TK 7870.2 .D377 1995

Davidson, Homer L.

Electronic troubleshooting
and repair handbook

**NEW ENGLAND INSTITUTE
OF TECHNOLOGY
LEARNING RESOURCES CENTER**

Electronic troubleshooting and repair handbook

Electronic troubleshooting and repair handbook

Homer L. Davidson

NEW ENGLAND INSTITUTE
OF TECHNOLOGY
LEARNING RESOURCES CENTER

TAB Books
Division of McGraw-Hill

New York San Francisco Washington, D.C. Auckland Bogotá Caracas Lisbon London Madrid
Mexico City Milan Montreal New Delhi San Juan Singapore Sydney Tokyo Toronto

©1995 by **McGraw-Hill, Inc.**
Published by TAB Books, a division of McGraw-Hill, Inc.

Printed in the United States of America. All rights reserved. The publisher takes no responsibility for the use of any materials or methods described in this book, nor for the products thereof.

hc   2 3 4 5 6 7 8 9 10   DOC/DOC   9 9 8 7 6

Product or brand names used in this book may be trade names or trademarks. Where we believe that there may be proprietary claims to such trade names or trademarks, the name has been used with an initial capital or it has been capitalized in the style used by the name claimant. Regardless of the capitalization used, all such names have been used in an editorial manner without any intent to convey endorsement of or other affiliation with the name claimant. Neither the author nor the publisher intends to express any judgment as to the validity or legal status of any such proprietary claims.

Library of Congress Cataloging-in-Publication Data
Davidson, Homer L.
    Electronic troubleshooting and repair handbook / by Homer L. Davidson.
        p.   cm.
    Includes index.
    ISBN 0-07-015676-X (H)
    1. Electronic apparatus and appliances—Maintenance and repair-
-Handbooks, manuals, etc. I. Title.
TK7870.2.D377   1995
621.3815'4—dc20   95-9908
                                                              CIP

Acquisitions editor: Roland S. Phelps
Editorial team: Joanne Slike, Executive Editor
              Andrew Yoder, Book Editor
Production team: Katherine G. Brown, Director
                Jan Fisher, Desktop Operator
                Linda L. King, Proofreading
                Jodi L. Tyler, Indexer
                Jeffrey M. Hall, Computer Artist
                Janice Ridenour, Computer Artist
                Toya B. Warner, Computer Artist
                Brenda Wilhide, Computer Artist
Design team: Jaclyn J. Boone, Designer   015676X
            Katherine Stefanski, Associate Designer   EL3

This book is dedicated to all of the electronics technicians who service consumer electronic products and place them back into operation and to the many consumer electronics manufacturers who provide electronic products in the entertainment field. And last, but not least, this book is dedicated to my wife, Dolores, who has been by my side for over 53 years with love, support, and encouragement.

Contents

Acknowledgments xxv

Introduction xxvii

1 Basic electronics 1

Identifying important components 2
Power (watts) 25

2 How to read schematics 27

Symbols 27
Partial schematics 35
Pictorial diagrams 35
The block diagram 37
Tie or terminal points 37
Identifying parts on the PC board 40
Information 44

3 How to use the digital multimeter (DMM) 51

Basic operation 52
DMM precautions 54
Auto ranging 56
Continuity tests 57
Resistance tests 59
Voltage measurements 62
Current tests 63
How to test small capacitors 65
Frequency tests 66
Diode tests 67

Zener diode tests 68
Bridge rectifiers 69
Detection diodes 70
Checking transistor function with diode tests 70
Transistor tests 71
NPN or PNP 72
Inductance measurements 73

4 Required electronic tests 75

Critical voltage measurements 75
Supply voltages 78
Critical resistance measurements 79
Checking transistors in the circuit 82
How to use the VOM 83
The VTVM 84
Required test instruments for the beginning technician 85
How to make tests in transistorized circuits 87
Making simple integrated circuit tests 89
Identifying the defective part 90
Why components break down 92

5 Testing solid-state components 95

Locating defective semiconductors with voltage tests 97
Diode voltage tests 100
IC and microprocessor voltage measurements 101
Critical resistance tests of solid-state components 102
Leaky diodes 105
IC resistance measurements 105
Transistor junction-diode tests 107
The open transistor 110
Horizontal output transistor with damper diode 112
DMM diode tests 112
Duo-diode tests 114
Bridge rectifier tests 115
Damper diode tests 117
High-voltage diodes 118
Testing solid-state components out of circuit 120
Removing defective components 120

6 Checking surface-mounted components 125

SMD transistors 126
Surface-mounted diodes 127
Surface-mounted ICs, LSIs, and microprocessors 129
Surface-mounted resistors 130
SMD capacitors 131
Special tools 133
Testing SMD components 133
Locating a component on the PC board 134
Electrostatic-sensitive devices 135
Removing chip capacitors and resistors 136
Removing ICs and SMD processors 138
Replacing SMDs 138
Part substitution 141
Intermittent board connections 141
Patience 142
SMD parts layout 142
Where to find replacement 143

7 Learn with kit building 145

Learn to solder 146
The different soldering tools 148
Read the directions 150
Check all parts 151
How to assemble components 153
Transistor and IC mounting 154
Putting the resistor color code to work 156
PC board layout 157
From schematic to board mounting 159
Mounting panel components 159
Transformer connections 162
Part checkoff 164
Check-out time 164
Testing 166

8 Build four different test instrument kits 167

The signal injector-tracer kit (ST-751) 167
Transistor test kit (EG-DT100K) 171
LCD meter kit (M-2665-K) 175

Parts identification 175
Deluxe regulated power supply kit (XP-620K) 178
Conclusion 181

9 Shock hazards in electronic repair 185

Power line problems (ac) 185
Check that power cord 186
The hot chassis 187
Isolation transformer 189
Grounded work bench 190
Grounded test equipment 192
Hazardous work areas 193
Heat problems 193
High-voltage precautions 194
Replacing the CRT 196
Charged capacitors 197
Microwave oven shock hazards 197
X-ray radiation 199
CD laser eye damage 200
Electrical shocks in test equipment 201
Protecting the customer 202
Conclusion 202

10 Simple electronic circuits 205

Amplifiers 205
Detector circuits 221
Oscillator circuits 222
TV high-voltage circuits 229
CRT circuits 231
Tape head circuits 232
Phono pickup 234
Conclusion 235

11 Isolate, locate, and repair 237

What's the symptom 237
Time is money 238
The different symptoms 239
How to isolate the different circuits 243
Pinpointing the defective circuit 244

Outside of the suspected circuit 248
Locating the defective component 249
Critical voltage measurements 251
Critical resistance tests 253
Critical waveforms 256
Replacing the defective part 256
Universal replacements 259
Heatsinks 260
Tests after repairs 261

12 Servicing low-voltage power supplies 263

Play it safe 264
Power-supply components 265
The rectifier 267
Voltage-doubler circuits 267
Filter capacitors 268
Filter chokes 269
How the power supply operates 270
Voltage regulators 271
IC regulators 272
Fuses 273
Circuit breakers 275
Power-supply systems 275
Critical voltage and resistance tests 276
Power wall adapters 280
Portable radio and cassette player power-supply circuits 280
Auto radio power-supply circuits 281
Boom-box power-supply circuits 282
Conclusion 283

13 Servicing simple audio circuits 285

Class-A amplifier 287
Class-B amplifier 287
How the amplifiers perform 288
IC audio amps 288
Stereo audio circuits 290
Different sound problems 293
Weak sound 294
Dead 295
Intermittent sound 296

Distorted sound 296
How to signaltrace audio signals 297
Servicing audio circuits with DMM 299
Servicing table radio audio circuits 300
Servicing stereo audio circuits 300
Checking power audio circuits 301
Speaker problems 303

14 Checking and repairing PC boards 305

Making your own PC boards 306
PC board wiring problems 309
The intermittent board 309
Cracked and broken boards 312
Circuit tie wires 313
Surface-mounted PC components 314
Locating parts on PC wiring 315
Board replacement 315

15 Required test equipment and how to use it 317

The most-used test equipment 318
Additional test equipment 335
Special test equipment 340
Specialized test equipment 349

16 How to signaltrace various circuits 351

Indicators 352
Signal injection 352
Signaltracing 353
Pencil and noise generator tests 354
External audio amp signaltracing 357
Audio signaltracing 359
RF signal generator tracing 362
Scope waveform tests 364
Power-supply voltage injection 365
Random voltage measurements 367
Random transistor tests 368
Tracing signals in radio circuits 369
Signaltracing audio stereo circuits 370
Conclusion 371

17 Waveforms to the rescue 373

Waveforms test in audio circuits 374
Troubleshooting the radio circuits with the scope 377
Important TV sweep waveforms 378
Critical horizontal sweep waveforms 379
Vertical sweep waveforms 382
Important CD waveforms 385
RF or EFM signal waveforms 386
Tracing error signal (TE) 387
Tracking coil waveform 387
Focus error waveform 388
Focus coil waveform 389
PLL VCO waveforms 389
No motor movement 390
SLED motor waveform 390
Spindle motor waveform 392
Conclusion 392

18 Identifying and replacing unknown parts 395

IC body part numbers 396
Unknown transistors 397
Transistor working voltages 399
Power output transistors working voltages 401
Different model, same chassis 403
Unknown IC part number 404
Replacing the defective transistor 405
Replacing defective ICs 409
Large surface-mounted LSIs 409
Unknown diodes 411
Unknown bridge diodes 412
The solid-state manual 413
Comparison circuit components 413
Manufacturer data 414
Manufacturer parts depot 414
Photographing parts 415
Help from local technicians 415
Local electronics stores 416
Conclusion 416

19 How to locate mechanical defects and problems 417

In action 418
Cracks and breaks 418
Broken belts 418
Squeaky bearings 419
Lubrication 420
Motor noises 422
Unusual noisy operations 422
Mechanical problems within the cassette player 424
Mechanical problems in the phonograph 428
Mechanical problems with the CD player 431
Mechanical problems with the VCR or camcorder 437
Conclusion 443

20 Troubleshooting and repairing solid-state circuits 445

Critical voltage tests 446
Power supply voltage 447
Critical resistance measurements 449
Transistor tests 450
Transistor bias voltage tests 452
IC voltage tests 454
Troubleshooting various solid-state circuits 455
Conclusion 467

21 Correct part replacement 469

Original components are best 470
Universal replacements 471
Transistor replacement 472
IC replacement 475
Diode replacement 475
SMD replacements 476
Resistor replacements 477
Variable resistor replacements 478
Fuse replacement 480
Capacitor replacement 482
Belt replacement 484
Battery replacement 486
Switch replacement 487
Speaker replacement 488

Parts that can't be substituted 490
Incorrect part numbers 491

22 Troubleshooting and repairing radio circuits 493

The portable radio 494
AM circuits (amplitude modulation) 496
AM converter 496
FM RF and converter circuits 497
Intermediate-frequency (IF) circuits 497
FM MPX circuits 500
The clock radio 500
Shortwave radios 502
Variable capacitor tuning 504
Permeability tuning 505
Varactor or varicap tuning 506
Digital tuning 507
Typical auto stop circuit 510
Monophonic audio circuits 511
Dial cord problems 512
AM/FM radio symptoms 512
Typical AM/FM alignment 515

23 How to test and replace defective motors 521

Different motor problems 522
Voltage tests 523
Continuity tests 524
Intermittent motors 524
Overheated motors 525
Erratic motor operation 525
Squeaky motor bearings 526
Flat armature 526
The tap test 527
How to locate and replace the cassette motor 527
Test and replace auto cassette motor 531
Checking the phono motor 531
How to test and replace CD motors 532
Microwave oven motor tests 535
Checking camcorder motors 537
Motor test voltage supply 539
Low-voltage motor-test supply 540

24 Servicing the cassette and phonograph circuits 541

Circuit breakdown 541
Breakdown of cassette circuits 542
The tape head 542
Bias oscillator 545
Erase head 546
Erratic switching 547
Preamp circuits 547
AF amp circuits 549
IC output circuits 550
Stereo amp circuits 550
Recording circuits 553
Troubleshooting IC record circuits 557
Tape speed adjustment 558
Head azimuth adjustment 560
Phonograph electronic circuits 560
Transistor phono circuits 562
Phono motor circuits 563
Conclusion 565

25 Basic TV screen symptoms 567

Block diagram 567
The 14 most-common symptoms found in TVs 573
Other TV symptoms 582

26 Simple TV repairs 593

Check the fuse 595
Check those silicon diodes 596
On/off switch tests 597
Capacitor tests 598
Voltage regulator tests 599
Leaky horizontal output transistor 600
Damper diode tests 602
Driver transistor tests 603
Scope deflection IC 604
Vertical output tests 605
Replacing vertical output transistors 606
Yoke tests 607
Sub the speaker 608
Transistor output sound circuits 609

IC sound output circuits   610
Signaltracing audio circuits   611
Tuner tests   611
Picture tube tests   613

27 Replacing the picture tube   615

Poor tube socket connections   616
High-voltage problems   618
CRT components   619
Larger glass picture tubes   620
Picture tube circuits   620
Safety factors   622
Testing the CRT   624
CRT repairs   625
Removing the picture tube   626
Picture tube replacement   629
Degaussing coil   630
Poor purity   631
Typical purity adjustments   632
Large screen purity adjustments   633
Typical convergence adjustments   633
Convergence problems   634
Typical slotted shadow mask convergence   635
Typical black-and-white adjustments   636

28 Servicing remote-control devices   639

Basic transmitters   640
Infrared remote   641
Infrared remote tests   641
Battery tests   643
Infrared remote tester   644
How the tester operates   644
Laser remote tester   645
Universal remote control   646
Remote-control receivers   648
Testing remote transmitters   649
Testing infrared receivers   650
Servicing interface circuits   651
Troubleshooting the keypad   652
Standby circuits   653
Conclusion   655

29 How to locate and repair intermittent problems 657

Hot and cold applications 658
Moving components 661
Intermittent boards 662
Locate and isolate the section 666
Broken boards 669
Shotgun terminal approach 670
Intermittent components 671
Intermittent problems in radio circuits 672
Intermittent amplifier problems 675
Intermittent TV circuits 676

30 Troubleshooting directly coupled circuits 683

Directly coupled circuits 683
Directly coupled preamp circuits 685
NPN and PNP dc circuits 686
Transistor interaction 687
Directly coupled IC circuits 690
Solid-state in-circuit tests 691
Critical voltage and resistance tests 692
Troubleshooting cassette preamp recording circuits 693
Repairing directly coupled sound circuits 694
Troubleshooting directly coupled audio circuits in TV chassis 695
Servicing directly coupled vertical circuits 696
Troubleshooting directly coupled horizontal circuits 698
Servicing dc video circuits 699
Wholesale part replacement 700
External power source 700

31 How to service consumer electronics without a schematic 703

Visual inspection 704
Listen to the music 704
Part numbers 705
The new chassis 709
Different face, same chassis 711
Service notes 712
Schematic comparison 712
Locating the various circuits 712

Ten service tests 716
Off the main chassis 717
Part replacement 718
Chemical products 720
Conclusion 722

32 Servicing electronic imports 725

Troubleshooting foreign devices 726
Imported solid-state components 727
Correct part replacements 728
Identifying components on the PC board 729
Critical radio circuits and components 730
TV import circuits 732
Vertical import circuits 736
Domestic and import replacement components 737
Distributors of imported electronic parts 738

33 Tough repair problems 741

Early to bed 742
From a different angle 743
Multiple problems 743
Intermittents 744
Try it again 745
Defective new part 746
Noisy audio 747
Intermittent recording 752
Insufficient vertical height 755
TV chassis shutdown 757
Conclusion 761

34 How to repair microwave ovens 763

Microwave energy 764
Oven components 766
Discharge 774
Required test instruments 775
The magnameter 776
Circuit saver 777
Servicing the low-voltage circuits 778
Control board panel 781

High-voltage problems 782
Intermittent cooking 783
Critical resistance measurement 783
Oven leakage tests 784
Conclusion 784
List of microwave oven manufacturers 786

35 Troubleshooting and repairing CD players 791

Safety precautions 791
Laser beam protection 792
Required test equipment 793
Compact disc cleaning 794
Surface-mounted components 795
Laser pickup assembly 795
Focus and tracking coils 798
The block diagram 799
Signal circuits 800
Signal processor circuits 801
Auto focus and tracking error signals 801
Servo circuits 803
CD motor circuits 804
Low-voltage power supply 807
Audio circuits 810
CD player adjustments 811
Troubleshooting CD players 814

36 VCR and camcorder repair 825

VCR and camcorder block diagram 826
VCR VHS and Beta formats 826
Camcorder VHS, VHS-C, and 8-mm formats 826
VCR tape heads 830
Tape head cleanup 832
Automatic head cleaner 833
VHS tape-loading paths 833
Beta VCR loading path 835
Camcorder tape transport 835
Periodic check and lubrication 836
VCR reel table drive mechanism 838
Rubber belt problems 840
Camcorder camera sections 841

8-mm camera circuits   841
Electronic viewfinder   843
Camcorder video circuits   844
Flying erase head   845
VCR servo and control systems   847
Camcorder servo and system control   848
Problems within VCRs and camcorders   849
Pulling of tape   850
Cannot remove the cassette   850
Improper loading   851
Shutdown after loading   852
Capstan speed problems   852
Erratic camcorder zoom operation   853
No camcorder auto focus   854
Dead VCR or camcorder   854
Poor VCR recording   855
Poor or no head erase   858
No audio in playback   858
Poor VCR picture in playback   859
Camcorder motor circuits   860
Conclusion   860

37 Servicing TV low-voltage and regulator circuits   861

Line voltage circuits   862
Voltage-doubler power circuits   862
Bridge rectifier circuits   863
Low-voltage transformer supply   865
SCR switching low-voltage circuits   866
SCR tests   868
B&W TV low-voltage supply   868
Different voltage sources   869
Line-voltage power-supply regulators   870
Transistor and zener diode regulators   872
Keeps blowing the fuse   874
Check by the numbers   875
Filter capacitor problems   876
Black hum bars   878
If lightning strikes   879
Critical *B+* adjustments   880
Chassis shutdown   881

38 Repairing the vertical circuits 883

In the beginning   884
Vertical supply sources   885
Then there were none   886
Important vertical waveforms   887
Vertical countdown circuits   888
By the numbers   890
Vertical sync problems   890
Intermittent vertical sweep   891
Supply pin terminal tests   893
Vertical voltage injection   893
Vertical foldover   895
No work/no play   896
Inside outside   897
An unusual vertical problem   899

39 Troubleshooting the TV horizontal circuits 901

Countdown and deflection circuits   902
Horizontal driver circuits   903
Horizontal output circuits   906
High-voltage problems   908
Damper diodes   908
High-voltage hold-down problems   909
Flyback operation and problems   911
Scan-derived screen and focus voltage   913
Critical waveforms   914
Pulled in on each side   916
Horizontal foldover   917
Deflection yoke problems   917
HV shutdown   919
Horizontal disable circuits   920
High-voltage shutdown test   922
Keeps blowing the fuse   922
Keeps destroying the output transistor   924
Replacing hot transistor   925
Intermittent horizontal problems   927
Lines in the raster   928

40 Repairing special TV circuits 931

Repairing the color circuits 932
Critical test points 933
Servicing RCA's chopper and VIPUR circuits 936
Troubleshooting Sylvania's SMPS circuits 937
Comb and saw filter circuits 941
Computer tuning circuits 946
TV picture on-screen display 948
Mono audio circuits 950
TV stereo audio circuits 950
Troubleshooting stereo decoder circuits 952
Stereo headphone operation 953
Picture in a picture 953
Standby circuits 955
Pincushion problems 957

41 Keep that test equipment in tip-top shape 961

Simple maintenance 961
Test instrument schematics 962
Where to obtain parts 963
Instrument warnings 963
Repairing cords, plugs, and jacks 966
Internal VOM and DMM repairs 967
Power supply repairs 969
Servicing the external audio amp 972
Broken dials and knobs 972
Generator repairs 973
Transistor tester repair 975
Capacitor tester repairs 977
FET meter repairs 977
Portable color-bar generator repair 978
Cabinet repairs 979
Soldering iron repairs 979
Factory instrument adjustments 980

42 Taking care of business 983

The CET test 983
Business ethics 984
Radio and TV sales and service 984

TV repair and satellite business 985
TV and VCR repair 986
Medical Electronics Technician 987
Computer technicians 988
Other electronics repair fields 989
Getting into the business 989
Bookkeeping 991
Sperry pricing book 992
Local and federal tax laws 993
Conclusion 994

43 List of electronics manufacturers 997

Electronics book publishers 1010
Electronics magazines 1010

Glossary 1013

Index 1049

About the author 1073

Acknowledgments

I thank all electronics manufacturers who have contributed circuits, ideas, and schematics for this book (and previous books). To all those electronic technicians, friends, and editors who have shared electronics experiences, encouragement, and data for this book.

Forest Belt, editor and author
Jack Darr, author
Jack Hobbs, editor of *Electronic Technician/Dealer*
Glen Jochims, Radio Trade & Supply
Tom Krough, Krough TV Repair
J.W. Phipps, editor, *Electronic Servicing*
Julian S. Martin, editor, *Popular Electronics*
Nils Conrad Persson, editor, *Electronic Servicing & Technology*
Tom Rich, Tom's TV Repair
Fred Shunaman, editor, *Radio Electronics*

Special thanks to my electronics acquisitions editor, Roland S. Phelps, and Senior acquisitions editor, Kimberly Tabor, of Tab/McGraw-Hill for helping to make this book possible. A great deal of thanks to Robert Douglas, and Kathleen and Darrell Guymon who helped with the book manuscript. Finally, to all of my customers out there who supplied products for the service bench for over 38 years in the radio and TV business, and especially to those people who would say "thanks for a job well done." Thanks for those memories.

Introduction

This book will help you to learn basic electronics, how to apply it, and how to become an electronics technician. This book contains more practical data than theory. The book shows you how to recognize the different electronic components with lots of photos and drawings. It points out the various electronic circuits and how to repair them. Actual photos of the various electronic components in a wide range of electronic products are shown throughout the many pages. It also shows how to repair and service electronic products in the consumer electronic entertainment field.

The first 14 chapters start with basic electronics, how to read schematics, how to use the digital multimeter (DMM), how to test solid-state components, shock hazards, how to make required electronic tests, how to learn with kit building, how to locate and repair, how to service simple audio circuits, and how to make PC board repairs.

Chapters 15 through 28 begin with the required test equipment and how to use it, how to signaltrace the various circuits, how to take correct waveforms, how to identify unknown electronic parts, how to locate mechanical problems, how to troubleshoot solid-state circuits, how to make correct parts replacements, how to repair radio circuits, how to test and repair defective motors, how to service cassette and phono products, how to learn basic TV screen systems, how to make simple TV repairs, how to replace the defective picture tube, and how to service the defective remote-control devices.

Chapters 29 through 43 start with those pesky intermittent problems; how to troubleshoot directly coupled circuits; how to service products without a schematic; how to repair imported electronics; how to service microwave ovens; how to make CD player repairs; how to service TVs, VCRs, and camcorders; how to service TV power and voltage regulator circuits; how to service vertical and horizontal TV circuits; how to keep test equipment in tip-top shape; and how to take care of the business.

The electronics technician within the consumer electronics field keeps thousands of electronic products and equipment operating each year. The life of the electronics technician is not only a busy one, it can be very rewarding, and you learn something new each day. These electronic products include TVs, cassette and CD players, camcorders, radio receivers, audio, and hi-fi equipment.

Today, the electronics technician can branch out and expand into the exciting field of computer repair and maintenance, industrial electronics, and audio careers. Every year, something new appears over the horizon, such as high-definition TV (HDTV), digital audio tape (DAT), satellite TV, home and medical electronics, and on it goes.

Learning about electronics is fun. Servicing and repairing those electronic products is exciting and even more fun. You will have to make difficult repairs, but it's worth the whole trip. One day, you know you have just arrived as a professional electronics technician.

Basic electronics

THOUSANDS OF PEOPLE REPAIR CONSUMER ELECTRONICS (VCRs, TV sets, stereo equipment, radios, tape recorders, camcorders, etc.) as a profession. Today, the computer industry is wide open. Service technicians are needed to keep the music playing, camcorders recording pictures, and the TV bringing news, weather, and entertainment into the living room.

The electronics technician must be able to repair and service many different electronic products found in the home, on the playground, and in the factories (Fig. 1-1). Most electronics technicians have opened their own electronic service business and become their own boss. Whatever electronic field you decide to enter, electronics is still new, exciting, challenging, and presents another experience each day. Just about everything we touch is related to electronics.

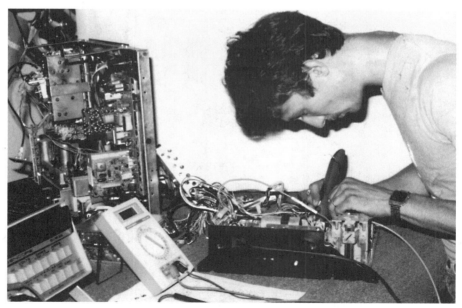

■ **1-1** *A service technician soldering components in an RCA TV chassis with a battery-operated soldering iron.*

In the early 1930s, electronics really began with battery-powered vacuum-tube AM and shortwave receivers. Then, the power line came to the home providing ac power for electronic devices. Later, the whole industry was changed with solid-state devices: transistors, ICs, and diodes. Today, the microprocessors and LSIs have generated another field of electronics.

The service technician's career might begin as an electronics hobby—soldering and wiring sets together, becoming an amateur radio operator, and repairing small items around the house. In the early days of servicing, electronics troubleshooting was self-taught. Just about everyone starts out as a beginner, moves up to the intermediate level, and finally ends up as the advanced or professional electronics technician. Let's start at the beginning.

Identifying important components

The TV or radio receivers consist of many components. When these parts are tied together, they form a circuit. The electronic circuit can be identified as a diagram or schematic (Fig. 1-2), which is an electronics "road map." The electronic circuit must be powered by a dc or ac voltage source.

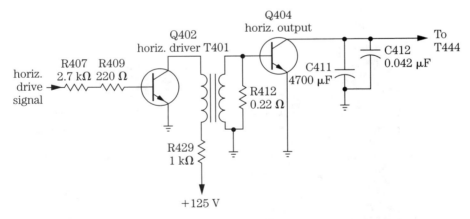

■ **1-2** *Electronic components are tied and soldered together in a circuit; they can be identified with the circuit diagram or schematic.*

The following electronic components are found in various products of the consumer electronics field:

ac voltage Ac voltages are carried by the power lines to receptacle outlets in the home. Ac voltages operate nearly any appliance or electronics gadget in your home.

power supply The power supply filters and changes the voltage of any power source (ac power outlet or dc battery). This power is then altered so that it is appropriate for the different sections to operate properly.

battery The dc battery generates electricity by electrochemical action. Usually there are several cells found in a 6- or 9-V battery. The 9-V battery powers small radios, meters, toys, and other hobby projects. Batteries eventually decrease in voltage and can be checked in a battery tester, VOM, or digital multimeter (DMM) (Fig. 1-3).

■ **1-3** *Small batteries are used in many portable units. They can be checked in a battery tester.*

capacitor The unit of capacitance is given in farads or microfarads (F) or (µF). µF is the abbreviation for microfarads and pF for picofarads. A capacitor has two conductors separated by a dielectric, as charges are stored between the conductors. The dielectric material of a capacitor might be air, vacuum, rubber, oil, paper, mica, glass, or ceramic. The electrolytic capacitor is found in filter networks in the low-voltage power supply. A bypass capacitor returns ac or RF signal to ground in electronic circuits. The coupling capacitor couples the ac or audio signal to another component and maybe a fixed or electrolytic capacitor. Variable capacitors are used for tuning stations in AM/FM and TV receivers. Electrolytic high-frequency/high-temperature capacitors are widely used in computers.

Electrolytic capacitors have an electrolyte paste film between aluminum plates to provide insulation and capacity. The dry electrolytic capacitor has electrolyte in a paste or solid form and the wet electrolytic capacitor contains a liquid as electrolyte.

The defective filter capacitor might dry up or go open and have less capacity, causing 60- or 120-Hz hum in the sound with hum bars in the TV raster (Fig. 1-4). Shorted or leaky filter capacitors can blow a fuse or destroy silicon diodes in the ac and dc power supply circuits. A low dc voltage might result from a dried up or leaky capacitor. Electrolytic film capacitors are also found in voltage decoupling circuits. Some electrolytic capacitors are used as coupling components in interstage and audio output circuits.

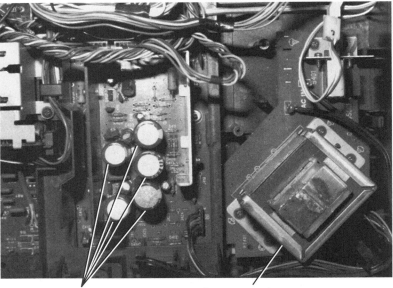

■ **1-4** *Several electrolytic filter capacitors might be used in the audio amp or tape player in the low-voltage circuits.*

Paper, ceramic, mica, tantalum, polyester film, mylar, metallized film, and nonpolarized crossover capacitors (Fig. 1-5) can be used as bypass and coupling capacitors. The NPO (nonpolarized) and silver mica capacitors are located in high-frequency controlled circuits. Memory backup capacitors are used in VCRs, computers, and other electronic equipment to maintain clock and other presets when power is removed from the unit. Dip capacitors are found in dual in-line packages.

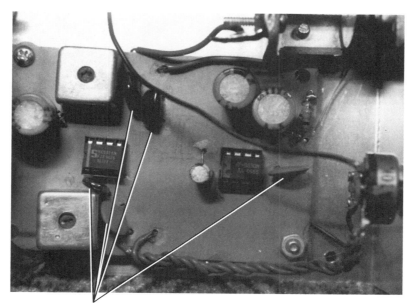

Ceramic bypass capacitors

■ **1-5** *Ceramic bypass capacitors and electrolytic capacitors in a table radio.*

Capacitors are available in either a radial- or axial-type mounting (Fig. 1-6). The radial capacitor has terminal leads out of the bottom of the capacitor and the axial capacitor leads come out each end. Surface-mounted capacitors (SMD) are mounted flat on the PC wiring and soldered at each end. A defective capacitor could open, lose capacity, or short between the elements. Check the suspected capacitor with voltage and resistance measurements and a capacity tester.

coils and chokes A coil consists of several windings of copper wire, which results in inductance. The coil might be wound in single or double layers or alongside another winding. The inductance of a coil or choke is the ability to oppose any change in current of an electronic circuit (Fig. 1-7). The unit of inductance is in henries. The coil form is the insulating support around which a coil is wound and might consist of air or self-supporting, ceramic, fiber, plastic, and glass forms. Coils are found in AM/FM radios, TVs, tape and CD players.

Choke coils restrict particular current or frequency and provide high impedance to alternating current while offering low opposition to dc current. The RF choke might be rated in microhenries (µH) and millihenries (mH). Check the defective choke coil with conti-

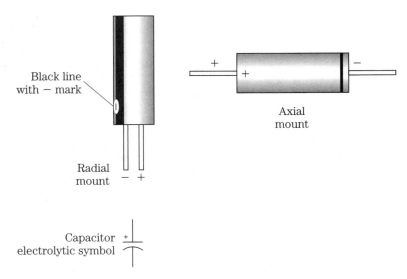

■ **1-6** *Radial electrolytic capacitors have leads at the bottom of the capacitor and axial types have leads at each end.*

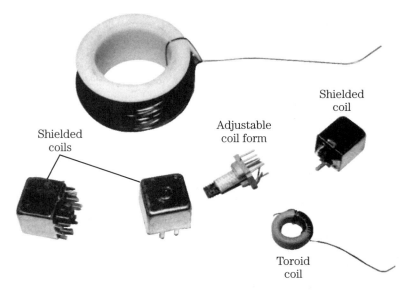

■ **1-7** *Adjustable shielded and ferrite core coils in a shortwave receiver.*

nuity measurements on an ohmmeter. The defective coil might open or become burned with a leaky component in the circuit.

dc voltage Dc voltages do not change in polarity and when scoped might appear as a straight line. Batteries and dc power supplies provide dc voltage to other circuits.

fuses Most electronic products are protected with a small fuse (Fig. 1-8). When current passes through a low-melting strip and exceeds the amperage of the fuse, the soft metal melts or blows open the circuit. Low-amp fuses are found in the ac input and horizontal circuits of the TV chassis. Stereo amplifiers might have a fuse in the transformer input, dc voltage, and speaker circuits for added protection. Check the fuse on the low ohmmeter scale or continuity tests of an ohmmeter. Replace the open fuse with one that has the exact current and voltage rating (Fig. 1-9).

Plug-in ac line fuse

■ **1-8** *Most electronic chassis are protected with a small fuse. Here, a clip-on fuse is located in the power line of a TV chassis.*

rectifiers A rectifier is a device that changes ac into dc, such as a rectifier tube, semiconductor, or mechanical vibrator. Various diodes found in electronic equipment are germanium, silicon, high-voltage, and bridge rectifiers (Fig. 1-10). The two- and four-element vacuum tube was used as a rectifier in the early radios, TV sets, and amplifiers. Germanium diodes are used to rectify RF and IF signals in the radio chassis.

Today, the silicon diode is found in practically every consumer electronic product operating from the ac power line. Silicon diodes have a low-resistance measurement in one direction and

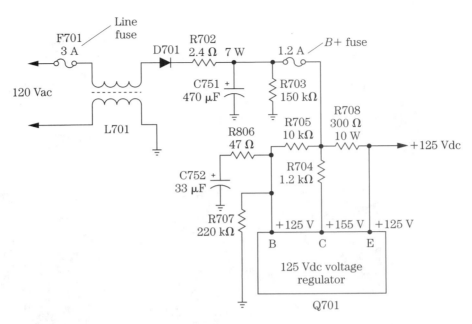

■ **1-9** *Many types of fuses are used in various chassis. Always replace the open fuse with one that has the exact current and voltage rating.*

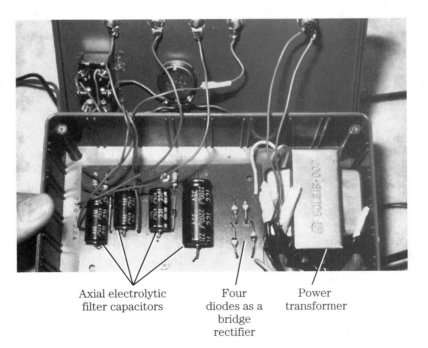

■ **1-10** *Four 1-A silicon diodes formed a bridge circuit in a low-voltage power supply.*

high-resistance measurement in the other direction, which provides higher voltages. You might find the silicon diode in full-wave, half-wave and bridge rectifier circuits in the line-voltage power supplies.

Selenium rectifiers were used in low-voltage power supplies of early electronic equipment, battery chargers, and stacked high-voltage diodes in the TV chassis (Fig. 1-11). The selenium is a junction diode processed of selenium material. Large plates of selenium and metal in the form of a stack take up more room within the chassis than a silicon diode of the same value. Selenium rectifiers were replaced with the silicon diode.

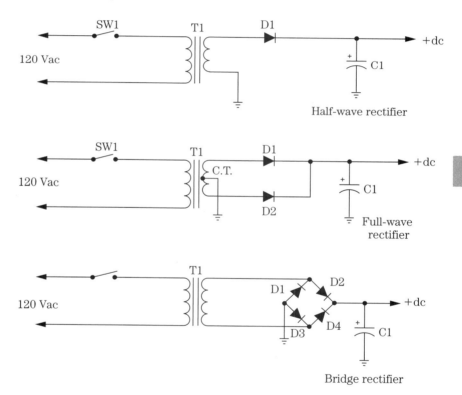

■ **1-11** *A schematic of a full-wave and bridge rectifiers, made up of single silicon diodes.*

Leaky silicon and selenium rectifiers can destroy the fuse and power transformer in ac operated equipment. The normal silicon diode has only a few ohms of resistance while the selenium rectifier has a high measurable resistance. Check the suspected diode or rectifier with an ohmmeter and diode tester. Most digital multimeters (DMM) have a separate diode test position.

resistors The resistor is an electronic part that offers resistance in a form with oppositions to current flow. The unit or value of resistance is measured in ohms. The value of the resistor is in ohms, kilohms, and megohms. Resistors consist of carbon, carbon film, and wire-wound material (Fig. 1-12). Fixed resistors have a set resistance and variable resistors have a wiper blade that rotates on the carbon material for a variable resistance. Surface-mounted resistors are chip types with soldered terminals at each end; they are placed directly on the PC wiring.

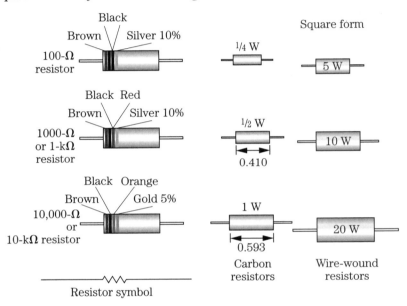

■ **1-12** *The different wattages of carbon and wire-wound resistors used in the electronic chassis.*

Variable resistors are used in circuits to control audio, volume, bias, voltage, and current. These potentiometers might be linear or audio tapered controls. Some potentiometers have a 2-W rating with low ohm values used in speaker circuits. Usually the variable resistor has three terminals with the center terminal as the wiping blade. Variable resistors might have a short or long shaft and screwdriver-type adjustment. Trimmer pots can be mounted horizontally, vertically, or flat upon the PC wiring.

The variable resistor can be held into position with a mounting nut or PC metal tabs (Fig. 1-13). A rear switch might be found on the rear of volume control. Variable resistors might become dirty, noisy, have burned or worn spots, and become open. Check the resistance with the ohmmeter. The digital multimeter makes accurate resistance tests.

10-W wire-wound resistors   15-W resistor

■ **1-13** *Large 15- and 10-W wire-wound resistors used in the low-voltage regulated power supply of TV chassis. Notice that the white rectangular resistor is 160 Ω at 15 W.*

Universal carbon resistors are available in power ratings from ⅛ to 1 W. Flameproof resistors are used in circuits under excessive heated conditions. High-wattage wire-wound resistors have low resistance and run from 2 W to 25 W (Fig. 1-14). Metal-oxide and noninductive resistors are below 1 kΩ, 20 W. The defective wire-wound resistor might show burns and cracked areas.

switches A switch opens and closes a circuit or connecting lines (Fig. 1-15). Switch types are: knife, toggle, slide, bat wing, rocker, pushbutton, rotary, and key types. A single-pole single-throw (SPST) switch connects one circuit in one direction. The double-pole double-throw switch (DPDT) connects two different circuits in both directions. Rotary switches might have more than two poles with several different switching positions. Several switches might be found in each piece of electronic equipment.

The volume control might have a rotary type on/off switch mounted on the rear side, used in amplifiers, radios, and auto receivers (Fig. 1-16). Function switches found on the CD player, TV set, VCRs, and boom-box players might be of the rotary type with several different poles and positions. Slide or dip switches might have multi-circuit contacts. Switch contacts might also appear in earphone jacks, antenna, motor, and relay circuits. Defective switches might

■ **1-14** *The variable resistor (center bottom) can be held into position with a lock nut, tabs, or self-supporting terminals.*

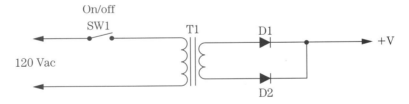

■ **1-15** *An on/off switch in a TV low-voltage power supply.*

have dirty, worn, burned, or open contacts. Take continuity and low ohmmeter tests across the switch terminals to locate the defective switch.

speakers and headphones A loudspeaker converts electrical energy into sound waves. Speaker size can range from 1 to 18 inches. The PM speaker has a permanent magnet over the voice coil (Fig. 1-17). Field coil speakers have spools of wires that form a magnetic field and are used as a choke in the early radio receivers. Speaker impedance might vary from 4 to 32 Ω. The typical speaker impedance is 8 Ω.

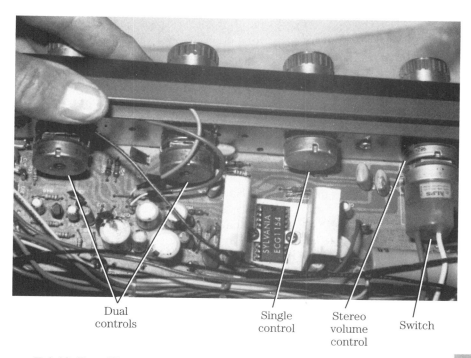

■ **1-16** *The different volume and tone controls with a rear-mounted switch in a boom-box player.*

■ **1-17** *The speaker has a large magnet mounted at the rear of the speaker basket in a boom-box player.*

Identifying important components

The speaker cone can be made of paper, coated paper, graphite, poly, carbon fiber/poly, etc. A voice coil is attached to the end of the cone and moves inside a metal and PM magnet grooved area. The PM magnet is mounted at the rear of the speaker. In deluxe auto and sound installations, you might find tweeters, mid-ranges, and woofers.

Headphones are headsets with built-in speakers that rest near your ears so that you can listen to something comfortably without bothering others. Permanent magnet (PM) headphones are small speakers enclosed in a plastic case with a headband to hold the earphones into position. Early headphones might have an impedance of 1 or 2 kΩ and the present-day PM types vary from 4 to 35 Ω (Fig. 1-18). Headsets might have a voice coil, or a magnetic or crystal diaphragm. Check the defective headset coils, cables, and plugs with continuity and low-ohm measurements. The defective headphones might have intermittent, open, or dead conditions.

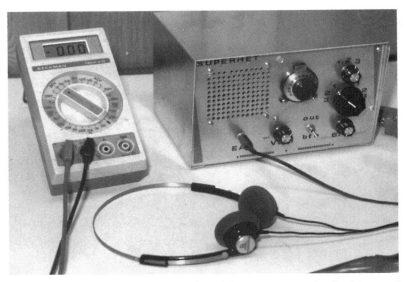

■ **1-18** *Here, a set of 4- to 32-Ω headphones are used with this small shortwave superhet receiver.*

transformers The transformer is a component that uses electromagnetic induction to transfer electrical energy from one circuit to another. Usually the transformer has a primary and secondary winding (Fig. 1-19). The auto-transformer has one large winding. When ac current flows through the primary winding, the magnetic flux of the core induces alternating current and voltage to flow in the secondary winding. Transformers can have air core or ferrite forms.

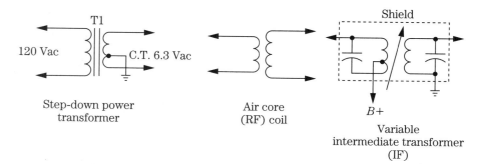

■ **1-19** *This simple step-down transformer has a primary and secondary with metal iron cores. IF transformers have an adjustable iron core.*

Different types of transformers are found in the consumer electronics products, of low-voltage power supplies, step-down transformers, step-up transformers, intermediate frequency (IF), start-up, audio output, driver, isolation, line, and matching transformers. The step-down transformer reduces the secondary voltage with fewer turns of wire (6.3 to 12.6 V). This component is found in modern receivers, tape and CD players, TV sets, amplifiers, and VCRs. A step-up transformer increases the secondary voltage with many turns of wire. This component is used in older radios, TV sets, and industrial equipment.

Although the IF transformer has been replaced with ceramic filters, they are still used and replaced in some AM/FM auto and home radios. Audio interstage drivers and audio-output transformers are found in the audio circuits of early radios and TV sets (Fig. 1-20). The TV horizontal output (flyback) transformer is fully described in the TV circuits.

Defective transformers might have an open winding, arcing between windings, or burned and shorted windings. Both primary and secondary windings of a step-down transformer have low resistance. The step-down transformer has a higher primary resistance than the secondary winding. Measure the resistance of each winding upon the low range of the ohmmeter. Sometimes shorted turns in some interstage transformers will not show up with resistance measurements.

Solid-state components

Semiconductor devices are solid-state components. A solid-state device is one in which the flow of electrons and holes are controlled in special blocks, wafers, disks, and rods. Semiconductors consist of a solid material that lies between conductors and insu-

Driver transformer   Start-up transformer

■ **1-20** *Audio, driver, start-up and interstage transformers are used in the TV chassis.*

lators, usually made of germanium and silicon. Transistors, diodes, integrated circuits (ICs), large-scale integration (LSI), and microprocessors are called *semiconductor devices* (Fig. 1-21).

The word *transistor* is a contraction of the words *trans*fer and re*sistor*. The transistor is a three-terminal semiconductor that is capable of switching, oscillating, and amplifying. Transistors have replaced tubes in receivers, TVs, and amplifiers. You will find bipolar, diffused-junction, field-effect, metal-oxide, germanium, unijunction, and silicon transistors in consumer electronic products. Transistors are also used in test equipment, which locate defective transistors.

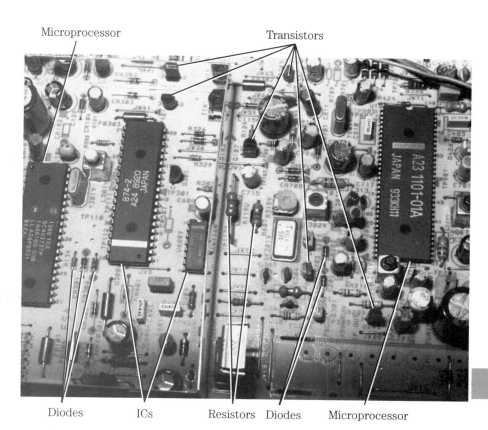

1-21 Transistors, ICs, and microprocessors are used in the TV and VCR chassis.

Transistors are PNP or NPN types (Fig. 1-22). The early transistors were constructed of germanium and later ones of silicon material. Transistors have a base, collector, and emitter terminals (B-C-E). Remember that the PNP type of transistor always has a negative collector voltage while the NPN transistor collector has a positive voltage. The bias voltage of a PNP transistor, between the base and emitter, is 0.3 V, and 0.6 V on the NPN transistor. Usually the transistor is normal when the correct bias voltage is between base and emitter elements.

Field-effect transistors (FET) have low power consumption and high impedance, similar to the vacuum tube. FET transistors are used as RF amplifiers and voltage amps in the AM/FM receivers. Metal-oxide semiconductors (MOS or MOSFET) are used in broadcast and shortwave receivers as RF amplifiers, converters, and IF and audio input circuits. Surface-mounted transistors have three soldered PC connections.

Identifying important components

■ **1-22** *Transistors are either PNP or NPN types. Most of the transistors used in today's electronic chassis are made of silicon. Notice the arrow is reversed in the NPN transistor.*

Check the suspected transistor with in-circuit transistor tests, out of circuit with the transistor tester, and diode-transistor junction tests with the digital multimeter (DMM). The defective transistor might be checked with critical voltage and resistance measurements. Intermittent transistors can be monitored with voltage and signal tests or subbed with new components. Transistors are mounted flat, horizontally, and vertically, and with self-supporting lead terminals.

diodes The solid-state diode can be constructed from germanium or silicon material (Fig. 1-23). These diodes are formed by joining a piece of P material with a piece of N material. The junction diode is used to rectify ac current while the point contact diode is for rectifying signals. Germanium diodes are found to rectify RF and IF signals in the radio receivers and have a higher measurable resistance. The silicon diode is found in low-voltage power supplies and switching circuits with low resistance in one direction and infinity in the other direction. Surface-mounted diodes might contain more than one diode in a miniature component. Check each diode with the low ohmmeter scale or with DMM tests.

varactor diode The varactor diode is known as a *capacitance diode* that will change capacity with a variable dc voltage applied (Fig. 1-24). The variable diode acts as a variable capacitor. The varactor diode tunes in a certain frequency with an inductor to obtain a certain frequency. These diodes are found in AM/FM/MPX receivers and TVs, instead of the variable capacitor that is today difficult to find.

zener diodes Zener diodes are a special type of diode and break down at a certain biased voltage. Universal zener diodes might operate from 2 to 56 Ω with 1- to 5-W rating in consumer electronic products (Fig. 1-25). Defective zener diodes become leaky or shorted and might become overheated with burn marks. These diodes look exactly like a regular silicon diode.

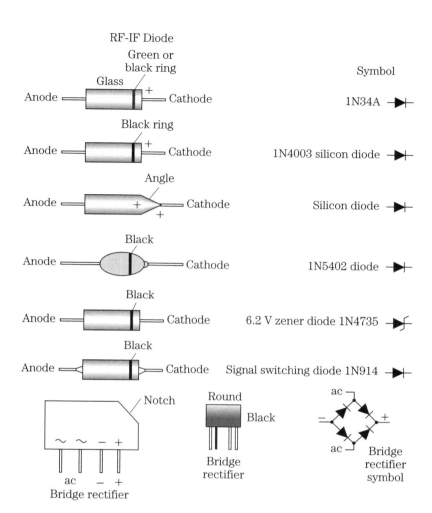

■ **1-23** *The various diodes used in the electronic chassis.*

other diodes The switching diode has low capacitance and fast recovery, found in computer circuits. The Schottky diode is a special semiconductor with a metal and semiconductor form with a PN junction and has high amperage capabilities. Surface-mounted diodes are miniature parts with soldered ends mounted on the PC wiring.

Integrated circuits

The integrated circuit (IC) consists of components and internal connecting wires inside of a silicon chip. Transistors, diodes, resistors, and capacitors form a circuit inside the chip with many terminals. The IC might look like a multi-legged transistor, flat pack,

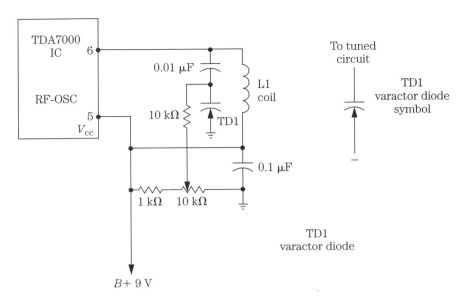

■ **1-24** *The varactor diode has a capacitance that is varied to change the frequency of (tune) the radio or TV set.*

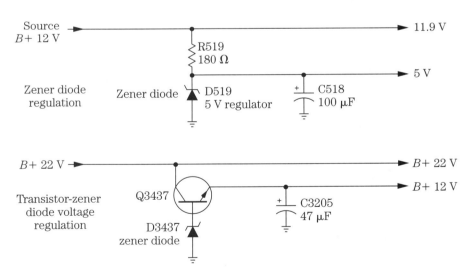

■ **1-25** *Zener diodes are used as voltage regulators. Sometimes they are used in a circuit by themselves or in conjunction with a transistor.*

dip, or dual in-line package (Fig. 1-26). ICs are found in most electronic equipment, from toys to computers. The surface-mounted chip has flat terminals to solder directly on the PC wiring. The IC might have more than one power source or terminals (V_{cc}).

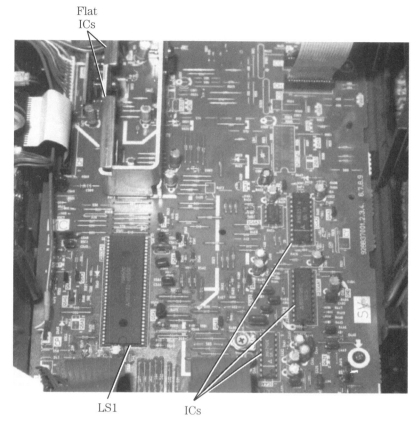

■ **1-26** *Several ICs, LSIs, and microprocessors are used in the VCR.*

microprocessor A large single chip has computer and control elements, processing memory, and logic functions. A microprocessor can be programmed, but cannot function by itself. Microprocessors can mount in a socket, pins can be soldered directly to PC wiring, or they can be mounted flat with gull-type wing terminals. Several microprocessors are often found in the TV chassis, camcorder, CD player, VCRs, and computers (Fig. 1-27).

op amp The operational amplifier is a stable linear and direct-coupled amplifier with IC capabilities. This op amp has high gains and can operate in frequencies up to 1 MHz. The op amp is constructed inside with transistors, diodes, and capacitors to function as a high-gain broadband amplifier. These op amps work in an open or closed-loop circuit (Fig. 1-28).

silicon-controlled rectifier The silicon-controlled rectifier (SCR) or diode is a special type of rectifier, like that of a thyratron. SCRs

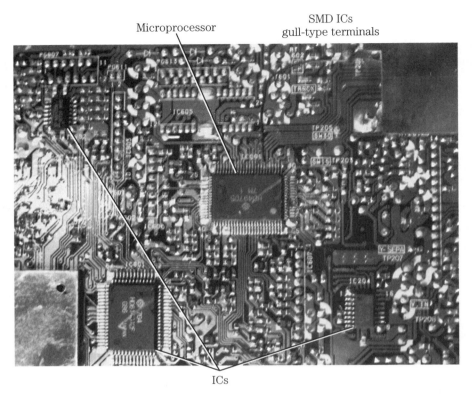

■ **1-27** Several SMD microprocessors are used on the PC board of this camcorder.

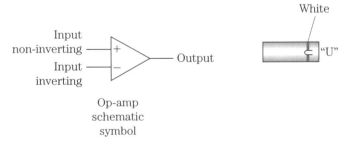

■ **1-28** The op amp is in an IC with a triangular symbol used in TVs, VCRs, CD players, and amplifiers.

are four-layer devices made up of NPNP or PNPN semiconductor materials. The SCR has an anode, cathode, and gate terminal (Fig. 1-29). When a small voltage is applied to the gate and cathode terminals, the resistance is controlled between anode and cathode, allowing current to flow through the SCR.

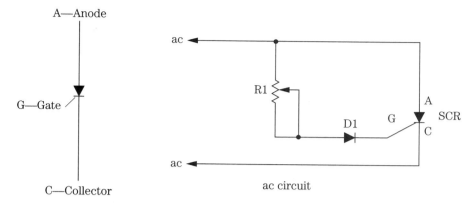

1-29 *The silicon rectifier (SCR) has three terminals, anode, cathode, and gate and is used to control ac circuits.*

triacs A triac is a three-terminal device (somewhat like the SCR) that is gate controlled with semiconductor ac switching. The triac has a gate and two anode (A1-A2 or MT1 and MT2) terminals. Triacs are used in TV sets, microwave ovens, and industrial electronic equipment (Fig. 1-30).

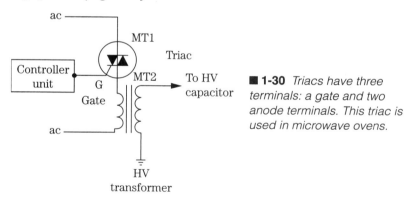

1-30 *Triacs have three terminals: a gate and two anode terminals. This triac is used in microwave ovens.*

surface-mounted components (SMD) Surface-mounted devices are actually transistors, ICs, diodes, capacitors, and resistors inside a miniature package, soldered directly to the PC wiring. These multi-layer ceramic and plastic chips have flat soldered ends to tie into the PC circuits. You might find more than one resistor or diode in one component. ICs and microprocessors have flat gull-type wing terminal connections.

These components are designed for miniaturized circuits and are found in today's consumer electronic products. You might find regular size parts upon the top side of the PC board with SMD parts soldered directly on the PC wiring (Fig. 1-31). Surface-mounted components require special care, patience, and equipment to locate, remove, and replace the small parts (Chapter 6).

■ **1-31** *Surface-mounted devices (SMD) include transistors, ICs, diodes, resistors, and capacitors in miniature components soldered directly on the PC wiring.*

Ohm's law By using Ohm's law, you can figure out the resistance, current, voltage, and wattage with a given resistor. Ohm's law is an easy formula to remember:

$$I = E/R$$

$$R = E/I$$

$$E = I \times R$$

I represents current in amps, E equals voltage, and R is resistance. For example, if the voltage is 200 V and the resistance is 25 Ω, the current would be:

$$I = 200/25 \text{ or } 8 \text{ A}$$

If you need a certain value resistor to drop the voltage to 100 V and the current was measured with the amp or milliamperes scale of the VOM or DMM of 1 A, use Ohm's law to get the correct resistance:

$$R = E/I$$

$$R = 100/1 \text{ is } 100 \text{ Ω}$$

Divide 100 W by 1 A; the result is 100 Ω.

Power (watts)

Wattage is electrical power. The wattage rating is the power dissipated by the component or device. The formula for finding the correct wattage of a resistor replacement is:

$$P = E \times I$$

P is in watts equals voltage (E) times the current (I). If the resistor you want to replace drops the voltage at 50 V and 100 mA of current is measured in the circuit, just multiply voltage times the current. First convert the 100 mA into A (0.100). Now 0.100 times 50 V equals 5.000 or 5 W. To be on the safe side, you might want to replace the 5-W resistor with a 10-W resistor so that it will not overheat.

Memorize the resistor code

Simply try to memorize the resistor code because you will use it every week as an electronics technician. The resistor code is standard and it is easy to mark the values of different carbon resistors. Different color rings are found around the body of a small resistor, representing numbers. Read the bands from left to right. Usually the color band nearest to the one end is the first number. Another method is to start at the opposite end of a silver or gold band.

The first band is number 1, the second band gives the next number, and the third band represents the multiplier. The fourth band spells out the tolerance of which silver is plus or minus 10 percent, gold is plus or minus 5 percent and the fourth band represents resistors of plus or minus 20 percent tolerance or accuracy.

For instance, a 100,000-Ω carbon resistor would have a brown band as number 1; the second band, black, is zero (0); and the last color band is yellow, four zeros (0000). The color coding of a 100-kΩ resistor is brown, black, and yellow (Table 1-1).

■ **Table 1-1** The resistor code.

0. Black
1. Brown
2. Red
3. Orange
4. Yellow
5. Green
6. Blue
7. Violet
8. Gray
9. White

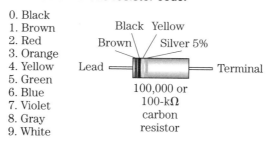

Foreign capacitor chart

Many foreign products are sold to local electronics and mall stores that must be serviced. Different parts and components might have different symbols than found in American diagrams. Foreign publications and schematics might have the capacitor marked in nanofarads (nF).

Simply translate nanofarads to microfarads when replacing a defective capacitor in the electronic chassis (Table 1-2). Now see how the various components are tied together on a schematic diagram. Learning the basics of electronics paves the way to an exciting field of servicing consumer electronic products.

■ Table 1-2 A foreign capacitor chart.

nF	µF
1.0	0.001
10.0	0.01
100.0	0.1
1000.0	1.0
10,000	10.0
100,000	100.0
1,000,000.0	1000.0

How to read schematics

SCHEMATIC SYMBOLS (FIG. 2-1) REPRESENT ELECTRONIC components in a circuit diagram. Circuit diagrams consist of a wiring diagram or schematic. Various circuits are found in consumer electronic products.

When parts are soldered together, an electronic circuit is formed. You must be able to read a schematic or wiring diagram before some electronic products can be repaired (Fig. 2-2). Of course, simple electronic projects, radios, and toys might be serviced by simply tracing out each component and seeing how they are wired together. Schematic diagrams help the electronics technician to isolate and locate the defective components. Wiring diagrams are must-have items when troubleshooting and repairing electronic equipment.

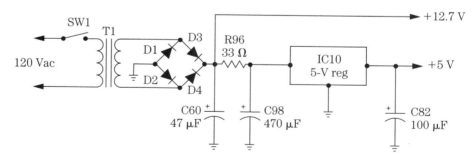

■ **2-1** *Several different symbols and parts are used in the low-voltage power supply of an AM/FM stereo receiver.*

Symbols

Every component in a schematic has a representative symbol. Try to picture each part as it is worked together in the schematic or circuit. These components must then be identified on the PC board or chassis.

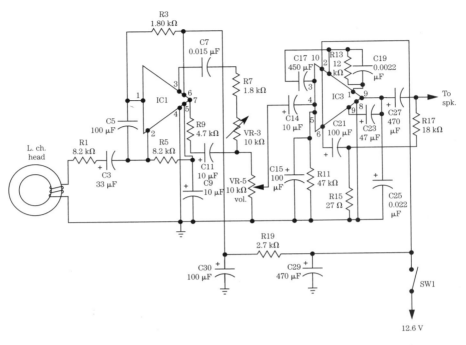

■ **2-2** *A complete diagram of the input and output IC amplifiers in a portable tape player.*

The most common electronic part symbols are found in Table 2-1. Semiconductor symbols are found in Table 2-2. Component parts with symbols are found in Tables 2-3 and 2-4, with semiconductor parts and symbols in Table 2-5.

The straight line

Straight lines represent the connecting wires between components on a schematic. Usually, the straight line starts from the terminal of a component and is tied into the circuit with a dot or intercepting line. Of course, most schematics are accurate and very helpful, although some have an occasional mistake.

For example, in Fig. 2-3, the major components in a TV vertical output circuit are the output IC and yoke assembly. The vertical countdown pulse begins at pin 5, goes through resistor R402, and connects to terminal 4 of IC500. Actually, the 500-Ω resistor is tied to pin 5 and 4 of each IC. The terminals of IC301 and IC500 can be soldered directly to the PC board and wiring. Sometimes small ICs plug into dual-line pin sockets.

■ Table 2-1 The most common symbols found in consumer electronics products.

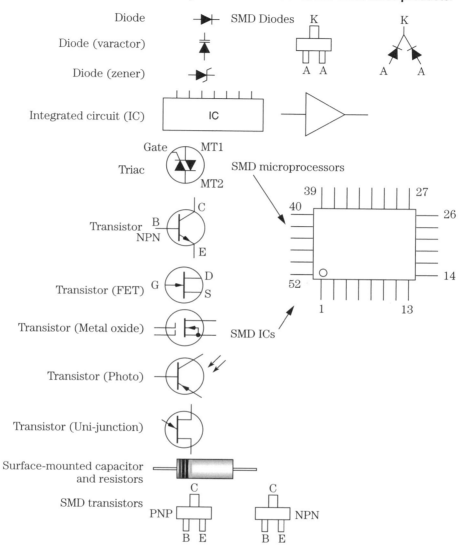

■ **Table 2-2** Most common symbols found in Consumer Electronics Products.

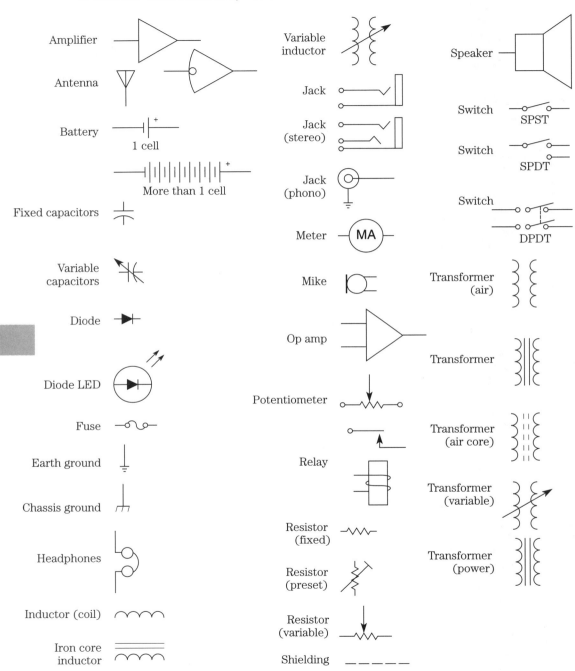

How to read schematics

■ **Table 2-3 Symbols of electronic parts and actual components found in the electronic chassis.**

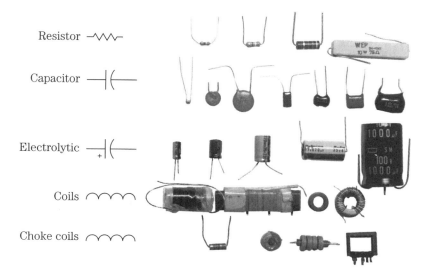

■ **Table 2-4 Electronics symbols and parts found in the electronic chassis.**

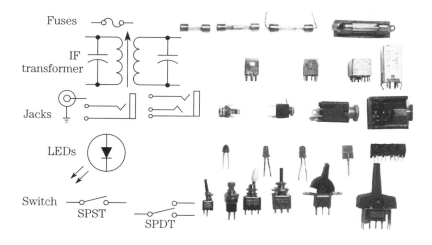

■ **Table 2-5** Semiconductor symbols and parts found in the electronic chassis.

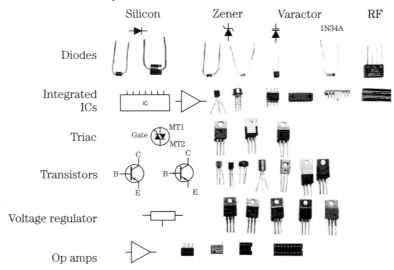

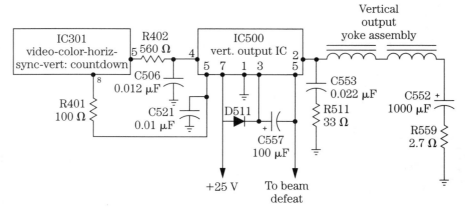

■ **2-3** *Notice the various straight lines that connect to the various components represent hookup wire, terminal connections, or PC wiring.*

Notice that a dot connects the line going to capacitor C506. The terminal of C506 is soldered to this junction on pin 4 of IC500. Where there is a dot, the component ties into that circuit. Notice that diode D511 and electrolytic capacitor C557 are in series; one end of D511 ties into the supply voltage (+25 V) and to pin 7 of IC500. The cathode terminal of D511 connects to the positive terminal of C557 and pin 3 of IC500. The negative terminal of C557 connects to pin 5 and to another outside circuit.

The vertical output pulse on pin 2 of IC500 drives the vertical deflection yoke winding. The yoke winding is soldered to pin 2 and the return wire connects to the positive side of electrolytic capacitor C552. The negative end of C552 solders to resistor R559 (2.7 Ω) and the end of the resistor connects to the chassis (common ground). Straight lines connect components in the circuit diagram and represent PC wiring or hookup wire. Arrows found in a partial schematic indicate the circuit ties into another outside circuit.

Dots and dotted lines

Dots on the schematic represent a terminal or connection. When two lines or PC wires intersect, they connect together (Fig. 2-4). Here, capacitor C521 terminal connects at the dot and is also connected to pin 5 of IC500 and resistor R401 (100 Ω). The ground terminal of bypass capacitor C521 is connected to chassis ground. A black line on the body of C521 represents the ground terminal and is soldered to common ground. Remember, dots connect circuits together in schematics.

Dotted lines often represent a shield or component container in a given area. Shielded transformers have dotted lines around the transformer symbol on the wiring diagram. These dotted transformer lines are usually grounded (Fig. 2-5). Notice that sound coil L202 is adjustable indicated by the arrow and is shielded with the dotted lines. Likewise, sound input transformer T201 is enclosed with dotted lines, which indicate that the winding is shielded and at ground potential. The inverted V indicates the input or beginning of a circuit at the primary winding of T201.

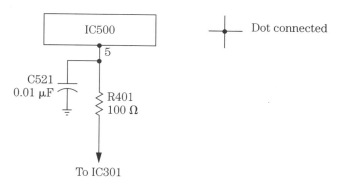

■ **2-4** *A dot on the schematic indicates that the circuits are tied or connected together at this point.*

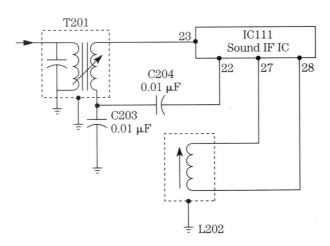

■ **2-5** *Dotted lines around the coils or transformers indicate that a shielded component has a shield tied to common ground in an IF TV sound circuit.*

Nonconducting lines

When two lines are drawn over one another without a dot, these lines do not connect, they go to other circuits. In some schematics, you might find an inverted or jumper-type line, which indicates no connection. Both of these nonconducting lines might be found in the schematic diagram (Fig. 2-6). They are drawn this way to connect to other components or circuits.

In this horizontal output transformer voltage source, you will find dotted lines and two types of lines that do not connect. Notice the bottom winding of terminal 1 goes to the center tap of windings 3 and 4. These dotted connections are made inside the transformer (winding) and are soldered together. When the winding terminal numbers are given, such as 1 through 8, these terminal connections are outside the transformer.

The line that connects terminal 1 of the transformer and goes down to A1 does not connect to any other transformer windings or terminals. These lines are shown jumped over the transformer connecting lines. Another type of nonconducting line from terminal 1 to A2 does not connect to any of the transformer sources. From pin 4 another nonconducting line goes to A3, and from pin 7 a straight line does not connect to any other line in the voltage sources. Terminals A1, A2, A3, and A4 (with arrows) are connected outside of the transformer circuits. Remember lines that go across, jump over, and intersect do not connect at that junction. The only lines that connect together are when a dot is used. The directly nonconducting lines that intersect and go across other lines are used in the latest circuits.

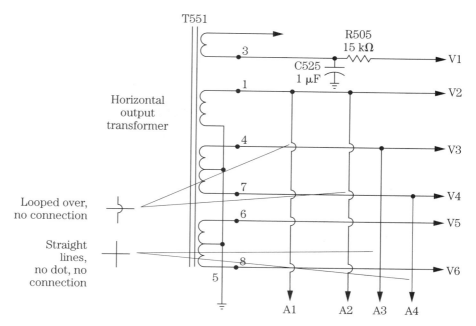

■ **2-6** *If a line jumps over or goes through another line without a dot to connect them, the circuit lines are not connected.*

Partial schematics

The partial schematic contains a small part or section of a given circuit. The schematic is broken down to help in troubleshooting, isolating, or pointing out certain components. By breaking the schematic down to a certain section, it is much easier to understand and locate the problem.

Partial schematics are used in electronics books and magazines to illustrate a single circuit. In TV schematics, you might find a partial or section-type schematic for each of the different sections (Fig. 2-7). Partial schematics are found in the chapter to help understand how circuits and symbols are wired together.

Pictorial diagrams

A pictorial diagram is a real drawing of components tied together. Pictorial diagrams are intended to help beginners in electronics to see how each part is placed and soldered together. The components are connected together with ordinary hookup wire or printed circuit (PC) wiring (Fig. 2-8).

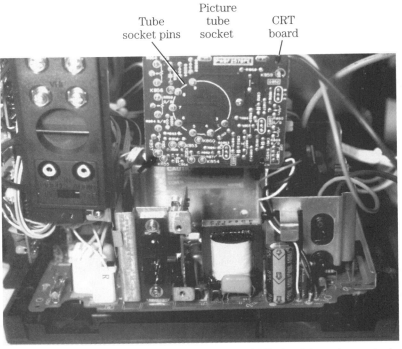

■ **2-7** *Here, a partial circuit might consist of a CRT circuit board mounted on the rear of the TV picture tube. Notice that the tube socket pin numbers are in a circle.*

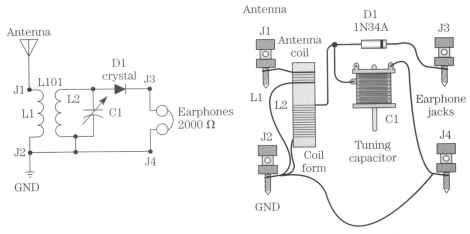

■ **2-8** *A pictorial diagram of a small crystal set with the schematic at A and a pictorial drawing at B.*

How to read schematics

The schematic of a simple crystal radio is shown in A and the pictorial diagram of the same circuit is shown in B. The outside antenna wire is inserted into jack (J1) and is wired to the primary winding of the antenna coil (L1). The end of the primary coil is soldered to J2 (ground). The secondary coil (L2) connects to the fixed crystal (IN34A) and to the variable tuning capacitor. The outside lug or metal ground of the variable capacitor is soldered to common ground (rotor). A terminal lug on the side of the stator terminal is soldered to the crystal and to the top side of L2.

Notice the cathode end (dark line) of the fixed crystal is connected to earphone jack J3. The other earphone jack (J4) is soldered to common ground. A 2000-Ω headset produces audio signal from a local tuned radio station. Notice the dots that connect the circuits together in A and how they connect in pictorial diagram B.

The block diagram

The block diagram is a simplified schematic of an electronic circuit drawn in boxes. Often, no detailed wiring or circuit is found. The block diagram can be used to help locate the defective component to a given stage. Just follow the symptom and locate the defective circuit and component in a blocked-off area.

Tie or terminal points

Electronic components or parts have connecting terminals that might have terminal wires, self-containing wires, and lugs (Fig. 2-9). The variable resistor has metal terminal pins or lugs that can be wired into a circuit with hookup wire or can be mounted directly on the PC wiring. The center terminal of the variable resistor varies the resistance of the control.

Fixed terminals of fixed resistors or capacitors come directly out of the component to be soldered into the circuit (Fig. 2-10). Terminal leads are cut off of fixed resistors and capacitors when mounted on a PC board. A poorly tinned terminal lead could cause a poorly soldered connection, resulting in a malfunctioning or intermittent circuit.

Fixed coils and small choke coils might use the original winding lead to connect into the circuit. Some choke coils might have an iron or plastic core. In the FM section of an AM/FM/MPX receiver, the FM coils are self-supporting and are soldered directly to the PC wiring.

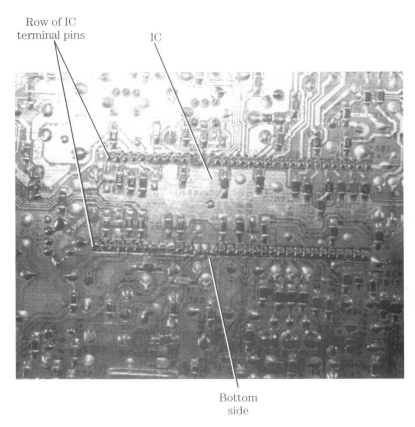

■ **2-9** *The pins of a large IC stick through the PC board chassis and are soldered directly to the PC wiring.*

Power and interstage transformers have terminal lugs that are connected with hookup wire or have PC lugs that solder directly to the PC wiring. The inside wiring of the transformer is soldered to a terminal lug for easy connections. Transformer PC terminal lugs are embedded into plastic or cardboard-fabric paper for self-supporting terminals (Fig. 2-11).

Transistors and ICs either plug into a socket for mounting or solder directly into the circuit. Power transistors must have larger mounting holes. ICs have long pins to fit into the respective holes of the PC wiring. Make good solder connections on transistors. ICs can plug into an IC socket that is soldered to the PC board or the ICs mount directly on the PC wiring (Fig. 2-12).

Ground tie points might be found from metal tabs of the chassis framework, holding the PC board into position. These chassis tabs

■ **2-10** *Wire terminals of a fixed resistor or capacitor support the part on the PC wiring. Transformer and coil wires might solder directly into the circuit.*

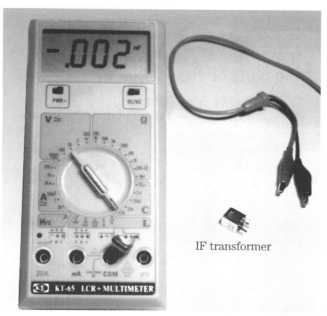

IF transformer

■ **2-11** *Small interstage or IF transformer lugs or terminals might solder directly into the PC board.*

■ **2-12** *The battery-powered soldering iron is ideal to solder small transistors and IC pins.*

or lugs stick through eyelets on the PC board. The eyelet is filled with liquid solder to tie the chassis and PC board grounds together and lock the board into position.

Identifying parts on the PC board

The PC board is a thin insulated board with copper strips of wiring etched onto the surface. The board consists of a plastic, epoxy, fiberglass, or phenolic base on which components and circuits can be connected together (Fig. 2-13). Each part terminal is stuck through a hole on top of the PC board and soldered on the etched wiring side. Some compact wiring might have PC wiring on top and bottom sides of the board, called a double-sided PC board.

Some PC boards have a trace of the PC wiring on the top side of the board. This traced wiring lies right over the exact PC wiring underneath the board. You can signaltrace the circuit and match up with the correct part (Fig. 2-14). Topside voltage and resistance measurements are easily made on the top side of the board.

Connecting cables or socket terminals are found on the etched wiring side of the board. Numbers or letters that correspond with

■ **2-13** *The PC wiring is etched on a plastic, epoxy, or phenolic base board with the components on top soldered into the different circuits.*

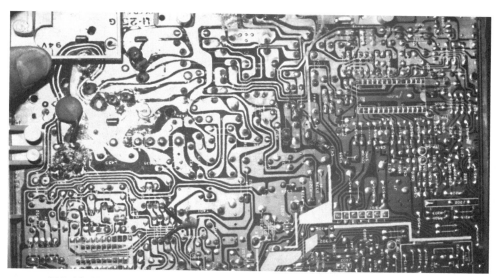

Notice white PC wiring lines

■ **2-14** *Notice that the white trace of wiring on the top side of a module, representing actual PC wiring under the board, makes signal or part tracing much easier.*

each connection are stamped on the board. You can find each component marked on the PC wiring. Long stamped white or black lines indicate where a resistor or capacitor connects within the circuit (Fig. 2-15).

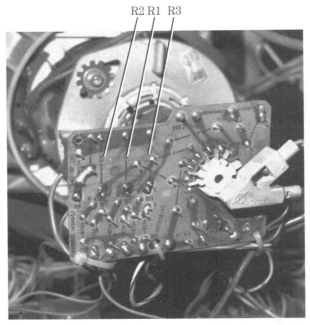

■ **2-15** *On the TV CRT board, the various resistors are mounted on top and black trace marks with the part number are located on the PC wiring side.*

Most parts are mounted on the top side of the board, except with surface-mounted (SMD) components. Usually, small parts are mounted close together to larger components, such as ICs, processors, and transformers. Small components use their own terminals to support, mount, and solder into the PC circuit (Fig. 2-16). Short part leads or terminals are used in RF and high-frequency circuits.

You might find the part numbers or symbols printed or stamped underneath of the component. Some manufacturers stamp components on the top side of the PC board. Voltage sources and connecting terminal wires might be marked on the circuit board. You might locate the different circuits stamped on the PC side of wiring for easy troubleshooting. Of course, this is not true in all cases. Sometimes you have to hunt and trace out each part on the chassis.

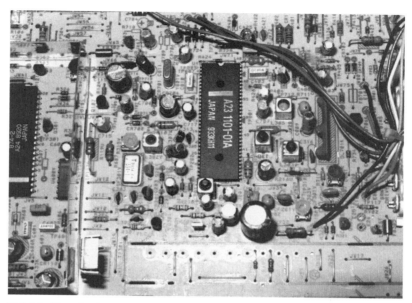

■ **2-16** *The single components on this TV chassis are held in position with soldered terminal leads. Use the component terminals to hold them into position on the PC side.*

Numbered parts

Each part in a wiring diagram must have a number or corresponding letter, identifying the part on the PC board and in the parts list. Notice in the partial circuit of audio IF and output stages that the capacitors have numbers (Fig. 2-17). All capacitors begin with a C letter, resistors with an R, transistors with Q, transformers with a T, diodes with a D or CK, ceramic filter network with a Z or CF, variable resistors with VR or R, waveforms with WF, SW for switches, coils with an L, and voltages with a number.

For example, the transistors are marked with a letter Q and one number for each transistor (in ascending order through the schematic). The number and order of the parts can vary, depending on how many transistors are in the circuit and how the manufacturer decides to number each one.

Notice that (in Fig. 2-17) all resistors start with an R and proceed with a 200 number. Each resistor has the correct resistance marked in Ω, kΩ, or MΩ along the side. Again, these numbers might begin with R1 on up in a small schematic or begin in the thousands in larger TVs and computer schematics.

Identifying parts on the PC board

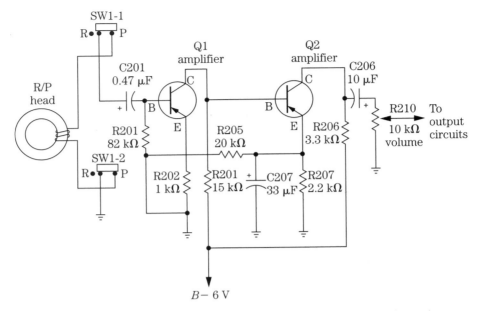

■ **2-17** *The symbols and parts are numbered and lettered in a schematic for identification and a parts list is included in this preamp stage of the tape player.*

Capacitors are marked in microfarads (μF). Here, the first capacitor is C201. Notice both resistors and capacitors have 200 numbers beside the component symbols. The last capacitor is marked as C207. Marking electronic components with corresponding letters and numbers makes them easy to locate. You might find some foreign circuits where the parts have a letter and number, except that the value is left off.

These same part symbols, letters, and numbers are found in the parts list (Fig. 2-18). Usually, the parts list provides the actual component number that must be replaced. Replace the part with exact part numbers or with a universal replacement.

Information

Terminal guides

Usually, each electronic semiconductor and transformer must have a terminal guide for correct connection in the circuits. These guides might be rough drawings of the different components that share guide terminal numbers or letters (Fig. 2-19). These same numbers appear on the schematic. You must know what terminal

■ **2-18** *Notice the different part numbers, symbols, and voltages stamped on the PC wiring of this portable TV set.*

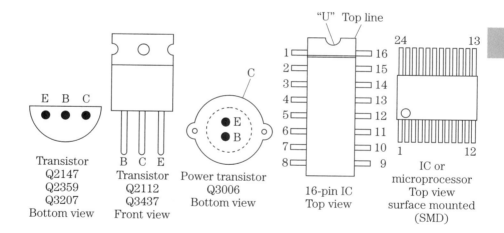

■ **2-19** *The terminal guides of the various semiconductors help to identify each terminal number with symbol letters and numbers.*

is connected when it is in the circuit for voltage and transistor tests. Terminal guides are found with most schematics in magazines, books, and service manuals.

Electronic hobby and project schematics

Usually, electronic projects have very simple schematics that most anyone can follow (Fig. 2-20). These projects consist of one or two

Information

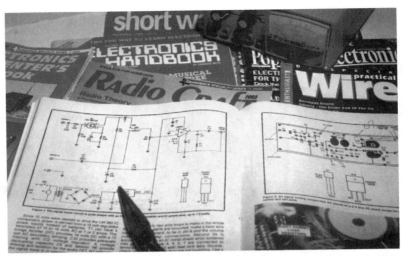

■ **2-20** *Very simple schematics are used for projects constructed in hobby or electronic magazines. Each symbol has a part number that is used in the parts list.*

transistors, ICs, or a combination of both. Again, resistors and capacitors begin with number 1 and the letter R or C. Likewise, if two ICs are used, they might be marked IC1 and IC2. These projects might be handwired or mounted on an etched PC board.

The terminals of transistors and ICs can be found within the schematic or in a schematic key. The IC symbol might have the numbers around the triangle or rectangular component with numbers in any order. Often, an underside view of the IC terminals on the etched board might start with pin 1 marked on the PC board (Fig. 2-21). Sometimes the IC is numbered from the top with the "U," and line and terminal number 1. You can count off the numbers around the in-line socket or etched base terminals.

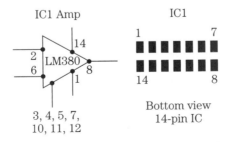

■ **2-21** *IC1 has 14 terminal connections for the IC amp; the different terminals are used on the PC wiring.*

Electronic Servicing & Technology (EST)

Besides the electronic hobby magazines, electronics servicing magazines sometimes have complete or partial schematics stapled inside of the magazine (Fig. 2-22). These schematics are reprints of the actual schematics with current model numbers. *Electronic Servicing & Technology (EST)* magazine has several schematics placed in each issue.

■ **2-22** *The large extra issue of* Electronic Servicing & Technology *magazine features schematics for various consumer electronic products.*

A larger copy of the regular magazine is issued at the end of the year with every subscription of *EST*. This 1993/1994 copy has 12 different manufacturers' schematics, in a large drawing that can be easily seen and traced. This issue also contains directories of consumer electronics, manufacturers, distributors, and state associations.

Electronic Servicing & Technology is edited for professionals who service electronic equipment. This includes electronic service technicians, avid servicing enthusiasts and field service technicians, who service and repair audio, video, computer, and other consumer electronic equipment. This monthly service magazine can be obtained by subscription only, not at the newsstands.

Electronic Service & Technology
76 N. Broadway
Hicksville, NY 11801

Howard Sams Photofacts

Howard Sams Photofacts covers the majority of consumer electronics schematics manufactured each year (Fig. 2-23). When schematics are not available from electronics manufacturers, these Photofacts are a must-have item. You can purchase these schematics from your local electronics supply house.

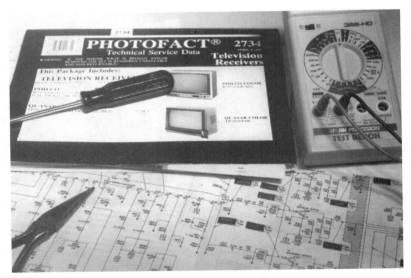

■ **2-23** *Howard Sams Photofacts feature waveform and voltage measurements of the TV, VCR, computer, and classic radios.*

Photofacts has an annual catalog-style index that lists all of their schematics by manufacturer and model number. When a certain model number is not listed, look up the number next to it and compare the circuits. A lot of manufacturers use the same circuits year after year.

In Sams TV Photofacts, you will see partial schematics or broken down sections of the TV chassis of the low-voltage power supply, CRT circuits, color output, tuner, stereo sound, and special circuits. The main TV schematic might fold out to a larger size for easy servicing. Besides regular schematics, pictorial and hookup drawings are shown in detail.

Special photos of the TV chassis are found with components marked out for easy location. Photos of a given chassis might include the audio board, cabinet rear-view, control board, CRT neck assembly, main board, PC board, rear control board, and RGB board. Each board might have quick-checks of troubleshooting with grid-trace part location. Troubleshooting tips and miscellaneous adjustments are found on page 1. Safety precautions of working on the TV chassis are given.

Service information, servicing in the field, and proper test equipment are important items. Test jig hookup and terminal guide with notes saves the electronics technician valuable service time. Critical voltage, resistance, and waveforms are found throughout the main schematic. Howard Sams Photofacts and consumer electronic service manuals can help locate those tough problems. The parts list with various universal part replacement is worth the price of each Photofact.

Howard W. Sams & Co.
2847 Waterfront Pkwy.
East Dr., Ste. 300
Indianapolis, IN 46214

How to use the digital multimeter (DMM)

THE DIGITAL MULTIMETER IS THE MOST VERSATILE TEST instrument found on the service bench (Fig. 3-1). You can check ac or dc voltage, resistance, continuity, ac and dc current, diode-transistors, transistors, capacitance, inductance, logic probes, and check the frequency in the deluxe DMM. The accuracy of dc voltage can be from 0.5 to 0.1 percent.

Digital multimeters are available in hand-held, portable, and bench models. Most DMMs have table cases or a shock-resistant heavy-duty case with rubber holster and tilt stand. You can pur-

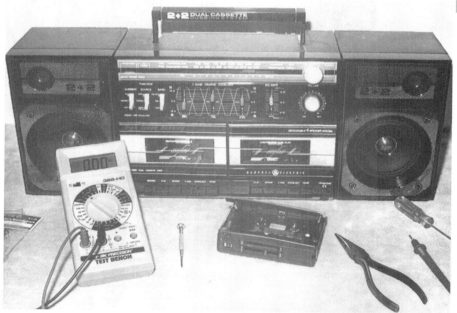

■ **3-1** *The small digital multimeter is one of the most valuable test instruments on the service bench.*

chase a DMM from $19.95 up to $299.95. Of course, the low-priced models might have only a few features.

The DMM is very accurate in taking resistance and voltage measurements. Troubleshooting is fast and accurate—with no guesswork. Of course, no test instrument is any better than the person operating it. Just learning how to operate and use the DMM might save you time and money.

Basic operation

Most digital multimeters have a liquid-crystal display (LCD) that can be seen in very dark rooms or very bright sunlight. Choose one with a large easy-to-read 3¾-digit LCD display. The small meter is ideal for the electronic service technician, engineer, hobbyist, or home owner who demands a very accurate and reliable test instrument. This meter operates well on the service bench, laboratory, or in the field (Fig. 3-2). Some meters are equipped with a rotary function switch or pushbutton operation.

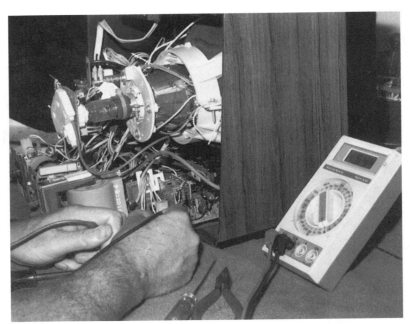

■ **3-2** *The DMM can be used on the service bench, in the lab, or in the field.*

These meters are very rugged and can last for years of trouble-free operation under heavy-duty servicing. You might have to replace the test leads several times during the life of the meter because it is used so much. The plug-in connection of the test leads wear and spread after several years until they need to be replaced. Most meters have built-in overload protection for the voltage and resistance ranges.

Many of these multimeters can operate with the same battery for over 2000 to 3000 hours because the circuits and the liquid-crystal display have a very low current drain. Even when the battery becomes very weak, you can still see the digital display with very light numbers. In some models, the decimal point in the digital display blinks when 200 hours of battery life remain or the "Lo Bat" and marks are displayed on the LCD panel. Most meters operate from a 9-V battery.

Maximum display readouts are always one digit less than the marked range. For example, the 200-Ω resistance range reads between 0.000 and 199.9 Ω (Fig. 3-3). When higher resistance is measured on the range, the OL or 1 (overrange indication) will show in the display. Likewise, with the 2-Vdc range, the voltage starts at 0.000 and measures to 1.999 V. You do not have to worry about a high voltage damaging the needle pointer or meter of a regular VOM. Simply rotate the function switch to a higher range.

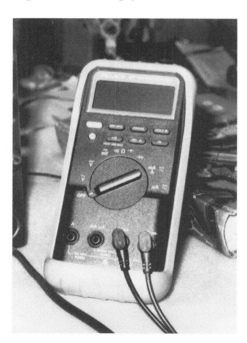

■ **3-3** *The digital multimeter will check voltage, current, resistance, diodes, and capacitance.*

The dc voltage and current readings have automatic positive and negative polarity symbols. Don't worry about switching test leads or accidentally selecting the wrong polarity. Always observe the correct voltage polarity as shown in front of the voltage reading. In some models, an instant Ω sign appears in the top left corner, which indicates correct resistance tests. When the Ω sign is erratic, poor probe contacts or intermittent tests are noted. Other models have an audio tone to alert the operator when the resistance measurement is erratic.

DMM precautions

When replacing the battery or fuse, remove the test leads and make sure the snap terminals are correctly seated. Select the highest proper measurement range and progressively select lower ranges until the measurement falls within the correct range. Most DMMs will not measure voltage above 1000 Vdc or 750 Vac. Do not try to measure high voltage in the TV chassis or microwave oven (Fig. 3-4) with a DMM.

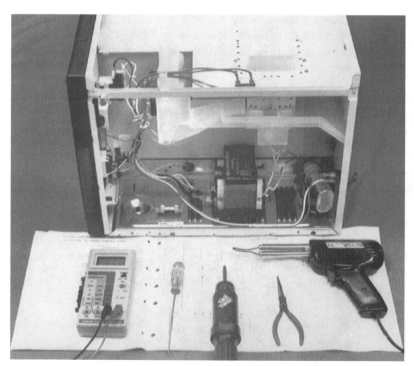

■ **3-4** *Use the DMM to make continuity tests on the microwave oven with the power cord pulled. Do not take the high-voltage test on the microwave oven.*

Be careful when taking high current measurements. Sometimes the 20-A range might be unprotected and have a low internal resistance. Do not attempt to make unknown current measurements above 20 Adc. The common terminal on some meters should not exceed 500 V measured to ground. The mA ranges are usually fuse protected. To avoid possible electric shock, instrument damage, and equipment damage, always first select the highest current range. Remove test leads from the circuit when changing the measurement range.

Before changing ranges, always re-energize the circuit completely. It's best to discharge filter capacitors before trying to take resistance measurements on the low-voltage power supply circuits (Fig. 3-5). Remember, on some meters, an open circuit exists between the test leads during range changing of the multimeter.

Completely discharge the circuit or device in which the resistance is to be measured. All resistance and diode measurements should

■ **3-5** *Always discharge electrolytic capacitors before checking them with the ohmmeter scale of DMM.*

be taken in de-energized circuits only. Do not connect the COM (common) terminals to potentials exceeding 500 V to ground.

Always depress the correct function and range pushbuttons or rotary switch before taking measurements. Do not exceed the maximum voltage or current range at the input terminals. Always switch the power off when the instrument is not in use. Remove the battery if the instrument will not be used for a long time.

Auto ranging

Some DMMs have an autoranging method to determine the range of the meter. When the meter is turned on, the DMM is in the autorange mode. A switch selects the correct function to be measured. When the test leads are connected, the overrange (OL or 1) indicator lights if the autorange is not high enough (Fig. 3-6). Push the autorange button until a reading is noted on the display. The overrange display (OL or 1) will go off when the correct reading is found.

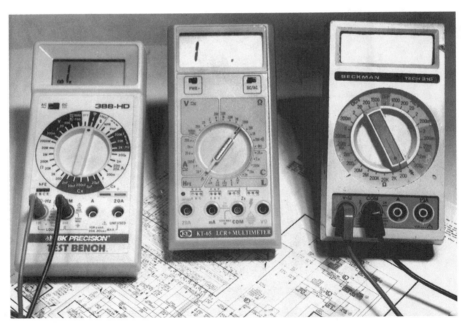

■ **3-6** *The overrange of the DMM might be OL or 1 symbol on the different meters.*

Separate range scales are not switched into the DMM circuits, like other meters. The rotary switch or pushbuttons select any voltage, current, ohmmeter, and diode tests. The correct measured value is selected by the overrange feature. For example, if a dc voltage is to be measured, turn the function switch to the dc voltage range. Now, measure the voltage (Fig. 3-7). If the voltage is too high, the overrange (OL or 1) display will be seen. Now press the overrange button for a higher reading.

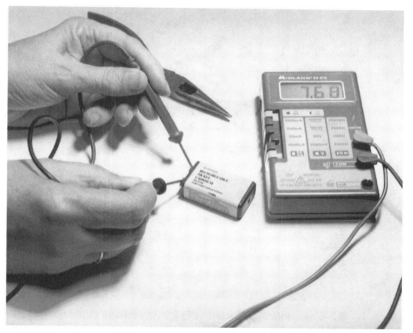

■ *3-7* Checking a 9-V battery with the DMM. If the battery is checked on the 2-V scale, the overrange symbol will appear.

Continuity tests

Continuity tests indicate whether the component or circuit is open or closed. Besides the ohmmeter range of a VOM or DMM, continuity tests can be made with a battery and lamp bulb, or a battery and buzzer.

A constant tone or noise can be heard in some DMMs when taking diode and continuity tests. The tone test can quickly indicate intermittent, open, and poor test lead connections. The sound starts and stops quickly. No beeping sound is heard with open or high-resistance circuits. Often beeper sounds are heard under 100- to 200-Ω measurements.

Continuity tests are quick (Fig. 3-8). They are usually made on a low-resistance component, broken wires, and pc wiring. Rotate the DMM to the lowest Ω scale. Make sure the test leads are plugged into the Ω and common (COM) terminal of the meter. Touch the test leads together for a shorted measurement (000). You should hear a tone or beeper sound if the meter has continuity beeper or tone features. If the overrange (OL or 1) display comes on, recheck the meter settings and test leads.

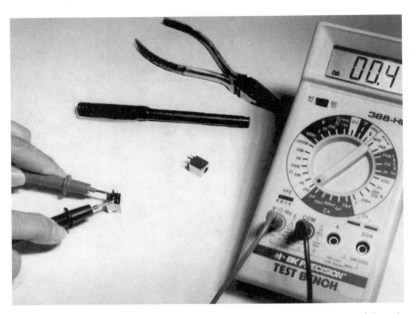

■ **3-8** *A continuity test with the low-Ω scale of DMM will indicate if the circuit or part is open or has continuity.*

Suspect a broken connection or circuit if the sound stops. The intermittent beeping sound might indicate an intermittent or erratic connection. Some meters have a beeping noise on the diode test.

Take a look at the typical symptom of a dead AM/FM stereo receiver. No sound—no panel lights indicate problems within the low-voltage circuits. The power switch and transformer are suspect with no panel lights. Check the on/off switch by taking a continuity test of the switch terminals (Fig. 3-9). Remove the power plug from the ac receptacle. Rotate the DMM to the 200-Ω range. Place the test probe across the switch terminals.

Rotate the switch to the on position. The normal switch will show a shorted measurement. No reading indicates open switch terminals. Erratic or a low-Ω measurement indicates a worn or dirty switch

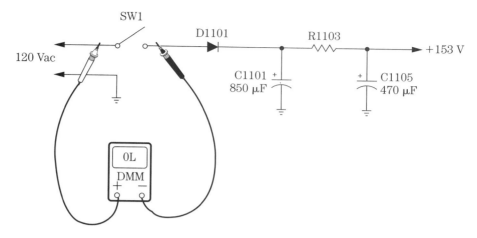

■ **3-9** *The continuity of an on/off switch is found by starting on a low-Ω range and switching the switch off and on.*

contact. Replace the defective switch. Remember, continuity measurements are made on low-resistance circuits or components.

Resistance tests

Most resistance ranges on the average DMM are 200 Ω, 2 kΩ, 20 kΩ, 200 kΩ, 2 MΩ, 20 MΩ, and 2000 MΩ (Fig. 3-10). On the deluxe mod-

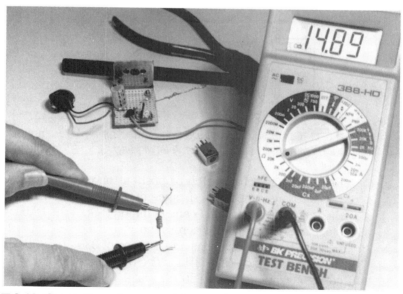

■ **3-10** *Check the unknown resistance of a resistor by starting at a high-resistance range and switching down to the correct scale of DMM.*

els, the Ω range might start at 200 Ω and go up to 4000 MΩ. You might find a musical tone symbol at the low (200 to 400) Ω range.

Accurate resistance tests can be made with the DMM taken in or out of the circuit. Measuring resistors in circuits containing transistors and diodes is quite accurate with most meters. The digital multimeter quickly and accurately checks critical tolerance of emitter resistors in transistorized circuits. The resistance measurements of coil windings found in the receiver or TV circuits are very valuable. Comparable resistance measurements of stereo tape heads in the cassette or tape player might determine if the tape head is defective or not (Fig. 3-11).

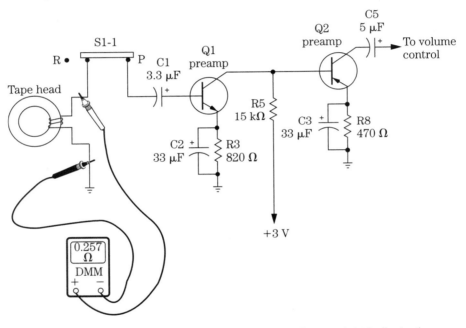

■ 3-11 Continuity of the tape head in the tape player, might indicate the winding is open or has a complete winding.

Rotate the selector switch or pushbutton to the correct resistance range to measure the resistance of a given component. Make sure the test leads are in the Ω and common jack of the DMM. For example, place the probe tips across the suspected resistor or leaky component. If the resistor has a resistance of 330 Ω, rotate the selector to the 2-kΩ range (Fig. 3-12). The overrange display (OL or 1) will come on when using the 200-Ω scale. The resistor is either open, or increased or decreased in resistance if the average display lights. Suspect a burned resistor if the reading is less than 300 Ω. Remember a 330-Ω 10% resistor might be off 30 Ω and still be normal.

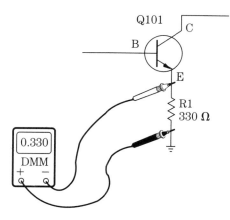

3-12 Measuring the emitter resistor of Q109 to check for correct resistance. Remember that the resistance might be off 5 or 10%.

Remove and test the suspected resistor out of the circuit for a change in resistance. Make sure that no coil or diodes are found in parallel with the resistor to be measured. If in doubt, remove the terminal lead of the resistor and make another test. You can take resistance measurements of transistors, ICs, diodes, triacs, SCRs, etc. (Fig. 3-13).

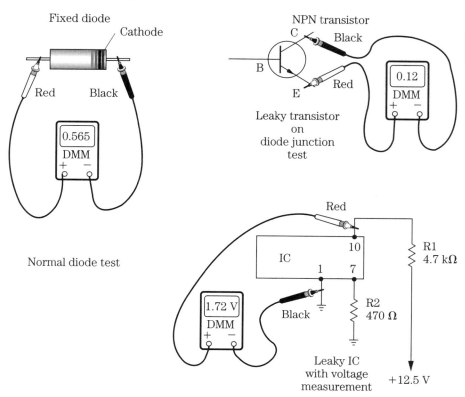

3-13 Check the diode, transistor, and IC for shorted or leaky conditions.

Resistance tests

Voltage measurements

The DMM is ideal for measuring critical voltages in semiconductors and solid-state circuits. Low critical voltages determine if a transistor is defective or normal. A normal silicon transistor might have a forward reading of 0.6 V, but the normal germanium transistor has a forward reading of 0.3 Vdc. The typical VOM will not measure voltages less than 1 Ω accurately. You can quickly determine if a transistor is open or leaky with a forward bias voltage reading between base and emitter terminals (Fig. 3-14).

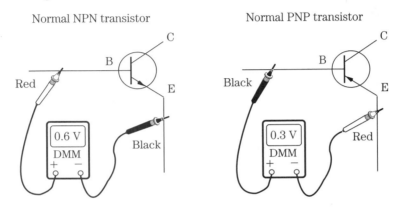

■ **3-14** *Determine if the transistor is defective with a very low voltage test between base and emitter terminals.*

Accurate voltage measurements can be made in transistorized circuits. Check the voltage measurements on the schematic diagram when taking voltage measurements. Compare these voltages with the one found on the DMM. Rotate the function switch to the desired voltage range. If the voltage is not known, start at the highest scale and turn down to a working voltage. Any time that the average display (OL or 1) comes on, turn to a higher voltage scale. Most DMMs will indicate the correct voltage polarity without changing test leads (Fig. 3-15). Both ac and dc voltages can be checked from the same test jacks with selected voltage settings or separate ac/dc switch.

For example, if you are taking voltage measurements within the horizontal output and drive transistor circuits, rotate the voltage to a higher scale. Place the selector to the 200-V range. Clip or place the black lead to the chassis (common ground). Touch the red probe to the collector (C) terminal of the driver transistor (Fig. 3-16). If the voltage is normal, the horizontal oscillator and drive circuits are okay. If the voltage is very low, suspect a leaky driver transistor, increased voltage at the dropping resistor, no drive signal, or an improper voltage source.

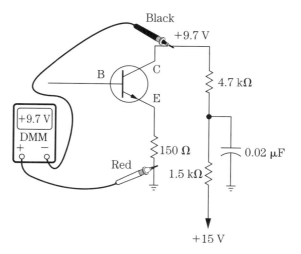

3-15 You do not have to change test leads to measure voltage in a circuit with the DMM. It will spell out the correct polarity on the display.

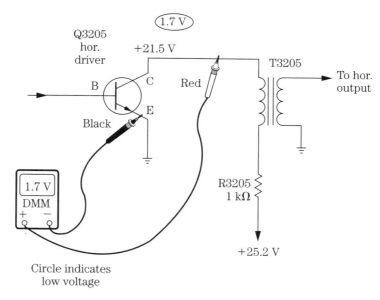

3-16 Touch the DMM red probe to the collector terminal of an NPN transistor to determine if the transistor is leaky or if there is no supply voltage.

Current tests

The current test measures the actual flow of current in a circuit. A current drain might be caused by a leaky or overloaded component in a circuit. You can measure the current flow by placing a current meter in series with the voltage source and load

(Fig. 3-17). The current is measured in amperes or microamperes. Current scales might start at 200 or 400 µA up to 10 or 50 A, with a separate jack for high-ampere measurement.

In battery-operated circuits, you can measure the current drain by connecting the terminal probes in series with the battery or across the on/off switch. Simply leave the project or product turned off and measure the current (Fig. 3-18). Likewise, the current drain of a TV set can be checked for overload condition by clipping the ac amp meter across the ac switch. Always remember in dc circuits to flip the current meter to dc milliamperes and when in ac current, to ac amps. Do not connect a current meter across the power line on dc voltage sources. Be very careful when measuring ac parallel line voltage and current.

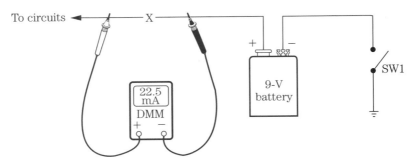

■ **3-17** *Perform the current test on an electronic project by inserting the meter leads between the battery or load.*

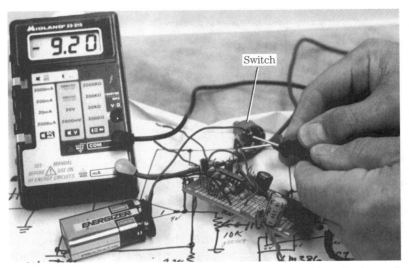

■ **3-18** *Measure the current of a small audio project by placing milliampere test leads across the switch with unit turned off.*

How to test small capacitors

Suspected capacitors should be removed from the circuit to provide accurate measurements on the DMM. Discharge large filter capacitors before attempting to measure. Electrolytic capacitors can be checked with the 20-kΩ range of some digital multimeters. Capacitors have a tendency to charge up and discharge through the meter (Fig. 3-19). Often, the reading will increase until the overrange (OL or 1) appears on the meter. Then afterwards, the Ω symbol will disappear.

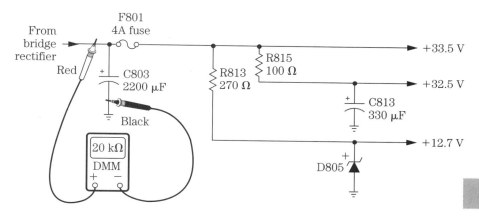

■ **3-19** *When checking the electrolytic capacitors, the capacitor will charge up and discharge on the DMM.*

Now, reverse the test leads across the capacitor. The reading might go to zero and charge up again until the overrange readout appears. The charging and discharging of the electrolytic capacitors with the insta-Ω symbol and overrange readout might indicate the condition of the capacitor. A capacitor with an open or dried-up condition might not charge or discharge. Of course, smaller capacitor values have a shorter charge and discharge rate. They discharge very fast.

Small electrolytic capacitors (under 1 µF) might not produce a quick indication on the DMM, except that the meter Ω symbol might stay on a few seconds. Very small bypass capacitors (0.01 µF) might not show a reading on the Ω symbol of the DMM. The condition of a normal electrolytic capacitor might show the charge and discharge rate with the presence of the insta-Ω symbol.

Place the function/range switch in the 20-kΩ range position. Connect the red test lead to the voltage Ω input jack and the black lead to the common test jack. Discharge the capacitor. Now, place the test leads across the capacitor (Fig. 3-20). If the insta-Ω symbol appears briefly and then disappears, the capacitor is accepting the charge. The capacitor is open if the insta-Ω symbol does not appear. A low resistance reading will be displayed if the capacitor is leaky.

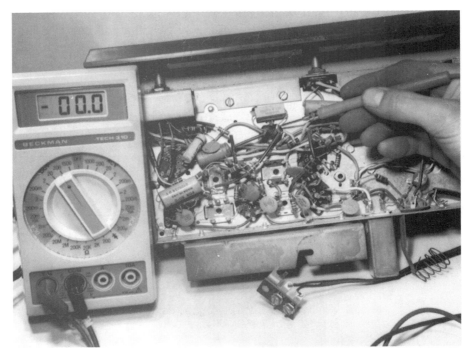

■ **3-20** *Check the voltage across the filter capacitor, then discharge it, and take a resistance measurement with a very low voltage on the capacitor terminals.*

Besides making resistance tests on a suspected capacitor, some DMMs have a capacitor test from 2 nF to 20 µF. The deluxe models might go from 4 nF to 40 µF. Insert the meter leads into the red and common jacks and rotate the function switch to the correct (CX) band. Place probe tips on each capacitor terminal (Fig. 3-21). Very seldom can these small digital multimeters accurately measure large electrolytic capacitors. Choose a regular capacitor tester for larger capacitance values.

Frequency tests

The deluxe DMM might have a frequency counter covering from 0 to 200 kHz, with a resolution of 10 Hz. Auto ranging might be

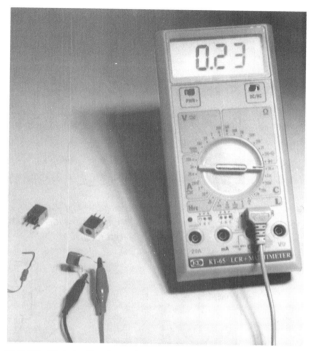

■ **3-21** *Rotate the function switch to the capacitor test and place the probes across the capacitor terminals. Rotate the switch until the right scale is indicated.*

found in the more expensive DMMs. Plug the test leads into the red jack and common lead to the black jack. Start at the highest band and switch down to the correct frequency found on the LCD. Measure for the correct frequency.

Diode tests

Rotate or push the selector switch to the diode test range. Some DMMs have an audible tone for the diode test position. Place the black probe (−) to the cathode terminal of the diode and the red to the anode (Fig. 3-22). The cathode can be marked with a black or white line at one end of the diode. A normal silicon diode reading will indicate only an overrange measurement (OL or 1) with reverse test leads.

No measurement reading indicates that the diode is open. A leaky or shorted diode will show a low measurement in both directions. Always reverse the test probes to see if a reading can be obtained (Fig. 3-23). Normal silicon diodes have a low reading in one direction and germanium diodes have a higher resistance.

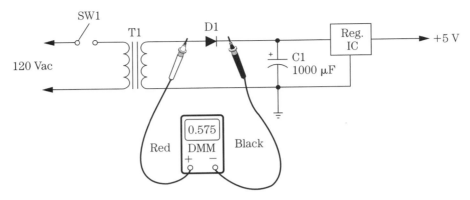

■ **3-22** Place the black probe lead to the cathode and red lead to the anode terminal for a normal diode test. If the diode is shorted, a low resistance is measured with reversed test leads.

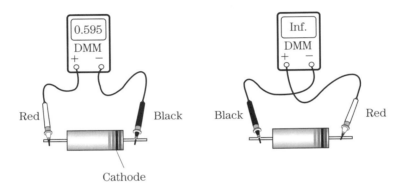

■ **3-23** The normal reading of a silicon diode taken with the DMM.

Diodes can be checked in the circuit. If in doubt, remove one lead from the circuit. Discard the diode if a low measurement is found in both directions. Go a step further if the diode only shows a measurement with the negative terminal at the collector terminal. Switch the function back to the 2-kΩ resistance scale. Notice if a reading is found with the negative probe at the anode, the negative terminal of the diode. Rotate the diode to the 20-kΩ scale. Replace the diode if a reading is found in both directions on the high-Ω scale.

Zener diode tests

The zener diode can be tested in the same manner as any other diode. When leakage is found in both directions (switched probes), remove one end from the circuit for accurate continuity tests. You can test the zener diode with the diode test and a 20-kΩ resistance scale of the DMM. If in doubt with the reverse measurement, replace the suspected diode. Most zener diodes found in consumer electronic products are 1- or 5-W units. Most defective zener diodes are found leaky and operate quite warm.

Bridge rectifiers

The bridge rectifier consists of four silicon diodes connected in a bridge circuit. Two diodes in each end of the unit provide half-wave rectification. The bridge rectifier permits full-wave rectification. Usually, a bridge rectifier is one unit with four terminal connections (Fig. 3-24). You can also make a bridge circuit with four separate silicon diodes.

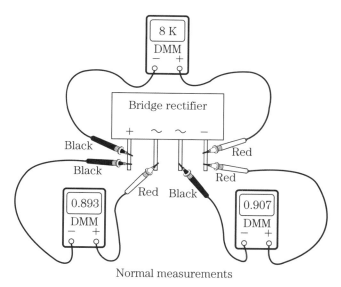

Normal measurements

■ **3-24** *The normal measurement of a bridge rectifier is taken with the diode test of the DMM.*

The bridge diodes can be checked with the ohmmeter or with a diode test. With the DMM diode test, a resistance measurement can be made in one direction. If a very low reading is made with the probes in both directions, the rectifier is defective. The positive terminal of the bridge rectifier connects to $B+$ and large filter capacitor while the negative terminal is at ground potential. An ac voltage is applied to the sine-wave symbol terminals (Fig. 3-25). If a section is leaky, the beeper will sound on testers that have this feature.

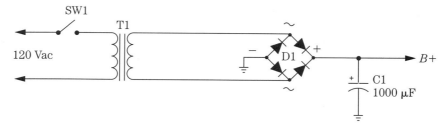

■ **3-25** *Notice that the sine-wave symbol marked on the bridge rectifier connects to the ac leads of the power transformer.*

Detection diodes

Diodes found in the early detection circuit of a TV or radio are often of the germanium type. You will find silicon diodes in modern electronic equipment (Fig. 3-26). Common detector diodes are the 1N34, 1N60, and 1N270. Check the diode with the diode test. Notice the resistance of a detection diode might have a higher reading than a regular silicon diode. If in doubt, remove one end for an accurate measurement.

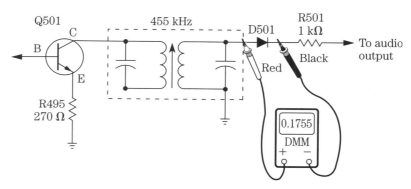

■ **3-26** *The detection diode (1N34) will indicate a higher resistance measurement than a silicon diode on the DMM.*

Checking transistor function with diode tests

The transistor function test can be made on the DMM diode test. Transistors can be tested in a circuit rather quickly with this method. Switch the DMM to the diode test. Leave the red and black plug in respective jacks. With a silicon NPN transistor, place the red probe to the base terminal. Move the black probe to the collector and then to the emitter terminal. A normal transistor will show a comparable reading at each terminal (Fig. 3-27). Remember that the base terminal is common to the emitter and collector terminals.

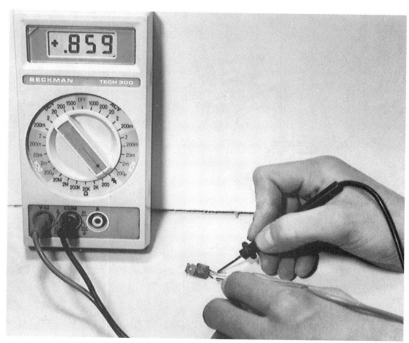

■ **3-27** *The normal transistor will have a comparable measurement from the base to the collector and from the base to the emitter on the diode junction test of the DMM.*

To determine if the transistor is a PNP type, place the black probe (−) to the base terminal. If you do not know what any one of the terminals are, the base terminal will have a reading common to the collector and emitter terminals. When you receive a reading with the negative terminal common to the other two, you can assume that you have located the base terminal and that the transistor is a PNP type (Fig. 3-28).

An NPN transistor will have a common measurement if the red probe (+) is placed on the base terminal. Remember the collector terminal of an NPN will have a positive voltage reading, but the PNP transistor has a negative voltage measurement on the collector terminal. Suspect a leaky transistor if all three terminals have a low reading between any two terminals.

Transistor tests

Some larger DMMs have a separate transistor socket and HFE measurements. Remove the test leads from the instrument. Rotate the range switch to HFE position. Use the appropriate (NPN or

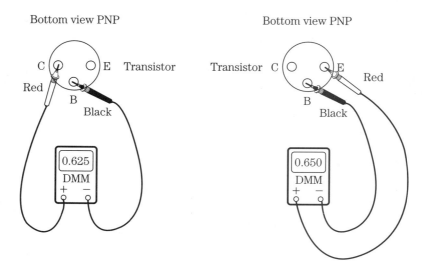

3-28 *Normal diode junction test of a top-hat transistor on the diode tester DMM. Notice that the black probe at the base terminal indicates that the transistor is a PNP.*

PNP) setting. Insert the transistor into the EBC socket. Make sure correct transistor terminals are inserted in respective holes. Read the common emitter forward gain on the LCD.

NPN or PNP

You can quickly determine if a transistor is a PNP or NPN with the diode test of the DMM. Switch to the diode test. Because most transistors are NPNs, place the red probe to the terminal that you think is the base terminal. Now check the other two terminals with the black probe. If you find a comparable reading on the other two terminals, you have spotted the NPN base terminal (Fig. 3-29).

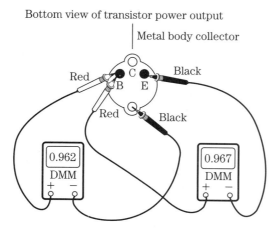

3-29 *The normal diode junction test of an NPN transistor on the diode test of the DMM. Notice that the base-to-collector reading is always lower than the base-to-emitter terminal reading.*

The lowest measurement between the base and another terminal is the collector terminal. The emitter terminal will have a higher measurement between base and emitter. Remember that the base terminal is common to the other two terminals. If the black probe is common to the other two terminals, the transistor is a PNP type.

Inductance measurements

Only a few high-priced digital multimeters have a separate inductance range. You might find another jack to place the red probe into, for inductance measurements, with the common black terminal jack. Rotate the selector to the LX settings on the inductance scale. If the inductance range is not known, set the selector switch to the highest range. The inductance can be measured from 2 mH up to 40 H (Fig. 3-30). Mastering the digital multimeter is one of the most important jobs in electronic servicing. The electronic technician who masters the DMM can troubleshoot most any electronic circuit.

■ **3-30** *The correct inductance of a small choke and coils can be used with the LX range of the DMM.*

4

Required electronic tests

THE NOVICE ELECTRONICS TECHNICIAN SHOULD BE ABLE to locate the defective component with critical voltage and resistor tests, transistor in-circuit tests, and isolate the leaky or open IC (Fig. 4-1). Troubleshooting electronic circuits is easy with the proper test equipment and wiring diagrams. You can then locate and isolate the defective part with only a few test instruments. How to remove and replace the defective component is just as important as how to locate one.

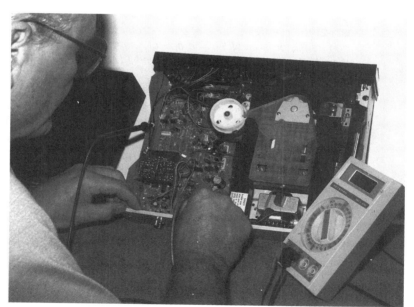

■ **4-1** *Technician taking the voltage measurement of a voltage source in the low-voltage power supply circuits of a compact disc player.*

Critical voltage measurements

The digital multimeter (DMM) can measure low voltages (under 1 V), which are found in most electronic chassis. Besides checking the forward bias between emitter and base (silicon 0.6 V and germanium 0.3 V), critical base and emitter voltages can help you to locate a defective transistor or open bias resistor. Remember, the collector voltage on an NPN transistor is positive and is always higher than the base or emitter (Fig. 4-2). Base terminal voltages are also always higher than the emitter.

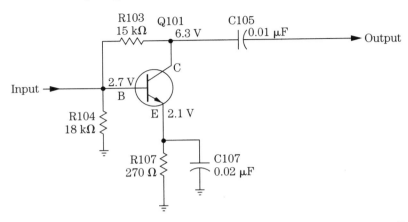

■ **4-2** *The positive voltage at the collector terminal indicates that the silicon transistor is an NPN. The collector terminal voltage is always higher than that of the base and emitter in the NPN transistor.*

In PNP transistor circuits, the collector has a high negative voltage compared to the other two elements. Although, you might find some circuits that have higher positive voltages on the emitter and base terminals, which results in a higher negative collector voltage. In other words, the collector is negative compared to the base and emitter terminals. In older radio and TV chassis, the PNP transistor was used extensively with a high negative voltage source on the collector terminal. Voltages are measured from transistor terminals to common ground.

Today, you might find a PNP and NPN transistor directly coupled in electronic circuits. For example, in a video transistor circuit, a PNP transistor is used as a video amp and an NPN transistor as an emitter-follower (Fig. 4-3). Notice that the forward bias voltage between base and emitter of Q101 is 0.3 V, indicating a PNP transistor, while in the emitter-follower (Q103), the difference voltage is 0.6 V. Remember, the arrow of a PNP transistor points inward and the NPN arrow points outward.

The collector voltage of Q101 is 5.3 V or 1.5 V more negative than the base terminal and 1.8 V more negative than the emitter terminal. Notice that the base is 0.3 V more negative than the emitter

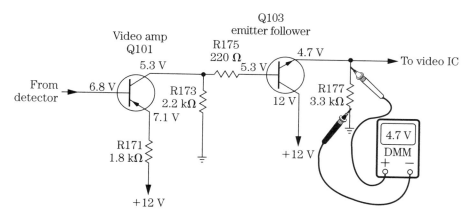

■ **4-3** *Here is a TV video circuit, Q101 is a video amp, and Q103 is the emitter-follower transistor. Notice that the output signal is taken off of the emitter terminal.*

terminal and a high positive voltage is applied to the emitter, resulting in a collector voltage that is higher negative than to the other two elements.

Emitter-follower Q103 has a high positive supply voltage tied to the collector, with a 0.6 V between base and emitter. Notice that the NPN transistor emitter terminal is tied to ground through resistor R177, making the emitter the lowest voltage element in the circuit. Likewise, in an NPN transistor, the collector terminal operates at the highest positive voltage.

Now let's look at an audio amplifier circuit using two PNP transistors (Q111 and Q112). The collector terminal of Q111 is less positive than the other two elements, making the collector more negative (Fig. 4-4). The bias voltage between base and emitter is 0.3 V, indicating that it is a PNP germanium transistor. Notice the collector terminal is tied through a 15-kΩ resistor to a high supply voltage (–6 V).

Q112 has a higher negative voltage (–3 V) than the base or emitter. The collector resistor (3.3 kΩ) is tied to the –6-V source. Also the base resistor (15 kΩ) is connected to the –6-V supply. Notice the difference in the two values of resistors, with only a 3.3-kΩ resistor tied to the collector terminal. This terminal should have a higher negative voltage because less resistance is in the circuit.

Critical low voltages (less than 1 V) can be found in pocket radios, cassette players, and amplifier circuits. A small cassette tape player might operate from two AA cells (3 V), providing low volt-

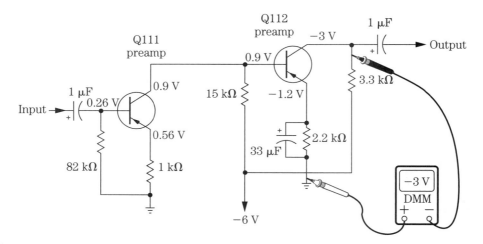

4-4 *The collector terminal of a PNP transistor is always more negative than the base or emitter terminals. The base-to-emitter bias voltage is 0.3 V in a PNP transistor.*

ages in the input and audio amp circuits. The DMM is very accurate in measuring low voltages found in transistor and IC circuits. In Figs. 4-3 and 4-4, you can see how the PNP and NPN voltages are supplied in either a negative or positive voltage source.

Supply voltages

Usually the supply voltage source is the highest voltage found on the IC component. The supply pin might be marked with a (V_{CC}) letter alongside or inside the IC symbol (Fig. 4-5). You might find more than one voltage source on large microprocessors or ICs. A lower-than-normal supply pin voltage might indicate a leaky IC, improper supply voltage, or a defective low-voltage power supply.

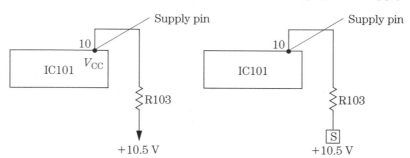

4-5 *The supply voltage source of an IC or microprocessor might be marked with V_{CC} or with the highest voltage terminal, coming from the low-voltage power-supply circuits.*

Voltage sources might be found with arrows and marked voltages, boxed letters and numbers found throughout the schematic. Boxed numbers with a letter alongside, indicate the same voltage, but from another source. These voltage sources might be powered with batteries or dc voltage from the low-voltage power supplies. Secondary-derived voltages are found in the horizontal output transformer circuits of a TV. It is much easier to read or find the different voltage sources with boxed numbers as they eliminate additional drawn lines (Fig. 4-6). In large schematic diagrams, it is often difficult and confusing to locate the defective components.

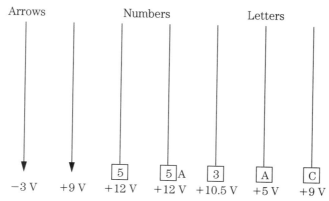

4-6 *The different voltage sources in a large schematic might be marked with arrows, boxed numbers, and letters.*

You must be extremely careful when taking voltage measurements on the IC terminals, to prevent shorting out the adjacent terminals. It's best to take voltages and resistance measurements on ICs on the PC board terminals. Sometimes these voltages can be taken from a component tied to each pin number, preventing shorting of terminals. All IC voltage tests are taken from pin numbers to common ground.

Be very careful when taking voltage measurements on the bottom leads of a transistor. Shorting the terminals together might further damage the transistor (Fig. 4-7). Make sure that the voltage probe tips have a sharp, fine point to provide a good contact on pin numbers and are not sliding off into another circuit.

Critical resistance measurements

Critical resistance measurements and component continuity help to locate defective parts. Accurate resistance tests can be made

4-7 Be careful when taking voltage measurements on the transistor or IC terminals. Don't short out other terminals and destroy this component and other parts.

with the DMM in or out of the circuit. Measuring resistors in circuits containing transistors or diodes is quite accurate with most meters. The digital multimeter quickly and accurately checks the critical tolerance of emitter resistor transistor circuits (Fig. 4-8). The continuity measurements of coil windings found in the TV and radio chassis are very valuable. Comparable resistance measurements of stereo tape heads in the cassette player determine if the winding is open or not.

Rotate the selector switch to the correct resistance range to measure a resistor or component. Resistance measurements on the elements of a transistor will help you to locate a leaky or open transistor. Check each capacitor with a resistance test across the terminals. Make sure the low measurement or leaky reading is not caused by a shunted diode or low-ohm resistor.

Take low resistance measurements from each IC terminal pin to common ground (Fig. 4-9). Check each reading after inspecting the schematic for ground or low resistors in the circuit. If the low-resistance measurement remains, suspect that an IC is leaky. Remove one terminal of the suspected component and take another reading. If removing a resistor, diode, or coil from the circuit does not make any difference, suspect a leaky IC. Check the resistance measurement between any two terminals. Make sure that the two

4-8 The DMM will quickly locate faulty resistance, voltage, and continuity measurements on the various electronic components.

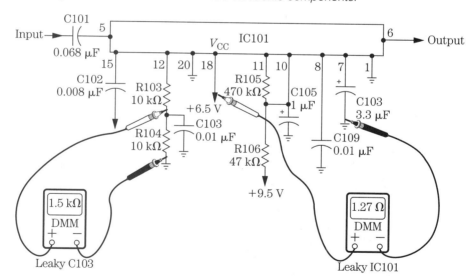

4-9 Before replacing an IC that you suspect is faulty, take critical voltage and resistance measurements on each terminal.

pins are not soldered or tied together with PC wiring. The accurate resistance measurement might locate a leaky capacitor or component tied to the IC terminals, instead of a faulty IC.

Notice in Fig. 4-9, with a low-resistance measurement across C103 off of pin 12 indicates a leaky 0.01-µF capacitor. If the supply voltage is low on pin 18 and the resistance to common ground is low, suspect a leaky IC101. You will find that pins 1 and 20 are at ground potential and should show a short to ground. Any other terminal with low resistances pin to ground, except supply (V_{CC} 19), might indicate that the IC is leaky.

Accurate resistance measurements from IC pins to common ground might indicate that the IC is leaky or that a nearby component is defective. Make sure any connecting component is defective before removing the IC from the circuit. If not, you might find the same service problem symptom still exists after replacing the suspected IC.

Checking transistors in the circuit

All transistors can be checked with a diode-junction transistor test while in the circuit (Fig. 4-10). Transistors can also be checked in the circuit with beta transistor tests. Rotate the selector to the diode test. Suspect a leaky transistor or a low-resistance component within the circuit if a low measurement is between the base and the other two elements. The transistor is definitely leaky if a comparable junction reading appears between all three elements. A low reading can be obtained when a coil, low-ohm resistor, or diode is found in the path of any two elements with a low-diode reading.

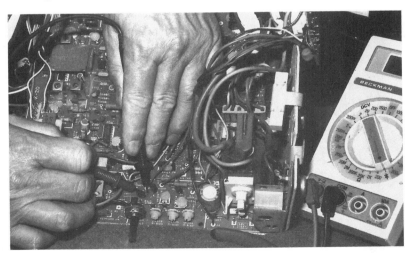

■ **4-10** Check the transistors in the circuit with the diode-junction test of the digital multimeter (DMM).

If there is no measurement with the base terminal common to the emitter and collector, suspect an open junction between these two elements. A higher resistance of 1 kΩ between the two elements and not the other indicates a high-resistance junction. Replace the suspect transistor. Always check the removed transistor out of the circuit for open or leaky conditions.

Quickly, rotate the meter to the 2-kΩ scale and take another reading. If there is no measurement, proceed to the next transistor for test. In case the reading is low, remove the corresponding element lead from the circuit. Check the schematic for a low-resistance path between the two terminals. For instance, if a 680-Ω measurement is found between base and emitter terminals, remove the emitter terminal from the circuit with solder wick. The transistor can be removed and tested out of the circuit.

A quick method is to remove only the corresponding leaky element with the base terminal. Now, just re-solder the emitter terminal with no leakage measured between the emitter and base terminal. Replace the leaky transistor if the same low-resistance measurement is found with the emitter terminal removed from the circuit.

Remember that intermittent transistors might snap on and test good in the circuit with meter probes attached. If in doubt, remove or replace the suspected transistor. A quick diode-junction transistor test on each transistor can be performed in seconds (Fig. 4-11). Check all transistors within that circuit with the same method. Transistors can also be checked by taking bias voltage test between the base and emitter terminals. This quick method of checking transistors in the circuit locates the defective transistor in several minutes.

How to use the VOM

The volt-ohm-millimeter (VOM) was the first test analog instrument used in servicing electronic circuits. The VOM will measure voltage, resistance, and current in electronic products. The Simpson model 260 is one of the most rugged testers found on the service bench. The meter is quite accurate compared to the inexpensive models.

Most early VOM testers were used for continuity and voltage measurements. These meters have a tendency to load down the circuit that provides different critical voltage and resistance measurement compared to the digital multimeter. The VOM is an ideal indicator in ac or audio alignment, whereas the DMM changes

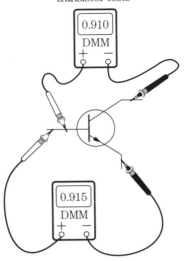
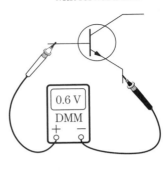

4-11 *Quick diode-junction tests on the transistors in the circuit can easily locate a leaky, open, or normal transistor.*

numbers very rapidly. You can see the position of the meter hand during critical alignment procedures.

Usually the VOM has different jacks for the various tests. Today, varactor or diode protection is provided in the higher priced models to prevent meter hand damage. You must determine the voltage polarity before applying the test leads in the circuit. Often, the voltage, current, and resistance ranges are limited in testers that are priced lower than the DMM.

Rotate the selector knob to the desired range. Plug in the test probe in respective jacks. Always observe correct polarity when taking voltage measurements. Do not accidentally place the test probes into a circuit with voltage while placed on the resistance range. The expensive VOM might have overload protection, including a reset pushbutton when overload occurs. The meter will not reset until the overload is eliminated.

The VTVM

The vacuum-tube voltmeter (VTVM) is a test instrument that uses tubes as an amplifier. The electronic voltmeter might be called an *FET voltmeter* in the solid-state field. The VTVM will not load down the circuit as the VOM and provides accurate measurements. No meter hands can be damaged with high ac or dc voltages if it is accidentally touched on the resistance scale. VTVM and FET voltmeters have a high internal and input resistance combined with excellent sensitivity and amplification.

In the early service days, the VTVM was used extensively in alignment, voltage, and resistance measurements. Most VTVMs have a large meter dial (Fig. 4-12). Besides normal voltage measurement, a high-voltage probe can be attached to measure the anode high voltage on the picture tube.

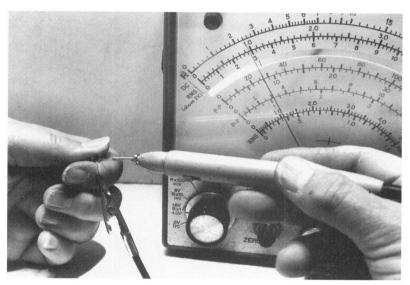

■ **4-12** *The VTVM or FETVM tests will not load down the circuit and will offer a higher resistance input measurement.*

The FET voltmeter has an FET transistor input amplifier circuit that has high-resistance input features. The analog FET voltmeter might have an input impedance of 10-MΩ overloaded protection for both meter and internal circuitry. The voltage range might go up to 1200 V on ac and dc ranges. Resistance can be measured from 1 Ω to 1 MΩ. Of course, the FET voltmeter operates from batteries, and the VTVM plugs into the power line.

Required test instruments for the beginning technician

When learning all about electronics, the beginner or low-level technician should know how to operate a VOM or DMM, be able to take various resistance and voltage measurements in simple electronic circuits, read circuits, know how to solder, and know the various components and symbols found in electronic circuits. The best way to take resistance, voltage, and current measurements is to simply apply the small meters to a given circuit (Fig. 4-13).

■ **4-13** *Accurate voltage tests on the motor battery or terminals might indicate if the motor is defective in the cordless hand drill.*

Begin by checking batteries in the small radio, tape player, or cordless tool. Next, check voltage and resistance measurements in the low-voltage power supplies. When these circuits appear normal, proceed to the transistor and IC circuits. You learn how to test transistors with the simple transistor tests or by using the diode-transistor tests of the DMM. Voltage and resistance measurements can be performed on the IC with the VOM or DMM. You can signal-trace the simple audio circuits with a signal generator and audio amplifier.

Required tools and equipment:

1. DMM or VOM
2. Transistor tester
3. Signaltracer
4. Audio amplifier tester
5. Solder gun and equipment (Fig. 4-14)
6. Hand tools

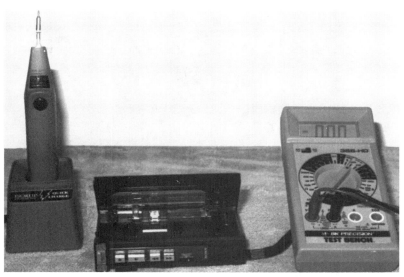

■ **4-14** *The battery soldering iron is ideal to resolder transistor or IC terminals after replacing the defective component.*

How to make tests in transistorized circuits

Before servicing any electronic product, you must have a symptom and several hints to help repair or solve the defective unit. For example, you might have a small portable tape player that is completely dead. The tape does not rotate and no sound is audible from the headphones or speaker. Naturally, the first thing you do is to check the batteries or power that operates the unit. Most electronics technicians make sure that the low-voltage power supply has the correct voltage sources before checking any other circuits.

If the batteries are normal or have been replaced, check the on/off switch. A leaf switch is found in most small cassette players (Fig. 4-15). Check the switch contacts and connecting wires with the low range of the ohmmeter. Defective batteries and switches often cause many problems with a dead tape player. You might have a defective motor that will not rotate the tape and a defective amplifier for no sound, but both will rarely occur at the same time.

The tape is now rotating after all of the batteries in the tape player have been replaced. Even if only one is defective, replace all batteries. This will prevent a call back. No sound at either the headphones or speaker indicates that the amplifier circuit is dead. Glancing at the small circuit diagram that came with the player, you might find two transistors (used as preamps) and one IC in the audio output circuits (Fig. 4-16).

■ **4-15** *A leaf switch is used in this portable cassette player to turn the unit off and on.*

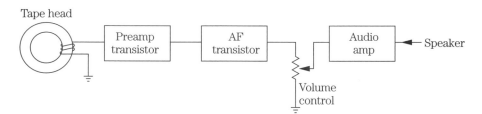

■ **4-16** *A block diagram of a simple tape player. The block diagram can help to isolate the defective stage with a given symptom.*

Divide the audio circuit in half by starting at the center terminal of the volume control. Rotate the volume control full on. Place your finger and metal screwdriver at the center terminal of the volume control. If the tape player makes a loud humming noise, the audio output circuits are probably normal. You can check the amplifier circuits by injecting a tone signal from a small generator and use the speaker as an indicator. Sound and hum is heard in the speakers, so proceed to the input circuits.

Another method is to check both transistors in the circuit. Use either a transistor Beta tester or the diode-junction transistor tests. Remember, in directly coupled circuits, the readings of one transistor might upset the other transistor tests. Remove the base terminal of the transistor and proceed with another test. If in doubt, remove the transistor and test it out of circuit.

You can spot a defective transistor with voltage measurements. Measure the voltage between emitter and base terminals. If the voltage is 0.6 V, you know the transistor is an NPN type and normal (Fig. 4-17). Take a test from collector to common ground. Likewise, check the emitter and base elements in the same manner. Suspect that the transistor is leaky if there is a very low collector, base, and emitter voltage. Be careful not to short out the transistor terminals. Remember, the collector voltage is always the highest in the audio amp circuits.

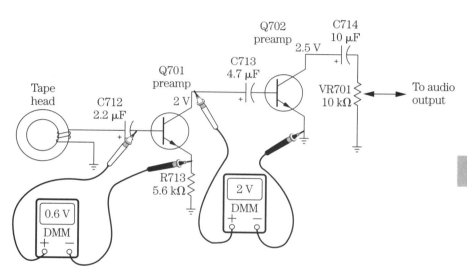

■ **4-17** *The critical voltage test of the forward bias and collector voltage might indicate that a transistor is defective. The NPN transistor has a forward bias of 0.6 V.*

Making simple integrated circuit tests

For another example, the hum tests did not indicate in the small speaker of the tape player, resulting in a defective audio output stage. Because only one small IC is found in the audio output circuits, signal and voltage tests can locate a defective integrated circuit. Locate the supply terminal of the output IC. If the voltage is very low, suspect that an IC is leaky. Do not overlook the possibility of low voltages from the voltage source or batteries.

Apply the audio signal from a generator or audio signal to the input terminal (Fig. 4-18). The audio signal could be from another

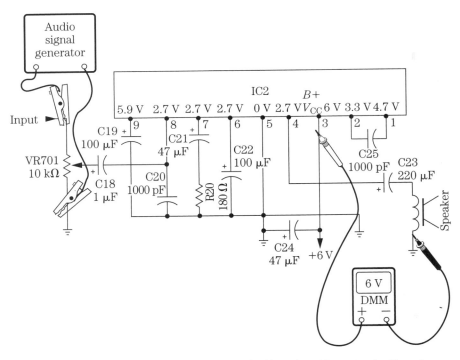

4-18 The audio IC amp can be tested with a signal in and out with a signal generator. A critical supply voltage on the supply pin of the IC (V_{CC}) might indicate that the IC is leaky.

radio or tone source. An audio oscillator signal should be heard in the speaker when injected at the top side or center terminal of the volume control. Weak or no signal at the speaker might indicate an open or leaky IC. Then, take voltage and resistance measurements on each terminal to common ground, to eliminate a leaky capacitor or defective resistor tied to one of the IC pin terminals.

Remember, ICs can be tested with a signal. If the speaker is suspected, clip another across the defective one. Check the supply voltage source terminal of the suspected IC. If the voltages are very low compared to the schematics, you might find a defective IC.

Identifying the defective part

The professional technician takes a good look at all components found on the circuit board to see if burned, overheated, or poor wiring can be located. Remove all dust and dirt so that each section can be inspected. You might find burned bias resistors, shorted silicon or zener diodes, overheated marks on transformers or a change in color of the body of a transistor. Sometimes a poorly soldered joint can be seen at the junction of an overheated component.

Leaky bypass capacitors will lower the voltage in that circuit and a leakage test across the capacitor terminals indicates a leaky capacitor (Fig. 4-19). A leaky bypass C11 will short the input signal to ground and drop the base terminal of Q101 to near zero. No voltage at emitter pin might indicate that Q101 or R10 is open. Measure the resistance across R10 from the emitter to ground and if it is normal, the transistor is open.

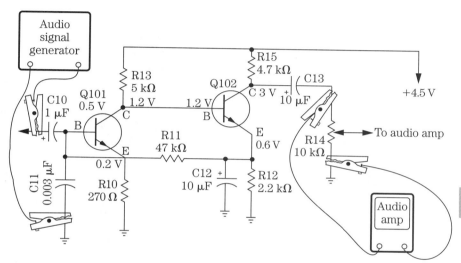

■ **4-19** *The audio signal in and out test of each preamp stage might help you to locate the defective part. After locating a weak or distorted stage, take critical voltage and resistor measurements.*

Leaky C12 (10 µF) might cause loss of volume and distorted sound. Take a resistance test across capacitor C12 terminal. If resistance is lower than R12 (2.2 kΩ), replace the capacitor. If in doubt, remove one end of the capacitor and test it for leakage.

A higher voltage (+4.5 V) on the collector terminal might indicate that Q101 is leaky. If the voltage is lower than 1.2 V, suspect that a transistor is leaky. Remember that if transistors become leaky or shorted, it will be between the collector and emitter terminals.

Weak volume or signal might result from a leaky C102, increase in resistance of R12, or dried up C13. The weak signal can be signal-traced with an external audio amp or voltage, resistance check, or part substitution. Check Q102 in the circuit for leakage or beta tests. Measure the resistance from emitter terminal to common ground to see if R12 is defective. Shunt another electrolytic capacitor across C13 to see if the volume increases. A leaky C13 will lower the collector voltage of Q102 and might produce distortion and weak volume.

A defective component will often appear discolored. Leaky capacitors will have a lower voltage and resistance in the circuit (Fig. 4-20). Open capacitors will create a lower or dead signal. Resistors have a tendency to increase in value and the body might be burned. The defective transistor might be open or leaky, which would produce a dead or weak stage.

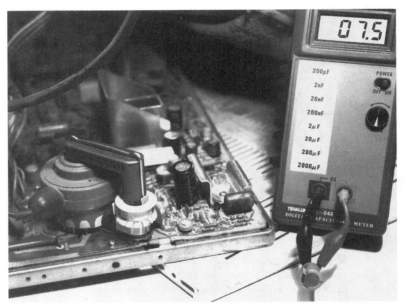

■ **4-20** *Leaky capacitors can be checked with voltage and resistance tests. Remove the suspected capacitor and test it out of the circuit.*

Why components break down

Components might break down with too much voltage applied, other leaky components, and become overheated. Solid-state components might destroy themselves. The leaky transistor might become leaky internally and take out collector resistors, diodes, and bias resistors. The leaky IC might cause bias or load resistors to become overheated with a low-voltage supply source (Fig. 4-21). Solid-state components might short internally, become leaky, or appear open. The connected IC or transistor might break down with a leaky solid-state component tied into the circuits.

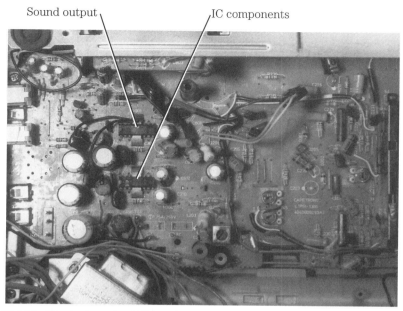

■ **4-21** *The leaky output IC in a boom-box player might cause the IC to run warm and burn the bias resistors.*

Voltage-dropping bias and load resistors might become quite warm and burn with a leaky IC transistor or capacitor in the circuit (Fig. 4-22). Often, carbon resistors will increase in value, upsetting the circuit. If too much current is pulled or passed by a small

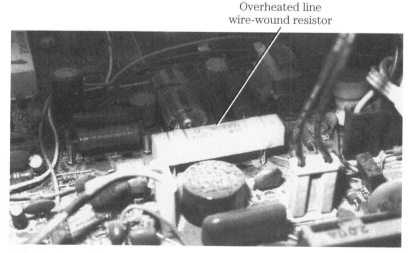

■ **4-22** *An overheated and cracked 10-W resistor in the input of the low-voltage power supply of a portable TV chassis.*

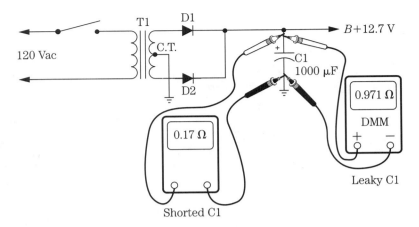

■ **4-23** *A leaky filter capacitor will have a resistance under 1000 Ω and a shorted capacitor might have resistance under 100 Ω.*

resistor, it might overheat and burn or crack. Remove one end of the resistor for correct measurement.

Filter capacitors break down between layers and become leaky or shorted. A shorted capacitor might have a resistance less than 100 Ω, and a leaky capacitor might have a resistance under 1000 Ω (Fig. 4-23). The dielectric compound in electrolytic capacitors might dry up and lose capacitance, producing a loss in volume or a hum in the audio. A leaky coupling capacitor might produce a loss in volume and distorted sound. Open coupling capacitors might cause a loss in volume or audio. Learning to make electronic tests is quite rewarding and is required in the consumer electronics field.

Testing solid-state components

THE THREE S's MIGHT BE YOUR MOST IMPORTANT TOOLS when electronic troubleshooting. Sight, sound and smell solve a lot of TV and stereo problems. Look over the entire chassis for burned resistors, a burned flyback transformer, or other lightning damage (Fig. 5-1). Cracked or overheated connections of a 15-W or 20-W resistor might indicate possible trouble. Burned bias resistors around an output transistor might indicate a leaky transistor. You might find a leaky electrolytic capacitor with a white or black substance oozing out at the bottom connections. Above all, your eyes identify the trouble symptoms from the front of the picture tube.

■ **5-1** *Look the chassis over completely to locate burned resistors, overheated transistors, and diodes in the defective chassis. A defective IF transformer is pointed out in the AM/FM/MPX radio receiver.*

You might smell an overheated voltage-dropping resistor or degaussing thermistor. The acrid smell from a power transformer might indicate shorted turns inside (Fig. 5-2). The smell of ozone might be traced to a screen-focus assembly, high-voltage anode lead, flyback transformer, and tripler unit. Not only can you smell an overheated transistor, diode, or resistor, but you can feel them.

Power transformer

■ **5-2** *A red hot power transformer might have shorted turns and windings, which will produce an acrid, burning smell.*

Lightly feel the overheated large power resistor for overloaded conditions. A light touch on a power output transistor might indicate a leaky transistor or defective bias resistor or diode. Do not touch any component while the unit is in operation and be sure to discharge any electrolytic capacitors because they can hold a charge indefinitely.

Intermittent or distorted sound might be traced to a defective sound stage. You might hear the tic-tic sound of the flyback transformer with high voltage or chassis shutdown. Arcover at the picture tube or horizontal output transformer might be heard. Noisy or improper horizontal sweeping might be heard from the flyback transformer. Some TV technicians can hear the 15,750-Hz sound of the flyback transformer, which indicates that the horizontal stages are performing (Fig. 5-3). Use your senses in electronic troubleshooting.

■ **5-3** *The horizontal output transformer in the TV chassis might have a frequency squeal that can be heard above 15 kHz.*

Locating defective semiconductors with voltage tests

The defective transistor can be located with voltage and transistor measurements. Voltage tests on each collector, base, and emitter terminal with common probe to ground can indicate a defective transistor. The open transistor can have a higher-than-normal collector voltage and no voltage on the emitter (Fig. 5-4). An open emitter resistor or terminal can have zero voltage reading.

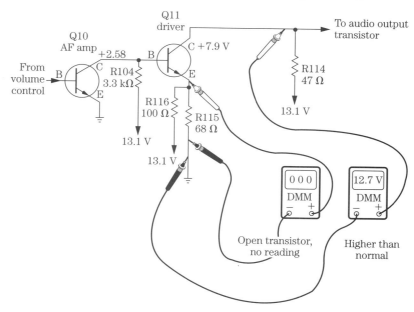

■ **5-4** *The open transistor might have high dc voltage on the collector terminal and zero voltage at the emitter.*

The leaky or shorted transistor might have similar voltages on all terminals (Fig. 5-5). Most transistors that become leaky do so from emitter to collector. Check the transistor in the in-circuit transistor tests and then remove it from the circuit and take another leakage test. Do not be alarmed if the transistor tests normal out of the circuit. Sometimes while unsoldering the transistor terminals, the transistor might heal itself. Replace the transistor if you suspect that it is faulty.

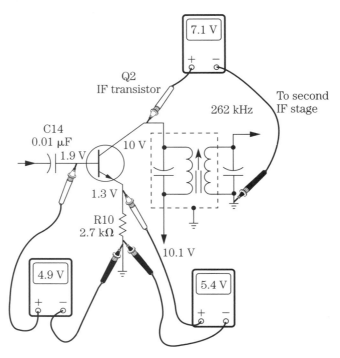

■ **5-5** *The leaky transistor might have close voltage measurements on all terminals.*

As mentioned before, another method to determine if the transistor is normal, is to measure the bias voltage from the base to emitter terminals. Usually, the transistor is normal if either 0.6 or 0.3 V is measured (Fig. 5-6). The NPN silicon transistor has a 0.6-V and the PNP germanium transistor has a 0.3-V bias voltage. The difference in both voltage measurements from base to ground and emitter to ground should equal the bias voltage of a normal transistor (Fig. 5-7).

The suspected intermittent transistor can show a different identical voltage measurement on collector and emitter terminals. The

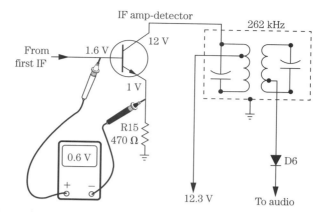

5-6 *The normal silicon NPN transistor will have a forward bias voltage of 0.6 V between the base and the emitter terminals.*

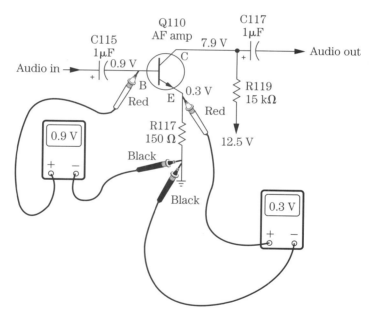

5-7 *The difference between the base-to-ground and emitter-to-ground voltage should equal the 0.6 V of forward bias of a silicon transistor.*

voltage can quickly change if the transistor is intermittent. Sometimes you can measure the voltage when the transistor is in the intermittent mode. Other times, you can't. When the probe touches the terminal, the transistor returns to normal, which shocks the transistor. Intermittent transistors might test open or leaky in the circuit and test normal when removed from the PC board. Replace any transistor that you suspect is faulty.

Diode voltage tests

Usually, regular silicon diodes found in the low-voltage full-wave or bridge circuits might blow the fuse or damage the power transformer before voltage can be measured (Fig. 5-8). The open diode will show no signs of voltage at the cathode terminal. It's best to take a leakage resistance test for power supply diodes. Of course, a voltage test across the main filter capacitor will indicate if the diodes and transformer are defective (Fig. 5-9).

■ **5-8** *Check the silicon diodes in the low-voltage power supply if a fuse is blown and there is possible transformer damage.*

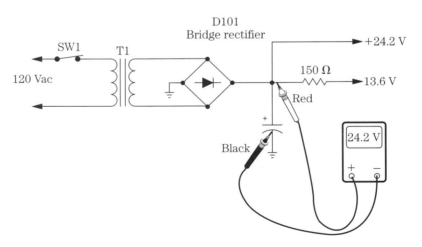

■ **5-9** *Measure the voltage across the electrolytic capacitor to determine if the missing voltage is caused by leaky diodes in low-voltage power supply.*

Voltage tests can be made across a zener diode in the low-voltage source. If the source voltage is quite low, suspect that a zener diode is leaky. These zener diodes have more of a tendency to appear leaky than go open. Look for signs of burned marks or overheating of the zener diode. Do not be alarmed if the voltage source is a few volts higher than rated by the zener diode (Fig. 5-10).

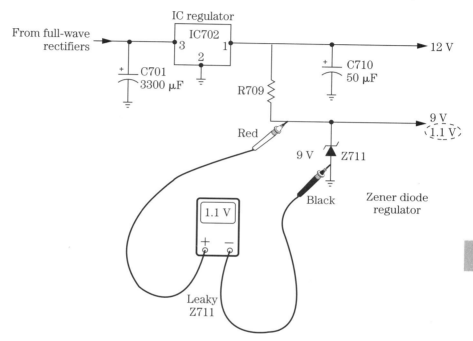

■ **5-10** *The burned or leaky zener diode might drop the voltage source, which can be checked with the diode test of the DMM.*

Small-signal diodes provide distortion in the sound of the detector stages in a radio receiver. No measurable voltages are found on the signal or detection diodes. Simply take a diode resistance measurement across the diode terminals. Replace the diode if both measurements are low with reversed test probes. Remove one end of the diode for accurate tests.

IC and microprocessor voltage measurements

The common IC configurations are the single in-line package (SIP), the dual in-line package (DIP), and the flat pack. Critical voltage measurements on the IC or microprocessor can indicate a defective IC (Fig. 5-11). Look for the supply voltage pin (*B+* or

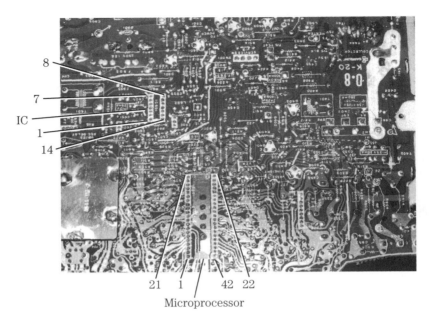

■ **5-11** *Locate the supply voltage pin and take critical voltage measurements on the suspected IC.*

V_{CC}). This voltage is always the highest voltage found on the IC terminals. Measure the supply voltage and compare it to that which is listed on the schematic. If the schematic is not available, compare the voltage to the good channel when servicing the stereo amplifier. Measure the voltage at the power supply source and measure the voltage that feeds the IC.

If voltage is lower at the IC terminal, suspect that the IC is defective. A low supply voltage can indicate that an IC is leaky (Fig. 5-12). Remove the supply pin from the circuit with solder wick and iron. Just remove excess solder around the pin. Flick the pin with a pocketknife or a screwdriver blade to make sure it's free and not connected. Notice if a big increase in voltage is found at the PC wiring connection. Replace a leaky IC if the voltage increases when the supply pin is removed.

Critical resistance tests of solid-state components

Transistors can be checked with a transistor tester, diode test of the DMM, and resistance measurements (Fig. 5-13). The resistance tests with a VOM, VTVM and FET VOM might appear a little different than with the digital multimeter. Quick resistance tests

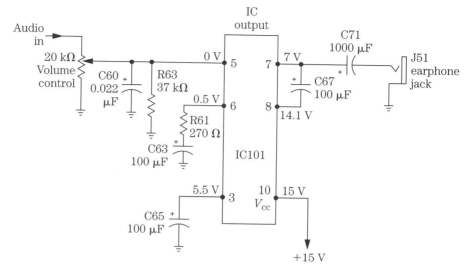

■ **5-12** *A low supply source can indicate that an IC is leaky or that a voltage source is improper.*

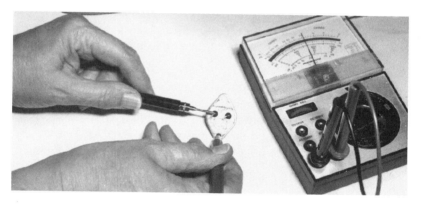

■ **5-13** *Check for the resistance between terminals of a suspected transistor with the low ohmmeter range. With voltage measurements, you can spot a defective transistor.*

can be made while in circuit for leakage tests on transistors, but outside the circuit tests are best for the VOM tests. Resistance tests taken within 20,000 Ω/V VOM meter might indicate a resistance measurement from base to collector and base to emitter. When the probes are reversed, a normal transistor will have an infinite measurement (Fig. 5-14). For instance, by placing the positive probe of a VOM to the emitter terminal and the negative probe to the base terminal of an AF NPN transistor, the resistance is around 12 kΩ. When probe tips are reversed on the same two ele-

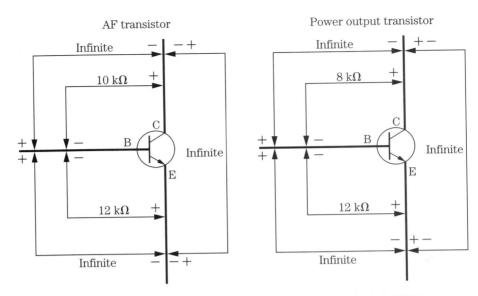

5-14 *A resistance test across the transistor terminals with a VOM can help you to locate a leaky or open transistor.*

ments, an infinite measurement is found. Remember, these measurements are approximate and might vary from transistor to transistor.

The leaky transistor will have a low measurement in both directions, when you change test probes. Leakage can occur between any three elements, although most leaky conditions occur between the collector and emitter terminals (Fig. 5-15). An open transistor has infinite or no measurement with test probes in any position.

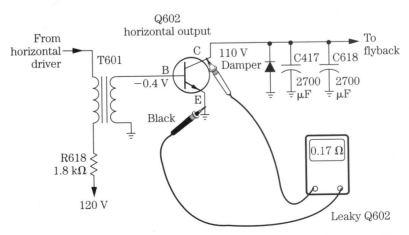

5-15 *Most transistors will appear leaky between the collector and the emitter terminals.*

Leaky diodes

The normal resistance measurement of any silicon diode might show a low measurement in one direction and with reversed test leads infinite or no measurement. When the positive terminal of a VOM, VTVM, or FET VOM is applied to the cathode terminal and the negative probe to the anode terminal of a fixed diode, you should have a low resistance measurement (Fig. 5-16). This is just the opposite of the digital multimeters (DMM). The normal diode with the DMM occurs with the positive probe at the anode and negative (black) probe at the collector terminal. Of course, the open fixed diode will not read in any direction.

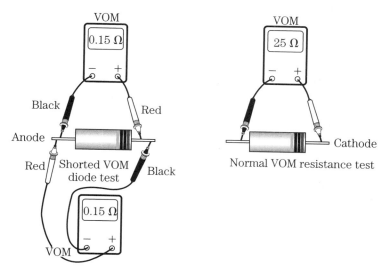

■ **5-16** *The different low-resistance measurements of a normal diode with VOM, FETVM, and DMM.*

The leaky diode will show a measurement in both directions. A shorted diode might have a resistance less than 10 Ω with reversed test probes (Fig. 5-17). The silicon damper or high-voltage diode should measure as any fixed silicon diode.

IC resistance measurements

Often, resistance measurements on ICs or microprocessors are not made from terminal to terminal, but terminal pin to common ground. When the voltage measurement is low on an IC pin terminal compared to the schematic, suspect a leaky IC or component (Fig. 5-18). Most ICs that short do it internally from supply pin

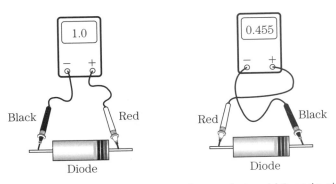

■ **5-17** *The leaky diode might have a resistance below 1 kΩ and a shorted diode might have a resistance less than 10 Ω.*

■ **5-18** *When the resistance is low on the supply source pin of the IC, suspect that an IC is defective.*

through a common ground terminal. Sometimes the IC must be replaced if the voltage and resistance are very low, and no signal is output from the IC.

First, take a resistance measurement of all terminals and mark them on the schematic. Go through the diagram and see what pins tie directly to ground or through a component to ground. Resistance measurements lower than 1 kΩ should be rechecked. Make sure that the test probe is on the correct terminal. When a low-resistance measurement is indicated on the meter, suspect that a component is leaky. Very low resistance measurements on the supply source pin, with a lower-than-normal voltage, indicate that the IC is leaky (Fig. 5-19).

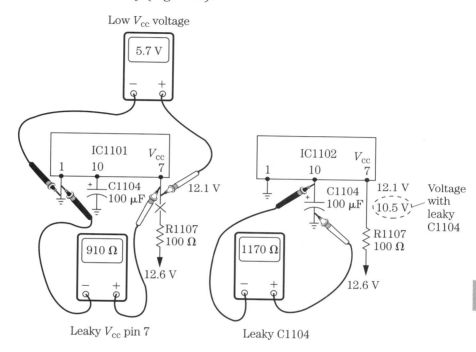

5-19 *Low resistance measured on the supply source pin after removing pin from the PC wiring indicates that an IC is leaky. A loss of voltage on the supply pin might be caused by a leaky component tied to one of the IC terminals to ground.*

Remove the pin of the IC suspected of leakage or remove a component tied to it. Remove excess solder with solder wick. Then take another measurement on the pin terminal. If the leakage has been removed, suspect that the part tied to that same terminal is defective. Remember, all resistance measurements of the IC or processor should be made from pin to common ground.

Transistor junction-diode tests

Set the function switch of the DMM to the diode test. All related tests are common with the base terminal of a transistor. With NPN transistors, place the positive (red) probe to the base terminal. While testing PNP transistors, place the negative (black) probe to the base terminal. Leave the red probe on the base terminal and take a resistance measurement between the base and collector and then the emitter (Fig. 5-20). Notice that the normal resistance junction test on the power output transistors is lower than the AF transistor.

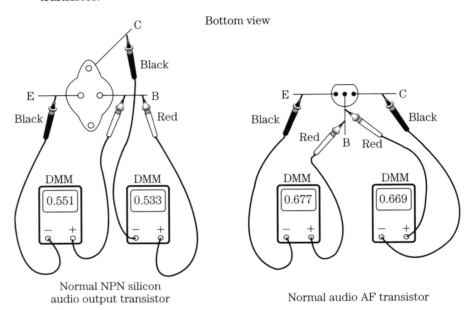

■ 5-20 *The normal transistor test of an AF and power output transistor in the radio or stereo chassis.*

Reverse the test probes on each test and notice if you receive a measurement. The normal transistor with reverse test leaks on the same elements will show an overrange symbol (1 or OL) (Fig. 5-21). When the probes are accidentally touched together in diode or transistor-diode tests, the DMM continuity buzzer will sound.

In the normal PNP transistor test, you place the black probe to the base terminal and collector, then to the emitter terminal. A comparable resistance test will indicate on the emitter and collector terminal. Notice that the base-to-collector measurement is lower than that at the base-to-emitter terminal (Fig. 5-22).

The leaky transistor will show a low resistance in both directions. If you have a resistance measurement under 1 Ω, the transistor is shorted and when the test probes are reversed, the measurement is the same, replace the shorted transistor (Fig. 5-23). The shorted transistor has only a fraction of an ohm short (0.017), but

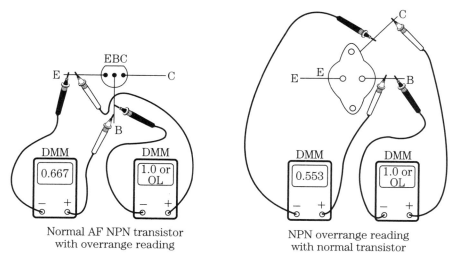

■ **5-21** *The normal transistor will show a normal measurement and with reversed test leads, it will show an overrange symbol.*

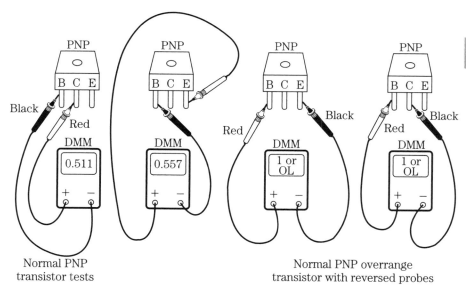

■ **5-22** *The normal PNP transistor tests with current tests and with reversed test leads. Notice that the black probe is attached to the base terminal of a normal PNP transistor.*

Transistor junction-diode tests

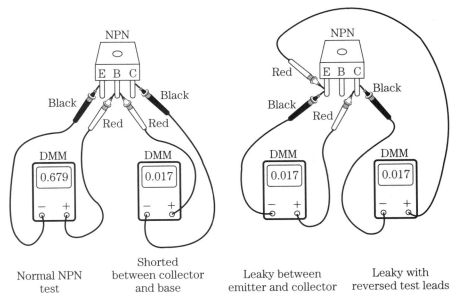

■ **5-23** *The leaky DMM diode test between the transistor terminals will show a low reading between these two terminals. The leakage should be the same with reversed test probes.*

a higher resistance can show a reading in both directions. The transistor can become leaky between any two elements or all three. Remember, the shorted or leaky transistor has a low-resistance measurement in both directions.

The open transistor

Besides finding a leaky or shorted transistor, you might locate an open transistor, which would be open between two elements. If the elements are open, no measurement is found between these two terminals (Fig. 5-24). This same transistor might be normal between the other two elements. For instance, an open measurement (infinite) is found between the base and emitter with normal measurement between the base and collector.

You might find a transistor with a normal measurement between the base and emitter terminals and a high reading between the base and collector (Fig. 5-25). Replace the transistor with a high-resistance measurement, which indicates a high-resistance junction. Remember, both normal measurements between the base and collector, and the base and emitter, should only be a few ohms apart.

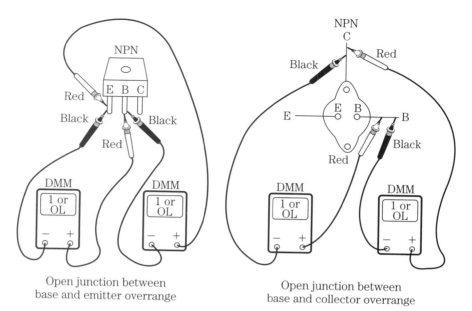

5-24 The open transistor will have no measurement between two terminals. An overrange symbol will appear on the meter or infinite resistance.

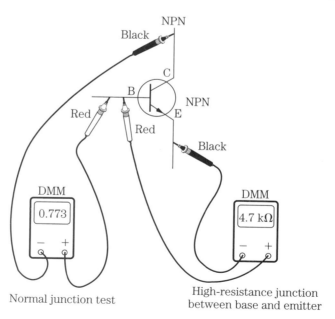

5-25 Here, a high-resistance junction is between the base and the emitter with normal voltage between the base and the collector. Replace the transistor.

Horizontal output transistor with damper diode

In the latest TV chassis, you might find the horizontal output transistor has a damper diode tied inside from collector to emitter terminal. This diode has a 1.5- to 1.8-kV rating. The cathode of the damper diode is tied to the collector and anode to the emitter terminal (Fig. 5-26). Before, the damper diode was found outside the collector terminal circuit to chassis ground.

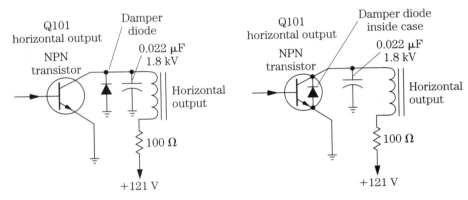

■ **5-26** *The regular damper diode might be found inside of the horizontal output transistor in newer TV chassis.*

When testing the horizontal output transistor in or out of the circuit, you will find a low resistance measurement in one direction, between collector and emitter terminals. A DMM diode test from the collector (metal body) terminal to chassis ground might indicate a normal transistor and damper diode. Before, a regular normal horizontal output transistor had no reading in any direction between collector and emitter.

Place the black probe to the metal body of the transistor and the red probe to the chassis (Fig. 5-27). Here, a normal diode measurement is taken on an NPN output transistor. If the measurement is below 10 Ω in both directions, suspect that the transistor is leaky between collector and emitter terminal. Doublecheck the horizontal output transistor out of circuit, before discarding it.

DMM diode tests

The normal diode test of the DMM will show a low-ohm measurement with the positive (red) probe at the anode and the negative (black) probe at the cathode terminal. An infinite or overrange reading is found when the test leads are reversed (Fig. 5-28).

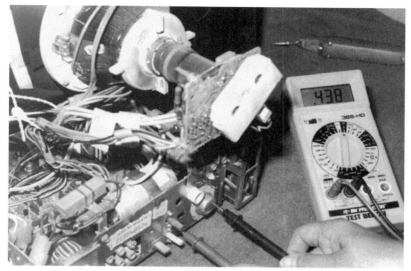

5-27 *A diode test from the collector (body) of the horizontal output transistor to chassis will indicate a normal damper diode measurement.*

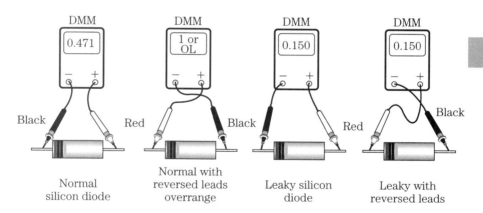

5-28 *The normal silicon diode test and a leaky diode located with the DMM.*

Set the function switch to the diode position of the DMM. Connect the red test lead to the V-O-Hz jack and the black lead to the COM jack. Touch the red probe to the anode and black probe to the cathode terminal of the diode for a normal diode test. The normal diode will show a reading in one direction and a shorted or leaky diode will read with reverse test leads in both directions (Fig. 5-29). The open diode will have no reading in any direction. Most defective diodes are leaky or shorted.

■ 5-29 *A normal silicon diode rectifier measurement taken on the diode test of the digital multimeter.*

Duo-diode tests

The duo-diode rectifier is found in the AFC control circuits of the horizontal oscillator in early TVs. This diode is usually of the selenium variety. The duo-diode is two separate diodes wired in a common cathode or series arrangement within one body (Fig. 5-30). You might find two separate silicon diodes in the same circuit of the later TV chassis.

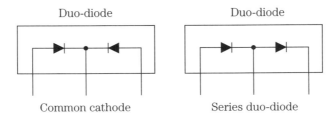

■ 5-30 *A common cathode and series duo-diode in the horizontal AFC circuits of an older TV chassis.*

Test the duo-diode as any other fixed diode. Place the black probe (−) to the common or positive terminal of the diode with the red probe (+) to the anode terminal. A very high resistance will be registered on the meter. No reading should occur when the test leads are reversed. A high or low resistance in both directions indicates that the diode is leaky. Replace the entire component.

Testing solid-state components

The regular resistance measurement of the duo-diode is a great deal higher than any silicon diode. After the regular diode test, check both diodes in each direction with the 20-kΩ scale (Fig. 5-31). You might have a measurement in one direction and not in the other, indicating that the duo-diode is normal. Now, check the diodes on the 2-MΩ scale. For example, if one is reading 130 kΩ and the other 65 kΩ on the 2-MΩ scale, doublecheck both diode readings of the duo-diodes. If the measurement is way off between the two diodes, replace it. If the measurement is normal, both diodes should have the same comparable resistance within a few ohms.

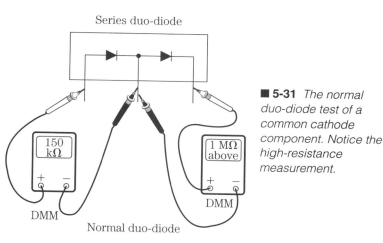

■ **5-31** *The normal duo-diode test of a common cathode component. Notice the high-resistance measurement.*

Bridge rectifier tests

Full-wave and bridge rectifiers can also be found in one package. Most diodes in the low-voltage supply circuits are of the silicon types and quickly tested with the DMM diode test. For a normal test, switch the DMM to the diode test and place the black probe to the positive (+) cathode terminal of the diode with the red probe at the anode (−) terminal. You should have a normal resistance reading on the meter. Now, switch the probe tips. A good diode will have a reading in one opposite direction (Fig 5-32). A very low resistance measurement indicates the diode is leaky in both directions.

Beware of leakage reading of a low-voltage diode when placed in a transformer full-wave rectifier circuit. A leakage reading might be obtained by feeding through one leg of the transformer winding. Remove one end of the diode for accurate diode tests. Each diode should test normally in a bridge rectifier circuit, unless one or two diodes are leaky.

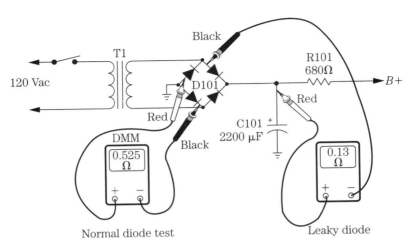

■ **5-32** *A normal and leaky diode in a bridge rectifier, taken with the DMM in the low-voltage power supply.*

You might find four separate diodes in one component of the bridge rectifier circuit. The bridge rectifier has four leads from a top hat or flat symbol. The normal diode tests in the bridge rectifier will have a reading only in one direction of two corresponding terminal leads (Fig. 5-33). All normal readings of each diode should be quite close in resistance.

■ **5-33** *The normal bridge rectifier test taken on a flat unit in the low-voltage power supply.*

A leaky bridge diode might have a shorted or low resistance reading in both directions (Fig. 5-34). Simply reverse the test leads when a low measurement is found between two terminals. A shorted or leaky bridge diode will have the same or comparable resistance in both directions. Replace the entire bridge rectifier when one leaky diode is located. If another bridge rectifier is not

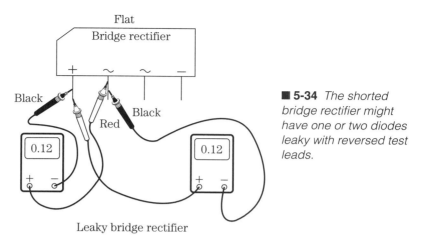

5-34 *The shorted bridge rectifier might have one or two diodes leaky with reversed test leads.*

available, four single 2.5-A diodes can be connected in a bridge circuit configuration.

Damper diode tests

The damper diode is found in the horizontal output transistor circuits (Fig. 5-35). In the early TV chassis, when the damper diode replaced the damper tube, the damper diode consisted of several layers of germanium diodes. Of course, this large diode will not show a regular diode measurement, except if it's leaky. This diode has a high peak voltage rating and must not be replaced with just any fixed diode.

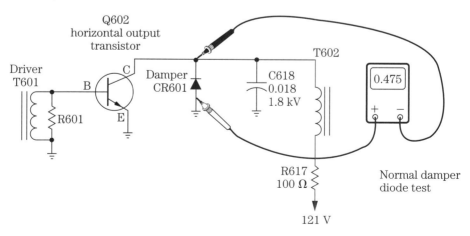

5-35 *The normal damper diode with the DMM in the collector circuit of the horizontal output transistor.*

Today, the damper diode is a silicon type and it can be tested as any fixed diode. Often the leaky damper diode will open the protection fuse of the horizontal or low-voltage power supply. Most damper diodes will become leaky or shorted. Very seldom do they open. You might find the damper diode built inside the horizontal output transistor in some TV chassis. If the diode goes open, the output transistor is destroyed.

High-voltage diodes

The boost diode rectifier is found to provide a boost voltage to the picture tube terminals in a TV set. The boost voltage can be taken from the top winding of the horizontal output transformers or from the horizontal output transistor (Fig. 5-36). An improper boost voltage to the picture tube (CRT) can be caused with a leaky boost diode or open voltage-dropping resistor. This boost voltage might vary from 150 to 480 Vdc, depending on the CRT circuit. Check the boost diode with the diode test of the DMM.

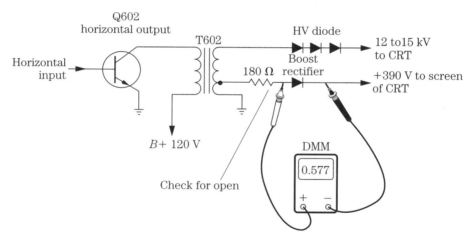

■ **5-36** *The boost diode in a flyback or horizontal output transistor circuit might become leaky and destroy the insulation or voltage-dropping resistor.*

A high-voltage rectifier stick found in the black-and-white TV set consists of layers and layers of germanium diodes. When one of these rectifiers arcover, it might have a sweet acrid smell. Check these diodes on the 200-kΩ range of the DMM (Fig. 5-37).

The high-voltage diode found in today's color chassis is molded inside of the flyback transformer winding. These diodes consist of high-voltage capacitors and diodes in series to produce the exces-

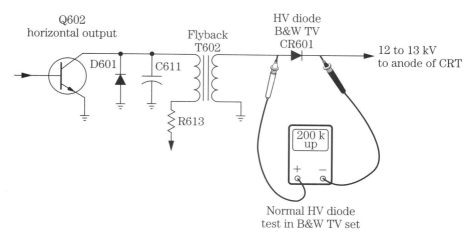

■ **5-37** *Check the high-voltage diode in the B&W TV chassis if no high voltage is at the anode terminal of the picture tube.*

sive high voltage found on the anode terminal of a picture tube (Fig. 5-38).

The high-voltage diodes have a tendency to break down and produce loud high-voltage arcing sounds. Arcing inside the molded flyback can be heard with a sizzling, popping, or banging sound.

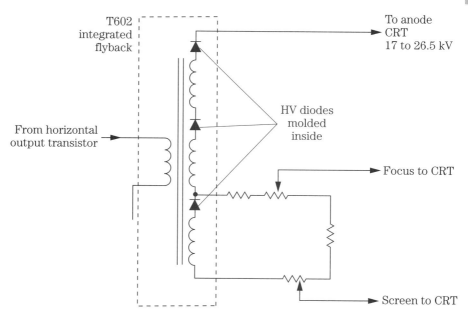

■ **5-38** *The HV diodes are molded into the flyback component of newer TVs.*

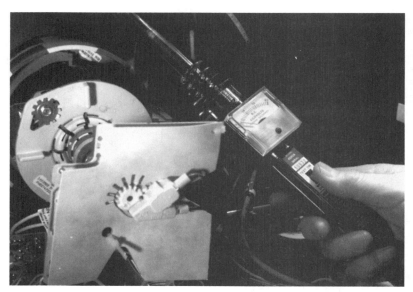

■ **5-39** *Check the high voltage at the CRT anode with a high-voltage probe voltmeter.*

Check the high voltage with an HV probe or meter (Fig. 5-39). Do not use a small VOM or DMM to check CRT anode high voltage.

Testing solid-state components out of circuit

Remove the suspected component after taking several different tests to determine if the part is defective. Again check the component out of the circuit to prove it is bad. Remember, transistors and ICs (even if they test defective in the circuit) can test normal out of the circuit. Replace it. Simply test all semiconductors out of the circuit, as in the circuit tests with the DMM and the transistor-diode tests (Fig. 5-40).

Test the replacement before installing it in the circuit. It will save you time and a lot of headaches. Also, remember that new components can be defective, right out of the package or carton.

Removing defective components

Most transistors and diodes can be removed from the PC board with solder wick and a soldering iron. Solder wick mesh comes in many different lengths and widths. With heat applied to the mesh material, it collects or soaks up the liquid solder and removes it from the PC board (Fig. 5-41). Use a low wattage iron or temperature-controlled soldering iron for IC and transistor work.

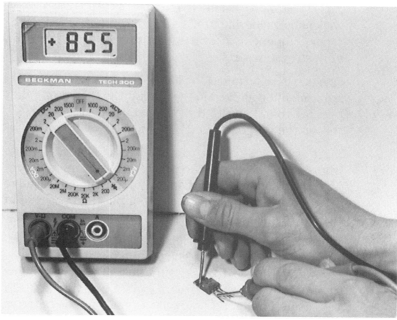

■ **5-40** *Test the defective component after removing it from the circuit and also test the new one before you install it.*

■ **5-41** *Removing excess solder from a defective part with solder wick and a soldering iron.*

After locating the defective IC, remove the old part with solder wick and a soldering iron. Use extreme care when working around other components near the IC terminals so that you don't damage or destroy any parts.

Check where terminal 1 is located before removing the IC. Terminal 1 might be marked in the IC or at the PC wiring side. If no numbers are found, locate the "U" section of the molded IC and mark this on the PC board.

Start at one side of the row of IC terminals with solder wick and iron. Go down the outside row of pins and remove excess solder. Likewise, go down the inside row of the same pins and remove solder. Now, go to the opposite row of pins and proceed in the same manner.

Carefully inspect each terminal. Apply the iron and solder wick on each pin until it is free. Flick the pin terminal with pocketknife or screwdriver blade, to make sure that all solder is removed. Pry up on the IC with a screwdriver, IC puller, or flat tool. Clean off all excess solder on the PC board after removing the defective IC.

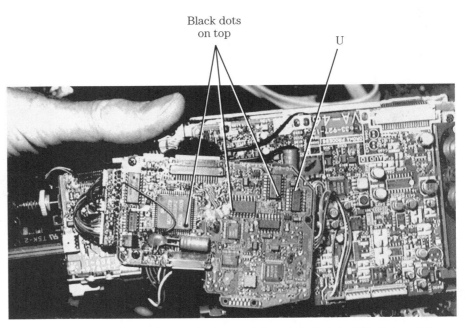

■ 5-42 Doublecheck for terminal one, white or black dot, white line or "U" at the end of IC to replace the new IC. Notice that the black dot and the white dot on the PC wiring identifies terminal 1 on a camcorder PC board.

Doublecheck for terminal 1 of the new replacement and for correct mounting on the PC board (Fig. 5-42). You might damage the new replacement if it is installed backwards. Bend over a terminal at each end of the IC so that it will not fall out when soldered.

Choose a 30-W soldering iron or a small battery-powered soldering iron to solder each terminal. Apply the small-diameter solder (0.032) on the opposite side of the iron tip. Make a good soldering connection, but do not leave the iron on the junction too long.

Clean around and between each terminal with the back edge of a pocketknife, screwdriver, and small steel brush. Run the blade between each terminal so that the connections will not be soldered together and so that the points of solder will not be touching another terminal. Finally, take a low-resistance measurement between terminals and corresponding tie points of the PC wiring. You might find a break between the foil and terminal connection when removing or replacing the defective IC.

Checking and testing semiconductor devices can be fun and interesting. Each day brings a new electronic venture into the field of electronics.

Checking surface-mounted components

SURFACE-MOUNTED DEVICES (SMD) ARE MINIATURE PARTS that are mounted on the surface of the PC wiring. These miniature components consist of ICs, microprocessors, LSIs, transistors, diodes, capacitors, rectifiers, and resistors. SMD components are now found in portable cassette players, radios, camcorders, VCRs, CD players, and the TV chassis (Fig. 6-1). They might appear on the wiring side as small black or brown chips.

■ **6-1** *Notice the small black dots and shapes on the bottom side of a PC board in a TV. SMD parts are found on the PC wiring side.*

Usually, the electronic components are molded inside a chip or plastic form with tinned ends that solder directly on the PC wiring. You might find regular-sized electronic parts on the top side of the chassis and SMD components underneath. SMD components provide a lot of parts mounted in a small space. This makes the electronic product smaller in size, but more difficult to service.

Servicing the SMD components might require special removal, installing and soldering equipment. Although expensive SMD soldering equipment might be found in the large service establishments, regular tools and soldering equipment can be used. Today most electronic technicians service the SMD chassis with a small soldering iron and miniature tools. A lighted magnifying glass is needed to locate the SMD devices.

SMD Transistors

The surface-mounted transistor appears in a miniature three-legged chip with three different tinned terminals. The transistor is mounted flat against the PC wiring (Fig. 6-2). Notice the PC wiring is very thin in SMD circuits. The collector terminal is found opposite the two bottom terminals. The transistor might be a PNP or NPN type. SMD transistors might be of the small signal, general-purpose, and fast-switching types.

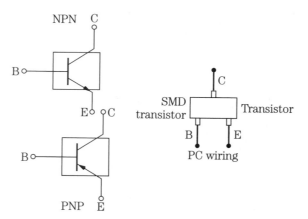

■ 6-2 SMD transistors are found in a flat three-legged component form with the collector terminal at the top, and base and emitter terminals at the bottom.

You might find digital transistors within the camcorder, VCR, and CD player. Although they look like a regular transistor chip, one or two small resistors might be included in the package (Fig. 6-3). The base terminal might have a resistor wired in series with the base terminal. In other digital transistors another bias resistor might be included from base to collector terminal.

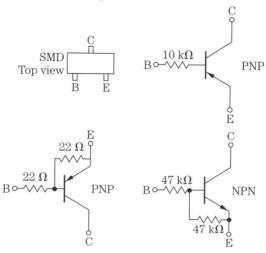

■ 6-3 *Digital transistors might have resistors built right in the transistor body. Resistors are found in series with the base and from base to emitter terminals.*

SMD transistors can be serviced in the circuit like the regular transistors with the DMM and transistor tester. Be very careful in checking digital transistors. Doublecheck the schematic and SMD component layout sheets to determine if the transistor is a digital type. You can make a test between base and collector and have an indication of a high-resistance junction with the diode test of the DMM. Likewise, a test between base and emitter might indicate a leakage between these two terminals when a bias resistor is included in the package (Fig. 6-4).

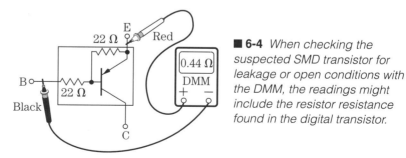

■ 6-4 *When checking the suspected SMD transistor for leakage or open conditions with the DMM, the readings might include the resistor resistance found in the digital transistor.*

Surface-mounted diodes

The SMD diode chip might appear like a capacitor or resistor with two or three terminals. Three-legged diodes might look like an SMD transistor. The two-legged diode might have terminals coming out each end or out of one side. Only one fixed diode is found in a single component that has two terminals. You might find two diodes in a single component with three terminals (Fig. 6-5). These can be mistaken for a transistor, if not careful. One end might be a solid-type connection while the SMD diodes have gull-type or curved terminals.

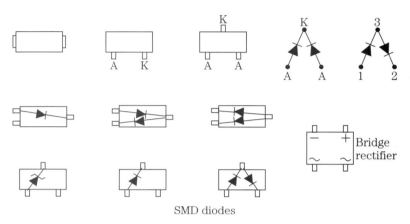

6-5 *The single SMD diode might be used in an end-to-end or a three-legged part form. Two diodes might be used in one component. The bridge rectifier has four diodes with gull-type wings.*

Small zener diodes are often contained in an SMD-type mounting. Often they look like a transistor three-legged device with gull-type terminals. These fixed diode terminals can be identified by letters, numbers, or a combination of both. The case style of a zener diode with solid ends might be an LL-34, LL-41, and D-64 mounting device. SMD-type diodes have gull-type wing terminal connections. Bridge rectifiers are now found in the surface-mounted component.

Low-amp diodes might appear in a two or three-legged chip up to 1 A. The voltage might vary from 50 to 800 PRV. You might find Schottky, fast-switching, and silicon diodes in the B-45 or SM case styles.

All of these diodes can be checked in or out of the circuit. If in doubt, remove the SMD component and make another test. Remember to throw away the surface-mounted devices after removing them—even if they test good. Often damage to the terminal or inside connection results when the small chip is removed. Manufacturers recommend destroying the removed SMD component.

You might find an SMD part that looks like a diode or resistor, but actually it is a solid-tee-feedthrough component. Naturally, the two terminals would test as a shorted component. Doublecheck the wiring schematic and parts layout from the manufacturer before removing the suspected part.

Surface-mounted ICs, LSIs, and microprocessors

Surface-mounted ICs and microprocessors are easily identified on the PC board with many terminals (Fig. 6-6). These flat-mounted chips have gull-type wing connections to the various circuits. Often, the IC might have terminals coming out of one or two sides while the microprocessor and large LSI components have wings or connections all around the outside area. These processors might have over 80 different soldered elements.

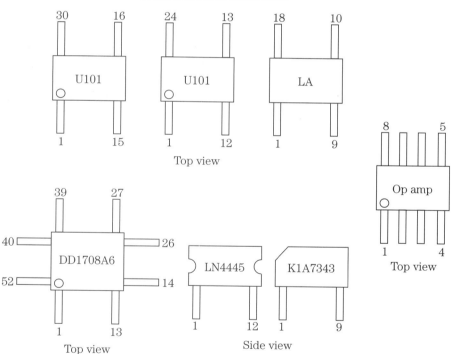

■ **6-6** *IC and microprocessors might have flat terminals on the two sides, but microprocessors and LSIs have gull-type and flat terminals.*

The surface-mounted IC or processor might have a round black or white dot that identifies terminal 1. Use a magnifying glass to get at the correct terminal. Simply count the various terminals from number 1 until you find the one you are seeking. You might find a white dot on the PC board for number 1. Sometimes the terminals on each end are numbered on the board for easy location (Fig. 6-7).

■ **6-7** *Here, the terminals are numbered at each end and marked with a white or dark dot for terminal number 1 in a portable CD player.*

Surface-mounted resistors

SMD thick film resistors and resistor networks look like black or brown carbon with flat soldered ends. These film chip resistors have tight temperature coefficients, excellent high-frequency characteristics, have excellent mechanical strength and electrical stability, and are flow solderable. These flat-type chip resistors can be of a 1% or 5% tolerance, a ⅛- or ⅒-W noted power. The small chip resistors can be purchased on a card of replacements, single bulk type, and reel rolls (Fig. 6-8).

Besides being a flat chip, some come in a round chip resistor and measure 0.14 inch in length. The flat-type resistor might be 0.83 and 0.122 in length with a width of 0.06 inch. As you can see, these resistors are very small.

You might find several resistors in one component (Fig. 6-9). A surface-mounted resistor network might look like a regular IC with several resistors inside. Of course, these resistor networks have

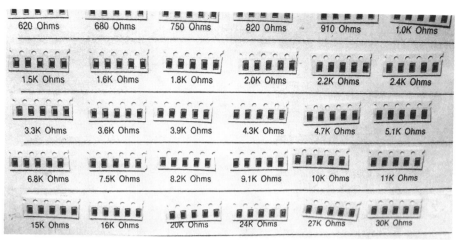

■ **6-8** *Universal SMD replacement of resistors might appear on a cardboard with the many different values.*

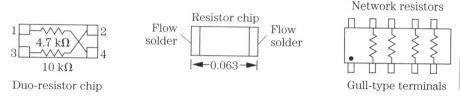

■ **6-9** *Resistor chips might appear with a single resistor end-to-end package or several in flat type mount. The network of resistors might look like an IC with gull-wing terminals.*

gull-type connections soldered directly on the PC wiring. The surface-mounted chip is quite miniature in size compared to other components found on the top side of the PC board in the TV chassis (Fig. 6-10).

SMD capacitors

Regular and NPO ceramic capacitors are found in many different values with 50-V rating. The SMD solid tantalum chip capacitors are small electrolytic capacitors in the 10-, 16-, 25-, and 35-V replacement. The molded SMD tantalum electrolytic capacitors are larger in size and have larger capacitance with 10 to 35 working voltage. SMD aluminum electrolytic capacitors are round and mount on a flat-chip platform for easy soldering. These aluminum capacitors might vary from 1 µF up to 100 µF. The SMD metallized

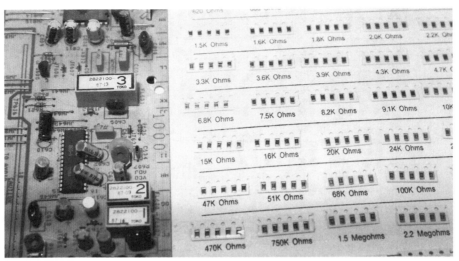

■ **6-10** *Notice the difference in size of the SMD parts to the regular resistor parts found on the TV chassis.*

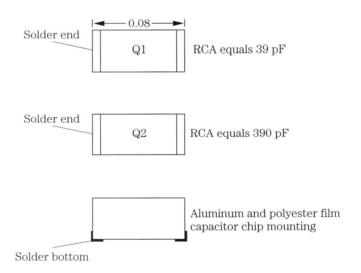

■ **6-11** *Fixed SMD capacitors might come in NPO, ceramic, tantalum, aluminum and metallized polyester film chip capacitors.*

polyester film capacitor is a larger component with 0.05- to 0.068-µF values (Fig. 6-11).

The surface-mounted capacitor might be checked in the circuit for leakage or open conditions with DMM. Check for correct capacitance with a capacitor tester. Check with the manufacturer for special SMD capacitor replacements.

Special tools

Most electronic tools and test equipment found on the electronic service bench can be used for locating, removing, and replacing the SMD chip. Sharpen up those test probes to a fine point to prevent slipping or shorting out the flat terminals. A small long-nose pliers with a sharp tip will help in removing and replacing SMD components. Select a sturdy surgical stainless steel tweezers and hemostat to clamp, hold, position, and twist the removable chip (Fig. 6-12). Small jeweler-type screwdrivers are handy when removing and replacing SMDs. Commercial flow-type soldering equipment is too expensive for the small electronic shop.

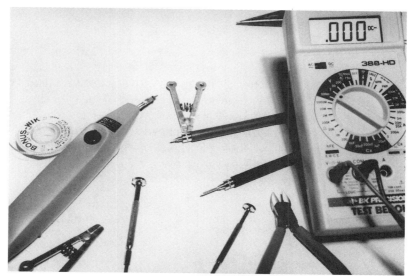

■ **6-12** *Only a few small tools are needed to handle, remove and replace the defective SMD component.*

Testing SMD components

ICs and transistors can be checked in the circuit with voltage, resistance, and instrument tests. Extreme care must be taken with LSI and sensitive microprocessors with voltage and resistance tests. Play it safe by wearing a grounded wrist strap. Make sure that you are working on the correct part and not another component. Check the manufacturer's literature for correct replacement.

Resistors, capacitors, diodes, and feedthrough SMD parts might all look alike. You can easily find a tie-feedthrough for a capacitor or diode and upon testing, the part shows a direct short. Several solid tie-feed terminals and connections might be found in the TV, VCR, and camcorder chassis. Three-legged transistors and diodes look somewhat alike and two diodes in one component might be taken for a transistor, resulting in improper tests (Fig. 6-13). Doublecheck with the manufacturer's service literature for correct parts location and take critical in-circuit tests.

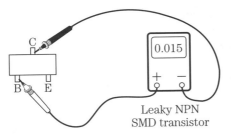

■ **6-13** *Check the suspected transistor like any regular transistor, with the diode-junction of DMM or transistor beta tester.*

Capacitors and resistors can be tested with critical resistance measurements. The normal capacitor should have no resistance or leakage. Then check it in-circuit with a capacitor tester (Fig. 6-14). Often in the electronic chassis, one end of the capacitor, resistor, or diode can be removed from the circuit and given an accurate test. The SMD units might be damaged trying to remove one end from the circuit. The whole part must be removed and tested. It should not be used again, so throw it away.

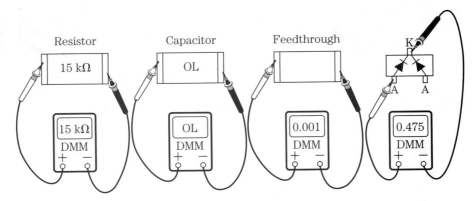

■ **6-14** *Resistors, diodes, capacitors, and feedthrough parts can be checked with the ohmmeter. Make sure that the component is either a resistor, diode, or capacitor before making tests.*

Locating a component on the PC board

It's not too difficult to locate small SMD parts on the PC board in tape players or radios. Try to locate semiconductors and part numbers in the same circuit. ICs and microprocessors are easily identified with many terminals. Some PC boards have the correct part symbol alongside of the board. Just follow the pin terminals to test a certain resistor or capacitor in the same circuit.

In camcorders, VCRs, CD players, and TV chassis, the SMD part is targeted with numbers and letters. The different sections of circuits might be outlined with black or white lines printed on the SMD and etched wiring side (Fig. 6-15).

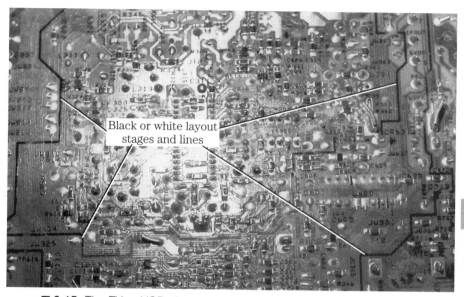

■ **6-15** *The TV or VCR chassis PC board might have the different circuits outlined and marked on the SMD PC board. Locate the circuit and then correct component.*

Manufacturers might have certain code letters and numbers to locate the part on the PC board. The numbers and letters are stamped right on the body of the SMD. For instance, RCA might use a Q2 as a 390-pF capacitor or Q1 as a 39-pF capacitor. Pentex might have transistor leadless identification numbers and letters, such as CD for a 2SA1122D and LD as a 2SC2462D transistor. Look for the manufacturer's leadless component chart for correct part identification numbers. A separate chart might be found for diodes, transistors, capacitors, and resistors.

Electrostatic-sensitive devices

Electrostatic devices might be damaged by static electricity. Some of these sensitive devices might be ICs, field-effect transistors, and

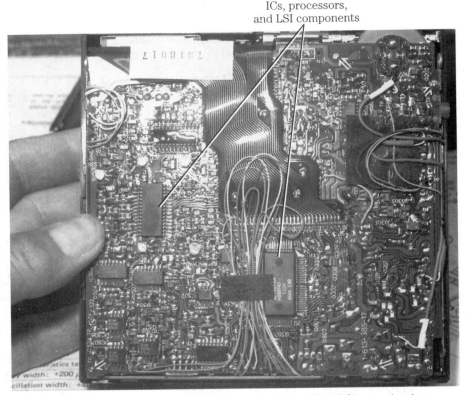

■ **6-16** *Electrostatic-sensitive devices such as ICs, LSIs, and microprocessors might be damaged by a charge from your hands and body.*

semiconductors (Fig. 6-16). Carefully handle all sensitive devices in their original electrostatic jacket or package. Do not unwrap or unpack them until you are ready to mount them.

Keep your body drained of electrostatic charges by keeping all clothing at ground potential. Wear a conductive ground wrist strap (Fig. 6-17). Sometimes a service technician might replace several sensitive LSIs and microprocessors without any danger. But the next time, the sensitive part might be destroyed. Make sure the electronic product and test equipment is grounded. Use a ground-tip soldering iron and solder removal tool to prevent static buildup. Prevent extra body motion that might create static.

Removing chip capacitors and resistors

To remove a chip capacitor or resistor, squeeze the chip with a pair of tweezers or sharp pair of long-nose pliers. Pull and twist upward

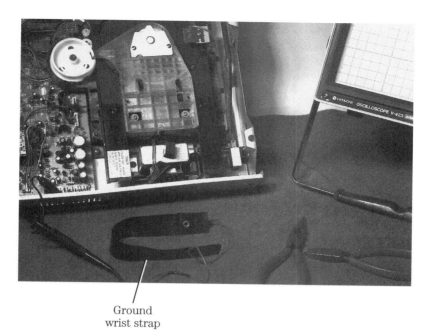

■ 6-17 *Wear a conductive ground wrist strap when working around these sensitive devices. Ground electronic products, test equipment, soldering irons, and your body when handling and replacing these devices.*

on both ends of the component while applying heat from the soldering iron (Fig. 6-18). Quickly melt the solder at both ends of the chip. Be careful not to pull up the etched PC wiring. Because the part must be replaced with a new one, damage to the SMD chip is not critical. Remove all excess solder with the anti-static suction solder device or with a controlled heat soldering iron. Because most SMD parts were glued to the board originally, heating and twisting the SMD might help to remove it.

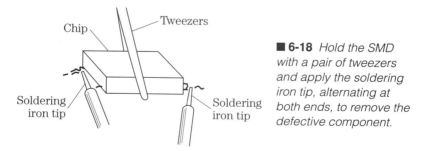

■ 6-18 *Hold the SMD with a pair of tweezers and apply the soldering iron tip, alternating at both ends, to remove the defective component.*

Apply solder braid or solder wick down along where old terminals were mounted to provide a flat board mounting. Clean off all excess solder and use a wire brush to remove excess solder flux, rosin, and sharp points of solder. Do not heat any soldered connections more than three seconds with the iron tip.

Removing ICs and SMD processors

To remove the flat pack IC or microprocessor, heat each terminal and gently pry up the lead as the solder is melted (Fig. 6-19). This is a much slower process than a commercial flow solder system. Make sure you know where pin 1 is soldered and check the position on PC wiring. Be very careful not to pull the actual PC wiring up from the board.

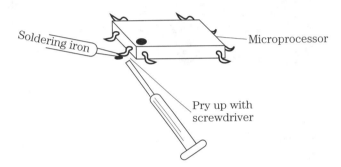

■ **6-19** *Remove the microprocessor or IC by applying heat at each terminal and prying up the gull wing or flat terminals.*

Another method to remove ICs or microprocessors that have gull-type wing terminals is to clip each terminal close to the body of the SMD with a sharp pointed pair of cutters (Fig. 6-20). Remove the SMD after all terminals are cut into. Then remove each terminal with soldering iron and solder wick or a soldering-sucking tool. Make sure all flat lead terminal pieces are removed from the chassis.

Replacing SMDs

Make sure the spot where the new SMD part must be placed is clean and level. Remove the epoxy bond or adhesive that was holding the part into position. Tin each wiring area or pad on the board before installing a new component. Recheck the new part for correct value and polarity. Carefully mount the new IC in place (Fig. 6-21). Doublecheck terminal 1. Press each IC lead against the foil and solder it.

■ **6-20** *In the small portable tape player, a flat mount and gull-wing mount ICs are used on the etched side of the PC board.*

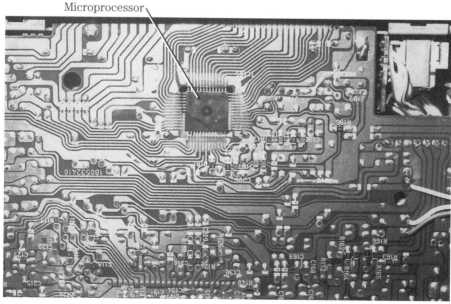

■ **6-21** *When replacing the microprocessor with many terminals, make sure that pin terminal 1 is at the right corner.*

Hold the small SMD with a pair of tweezers to anchor it in place while soldering each end. Keep the component flat. Do not apply heat for more than 4 or 5 seconds. You might destroy the new chip or PC wiring. Avoid using a rubbing stroke when soldering. Do not bend or apply pressure to the transistor or IC terminals. Use extra extreme care not to damage the new chip.

You can replace all SMDs except transistors, ICs, microprocessors by preheating with a hair dryer for 2 or 3 minutes. Re-tin or presolder the contact points on the circuit pattern or wiring where the resistor or capacitor mounts. Press down the component with tweezers and apply solder to the end terminal connections. Apply more solder over the terminal if it is needed for a good bond (Fig. 6-22). Do not leave the soldering iron on the transistor or IC terminals for too long. You might damage the internal junction of the semiconductor. The large IC or processor should be mounted flat and right over each correct terminal pad or connection.

■ **6-22** *After replacing the SMD chip, apply more solder at each terminal to make a good bond. Notice how close each SMD part is located in the small camcorder.*

Part substitution

Whenever possible, replace the SMD with an exact replacement. The biggest trouble in substitution of larger SMD parts is the size and shape, regardless of the correct value (Fig. 6-23). The IC and microprocessors must mount directly over the correct PC board terminals. Even with resistors and capacitors universal replacements the length should be taken into consideration for mounting.

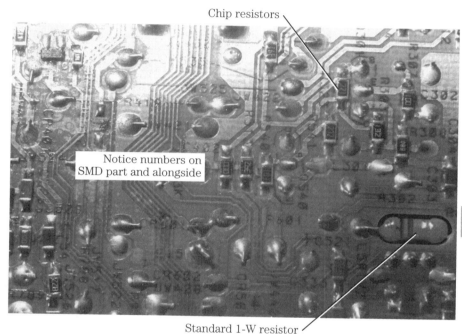

6-23 *Always, use original SMD parts numbers if they are available. Universal resistors, capacitors, and diodes can be subbed without too much difficulty.*

Intermittent board connections

Poorly soldered part terminals and board connections produce a lot of intermittent service problems in the SMD circuits. When PC wiring is found on both sides of the board, warping of the PC board might produce cracks or poor connections. Breaks usually occur at terminals or eyelets. Check feedthrough terminals for poor connections.

Inspect each component terminal for intermittent solder joints or broken connections. Sometimes touching up the SMD part terminal with a soldering iron might solve the intermittent problem. Pushing up and down with a pen or insulated tool on the board might cause the intermittent to act up. Moving the SMD part with an insulated tool will not move the part, but it might turn up a bad connection or an intermittent SMD component.

Patience

You must have a lot of patience and use extreme care when working with SMD components. These parts are quite fragile and easily damaged. Besides patience, you must have steady hands to test, remove, and replace the SMD. Do not use a large soldering iron or screwdrivers around them. A slip of the soldering iron might destroy the component (Fig. 6-24). By using a little patience, extreme care, and common sense, anyone can learn to remove and replace SMD components.

■ **6-24** *Be careful when soldering the small SMD parts on the PC wiring. You need patience and a steady hand.*

SMD parts layout

Look for SMD parts on the PC wiring side of the PC board. Often large components, such as transformers, electrolytic capacitors, and power rectifiers are mounted on the top side. When pulling off the bottom cover of a portable CD player, you can quickly spot these small components (Fig. 6-25). Do not slide or drag a TV chassis across the surface or you might damage the SMDs.

■ **6-25** *Look for a parts layout on the top and bottom PC boards in the service manual of each electronic product. Notice the thin PC wiring lines in the portable CD player.*

Check the manufacturers' parts layout chart for the location of SMD components. Usually the chart is found with the service manual. SMD parts are found where thin PC wiring is located. Most manufacturers have a PC board drawing of the SMD components on the bottom side and regular parts with PC wiring on the top side. Depending on the circuits, PC wiring might be found in only one side of larger PC boards. SMD parts might be mounted on wiring between large IC pin terminals that stick through the bottom side of the board in the TV chassis. They can be mounted just any place because they are so small in size.

Where to find replacement

Try to obtain original part numbers from the manufacturers' parts distributor, parts depot, and local electronics distributor. Check where the unit was purchased for nearest parts house. Small universal replacement SMD components can be found at large local electronic parts stores. Large electronic mail-order firms now handle many universal SMD components. The following mail-order firms provide universal parts:

Consolidated Electronics, Inc.
705 Waterveit Ave.
Dayton, OH 45420-2599

Digital-Key Corp.
701 Brooks Ave. S.
Thief River Falls, MN 56701

Fox International
23600 Aurora Rd.
Bedford Heights, OH 44146

Kevin Electronics
10 Hub Dr.
Melville, NY 11747

MCM Electronics
650 Congress Park Dr.
Dayton, OH 45459

Mouser Electronics
2401 Highway 287 N.
Mansfield, TX 76063

Parts Express International
340 E. First St.
Dayton, OH 45402

Premuim Parts Electronics Co.
P.O. Box 28
Whitewater, WI 53190

Learn with kit building

CHAPTERS 7 AND 8 SHOULD BE FUN AND EXCITING FOR YOU. There is no greater personal satisfaction and achievement in electronics than building projects and kit building. Besides, learning how to mount and solder the correct components can be a lot of fun (Fig. 7-1). Good soldering techniques are required in project building as in soldering replacement components in a TV set. Kit building is a good way to start learning electronic basics for the beginner, intermediate, and professional electronics technician.

■ **7-1** *Kit and project building teaches how to solder, mount parts, and connect wires, besides, you can learn to build, have fun, and gain a great deal of satisfaction.*

You can learn what electronic parts look like and how they are mounted. You will learn the color code of resistors, where they are mounted, the value of resistors in the circuit, and what they do. You must learn how to check each electrolytic capacitor, diode, IC, and transistor for correct polarity.

You can learn how the circuits are tied together with hookup wire or PC board wiring, how to follow the circuit and color code of each wire, and how to make the inside wiring look as sharp as the finished front panel. Soldering each component provides experience down the road in electronic repair. Kit and project building can teach discipline, determination, and patience. The professional electronics technician requires a lot of patience in today's consumer electronic field.

Learn to solder

Making good clean solder joints starts with soldering, soldering, and soldering. Practice and experience is the best teacher of clean soldering techniques. Only two tools are needed to solder the PC board: a low-wattage iron and 60/40 rosin core solder. Never use acid-core solder on any electronic connection. Use rosin-core solder, sold at most electronics stores and Radio Shack, for all electronic connections.

You can learn by simply soldering a piece of hookup wire to an insulated terminal standoff. When connecting wires to a soldering lug or terminal, wrap the bare wire around the lug tightly (Fig. 7-2). Place the tip of the iron on the lug and the wire connection. Apply solder to the lug and bare wires. Do not place solder on the iron when soldering extension wires or cables. Heat the joint first and push solder into the connection.

A good clean soldered joint should be clean and bright. The poorly soldered connection may have a dull appearance with solder in bunches. This might be called a cold soldered joint. The connection might perform now, but after a few months, you might have an intermittent joint. Do not pile solder up in a blob. Let the solder flow to the connection.

Soldering transistors in the circuit or on the PC board requires steady hands to place the correct terminal in the right hole. Know where terminals B, C, E go before mounting. The kit manufacturer might use the flat side of a transistor outline on the PC board for correct mounting (Fig. 7-3). This might apply to some transistors but not with others. Doublecheck the mounting of the transistor. If

■ **7-2** *Keep the iron tip clean and apply it to the material or joint to be soldered. Apply solder into the wires or junction.*

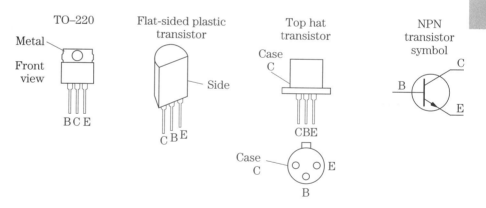

■ **7-3** *Check where transistor terminals BCE connect and check the position that the transistor is mounted.*

you connect a transistor backwards, it can be damaged in microseconds. Use a heatsink or a long-nose pair of pliers on each terminal to protect the transistor from overheating. Too much heat from the soldering iron can damage the internal junction of the transistor.

For ICs, check terminal 1 before soldering them into position. Some ICs are mounted into a socket and others are mounted directly on the board. Check for a dot or "U" and white line to determine where terminal 1 mounts. If the IC is mounted backwards, the project will not operate and the IC might be damaged. Be careful when soldering the IC terminals (Fig. 7-4). Make a good connection, but do not leave the iron on the terminal too long. Place the IC in the socket correctly. Looking down on an IC, the number 1 terminal is always to the left top corner. Recheck terminal 1.

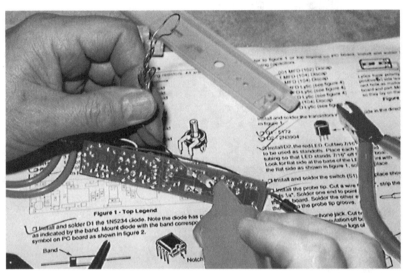

■ **7-4** *Solder the IC and transistor terminals with a regulated or battery-powered soldering iron.*

The different soldering tools

Choose a low-wattage soldering iron (25 to 40 W) for electronic kit building (Fig. 7-5). Rosin-core solder is available in a 60/40 mixture standard and different gauges. These solder rolls have a rosin-core center for making better connections. The small-diameter solder is ideal when soldering IC or microprocessor terminals. Rosin-core paste can be used on poorly tinned terminals or on large metal surfaces. Most kits have more than enough solder to complete the electronic project.

You might want to purchase a temperature-control iron for kit building and future repairs (Fig. 7-6). When the technician enters the world of troubleshooting and repairing electronic products, that soldering iron will be used most of the time. The battery-operated soldering iron is ideal when soldering transistor or IC terminals. Besides, you have no cord to pull or tug around. For heavy-duty soldering jobs, select a 150- to 350-W soldering gun. Always keep the

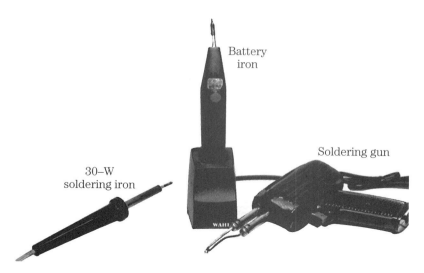

■ **7-5** *A low-voltage iron (25 to 40 W) can solder parts in a kit. The 150-W iron is used on large junction surfaces.*

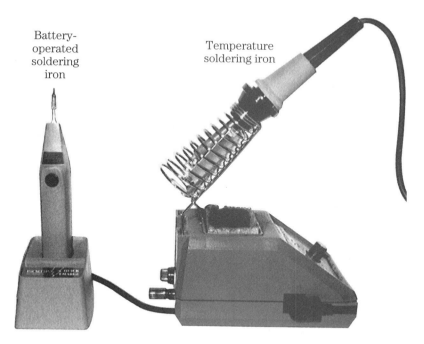

■ **7-6** *The temperature-controlled iron has a sharp point and is easy to keep clean. The battery-powered iron is excellent for soldering small wires and IC terminals.*

The different soldering tools

dering jobs, select a 150- to 350-W soldering gun. Always keep the soldering iron tip clean to make a better soldered joint.

Besides the soldering iron and solder, the only other equipment needed is some means of sucking or removing the excess solder. If, by chance, you installed a part in backwards or in the wrong place, the component must be removed. After you locate a defective part in the radio or TV chassis, it must be removed. A cheap method to remove solder is with a braided mesh material (Fig. 7-7). Solder wick is available in many different sizes or widths. The large width is used to remove large areas of solder and the small size is used around transistor and IC terminals. But it's cheap and it does a nice, quick, and easy job. Solder wick is used every day by the professional electronics technician.

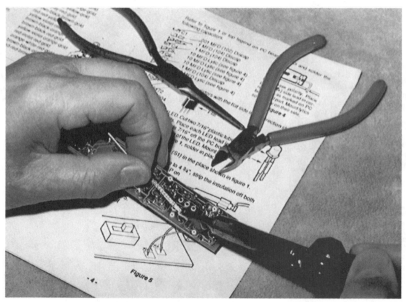

■ **7-7** *Solder wick is a mesh material with solder flux that, when heated, picks up excess solder.*

Read the directions

Before starting any kit building project, it's best to read the assembly operation steps. The directions will also tell you how the circuits operate. Knowing how the circuit performs, you learn more as you proceed along. Check each page of instructions over at least twice and start at the beginning.

Sometimes you might think that you can insert all the parts in the circuit holes and everything works out. Read the instructions over carefully and take each step as the manufacturer has produced many of these kits (Fig. 7-8). Carefully identify each component and verify that all parts are included. Lay them safely aside. Take your time. Check off each step of instructions. Avoid mistakes at the beginning and the project will operate when finished. Make good soldered connections.

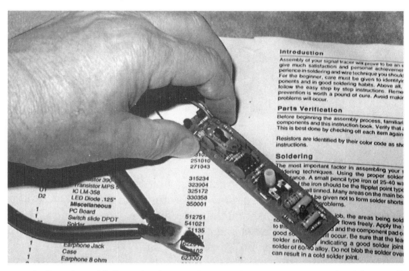

■ **7-8** *Read and follow the instructions on how to mount and solder each component.*

Check all parts

Carefully unwrap the enclosed kit parts and familiarize yourself with each component. Check off each part against the parts list. Do not forget parts that sometimes drop into the packing material. Lay all parts out flat for a checkoff. If a part is missing, most manufacturers have a part missing card that can be mailed back to the parts department (Fig. 7-9). The damaged, defective, or missing parts should be listed by the manufacturers' part number and description. Do not return the defective part unless the manufacturer requests it.

Usually, when unpacking components, you might find several clear packages of different components. The resistors might be found in one bag, and the mounting hardware and small electronic parts might be found in another. These parts can be checked off by viewing through the plastic bag and identifying each component

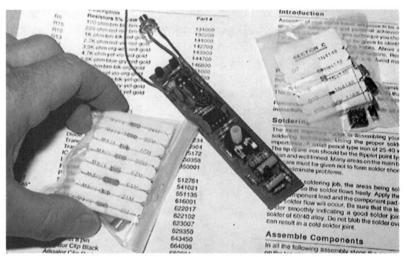

■ **7-9** *Lay each bag of parts out so that you can check each part off the list. Some kits contain a missing part card that can be mailed in to replace missing, defective, or damaged components.*

(Fig. 7-10). Besides learning what each part looks like, you can put the color code to use on resistors and capacitors. Bulky parts, including the case, might come in one large package. Mark down the number of plastic bags, in case one gets misplaced.

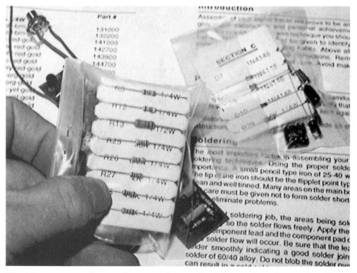

■ **7-10** *Most parts do not have to be pulled from the clear plastic bag to cross them off.*

How to assemble components

Small resistors can be mounted flat and pushed through the correct mounting holes. Recheck the color code and symbol numbers where it mounts on the PC board. If in doubt, doublecheck the resistance with the ohmmeter. High-wattage resistors should be mounted ¾ of an inch above the board to prevent burning or damaging the board. Often larger components are mounted last. Make sure each part is in the correct spot.

Bypass capacitors can be mounted ¼ inch above the chassis. Observe electrolytic capacitor polarity—the lead beside the minus-sign stripe on the capacitor is the negative or grounded side. Be sure that the positive lead is in the correct hole that is shown on the parts layout (Fig. 7-11). Diodes have a polarity band; white or black is the positive end and it should be mounted correctly. Transistors might be mounted with the flat side as reference in some project kits. Leave at least ¼ inch between the part and the PC board. Doublecheck all polarity signs of electrolytic capacitors, diodes, transistors, and ICs (Fig. 7-12).

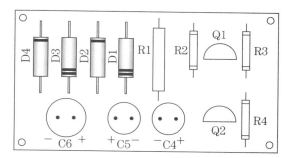

■ **7-11** *The PC board might have all of the part outlines shown on the board with each part number.*

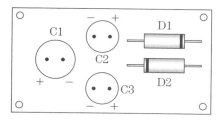

■ **7-12** *Doublecheck electrolytic capacitors, diodes, transistors, and ICs for the correct polarity.*

Most kit manufacturers make a board layout chart where each part symbol is mounted. This chart includes components mounted with polarity signs. If four or more diodes are placed side by side, make sure that the polarity bands are like those in the chart. Often, all components are mounted on the board. Parts can be mounted in a check-off group or those that are outlined. Some kits are put together by the number.

Bend the ends over of all components, such as resistors, diodes, and bypass capacitors (Fig. 7-13). This will keep the parts from falling out when the PC board is turned over for soldering. Some capacitors can be bent over on their sides for extra room on the board. You can mount each small part in the correct holes and solder them at once. If more than one component is mounted through a combing hole or pad, leave the soldering of that pad until both parts are mounted. You might have to remove excess solder if both holes are covered with only one component lead mounted and soldered. Clip the excess leads close to the PC board with diagonal cutters.

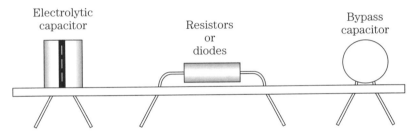

■ **7-13** *Push the terminals of each component in the right holes over the outline, bend the terminals back toward the board so that they will not fall out while you are soldering them to the PC wiring.*

Check off the parts listings as the components are soldered into the circuit (Fig. 7-14). Usually a small box is found alongside of each part for easy checkoff. Recheck the part for correct resistance, capacity, and terminal leads. Make sure the right part is in the holes. Again, check for correct polarity before leaving that group of components.

Transistor and IC mounting

When mounting transistors with terminals in a line, check the flat side so it faces the way drawn on the chart layout of the PC board. The same applies to mounting IC parts with a notch at the end. The IC socket can be mounted in any direction, just mount the IC correctly. The PC board might have the elements of transistors marked on top of the board. Of course, on the top side of the PC board, the symbol letters and the numbers of components referred

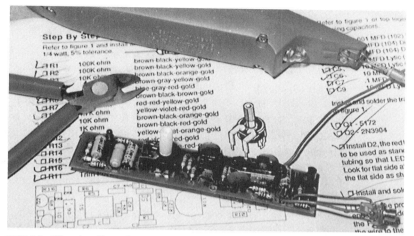

■ **7-14** Check off each component as you mount it and solder it onto the PC board.

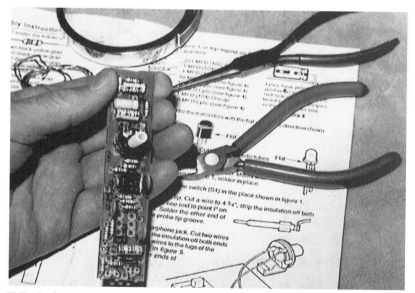

■ **7-15** Some kit manufacturers have the outline of the part, the part number, and the correct polarity marked right on the board itself.

to in each parts list will be marked on the top side of the board (Fig. 7-15).

Be careful when mounting components in small areas. Make sure that the leads are the right length and that there is plenty of room for the parts to lay down or stand up. Sometimes the part numbers might be difficult to see. Use a magnifying glass to locate the correct holes.

When mounting power or TO-220 transistors on the heatsinks, make sure the piece of mica insulation is between the transistor and the heatsink. Transistors that are mounted on a heatsink in the middle of the board do not need a piece of insulation. Place silicone grease on both sides of the mica insulator before bolting into position (Fig. 7-16). A plastic bushing must be in the hole to insulate the bolt and nut from the metal heatsink. Snug up the nut, but not too tight. Bend the transistor terminals upward so that when they are soldered they will not touch the metal heatsink. With most TO-220 transistors, the metal hole mounting and metal backside is the collector terminal and it must be insulated. IC regulators with TO-220 mounting have the metal hole and metal backside at ground potential. They should be grounded directly to the heatsink.

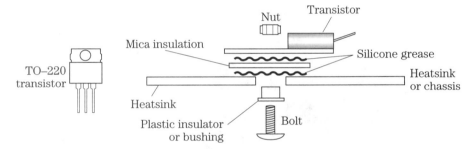

■ **7-16** *Place silicon grease on both sides of the mica insulator before mounting the power transistor. Take continuity checks between the metal of the transistor body and the metal heatsink to make sure that the transistor is insulated from the metal heatsink.*

Putting the resistor color code to work

Selecting the correct resistance to be placed in the proper holes is where the resistor color code goes to work. Although some kits might have a list of resistors by the number, the color code is shown off to the side (Table 7-1). This makes it easy to replace each resistor. Be careful in selecting the different colors. Sometimes the red, orange, and brown colors might all look alike. If you are colorblind, measure the resistance with the ohmmeter.

■ **Table 7-1 Parts list of resistors with the color code alongside.**

R1	100 kΩ	Brown/Black/Yellow/Gold
R2	180 kΩ	Brown/Grey/Yellow/Gold
R3	100 Ω	Brown/Black/Brown/Gold
R4	220 kΩ	Red/Red/Yellow/Gold
R5	4.7 kΩ	Yellow/Violet/Red/Gold
R6	10 kΩ	Brown/Black/Orange/Gold

PC board layout

Most kits have an outline of each component laid out on the top side of the PC board. Each part should have its part number alongside each outline (Fig. 7-17). Notice the positive and negative polarity of the electrolytic capacitors. Sometimes a diode might have a positive (+) symbol at the end of the diode. Most boards have a black or white line at the end of the diode, which indicates the positive or collector terminal. When several different diodes with different amperages are found in the same project, check the length of the diode against the outline on the PC board.

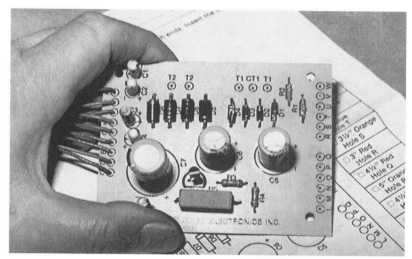

■ **7-17** *The parts layout on a low-voltage power-supply kit.*

Notice that small resistors do not have a line drawn at one end. Small diodes do. Doublecheck the resistance value, part number, and place it in the correct set of holes. These components might be placed close to one another and easily switched if not careful.

If a transformer is used in the circuit, the PC board might have different leads with different numbers. Of course, the primary color-coded lead wires are usually black. The secondary winding might connect directly to the PC board. Here, the transformer wires might be T1, T1, CRT, T2, and T2 (Fig. 7-18). Check the transformer color-coded schematic with the proper terminals. Here, T2 and T2 are the red secondary leads of 17 Vac. T1, T1, and CT1 terminals refer to the blue secondary leads with black as center-tapped (CT1). If the wires are not marked, measure the ac voltage of the same color of wires or take a continuity check with the ohmmeter.

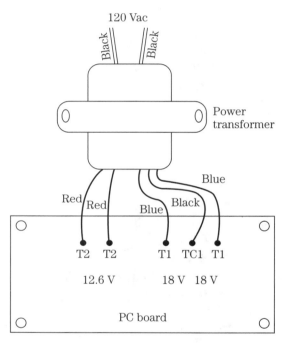

■ **7-18** *How the two different color-coded transformer secondary leads are soldered to the PC board.*

The primary winding will have the most resistance when measured with the ohmmeter. These wires will be the same color. The secondary terminal wires have the same color and low resistance. The center-tapped secondary winding will have a different color, but when checked for resistance with one of the secondary wires, it should have one half of the resistance of the whole secondary winding.

Transistors and ICs are mounted with the outline of each part on the PC board. Often, the transistor flat side is mounted one way.

The "U" or identification mark of the IC is inserted in another direction. By following the drawn outlines on the PC board, parts layout schematics and wiring diagrams, mounting and soldering each component makes kit building easy to follow.

From schematic to board mounting

When you build kits, you learn what the component looks like when you install it and you learn how each part functions in the schematic. When replacing four silicon diodes into the PC board, check and see what circuit or how the part functions in a particular circuit (Fig. 7-19). Likewise, when installing the transistors or ICs, notice where each terminal goes and how it ties into the project circuits. You learn by repetition and simply doing it.

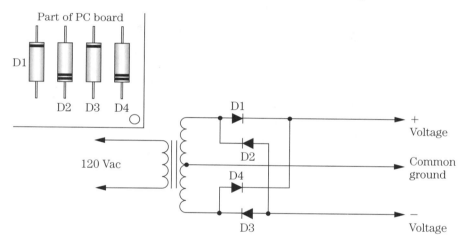

■ **7-19** *When installing parts, such as the silicon diodes, check both the board and the schematic to see how they are mounted and tie into the circuits.*

After a while, you can visualize the component symbol in the schematic diagram, pick it up, and mount the part in the right place on the PC board. Knowing how the component performs in a given circuit is half the battle in troubleshooting and repairing consumer electronics. Connecting the PC board with hookup wire to outside parts or panels, you learn how to tie the different parts and circuits together.

Mounting panel components

Installing controls, binding posts, controls, switches, jacks, and terminal lugs on the front and rear panel are fairly easy with a pictorial diagram. Most controls should be mounted with star lock washers. This holds the control from turning. Another method is to drill a small hole for the tab of the control in the back side of the front panel (Fig. 7-20). Do not drill the hole through the front panel; just enough to push the metal tab into. Controls and switches can be held in position with lock nuts. A dab of paint or fingernail polish on the nut will keep it from loosening. Wipe up the excess paint or polish and do not let it drip down on other parts.

■ **7-20** *Parts mounted on the front panel of a kit or project should have lock washers around the controls so that they will not loosen and rotate.*

Rocker and sliding switches can be held in place with pressure springs or bolts and nuts. Snug up all lock washers and nuts, then place fingernail polish over each nut to keep it from loosening (Fig. 7-21).

Mount all parts on the back panel with the same method. Often, the fuse holder and ac power cord are mounted on the rear panel. Always place the ac cord in a slotted-strain relief insulator and squeeze the two sections together with pliers. Then insert the strain insulator into the hole. When the strain relief expands, the ac cord is locked in place.

Usually the power transformer is mounted on the bottom area of the metal or plastic cabinet. Bolt each tab of the transformer in place. Tighten each bolt or screw it in securely (Fig. 7-22). Connect and solder each color-coded wire to respective pads on the

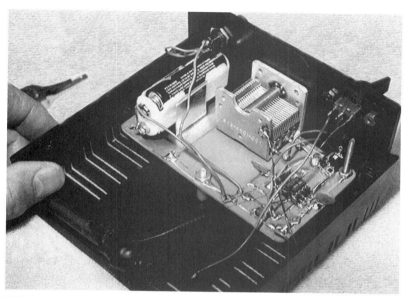

■ **7-21** *Place fingernail polish or a dab of paint over the small nuts and bolts to prevent them from loosening.*

■ **7-22** *Tighten all locknuts, bolts, nuts, and screws that hold the parts on the front or bottom panel.*

PC board. Solder one side of the ac cord to the fuse holder. Connect other ac cord wire to the rocker or on/off switch. Slip shrink-tubing over the bare wire leads or tape them to prevent shorting and producing a shock hazard.

Mounting panel components

Make all soldered connections to the PC board before bolting into position. Recheck each wire terminal. Inspect the bottom side of the board for poorly soldered joints or blobs of solder. Make sure that all terminal posts are insulated that might carry voltage or antenna input terminals. Recheck each extended wire that might connect to the PC board from transistors and ICs mounted off the board or on heatsinks.

Transformer connections

Small transformer leads have a color-code that identifies the correct wire and where it ties into the ac circuits. Often the primary wires are both black. The secondary leads might be red, green, blue, and white. Usually the red leads are the highest ac voltage winding. Blue leads might be a lower ac voltage, compared to the red leads. This might not be true all the time. The center tap of either secondary windings can be black, or a combination of red and green. When two color leads are the same color, they tie to the same coil winding (Fig. 7-23).

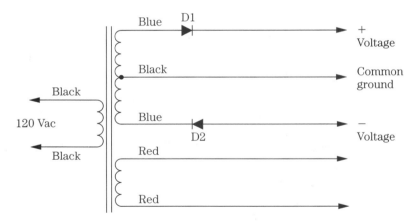

■ **7-23** Follow the color code of each transformer wire as it is connected to PC board and parts.

Take a continuity test of the windings if you are not sure where each wire is connected in the circuit. Remember that the primary winding has more resistance than the secondary windings in a step-down power transformer. Set the ohmmeter to the RX1 scale. Measure the resistance of the two wires that have the same color. Find the center-tapped wire with continuity tests between the same two colored wires. Mark down the center-tapped color wire where it ties into the circuit. In the low-voltage power supplies, the center-tapped wire goes to ground (Fig. 7-24). The other two colored wires (blue) are soldered to a pair of silicon diodes. It does not matter if the blue wires are interchanged with one another so long as the center-tap is connected to ground.

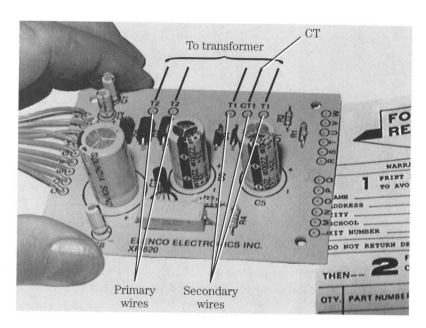

■ **7-24** *Make sure that the common ground wire is soldered to the ground terminal on the PC board. Keep all wires orderly.*

If in doubt about the secondary lead color-code, take an ac voltage measurement of both pairs of secondary wires. Temporarily connect the ac power cord to the two primary (black) leads. Wrap tape around each connection. Plug the ac cord into the power receptacle. Set the ac meter range to the 200-Vac scale. The ac voltage range can be lowered on low ac voltages. Measure the ac voltage across the red leads (12.6 V). Now measure the voltage across the blue wires (18 Vac).

To find the center-tapped lead, it should be connected to either voltage source. The ac voltage should be one half of the voltage measured (Fig. 7-25). Remember that the center-tapped lead might be any color. When the black lead has continuity with any one of the blue leads, you know it connects to both blue wires of the same winding. You might find that the center-tapped ac voltage might not be equal on both sides, but it is close enough for the

Transformer connections

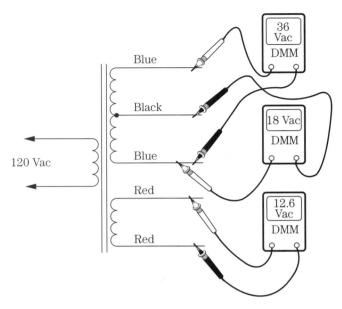

■ **7-25** *Take ac voltage measurements when you do not know where they go or if the wires are color coded differently.*

low-voltage power supply. Remember, the highest ac voltage ties into the circuit with the highest dc voltage source.

Part checkoff

Do not forget to check off each part as you solder components into the circuits. For instance, if you solder several resistors or diodes in a row, check off each part in that group. Make sure that each component has the correct part number or symbol in those sets of holes. Take your time. Do not rush the soldering process. Avoid making accidental mistakes and enjoy how to assemble the kit building project.

Check-out time

Go over each component and part connection at least twice before firing up the chassis. Doublecheck all parts with correct polarity, such as diodes, electrolytic capacitors, transistors, and ICs. Compare the wiring of the PC board to components mounted on the front and rear panels (Fig. 7-26). If in doubt, check it again. There is nothing more discouraging than having a project that will not

■ **7-26** *Doublecheck each wire lead from the PC board to the front-panel controls and make sure that they are on the right terminals.*

■ **7-27** *Test the finished product with voltage output measurements on the meter or measure the power supply voltages with the DMM.*

Check-out time

function the first time it is turned on. So you check and doublecheck, then sit back and enjoy the rewards of kit building.

Testing

When the project is turned on with no sound or indications, turn the project off at once. For example, you might have just wired up a dc power supply. Do not put on the front or back covers until the project is checked. Connect a dc voltmeter to the dc voltage output terminals. Make sure that the black post and black probe of the meter are connected to the same terminals (Fig. 7-27). Connect the red probe to the positive post or terminal. Now turn on the switch and watch the meter hand or numbers come up to the correct voltage. Viola! The power supply operates. Who said project building is a lot of fun? Now put it to work.

8

Build four different test instrument kits

THIS CHAPTER COVERS FOUR DIFFERENT KITS THAT YOU can assemble and use as test instruments. The signal injector kit (ST-751) will signal time and input signal into the radio, amplifiers, stereo, RF, and IF circuits. The transistor tester (EG-DT100K) will test diodes and transistors, in or out of the circuit. The digital multimeter (DMM) is the most handy test instrument (M-2665-K) to make voltage, resistance, current, capacitance, diode test and transistor hFE measurements. The dc power supply kit (XP-620K) provides output voltage of a positive dc variable source of 1.5 to 15 Vdc at 1 A and a negative dc variable source of 1.5 to 15 Vdc at 1 A. A positive (+5 V) dc source at 3 A can be used to operate electronic projects on the service bench. These four test instruments are from Elenco Electronic kits, available at many electronics stores and mail-order firms.

Besides learning the value of components, how to solder, polarity mounting, testing, and using the finished product, you will have a lot of fun. A list of kit manufacturers and addresses are found at the end of this chapter.

The signal injector-tracer kit (ST-751)

The signal injector-tracer operates around a dual op amp (LM-358). The first section serves as a square-wave oscillator, providing a 1-kHz signal with many harmonics that can inject a signal into the AM, RF, and IF stages of the AM band, and IF circuits in most FM receivers. The radio speaker serves as indicator with signal-injection methods (Fig. 8-1).

A second part of the op amp serves as an audio amplifier with a high-impedance input. An earphone is plugged into the phone jack to hear the gain and loss of sound in the audio section of a radio or audio amplifier. S1 switches in the probe tip to inject the signal or

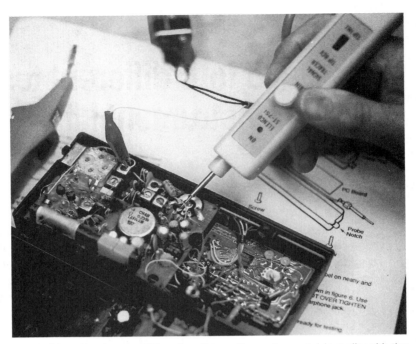

■ 8-1 *Signaltracing the RF and audio sections of a portable radio with the signal injector-tracer kit.*

pickup audio with the earphone. This signal injector test operates well with a 9-V battery.

How it works

Connect the red alligator clip lead to the positive voltage of the 9-V battery and the black lead to the negative voltage. Notice that the red LED should light. Clip the green lead to the probe tip and insert the earphone plug into the jack. You should hear a tone in the earphone. Rotate the gain control and the sound should get either louder or softer. The signal injector is used to input signal to the RF and IF stages of a radio receiver to determine what stage is not functioning.

Now switch the signaltracer to trace the audio signal from the volume control to the speaker. Connect the green lead to the input of the amplifier. The probe tip will pick up the audio signal. Start with the suspected preamp or volume-control circuit and trace to the speaker. This small instrument will inject a signal and trace the audio through the different stages of a radio receiver.

Verify all parts

Make sure that all parts are in the kit and check off each one on the parts list before attempting to mount each component. Read over the construction notes and how to put the kit together. The location of most parts can be verified by looking through the plastic bags. Resistors can be identified with the color code or by checking them on the ohmmeter. Q1 (5172) is black plastic with no numbers on the body of the transistor. Q2 is identified with 2N3904 number marked on the flat, gray side of the transistor.

Electrolytic capacitors have the value and polarity dash ground line on the body of each. The "U" shape or notch in IC (U1) provides correct mounting on the socket and PC board. Other single electronic components can be identified and mounted with instructions inside the booklet.

Assembly instructions

First, mount all components on the PC board (Fig. 8-2). Check the PC board layout. Insert each lead through the correct hole and bend it over. Solder each lead to the foil side. When each part is mounted, solder it into position. Check off the box of each component. Make an X or check mark in the box ahead of each part. When you end up with a different resistor at the end, you have misplaced the wrong color-coded resistor.

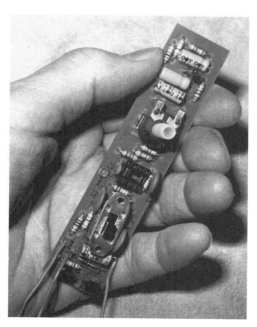

■ **8-2** *Mount all parts on the top side of the small PC board and follow the text and printer layout of parts on the PC board.*

Start with R1 through R11. Mount and solder each part in place. Make sure that trim pot (R11) is level before soldering in the body tabs. Next install D1 and the IC's (U1) 8-pin socket to the PC board. Be careful when soldering each pin; don't solder the pins together. Check off each box. Pass a screwdriver or knife blade between each pin terminal to clean out the solder.

Mount all bypass and electrolytic capacitors. Place the positive side of the electrolytic capacitor in the (+) hole. Bend electrolytic capacitors over so that they lay flat and will fit inside of the flat instrument case. C1, C2, C3, and C7 can be mounted flat and with any terminal because they are not polarized.

Install both transistors (Q1 and Q2) with the flat side, as shown on the top side outline of the PC board. Be very careful when soldering each transistor lead. Use a pair of long-nose pliers as a heatsink on the transistor terminals. This will prevent excess soldering iron heat from destroying the solid-state internal junction. Install the red LED (D2) with two 7/16" plastic tubes to be used as stand-offs on the terminal leads. Check the flat side of D2 and make sure that it mounts with the outline on the PC board (Fig. 8-3).

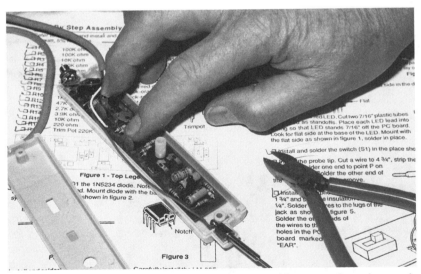

■ **8-3** *The inside view of the small signaltracer with connecting wires and components.*

Next, mount and solder the signal switch (S1). Install the probe tip. Apply a piece of hookup wire to one end and solder it to a point on the PC board. Install an earphone jack with two hookup wires from earphone terminals to the board terminals. The inside wire lead of the earphone jack should be soldered to the ground side of the PC board. It does not matter which earphone wire goes, except on the two flat rectangular boxes on the PC board. It's a good idea to go over component wiring and mounting after each page is checked off.

Signaltracing

Doublecheck the wiring and mounting of all components after finishing the project. Replace the top cover and place on the label. Connect the 9-V battery, observe the polarity, and notice if the red LED lights. Clip the green lead to the probe tip and insert the earphone. Rotate the gain control. You should hear the 1-kHz tone in the earphone.

Test the signaltracer out by setting the slide switch to the inject position. Inject the signal into a portable battery radio starting at the volume control. Use the speaker as indicator. If the tone is heard in the speaker, proceed to the first base terminal of the transistor or IC. Keep working toward the front of the receiver until the dead or intermittent stage is found (Fig. 8-4). Keep the volume control low as possible. Likewise, switch S1 to signaltrace position and listen to the audio at volume control to each AF transistor and output IC until the audio quits. Then you have located the defective stage.

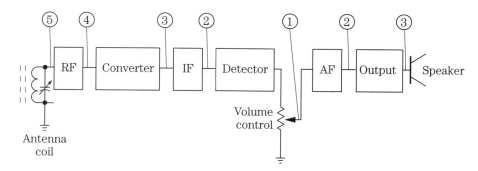

■ **8-4** *Troubleshooting the radio circuits by the number with the signal injector-tracer.*

Transistor test kit (EG-DT100K)

The small diode-transistor tester will check all germanium, silicon, power, LEDs, and zener diodes. The diode test will also indicate the cathode and anode leads of diodes. Various germanium, silicon, power, RF, audio, switching, and FET transistors can be tested. This transistor test will identify NPN or PNP types. In-circuit transistor

■ **8-5** *Taking the in-circuit transistor test of a transistor on a TV chassis with the Elenco EG-DT100R transistor tester.*

tests can be made in circuit with base and collector resistors as low as 100 Ω (Fig. 8-5).

Mounting components

Mount all small components on the PC board and check each one off as it is soldered. Doublecheck the mounting of the transistors and of IC1. Make sure that all transistors, ICs, LEDs, diodes, and electrolytic capacitors have the correct polarity. Wire SW1 and SW2. Make sure each wire is on the correct switch terminal. The switches, transistor socket, test switch, and LED indicators must be mounted flat on the PC board. The switch-mounting screws and test switch locknut holds the PC board to the front panel (Fig. 8-6). Make sure that all IC and IC socket pins are in the correct holes and are not bent over. When soldering the IC terminals, do not use too much solder or the terminals will connect together.

Diode tests

After installing a 9-V battery, select a silicon diode and test it. To test the diode-transistor tester, short black and red leads together and push the test button. The diode LEDs will alternate on at 1-Hz rate. Place the diode switch in the diode position and connect the diode to the red and black leads. Push in the test switch (red but-

■ **8-6** *Notice that the PC board is held into position to the front cover with screws in the top of the switches.*

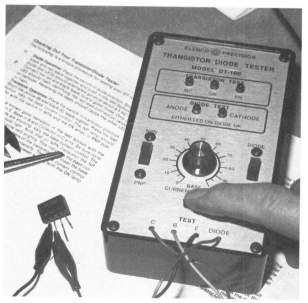

■ **8-7** *Testing a silicon diode out of the circuit with the transistor tester kit.*

ton). One diode LED should blink no matter how the leads are connected (Fig. 8-7). If both diode LEDs blink, the diode is leaky or shorted. When neither LED lights, the diode is open.

Transistor tests

For out-of-circuit tests, place the transistor in the blue socket or attach the leads to the C, B, E leads. The green clip is the collector (C), the base is the yellow terminal lead, and the emitter terminal lead is black. Push in the test button. Adjust the base control so that the red OK LED lights. This indicates a normal transistor. If the OK lamp does not light, turn the control up so that either the NP or PN LED lights, indicating if the transistor is an NPN (NP) or a PNP (PN) (Fig. 8-8).

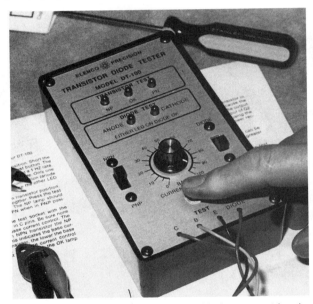

■ **8-8** *Testing a transistor out of the circuit with the correct leads clipped to each transistor terminal.*

When no LEDs light, the transistor might be open or it does not have the correct base terminal. If the OK LED lights, the base indicates a beta transistor. The lower the base control setting, the higher the beta.

In-circuit testing of a transistor in the chassis is done in the same manner. Always keep the power switch off or ac plug pulled when making in-circuit transistor or diode tests. Clip the colored alligator clips to the correct B, C, and E terminals of the transistor to be tested. Remember, a leakage test can be indicated if a resistor, coil, or component under 100 Ω is found in the base or collector terminals. Remove the base terminal from the circuit and take another test.

LCD meter kit (M-2665-K)

The DMM kit will measure dc voltages up to 1000 V with ranges of 200 mV, 2 V, 20 V, 200 V, and 1000 V. A maximum ac voltage range starts at 200 mV, 2 V, 20 V, 200 V and 750 Vac voltage measurements (Fig. 8-9). Resistance measurements start at 200 Ω, 2 kΩ, 20 kΩ, 200 kΩ, 2 MΩ, and 20 MΩ (megohm) ranges. The ac and dc current ranges start at 200 μA, 2 mA, 20 mA, 200 mA, and 20 A. Capacitors can be tested on a 2 nF, 20 nF, 200 nF, 2 μF, and 20 μF range with a test frequency of 400 Hz. Besides these tests, the LCD meter will also test transistors and diodes.

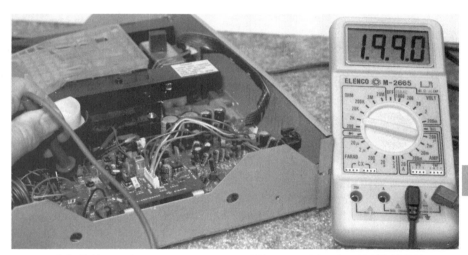

■ **8-9** *Taking voltage measurements within a combination AM/FM/MPX receiver, cassette player, and CD player.*

Parts identification

The resistors and diodes are mounted on a card for easy identification. You can take off the required part according to the check-off boxes. Mount all small parts on the meter display circuit board. Notice that the PC board is a double-sided board (Fig. 8-10). Section A is the meter display circuit and bottom side is section B. Section B has the dc voltage and current circuits. Section C has the ac voltage and current circuits.

Although PC wiring is on both sides of the board, soldering is done on the gold PC side of the board. Each hole or pad that is found on either side is tied to both sides of the wiring. All small parts mount on the side where the part numbers are stamped in white. The white part numbers might not show up under a strong light. Just tilt

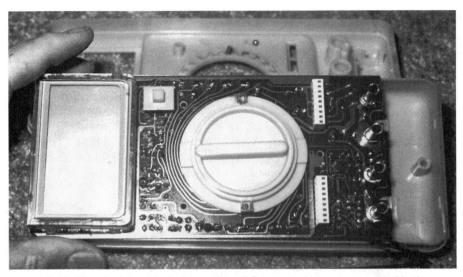

■ **8-10** *The double-sided wiring of the PC board with parts mounted in the deluxe 3½" LCD digital multimeter kit.*

one side of the board upward to take the glare off of the PC wiring. A magnifying glass will help you to see the small printed parts.

All small components are mounted on the side with part symbols printed in white. The ICs mount on the same side. Parts are soldered on the PC board side. Mount the LCD assembly on the PC wiring side where the terminals are soldered. Do not forget to place the zebra pad on the row of gold terminals inside the LCD housing, where housing fits in the small square holes. The rotary function switch and LCD are mounted on the same side.

Be careful when identifying resistor color codes because different color bands might appear to be the same color. If in doubt, check on another ohmmeter for correct resistance. Always, check off each component before starting to mount the parts. Each section has its own bag of parts.

Observe the correct polarity of diodes, electrolytic capacitors, transistors, and ICs. Notice that U1 mounts in the IC socket, but U2 is soldered directly to the PC wiring. Be careful when soldering the U2 pin terminals. Do not leave the iron on these pins too long.

Small capacitors are mounted ¼ inch above the PC board. Some resistors are mounted flat and others are mounted vertically. Likewise, certain diodes are mounted flat against the board and others might stand on end.

In section C, U3 is mounted in an 8-pin IC socket. In section E, U4 is mounted in a 14-pin IC socket. The only IC that is soldered directly to the PC wiring is U2 (IC 4030). Actually, U1 and U2 are mounted opposite the LCD housing assembly (Fig. 8-11). Use a 180- to 250-W iron to solder in the four input post sockets.

■ **8-11** *The inside view of the components mounted on the PC board with battery leads attached to the DMM kit.*

Last-minute checkoff

Doublecheck all parts and mounting of different assemblies. Make the final assembly by placing the PC board inside the top and bottom case. The battery will lay at the opposite end of the LCD assembly. Follow these mounting instructions closely, so as to not make a mistake or lose small springs and ball bearings.

Testing procedures

Comparison tests can be made with another accurate DMM. Measure the ac voltage with both meters and compare the readings. When the different tests do not appear as in the instructions, recheck the parts in that circuit. Check for cold soldered joints or misplaced components. If the meter is not working, perform the voltage and oscillator tests in the troubleshooting guide.

To take voltage measurements, set the selector switch to either ac or dc. Plug in the red test lead to the "V & Ω" input jack and plug the black lead to the "COM" jack. The ac and dc current measurements are done in the same manner with the red test lead at the "mA" input jack, up to 200 mA. For current measurements in the 20-A range, place the red lead into the "20 A" input jack.

For resistance measurements, rotate the selector to Ω. Plug in the red test lead to the "V-Ω" input jack and the black lead to "COM." Rotate the selector to the desired ohm range. All resistance measurements should be made with the power off. Discharge all large electrolytic capacitors. Compare the resistance measurement to a 5% known resistor.

To check diodes, set the selector switch to ohms, the red test lead to the "V-Ω" input jack and the black lead to the "COM" jack (Fig. 8-12). If the digital display reads overrange (1), reverse the lead connections. The red test lead is positive and the black lead is negative. A shorted diode will give a very low ohm measurement and if overrange (1) is displayed in both directions, the diode is open. All measurements can be compared with a known DMM for accuracy.

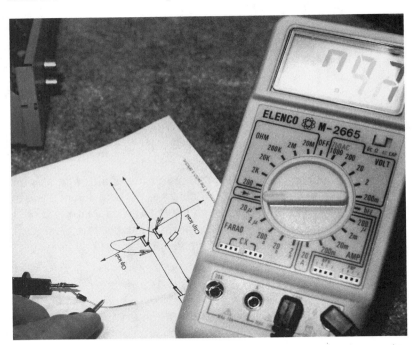

■ **8-12** *Checking a silicon diode with the DMM selector switch turned to the diode test range.*

Deluxe regulated power supply kit (XP-620K)

This deluxe regulated power supply provides a positive and negative voltage from 1.5 to 15 Vdc at 1 A. A 5-Vdc source supplies voltage up to 3 A. IC voltage regulation is found in the variable dc sources and a transistor and IC provide voltage regulation in the +5-V source. Full-wave bridge rectification is found in both sec-

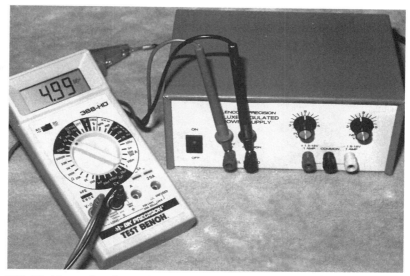

■ **8-13** *Measuring the variable output voltage of the regulated power supply kit with a DMM.*

ondary transformer windings. You can use this voltage-regulated supply for projects and a future service voltage source (Fig. 8-13).

Mounting components

All small components are mounted on a PC board with part numbers and letters outlined on the top side of this board. Notice that the transistor (Q1), electrolytic capacitors, and diodes are polarized. Make sure that the positive terminals of diodes and capacitors match those on the PC board. Mount the flat side of the transistor, as shown in the top outline.

Solder and check off each component as it is mounted. There are two different sets of silicon diodes. The largest diodes (1N5400, D5 through D8) and the small diodes (1N4001, D1 through D4) mount in the 15-V supply circuits. IC1 and IC2 are LM317 variable voltage regulators and transistors Q2 (2N6124) are all three TO-220 type mounting solid-state devices. Just look on the front side of these three-legged devices for the correct part number (Fig. 8-14).

Heatsink mounting

All four voltage-regulated ICs and power transistors are mounted on a separate heatsink. Bend up the transistor leads for easy soldering of connecting wire leads. Place silicone grease on the back side of each regulated IC. Notice that power transistor (Q2) and IC

■ 8-14 *The inside view of the parts connected to the front and rear panels, before replacing the top metal cabinet.*

regulators are insulated with a piece of mica away from the heatsink. Apply silicone grease to both sides of the mica and place it over the hole with a plastic bushing. Mount Q2 on top of a piece of mica insulation and bolt it into position. Make sure that you have the correct side of the metal heatsink before mounting the parts.

Check each IC and power transistor mounting with the drawn outlines on the metal heatsink. After mounting the transformer and all large components, connect the different leads of ICs to the PC board with colored hookup wire. Make sure that none of the ICs or power transistor terminals touch each other and the metal heatsink (Fig. 8-15).

Testing

Check the troubleshooting guide for possible problems if the supply blows the line fuse or does not supply output voltage. Notice

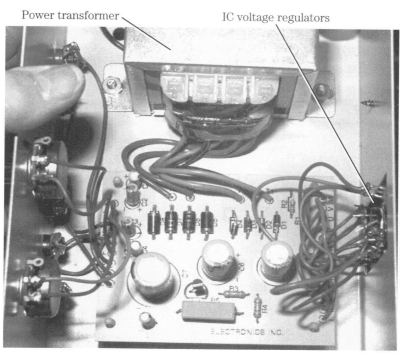

■ 8-15 *Notice how the power transistor and IC voltage regulators are mounted and connected on the metal heatsink.*

the different voltage measurements on transformer, transistors, and IC regulator terminals. Measure the voltage: +1.5 V to +15 V from the red and yellow post to the common black insulated parts. Rotate each part to vary the separate positive and negative voltage. Check the +5-V source across the red and black common terminal posts.

Conclusion

You can learn how circuits tie together and function in kit building. For a career in electronic servicing, you must learn how to make good clean solder joints. Mounting the different components on the PC board provides correct replacement of parts and required patience. Take your time when placing the electronic kits together. Besides having a lot of fun, you can use these test instruments to service consumer electronic products.

Here is a list of kit manufacturers and mail-order firms that have many different kits available to the beginning electronic technician:

624 Kits
171 Springdale Dr.
Spartanburg, SC 29302

All Electronics Corp.
P.O. Box 567
Van Nuys, CA 91408

Antique Electronic Supply
6221 S. Maple Ave.
Tempe, AZ 85283

Cal West Supply, Inc.
31320 Via Colinas #105
Westlake Village, CA 91362

Check Kit List Mfg. book:
Rutenber Engineering
38045 10th St. E. #1145
Palmdale, CA 93550

D.C. Electronics
P.O. Box 3203
Scottsdale, AZ 85271

Electronic Goldmine
P.O. Box 5408
Scottsdale, AZ 85261

Electronic Kits International, Inc.
16631 Noyes Ave.
Irvine, CA 92714

The Electronic Rainbow, Inc.
6254 LaPas Trail
Indianapolis, IN 46268

Elenco Electronics, Inc.
150 West Carpenter Ave.
Wheeling, IL 60090-6062

Gateway Electronics, Inc.
8123 Page Blvd.
St. Louis, MO 63130

Graymark International, Inc.
P.O. Box 2015
Tustin, CA 92681

Herback and Rademan
P.O. Box 122
Bristol, PA 19007-0122

Hosfelt Electronics, Inc.
2700 Sunset Blvd.
Steubenville, OH 43952-1152

Interactive Electronics
P.O. Box 913 Dept. K
Eulers, TX 76039

Kevin Electronics
10 Hub Dr.
Melville, NY 11747

Marcraft International Corp.
1620 E. Hillsboro St.
Tri-Cities, WA 99302

Mark V Electronics
8019 E. Slauson Ave.
Montebello, CA 90640

Oak Hill Research
20879 Madison St.
Big Springs, MI 49307

Ocean State Electronics
P.O. Box 1458
Westly, RI 02891

Pan-Com International
P.O. Box 130
Paradise, CA 95967-0130

Ramsey Electronics, Inc.
793 Canning Pkwy.
Victor, NY 14564

Shock hazards in electronic repair

THE ELECTRONIC TECHNICIAN MIGHT COME IN CONTACT with the power line, extremely high voltages, and shock hazards while working on electronic entertainment products. The professional technician is reminded each day to play it safe around the TV chassis, microwave ovens, and CD players. All electronic devices that operate from power line voltages are dangerous.

Surprisingly, very few technicians take the time to read the safety precautions, listed by the manufacturer on the service literature or manual. In every Howard Sams Photofacts, safety precautions and safety guidelines are given for color television receivers. Safety checks are made on fire and shock hazards, implosion, x-radiation, and tips for proper installation.

Beware of all shock hazards when servicing a new piece of electronic equipment. Read the manufacturers' safety precautions. Keep hands and test equipment out of the electronic product until you know what danger is involved. Play it safe. A shock might bring on a mild heart attack or cause painful burns on the hands and body. Or you could accidentally knock over test equipment and cause damage to it and more harm to yourself.

Power line problems (ac)

You must respect electronic products that receive power from the ac power line. When a defective unit blows the service bench fuses or main fuse in the building, expect a direct short in the ac circuits. Some ac commercial entertainment units are fused internally, and some are not. The ac cord might bring the voltage to a fuse holder or to an ac on/off switch for protection. Usually, a black or darkened ac fuse indicates a heavy overload.

While working with test equipment or metal hand tools around the chassis, the ac power line might be shorted out or provide a ground through the test equipment. Small VOM and DMM test equipment can easily be damaged if you are trying to measure high voltages. Improper grounding of test equipment might produce a shock hazard around ac-operated components.

Most manufacturers protect the consumer by bringing the ac cord directly to the power transformer primary winding without any bare wires or switches. This method is used in small radios, CD players, and tape players. Then, they switch the secondary winding or *B+* voltage to the desired circuits (Fig. 9-1). Other companies might place spaghetti or spot the ac switches with plastic dope to prevent shock damage. Remember, the ac power-line voltage is dangerous and should be respected and protected while servicing electronic products.

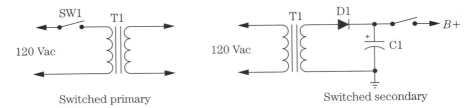

■ **9-1** *To prevent danger to the customer in some of the latest receivers, CD players, and tape decks, switch the secondary lower voltage of a transformer instead of the power-line input circuits.*

Check that power cord

Inspect every power cord of each product that is serviced. Ragged or cracked rubber ac cords might short out and start a fire in the customer's home. A bare section of power cord wiring could shock someone who touches the cord and TV set. Replace the defective cord if it shows signs of wear and tear, cracks and bare areas (Fig. 9-2).

Check where the power cord enters the chassis. Poor interlock contacts or a bare ac wire entering a metal chassis is very dangerous. Most ac cords arcover in the ac plug, if the cord is pulled out from the power receptacle. Bad plug connections can start a fire if they are close to curtains and cluttered newspapers. It's best to replace the power cord in one whole piece, plug and all. Replace it while the chassis is on the service bench. Check that power cord for possible shock hazards.

Interlock and power line cord (ac)

■ **9-2** *Check the interlock plug and cord of a TV chassis for wear and cracked line cords.*

The hot chassis

Just about every TV chassis has a hot chassis. No transformer is found in these chassis (Fig. 9-3). The hot wire of the power line might be attached to the chassis. Touching the chassis and ground at the same time might produce an accidental shock. You might touch the hot chassis and a grounded test instrument at the same time and receive a surprising shock. Do not fear the ac power line voltage (being fearful can create an even greater safety hazard), but respect it. Stand on a rubber mat to provide greater protection and to rest those tired feet.

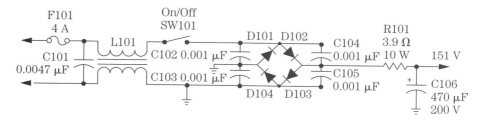

■ **9-3** *The hot TV chassis operates directly from the power line and it must be serviced with extreme care.*

The ac voltage of a TV set is wired directly to a protector fuse on one side of the power line and the other side might go to chassis ground or to a common point in the circuit. When the hot side of the power line is found at the ground side of the chassis, something must give. Remember, the TV chassis might have a hot and cold chassis.

Usually, the ac power line fuse will blow or open if a short occurs with a leaky diode or electrolytic capacitor (Fig. 9-4). In most chassis, the ac power plug is polarized so that it cannot be plugged in wrong. One plug spade is wider than the other and it can be plugged into the outlet in only one way. This same polarity should be observed when replacing the new power cord to the chassis (Fig. 9-5).

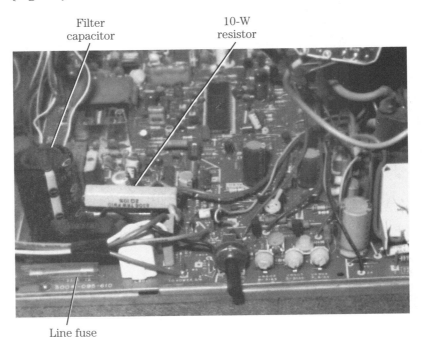

■ 9-4 *Often, the power-line fuse, high-voltage resistors, and silicon diodes are close together at the rear of the TV chassis.*

Keep the TV chassis and test equipment away from metal support posts, furnace ducts, heat pipes, and water pipes. You might get a shock from the TV chassis to grounded test equipment if it is not properly grounded.

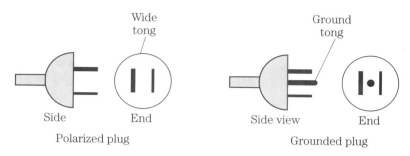

■ **9-5** *Two types of polarized plugs to prevent possible shock to the customer and technician.*

Isolation transformer

Today, before servicing any TV chassis, plug it into an isolation transformer. The isolation transformer provides a 1:1 ratio of ac voltage at the secondary. If 120 Vac is applied at the primary winding, 120 Vac is found at the secondary winding (Fig. 9-6). The secondary winding has a power ac outlet that you can plug the TV chassis into. With this method, the TV chassis is isolated away from the power line. Use the isolation transformer on every piece of consumer electronic equipment you service.

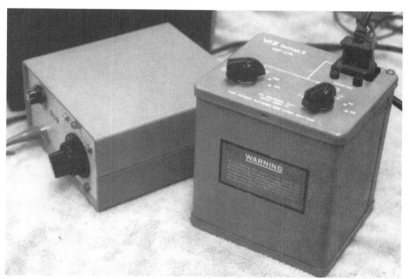

■ **9-6** *Always use an isolation power transformer while servicing the TV to prevent component parts and tester damage, and to prevent possible injury to you.*

You can damage test equipment, the TV chassis, and injure the electronics technician if an isolation transformer is not used on a hot TV chassis. Make sure that no test equipment is connected to the TV chassis while it is connected to the power line, unless connected through an isolation transformer. Remember, the TV might have the hot wire (120 Vac) connected to the chassis. When a test instrument ground is connected to the chassis, sparks fly and a fuse is blown. If you happen to touch the chassis and test equipment ground, 120 Vac is applied to your body.

Connecting grounded ac test equipment to the hot TV circuits might place the power line voltage across one of the diodes in the power supply, destroying the diode. Other damage to the TV set might occur like stripping grounded PC wiring and components. Besides, the electronics technician might be injured. Play it safe and use an isolation transformer on all electronic equipment that you service (Fig. 9-7).

■ **9-7** *Plug the isolation transformer into the wall outlet and plug the TV or electronic product into the transformer receptacle.*

Grounded work bench

Often, the electronic service bench power can be turned off or on when the technician arrives to work and when he or she finishes for the day. With this method, no electronic equipment is accidentally left operating. The power line voltage can be either 110 Vac or 220 Vac. For large service quarters and with several benches, 220-V circuits can be split and carry a heavier load. Actually, a double-pole, single-throw fuse power switch switches both sides of the 120-V power line (Fig. 9-8).

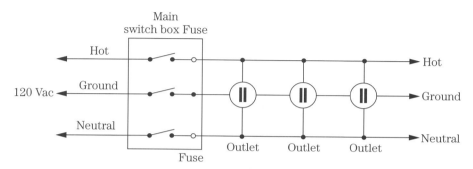

■ **9-8** *A DPST metal switch base or circuit breaker can be used to turn off the power each night before the technician leaves the shop.*

All outlets or bench power strips have polarized plug receptacles with a grounded terminal. When each outlet is grounded, less shock is found between the technician and the earth ground or grounded components. When the test instruments are plugged into the grounded receptacles, they are adequately grounded. No additional ground wires are needed. The brass or copper screw found on the receptacle ties to the common power ground wire (Fig. 9-9).

■ **9-9** *Electric power strips or outlets should be grounded for added protection while you service the electronic products.*

Grounded test equipment

All power line test equipment should be grounded through the polarized plug (Fig. 9-10). Some large test instruments have a separate ground clip that is screwed to the front plate of a power receptacle for grounding. The oscilloscope and large external power supplies should be properly grounded to prevent spurious spikes from entering the test equipment and applied to the unit that is being serviced. This will prevent shock to the service technician if he or she accidentally touches a grounded wire, metal post, or furnace duct.

■ 9-10 *A three-prong or polarized plug should be used to ground all test equipment.*

Notice when the ground clip of the test instrument is attached to the TV or radio chassis, if excessive sparking is noted. Then measure the ac voltage between the test instrument ground clip and TV chassis with a portable VOM or DMM. If the voltage is above 5 Vac, a possible shock hazard might exist (Fig. 9-11). Check the test instrument and the unit being tested for possible ac leakage.

When making a service call in the home, and when the antenna cable is connected, extreme sparking and arcing might occur. Check the TV for ac leakage. Remember that the outdoor antenna is properly grounded through the metal mast and ground wire to earth ground.

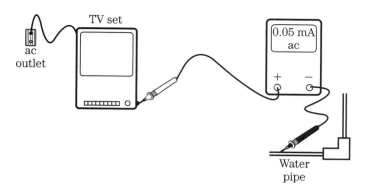

■ **9-11** *Check the TV, VCR, CD player, and tape deck for possible ac leakage with the ac voltmeter between unit and the earth ground.*

Hazardous work areas

You should never service or test electronic equipment in a damp or wet area. Keep open liquids away from spilling into ac outlets or wires. Stand on a rubber mat for added protection. Never service electronic equipment outdoors when it is raining or snowing. Play it safe.

Make sure that all power drills and tools operating from the power line are properly grounded. Do not cut off the ground terminal to make it fit in a power cord that has only two wires. Inspect extension cords for cracked and frayed areas (Fig. 9-12). Keep extension cords out of wet areas. Be careful where you drill in a wall so that you don't touch a hot ac cable.

Every year you see on the TV screen or in the newspaper that one or two people are accidentally injured or killed while repairing a TV antenna. Most antenna installers and service technicians use long aluminum ladders. Keep that metal ladder away from power lines. If an antenna could fall on a power line from a particular location, try mounting it in another place. Service technicians have been killed while installing antennas. Be careful.

Heat problems

You might accidentally place your hand or fingers on a heated soldering iron or you could drop molten solder on yourself. The soldering iron is dangerous if not handled properly.

Be very careful when operating a hot glue gun. You could get careless and move the component you are cementing and apply hot

■ **9-12** *Keep extension cords out of wet areas and replace frayed or damaged cords with polarized new ones.*

glue on your hand or fingers. It's natural to grab the hot glue and pull it away from burning flesh. Then you have two hands with hot glue stuck to them.

Do not touch hot output transistors to see if they are overheated (Fig. 9-13). You might burn your fingers. The same applies to hot 10- and 20-W resistors, which indicate overloaded or shorted components.

High-voltage precautions

Be careful while working around TV picture tubes. Remember, the high voltage found on the picture tube might exceed 32 kV. So, carelessness can be extremely dangerous. Most technicians become accustomed to the high-voltage lead and CRT anode when servicing the TV chassis (Fig. 9-14). The TV has very little current, but very high voltage.

The initial shock might not cause too much injury to the technician, but the damage done while pulling away can be costly. Hands and fingers can be smashed, test equipment can be knocked off the service bench, or tools can fly out of your hands and injure others.

■ **9-13** *Do not touch hot output transistors while the unit is operating.*

■ **9-14** *The anode lead of the picture tube might exceed 32 kV; it's a vicious shock hazard!*

High-voltage precautions

Extreme care should be taken when measuring the high voltage applied to the picture tube. Sometimes the high-voltage probe is difficult to get under the rubber anode lead, which pulls out of the high-voltage socket and tears the ground clip from the test instrument to chassis ground. The technician receives the full jolt from the high-voltage wire. Do not forget to clip the ground wire of the HV voltmeter to the chassis or you will receive a shock and you might drop the high-voltage probe and destroy a good tester.

Do not try to measure any high-voltage source with a VOM or digital multimeter. These instruments will be destroyed and if holding one of the probes, you might receive the full high voltage from the TV chassis. Take HV measurements with a high-voltage multiplier probe or with an HV probe connected to a VTVM. These voltages are extremely high and dangerous. The new high-voltage probes will measure up to 40 kV.

Replacing the CRT

Use extreme care when removing and replacing the color picture tube (Fig. 9-15). Besides being quite heavy, they can be dangerous if not handled properly. In fact, two men are required to lift a 31- or 35-inch TV picture tube.

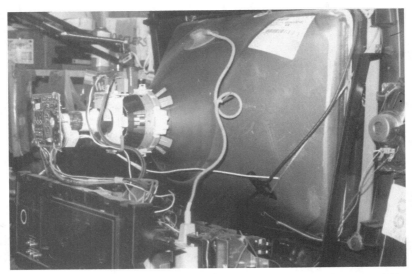

■ **9-15** *When removing the picture tube, use extreme care to prevent implosion or a broken CRT neck.*

How many service technicians have accidentally dropped the back TV cover down on the neck of the picture tube and snapped off the gun assembly? This seems to occur with most beginning technicians. It's very dangerous in the home or shop, embarrassing, and downright costly. It takes profit from several repairs to pay for that new picture tube. Picture tubes can implode and cause damage around it. Handle with care.

Before removing the picture tube, discharge the anode socket with a couple of long screwdrivers. Slip a pair of protective goggles over your eyes. Do not carry the tube against your body. Have someone help lift the tube in and out. Protect the neck of the picture tube. Lay the face of the tube down on a blanket or pad. Protect your eyes at all times.

Charged capacitors

Discharge all main electrolytic capacitors before taking any voltage and resistance measurements. Do not put your hands or fingers into the chassis until these capacitors are discharged. Some of these high-capacitance capacitors can hold the charge for months.

When replacing defective electrolytic capacitors, make sure that the correct polarity is applied. Check the foil terminal and place this at ground potential (Fig. 9-16). If large electrolytic capacitors are hooked up backwards, they can explode and throw pieces over the room. Doublecheck the polarity to prevent a capacitor from blowing up in your face and destroying your precious eyes.

Microwave oven shock hazards

Always discharge the high-voltage capacitor after removing the back panel of a microwave oven. Besides having high ac and dc voltages, the oven current is very high and can knock a person down. Each time that the oven is fired up, this HV capacitor must again be discharged before you touch any components or attach test instruments for certain tests (Fig. 9-17). This HV capacitor can charge up to several thousand volts (average 3.8 kV) with enough current to seriously injure or cause a possible heart attack.

When repairing the microwave oven, do not get distracted from what you are doing. Think each time you make a step in servicing these ovens. Do not put hands or test probes into the oven circuitry while the oven is operating. Pull the power plug and discharge the HV capacitor before clipping voltage or current meters

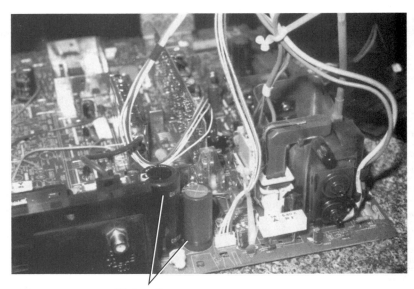

High-voltage
electrolytic
filter capacitors

■ **9-16** *Make sure that all electrolytic capacitors are installed with the correct polarity or they might overheat and explode.*

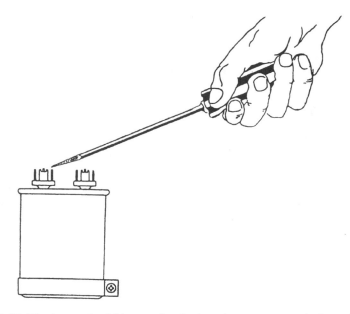

■ **9-17** *Discharge the HV capacitor in the microwave oven before attaching test equipment or replacing components. Discharge the capacitor each time that the chassis is fired up.*

to the oven circuits. Remember, this capacitor can hold a charge for months.

Keep all tools and test equipment away from the magnetron tube at all times. Look for the high-voltage signs (Fig. 9-18). Do not try to measure the high voltage of the microwave oven with a VOM or DMM. These meters will not go that high in voltage measurement (3.5 kV). They can be ruined within minutes. Attach only test instruments designed to work on and around the microwave oven.

■ 9-18 *Respect the high-voltage signs on the microwave oven and discharge the HV capacitor.*

X-ray radiation

Today, higher voltages are found on the picture tube in newer sets. The TV chassis is designed to shut down to disable the TV if higher voltages exceed the manufacturer's limits. In the early TV chassis, you could see that the HV had increased if the horizontal circuits were disabled and the screen showed horizontal lines. The TV set today is designed to disable the horizontal output circuits, preventing excessive high voltage and shutting down the chassis.

Always check the high-voltage setting and shutdown circuits before returning the TV to the customer. In most chassis, a couple of test points can be shorted to see if the chassis shuts down. Another method is to raise the dc voltage to the horizontal circuits and notice if the chassis shuts down. If not, check for defective components in the protection circuits of the TV. Make sure that the x-ray shutdown circuits are operating to protect the customer or operator.

CD laser eye damage

Your eyes can easily be damaged while servicing that small portable compact disc player. The laser beam is directly underneath the CD disc that rotates. When the disc is on the turntable and operating, the beam is reflected by the disc and will not harm your eyes. Remember, you cannot see this beam, it's invisible (Fig. 9-19).

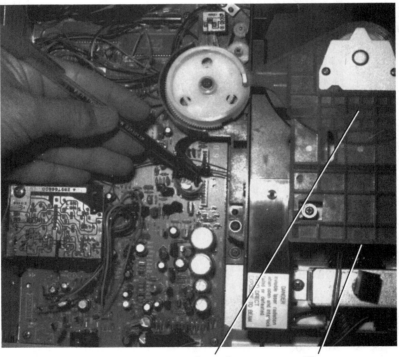

Laser lens assembly   Flapper

■ **9-19** *Do not look directly at the laser beam on a CD player while servicing it. The laser optical assembly is located under the CD disc.*

Take great care to be absolutely sure that the laser beam never enters the eyes, while repairing the compact disc player. Do not look directly at the laser beam coming up from the pickup. Do not allow it to strike against your fingers, skin, or arms with the power on. Keep your eyes at least one foot away from the laser lens area.

Keep the compact disc on the turntable at all times while servicing the CD player, if possible. Tape a piece of tin foil over the laser lens assembly if the unit must be turned sideways or upside down for servicing.

Clean the lens assembly only when the player is shut off. The disc changer or loading mechanism must be removed before getting to the lens assembly on table models. Of course, the lens assembly stares up at you, with CD removed, on the portable CD player (Fig. 9-20). Dust off the lens area with a camera lens airbrush or liquid and a cotton swab or stick.

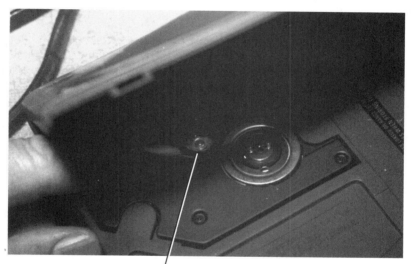

Laser lens assembly

■ **9-20** *In table-top models, the CD laser beam is covered with a flap or loading mechanism. In the portable CD player, the laser assembly points upward.*

Electrical shocks in test equipment

Make sure that all test equipment operating from the ac receptacle has no leakage. If ac leakage is found, it's possible to damage transistor and sensitive semiconductor devices with only a slight ac or dc voltage charge. Besides damaging components in the unit to be serviced, bench fuses might be blown and test equipment can be damaged. Make the same leakage test on the bench test equipment as in making a leakage test on the customer's product.

Protecting the customer

After completing the repair of each unit, make sure that the unit will not cause a fire or possibly cause shock to a person. Check the repaired TV or CD player for possible line leakage. Simply measure the ac leakage from chassis to a good earth ground or water pipe.

How many electronics technicians take the time to check the TV set, compact disc player, VCR, and receivers for ac voltage cold and hot leakage current checks before returning the unit to the customer? Make sure that all safety-related parts have been properly replaced. Check the manufacturer's star or triangle safety symbols that mark those special safety-related components.

Replace all protective shields. Make sure that all built-in protection devices have been restored to normal operation. Replace knobs with plastic and not metal knobs. Replace the back covers on the TV set, while waiting for parts. Keep the TV chassis isolated away from those in the house.

When replacing components, make sure they are not laid next to or on high-wattage resistors that can cause a fire. Remove all sharp points of soldering in the safety circuits. Make a good soldered joint. Check any component that looks overheated and might cause damage later on. Place rubber or silicone cement over possible high-voltage or power-line areas that might become shorted out.

For a chassis or antenna check, short the ac plug with clip wires. Make sure that the power cord is pulled. Rotate the on/off switch to on. Connect one lead to the ac prong and the other to the antenna terminal of the TV set. Check for exposed metal cabinet screws (Fig. 9-21). If the ohmmeter measurement is less than 1 MΩ, check for leakage in the TV set or repaired product.

Now make a hot current test. Plug the unit into the ac power line with the switch on, then turned off. Take a current test by touching metal screws, front panel, metal cabinet, and metal overlays with the leakage meter negative terminal grounded to a water pipe or earth ground. If the current exceeds 0.5 mA, check for component leakage within the set (refer to Fig. 9-11).

Conclusion

Always read the safety precautions in the front of the service manual on a new or unknown product before servicing. Look and think

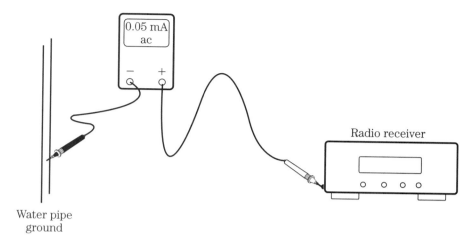

■ **9-21** *Take a hot and cold ac voltage and resistance test on the product before returning the repaired unit to the customer.*

before attaching test equipment. Keep wide awake at all times when working on a hot chassis or microwave oven. Keep fingers and hands out of dangerous places with the product operating.

This chapter is not placed in the book to scare the potential electronics technician, but to make you aware of the possibility of electrical and physical hazards that occur in the electronics field. The thoughtful, hardworking, wide-awake technician has and will continue to make repairs on these entertainment devices for years to come.

Simple electronic circuits

10

MANY BASIC CIRCUITS ARE CONTAINED IN RADIOS, TVS, amplifiers, CD players, and tape decks. Most radio receivers have six different stages: RF amp, converter or oscillator, IF amp, detector, audio driver, and audio amplifier. The TV chassis contains many basic circuits: RF tuner, IF amp, video, vertical, horizontal, sync, AGC, color, audio, high voltage, and picture tube circuits (Fig. 10-1). If the product in the consumer electronic field operates from ac power-line voltages, the low-voltage power supply and regulator circuits must be included.

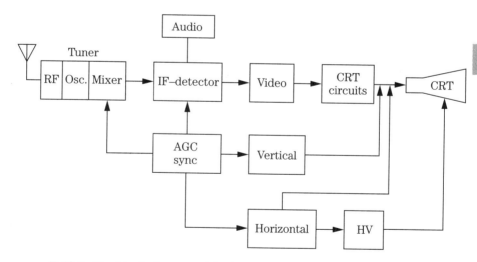

■ **10-1** *The block diagram of the basic circuit found in the TV chassis.*

Amplifiers

Many forms of amplifier circuits are contained in just about any electronic product found on the service bench. Amplifiers of some sort are found in the TVs, radio receivers, CD and tape players, VCRs, power amps, and camcorders. The radio receiver uses an RF, IF, and AF amplifier, and the TV set incorporates the RF, IF, color, horizontal, vertical, and AF amplifiers (Fig. 10-2). Although the RF

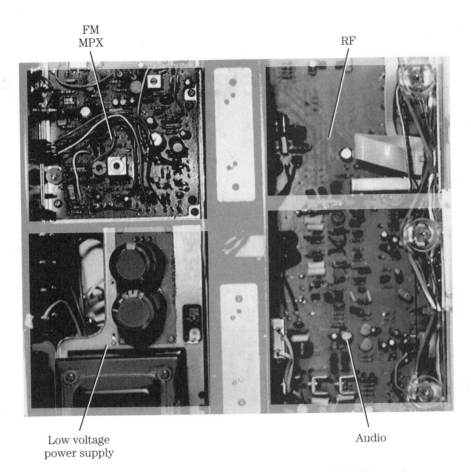

■ **10-2** *The different sections of a complicated AM/FM/MPX receiver on separate PC boards.*

amp of a radio receiver might amplify the radio frequency signals, the video amp in the TV set amplifies the video or picture signals.

The amplifier is a device that increases the amplitude of the applied signal. Usually, the amplifier receives an input signal and amplifies or increases the output signal. There are many different amplifiers, such as current, power, and voltage amps.

Years ago, these electronic products contained tubes and some solid-state transistors. Although tubes have almost disappeared from these circuits, they are still found in high-powered amps and electronic industrial machines. Today, transistors, integrated circuits (ICs), microprocessors, and LSIs are found in the electronic entertainment products. In the future, the simple transistor might become obsolete.

Audio amplifiers

Audio amplifiers do nothing more than amplify the audio signal. The audio frequency amplifier operates in the 20-Hz to 20-kHz audible range. For the early audio solid-state chassis, the audio stages might consist of PNP or NPN transistors (Fig. 10-3). The audio amplifiers in a tape deck might consist of a preamp, AF amp, driver, and an audio output amplifier that drives several speakers.

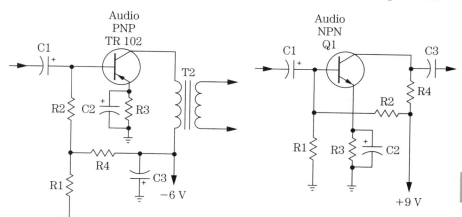

■ **10-3** *The PNP and NPN basic audio circuits with transformer and resistor-capacitor coupling methods.*

Notice in the PNP transistor audio circuit that the voltage applied is negative to the collector terminal. C1 couples the audio input signal and R1, R2, and R4 provide bias to the base terminal. R3 is the emitter bias resistor. T2 couples the amplified signal to the next audio stage.

In the NPN audio circuit, a positive voltage is applied to the collector terminal. C1 couples the input audio signal to the base terminal and isolates voltage from the previous circuit. R1 and R2 provide base bias voltage, and R3 is the emitter resistor. Instead of a transformer winding applied to the collector terminal, resistor R4 is used instead. This audio circuit is known as a *resistance-capacitance coupling circuit.*

Audio-frequency (AF) amplifiers might consist of resistance-capacitance, impedance, transformer, and direct-coupled circuits. The resistor-capacitor coupling consists of a collector and base resistor with an isolation-coupling capacitor between the two circuits (Fig. 10-4). The impedance coupling might consist of an inductance as the collector load with C1 coupling the signal to a resistor base circuit. Transformer coupling connects the primary winding to the collector circuit and secondary winding to the base terminal of the next circuit. Direct coupling is when the output of one transistor is wired directly to the input circuit of the following stage. In this case, an NPN and PNP transistor are used.

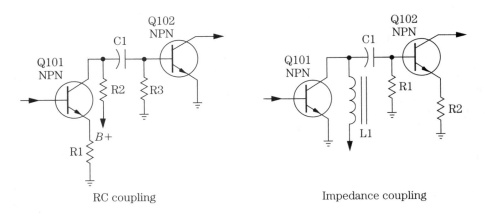

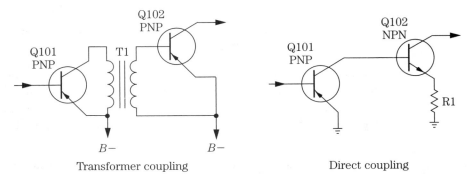

■ 10-4 *The early audio circuits consisted of transformer and impedance coupling. Today, resistance-capacitor (RC) and direct coupling is used between the audio stages.*

Today, you will find RC coupling, direct coupling, and some transformer coupling in the audio transistor circuits. Of course, the IC has taken over most of the audio amplifier circuits. You might still find a combination of preamps, AF amp transistors, and IC power output audio circuits. Even the small audio amplifier might consist of one IC as preamp and large power amp IC in the tape player. In the latest audio circuits, the entire audio signal is amplified by one large IC.

Preamp audio circuits

Tape and phono players must have a preamplifier stage because the input signal is so weak that it must be amplified, while the radio signal might work directly into the AF circuits. The preamp circuits might consist of two transistors in a direct or RC coupled circuit. Today, the preamp audio circuit might consist of one IC in the monaural or stereo input circuit (Fig. 10-5). Here, C501 couples the weak input signal to the input terminal 4 and C507 couples the amplified audio signal from pin 2 of IC501, to the next AF amp stage. Pin 5 is the voltage supply (V_{cc}) pin. IC501 might have another identical audio stage for the other stereo input circuit contained in one component.

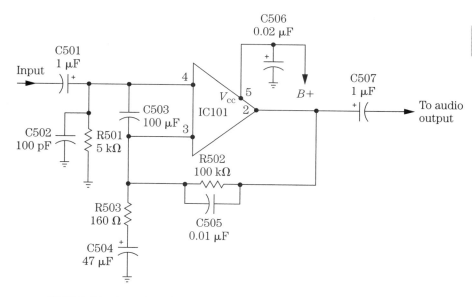

■ **10-5** *The tape head or phono pickup might connect to one preamp IC instead of transistors.*

AF audio circuits

The audio frequency (AF) stage might consist of a transistor or IC. The input audio signal is controlled by a volume control in the base circuit of the AF amplifier (Fig. 10-6). Notice that the output signal is taken from the emitter circuit.

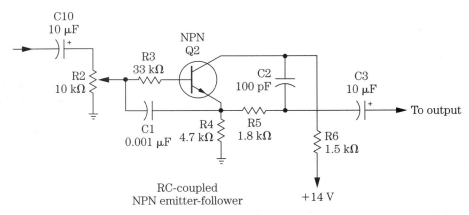

10-6 *RC coupling is used in the input and output circuit of an NPN transistor emitter-follower circuit.*

You might find a separate AF and driver transistor in the early transistorized audio circuits. The driver transistors are used to drive the audio output transistor with greater volume. Today, a complete audio circuit can be found in one IC. The AF and audio output circuits are located in one IC that might also be included in a stereo power output circuit (Fig. 10-7).

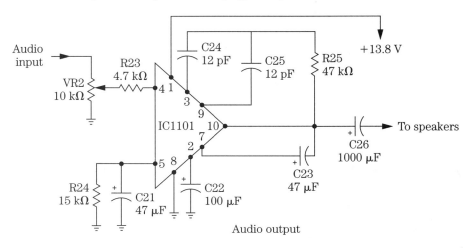

10-7 *One large IC might be used in the audio circuits in a boom-box or large portable cassette or tape player.*

The audio input signal is controlled by VR2 to the input terminal of pin 4. The amplifier audio output signal is capacity coupled to the speakers from pin 10 of audio output IC1101. Pin 4 is the supply voltage (V_{CC}) terminal. Notice that coupling capacitor C6 has a very high capacitance (1000 µF).

Stereo circuits

The stereo audio amplifier consists of two identical stages or circuits that produce audio signal from 20 Hz to 20 kHz to several speakers, large woofer, midrange, and tweeter speakers. The stereo amp is a two-channel amplifier for binaural reproduction. The stereo amplifier might be connected to the tape head, phonograph, boom-box, CD player, stereo receiver, VCR, and TV (Fig. 10-8).

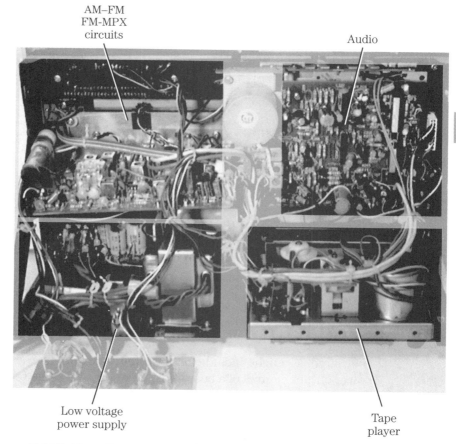

■ **10-8** *Here, the stereo amplifier circuits connect to the output of AM/FM /MPX receiver and tape player in the compact entertainment center.*

Identical preamp, AF, driver, and power output stages are found in the left and right channels of a stereo amplifier. Transistors were used in the early audio circuits. Now, ICs are found throughout the stereo amp circuits. In fact, in small amplifiers, one IC might include both stereo channels, and in larger amps, power ICs and transistors form the audio circuits.

The stereo preamp tape circuits might be included in one IC. The tape head signal is fed to input pins 1 and 8 of IC201, and through coupling capacitors C203 and C204 (Fig. 10-9). The left amplifier output signal is taken from pin 6 and the right stereo audio is from pin 3. C209 and C210 couple the output audio from IC201 to the top side of the volume control (VR-3A and 3B). The supply voltage is applied at pin 4.

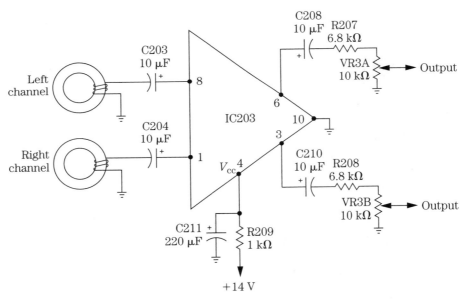

■ **10-9** *The preamp IC of the tape player might have RC coupling between the different audio circuits.*

One large IC might include both stereo channels in the audio power output circuits of a boom-box, tape, CD, and audio receiver. The stereo amp IC has two input circuits on pin 6 and 7 of IC205 (Fig. 10-10). Pin 9 is the supply voltage pin (V_{CC}). The left audio sound is amplified out of pin 2, and the right stereo channel is at pin 11. These two separate power audio channels provide signal to the respective speakers.

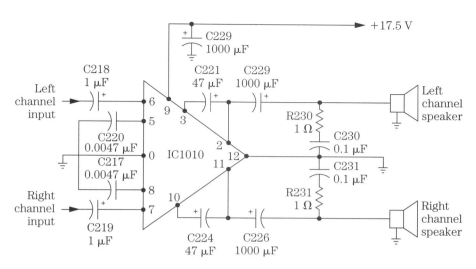

■ **10-10** *One large stereo audio output amp might often be contained in one IC.*

Transistorized power amps

Several power output transistors might be found in each stereo channel of a power amplifier or stereo receiver. Darlington transistors might be found in the driver stages. A high-powered amplifier might have 7 or 8 separate transistors in one channel (Fig. 10-11). Often, these transistors are directly driven. When one transistor fails, voltage might change in all transistors in the line up. Defective transistors in the audio circuits can be replaced with universal replacements. Follow the dark lines for several audio stages amplification.

Notice that the higher dc voltages are found on the output transistors with a positive and negative power supply (33.5 V). The bottom transistors (Q505 and Q508) operate from a +33.5 V and the top (Q504 and Q506) operate at a high negative voltage (−33.5 V). The voltage found at the 3-A fuse is zero. When either one of the output transistors Q524, Q526, Q507, and Q508 become leaky, the voltage at the fuse will rise in either a positive or a negative voltage, which will destroy the fuse. If a speaker fuse was not included, the voice coils of the woofer speakers might be damaged.

RF amplifiers

The radio-frequency (RF) amplifier increases the weak radio frequency from the antenna and supplies this signal to the converter or mixer stage in the auto, table, and stereo radio receiver. Radio

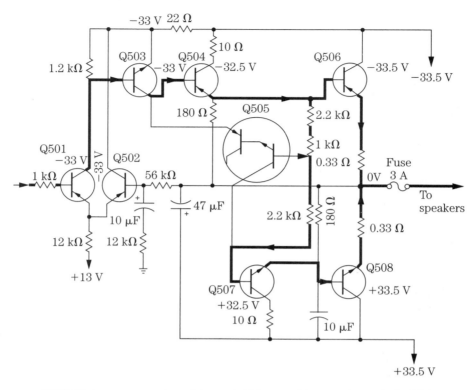

10-11 *Large power transistors might be used in one channel of a high-powered stereo receiver. Follow the heavy black lines to trace the signal through the various transistors.*

frequency is higher than the intermediate or audio frequencies. In the superhet circuits, the RF amp amplifies the incoming signal. The RF amp might be an RF transistor, FET transistor, MOSFET transistor, or RF IC. In the small table top or boom-box player, the RF amp might be eliminated with a converter circuit. The RF amp in the shortwave receiver operates at a much higher frequency than the average radio (525 to 1700 kHz).

The RF amp found in the RF AM circuits might consist of an RF or FET transistor. The RF amplifier picks up the radio frequency from the outside antenna or antenna loop and is transformer coupled with the RF antenna coil. The secondary winding of the antenna or RF coil (L105) couples the RF signal to the base of transistor Q152. The collector terminal of Q152 is connected to another coil fed to the mixer or converter circuits. The superhet receiver has an RF, oscillator, mixer, AF, and audio output circuits (Fig. 10-12).

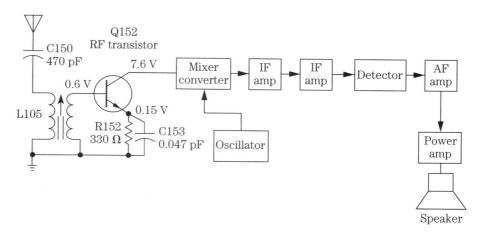

■ **10-12** *The superhet receiver circuit might have an RF transistor, mixer, oscillator or converter, several IFs, a detector, and audio stages.*

The FET has a high-impedance input like the vacuum tube and it can be found in AM and FM RF circuits. The only difference between the symbol of a regular transistor and an FET is an arrow at the input terminal (6). The FET has a gate, drain, and source elements, which represent a base, collector, and emitter elements in a transistor. The FET operates at higher frequencies with a high impedance input (Fig. 10-13).

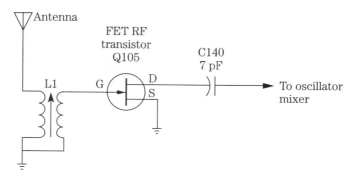

■ **10-13** *The RF field-effect transistor (FET) has a high input impedance, like that of a vacuum tube.*

The RF amp FET found in the FM circuits operates at a high frequency with critical components. The FM RF transistor operates the same as the AM circuits, except that they have shorter leads in the FM units. All defective FM transistors must be replaced with units that have the same lead length and must be mounted in the same spot. The RF amp might be tuned with a variable tuning capacitor, permeability tuned coils, or varactor diodes.

TV RF amp

In early TV receivers, the RF transistor, FET, and MOSFET were found in the rotary or turret tuner (Fig. 10-14). Today, the TV receiver might use an electronic tuning control with tuning modules or tuner systems. The tuning control system might provide different tuning voltages for a varactor tuning system. How the tuning control systems function is found in later chapters.

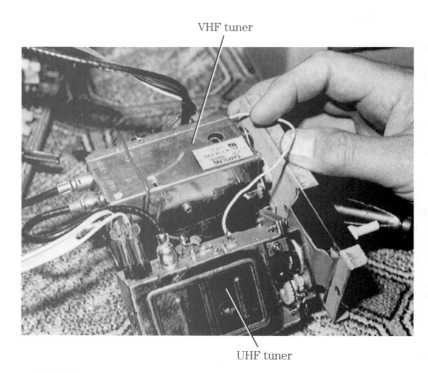

■ **10-14** *The early turret or rotary VHF and VHF tuners were transistorized with separate tuning knobs and antenna input cables.*

Converter amp stages

The converter transistor stage mixes two separate signals (RF input and oscillator) in a superheterodyne receiver. In smaller table model receivers, no separate RF stage is found because the RF signal is coupled directly to the converter circuits. The same transistor also serves as an oscillator (Fig. 10-15). The converter circuits are used in many different types of AM receivers.

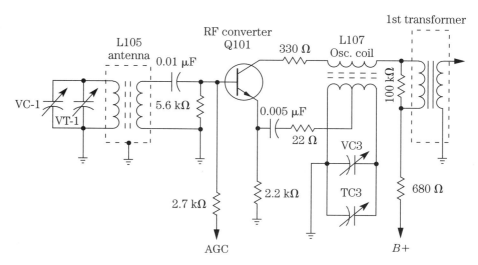

■ **10-15** *The RF amp and oscillator circuits might be called a* converter *stage in the superhet receiver.*

IF amp

The IF amp is found between the mixer, oscillator or converter, and detector stages in the superhet receiver. Also, the IF amp is found in TVs. This IF amp circuit boosts the IF signal in a fixed IF circuit. You might find one, two, or three different IF stages in a given receiver. The IF might be 455 kHz in an AM radio, 10.7 MHz in an FM radio, 262 kHz in an auto receiver, and 45.75 MHz in a TV (Fig. 10-16). The latest receivers might have a ceramic filter used in place of an IF transformer.

IC RF, IF, and mixer amps

Today, the AM radio might have only one IC that serves as an RF, mixer, oscillator, IF, and detector amplifier. The FM RF and converter circuits feed into IC101, providing FM IF, AM RF and converter, and AM mixer and IF stages (Fig. 10-17). The FM front-end signal is tied to pin 12 of IC101. The AM RF antenna coil operates at pin 13 and the AM oscillator circuits at pin 16. The 10.7-MHz filter is found at pin 8 and the 10.7 MHz IF is at pin 6.

Pin 9 ties to the first 455-kHz AM IF transformer and it is connected to the second IF 455-kHz transformer. The IF output amp signal is fed to pin 15 of IC101. The AM signal is detected by D124 to the audio circuit. Pin 4 feeds the FM signal to the FM/MPX IC. You might find a combination of transistors and ICs in the RF, IF, converter, oscillator, and mixer stages (Fig. 10-18).

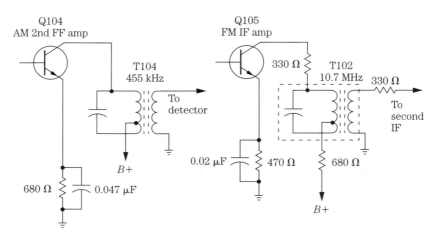

■ **10-16** *The intermediate frequency (IF) circuits of the AM receiver is tuned at 455 kHz, but the FM IF is aligned at 10.7 MHz.*

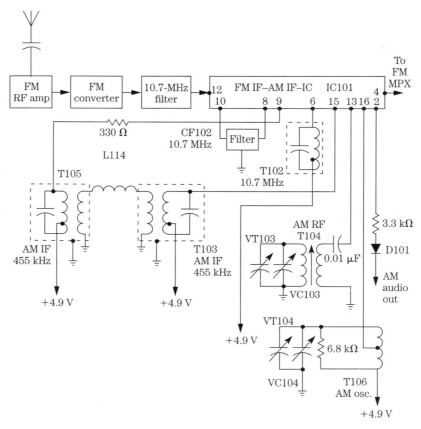

■ **10-17** *One large IC might provide AM RF, AM IF, and FM IF circuits. The FM front-end circuits are separate transistors.*

Simple electronic circuits

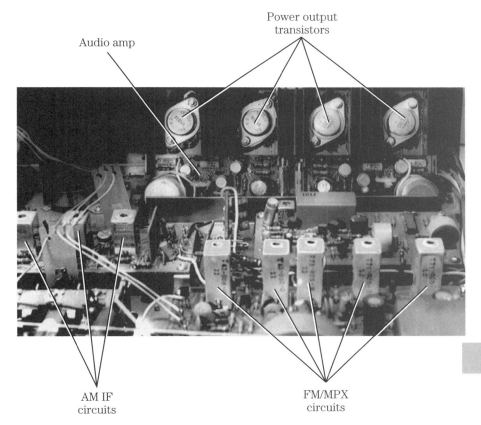

■ **10-18** *Separate transistors and IF transformers are used in the early AM/FM/MPX receiver circuits.*

Video amp

In the TV chassis, the video amp is a wideband stage that amplifies the picture signal and feeds it to the picture tube circuits. The video amp is found after the IF amp and detector circuit. Often, the first video amp has a 4.5-MHz sound trap that prevents audio from entering the video circuits (Fig. 10-19). The video amp might be directly coupled to several transistor video amplifiers and coupled to the picture tube circuits.

In the latest video amp circuits of the color TV set, the input 4.5-MHz sound trap might consist of a 4.5-MHz ceramic filter network. The ceramic filter is a resonant filter that uses piezoelectric ceramic material. No alignment is needed with ceramic filter networks; they are preset at a certain frequency. The 4.5-MHz filter network prevents the audio signal from entering the first video amplifier. The

Amplifiers

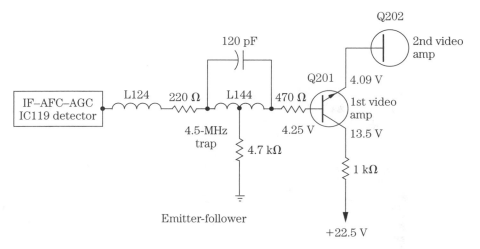

10-19 *The video amp circuit used in the TV chassis with a 4.5-MHz sound trap in the base circuit. The 4.5-MHz sound trap keeps the sound bars out of the TV picture.*

emitter-follower transistor (Q101) has a delay line tied to the emitter terminal and fed to the video/chroma IC203 (Fig. 10-20). Besides the TV chassis, the video amp is also used in VCRs.

Color amps

The color amplifier might be referred to as a *color output amplifier*. The TV chassis has green, blue, and red color amplifier transistors. Each color has its own amplifier stage. These transistors operate in a much higher voltage circuit and are fed through a re-

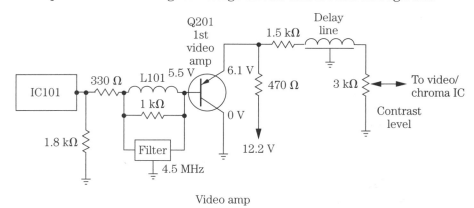

10-20 *The first video amp might have a 4.5-MHz filter network in the base circuit and the emitter terminal provide video through a delay line to the video/chroma IC.*

Simple electronic circuits

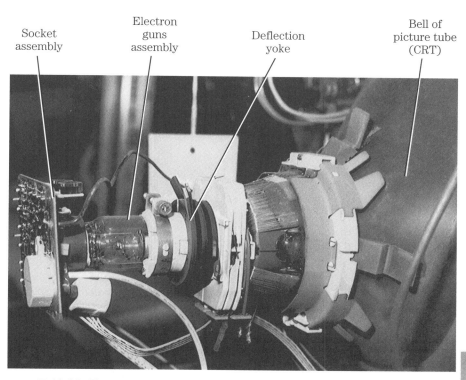

■ **10-21** *The color amps provide color and video drive to the cathode terminals of each color gun assembly in the picture tube. These color transistors might be mounted on the CRT socket board.*

sistor to the cathodes of the CRT. The color amps might be located on the rear of the picture tube socket board (Fig. 10-21).

The color amplifiers are exactly the same in each color circuit. The color signals are separated at the video/chroma IC and each color must be amplified to drive the picture tube color cathodes. The color signals are connected directly from the chroma IC to the base circuit of each color amp. The video signal is fed from the video amplifier to the emitter circuit of each color amp. The color transistor amplifies the color signal and is fed from the collector through a resistor to each cathode terminal of the picture tube (Fig. 10-22). When one color is missing from the picture, a color output amplifier is defective.

Detector circuits

In the superhet receiver, the AM detector might be a silicon or germanium diode. Likewise, the FM detector might consist of two

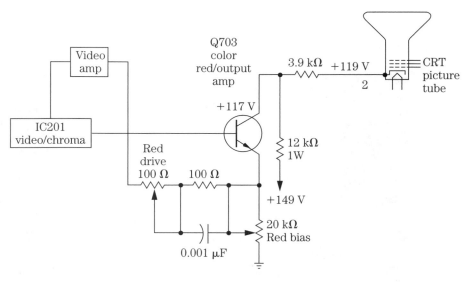

■ **10-22** *The color signal from the video/chroma circuits is applied to the base terminal of each color output transistor. Notice that the color output is directly coupled to the cathode terminal of the picture tube.*

diodes or detectors found in the discriminator and multiplex circuits. Today, video detection might occur in the TV IF section, and audio detection is located in the audio-detector amplifier IC. The audio signal is separated from the intermediate, RF, and video detection. The diode detector is located after the last IF stage in the AM radio circuits and is capacitor coupled to the function switch that switches to the audio circuits (Fig. 10-23).

Oscillator circuits

AM/FM oscillator circuits

The oscillator stage is tuned in with the RF tuner incoming signal. As stated earlier, the IFs are typically 455 kHz in the AM radio, 10.7 MHz in the FM receiver, and 262 kHz in an auto receiver. In larger AM radios, you might find a separate RF, mixer, and oscillator circuit. Smaller AM radios use a combination converter stage as oscillator and mixer circuit. The FM oscillator transistor tuned signal is coupled to the FM mixer stage with C110. The 220-pF capacitor couples the oscillator signal from the emitter circuit of Q103 to IC101 (Fig. 10-24).

The function of a frequency converter in a superhet receiver is to convert the RF signal to an IF signal. To obtain the change in fre-

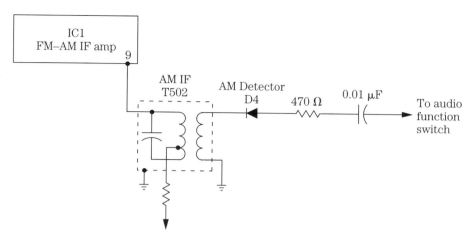

■ **10-23** *AM audio detection is used at the secondary winding of the IF transformer in the AM radio circuits.*

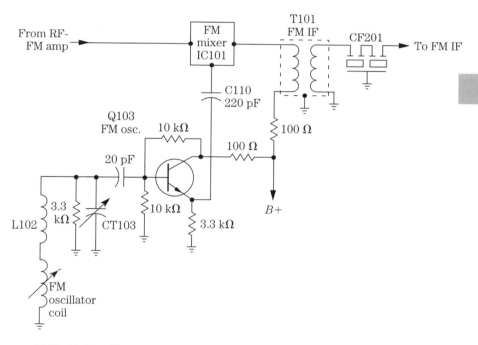

■ **10-24** *The FM oscillator circuit is coupled to the FM mixer (IC101) with C110. The resulting signal of the tuned incoming and oscillator signal is the IF signal.*

Oscillator circuits

quency, the incoming RF signal must be converted to a lower frequency. The mixer stage mixes the RF signal with the local oscillator (LO) signal. The resulting frequency is the IF signal amplified by the IF amplifiers or ceramic filter network. In the FM radio circuit, the RF, mixer, and oscillator stages might all use separate transistors.

Horizontal/vertical oscillators

The early horizontal oscillator circuit might have consisted of one transistor in a Colpitts oscillator circuit. The AFC signal is fed to the base circuit. The 15.575-kHz oscillator circuit is found in the base and emitter circuits (Fig.10-25). The ceramic iron core of L601 tunes or provides horizontal hold in the horizontal circuits. The horizontal hold control moves the picture horizontally. The collector oscillator signal is fed to the horizontal driver circuits.

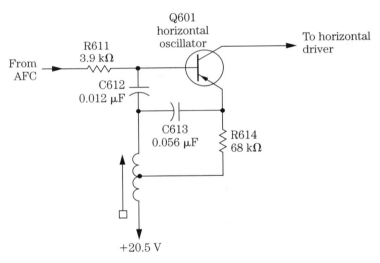

■ 10-25 *The early horizontal oscillator transistor circuit has a variable horizontal hold control in the base-emitter circuits.*

In older TVs, a tube blocking oscillator stage was used with a transistor horizontal oscillator in the solid-state chassis. Today, the horizontal and vertical oscillator circuit is found in a horizontal/vertical processor circuit. IC1001 contains the horizontal and vertical countdown circuits to drive a vertical output and horizontal driver circuit (Fig. 10-26).

Pin 3 of IC1001 provides a horizontal countdown and output driver pulse to the horizontal driver. Pin 14 provides a vertical amp drive

TV
large deflection
IC

■ **10-26** *One large IC might contain the horizontal and vertical deflection drive circuits.*

signal to the two vertical output transistors. When trouble is noted in either the vertical or horizontal stages, a quick scope of either vertical or horizontal waveforms at the horizontal/vertical processor will determine if the IC is normal (Fig. 10-27).

Horizontal driver circuits

The pulse waveform is fed from pin 3 of IC1001 to the base terminal of the horizontal driver transistor. The driver transistor is transformer coupled to the horizontal output transistor. Q500 is a high-powered driver transistor working in a high-voltage circuit (Fig. 10-28). The collector voltage is fed through driver transformer T500. R505 and the primary winding of T500 might appear quite hot if the horizontal driver transistor is leaky.

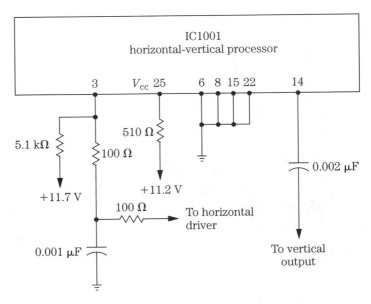

■ **10-27** *A waveform scope test made at pin 3 of IC1001 will indicate a horizontal drive signal and pin 14 a vertical drive signal.*

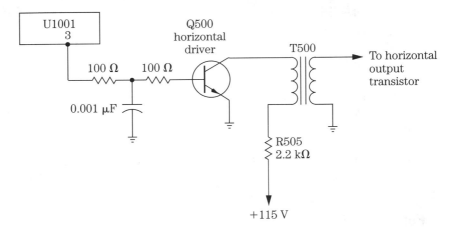

■ **10-28** *The horizontal driver transistor (Q500) receives a pulse drive from pin 3 of U1001, where it is amplified by Q500 and the transformer coupled to the horizontal output transistor.*

Horizontal output transistors

The horizontal drive signal from the secondary winding of the driver transformer (T500) is fed to the base terminal of the horizontal output transistor. This heavy power output transformer might have a separate heatsink or be mounted on the metal TV

■ **10-29** *The horizontal output transistor might use the metal TV chassis as a heatsink.*

chassis (Fig. 10-29). The horizontal output transistor is bolted to the heatsink with a piece of insulation and silicone grease on both sides of the insulation. Usually the metal body of the horizontal output transistor has more than 100 V applied.

The horizontal output transistor provides a large output pulse to the flyback or horizontal output transformer. The collector's supply voltage is fed through the primary winding of the flyback transformer. The dc voltage might or might not be fused applied to the primary winding (Fig. 10-30). The leaky horizontal output transistor might destroy the fuse and flyback winding. Remember, the damper diode might be found at the collector terminal to chassis ground or inside of the horizontal output transistor (Fig. 10-31).

Vertical output circuits

The vertical output circuit might consist of two output transistors or one IC in the vertical sweep circuits. The vertical output might have a top and bottom vertical transistor. The vertical driver might be found inside the deflection IC. A quick waveform test at the deflection IC vertical drive terminal indicates if the vertical input circuits are normal.

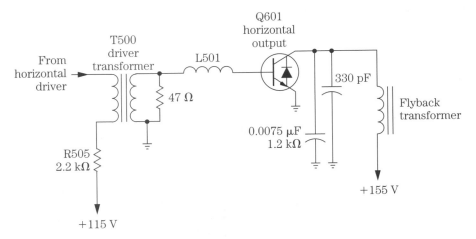

■ **10-30** *Driver transformers (T500) provide a drive signal or pulse to the base of the horizontal output transistor, which provides high voltage and different voltage sources in the secondary winding of the flyback.*

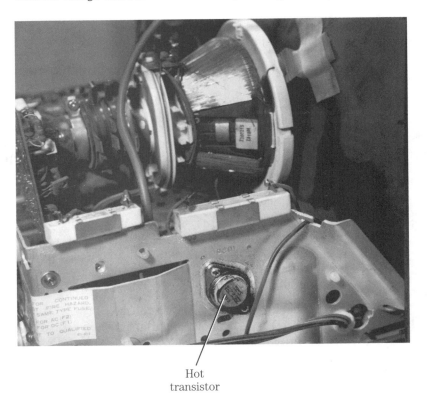

■ **10-31** *The damper diode might be used inside the same case as the horizontal output transistor.*

Simple electronic circuits

The vertical drive pulse is taken from pin 7 of IC501 and is directly coupled to the base of Q501. A high peaked output pulse is coupled from the vertical circuits with C517 to the vertical deflection yoke winding. If Q502, Q501 or R715 becomes open, a white horizontal line is found in the raster, indicating no vertical sweep signal (Fig. 10-32). An improper voltage on either transistor might affect the other.

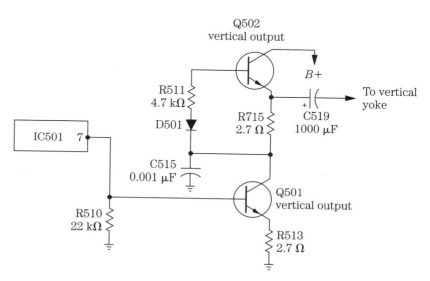

10-32 *The vertical drive output signal is used at pin 7 of IC501 to the base of Q501. The two vertical output transistors provide vertical sweep to the deflection yoke.*

You might find one power IC being used for the vertical output instead of two transistors in the latest TV chassis. The vertical output IC might be directly coupled to the deflection IC and it is driven with a low vertical input pulse or waveform. The vertical output waveform is taken at pin 5 of IC570. By taking an input waveform at pin 1 and output at pin 5, you can quickly determine if the output IC is normal or defective (Fig. 10-33). The supply voltage pin of IC570 is pin 9 (25 V).

TV high-voltage circuits

The horizontal output transistor provides drive to the horizontal primary winding, of which the horizontal output transformer provides several different voltage sources. With the integrated flyback, HV diodes are used in the HV winding to provide HV to the

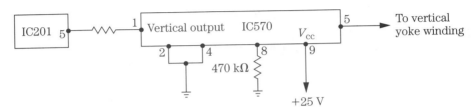

■ **10-33** *Check the input (1) and output (5) terminals of the vertical output IC (IC570) with the scope to determine if IC570 is normal.*

anode terminal of the picture tube. The secondary voltages are called *derived voltages*. The derived voltage might furnish different voltage sources to the CRT, video, color amps, boost, horizontal, vertical, IF, and chroma circuits. Once the horizontal circuits are functioning, the derived voltages will keep the horizontal circuits working. The HV components and windings are molded inside the flyback (IHVT) (Fig. 10-34).

■ **10-34** *The high-voltage winding and diodes are found molded inside the horizontal output transformer. Notice the HV anode lead out of top of IHVT.*

The high-voltage winding might furnish focus and screen grid voltage besides high voltage to the picture tube. A separate secondary winding might provide a dc voltage to the video color output transistors. Another winding provides three or more different voltage sources to other circuits within the TV chassis. Notice that each voltage source has silicon rectifiers to provide dc voltage to the various circuits (Fig. 10-35).

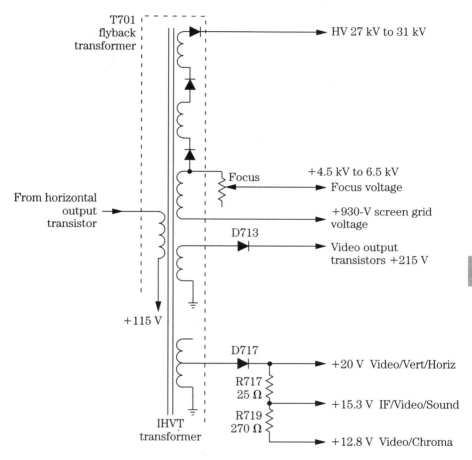

■ **10-35** *You might find various derived voltage sources that provide a voltage source to other circuits in the TV chassis.*

CRT circuits

The cathode ray tube (CRT) or picture tube must have different voltages applied to the various elements before a picture or raster can be seen on the screen area. The heater voltage is provided by a separate winding on the flyback. The three cathodes of each color gun assembly receive video and color drive signals from the color output transistors (Fig. 10-36).

■ **10-36** *Three different color gun assemblies are used in the neck of the picture tube. Each gun usually has its own heater and cathode elements.*

The anode voltage might vary from 15 to 32 kV to light the CRT. This anode socket is found on the top or alongside the picture tube. Be careful when working around the HV lead or socket of a CRT.

Variable screen and focus voltage is applied to the screen grid and focus grid elements of the picture tube. The focus control keeps the image in focus and the screen grid voltage provides adequate brightness to the viewing area (Fig. 10-37). The grid voltage is set at a determined dc voltage. Improper voltages found in the picture tube circuit might cause a weak, snowy, dark, or distorted picture.

Tape head circuits

When metallic tape passes the front of a tape head, the signal is picked up by the head and this is coupled to a preamp circuit. Because the signal is very weak, it must be amplified several times before it can be heard in the speaker. The portable, professional, tag-along, boom-box, and large cassette players all have a tape head. The tape head might be switched into the record or playback circuits.

The monaural tape player has only one head and a stereo cassette player might have two heads or circuits. Both stereo windings might be found in one tape head (Fig. 10-38). The audio from the

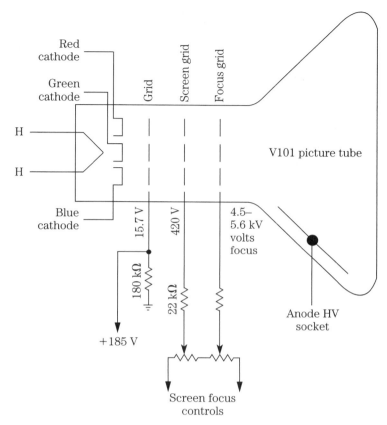

■ 10-37 *The CRT has three heater elements, three cathodes, a single grid, screen grid, and focus grid to control the electron beam in the picture tube.*

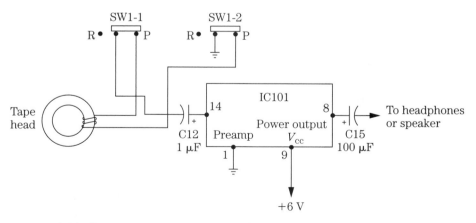

■ 10-38 *The magnetic tape pickup couples the audio signal on the tape surface to the preamp and audio output circuits.*

Tape head circuits

tape head might be coupled to the preamp circuits with a fixed electrolytic capacitor. One large IC might provide enough volume to drive a couple of speakers in a portable cassette player, but several stages of amplification are needed for the boom-box and console AM/FM/MPX cassette players.

Phono pickup

The phonograph cartridge pickup might be a crystal or coil-vane assembly. The crystal pickup needle is attached to a piezo wafer and when twisted, or moved up and down, it creates a voltage; when amplified, the music is heard in the speakers. The coil-vane or magnetic phono pickup has the needle assembly fastened to a metal vane. This movement of the vane changes the magnetic field resulting in music. The magnetic phono pickup must have an extra stage of amplification because the audio signal is very weak compared to the crystal cartridge (Fig. 10-39).

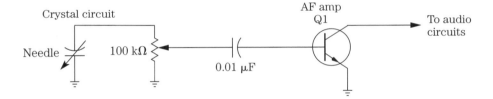

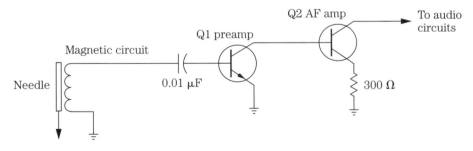

■ **10-39** *The phono needle might be attached to a crystal cartridge or magnetic-vane circuit that couples the audio to a preamp or AF audio circuit. The magnetic phono head must have a separate preamp circuit because the signal is much weaker than the crystal cartridge.*

Conclusion

Many simple basic circuits are found in products within the consumer entertainment field. These basic circuits are added to, as you proceed through pages of the book. Although everything might be a bit confusing at this point, the picture will clear up as more pieces of the electronics puzzle fall into place.

Isolate, locate, and repair   11

YOU MUST ISOLATE AND LOCATE THE DEFECTIVE PART before you can remove and replace it. How can you do that? By simply taking the symptoms found on the product, then isolate or locate it on the schematic and find the part on the chassis. Besides the symptom, use the three s's: sight, sound, and smell (Fig. 11-1).

■ **11-1** *Test the suspected vertical output transistor on the TV chassis after using the symptom on the raster of a white horizontal line.*

What's the symptom

The symptom of a TV raster with only a white line indicates no vertical sweep. You would check the schematic diagram and look on the schematic to find the component. The symptom in a cassette player might be distorted sound. Sound distortion in the amplifier might be a leaky transistor, IC, capacitors, or a change in bias resistors (Fig. 11-2). The symptom of the AM/FM/MPX receiver might be no FM with normal AM reception. Naturally, you would check the front-end FM circuits. The *symptom* is a change in normal operation.

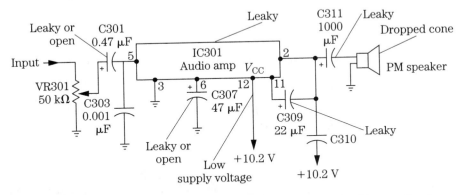

■ **11-2** *Check the components that might cause distortion in a small cassette amplifier.*

Anytime that the electronic product is not functioning properly, you have symptoms to help find the defective component. The symptoms of a common cold are runny nose, coughing, and a terrible headache. "Ailing" electronic products have symptoms much like these.

Likewise, take the symptom of the defective product, isolate on the schematic and chassis, and repair the unit (Fig. 11-3). The symptom can be isolated on the block diagram. Of course, an intermittent symptom or weak audio might take a lot more time and be more difficult to locate. Isolating and locating the defective component is more difficult than parts replacement. Most anyone can replace a defective component with the original part number. The time involved in locating the faulty component is the costly process.

Time is money

Time is one of the electronics technician's greatest tools. You can waste it or make money with it. The professional technician should be able to locate the electronic problem within one hour. The time spent servicing the tough dog or intermittent usually ends up in lost time and no profit. The more you know about electronics, how to

■ **11-3** *Take the symptom of the chassis, isolate it on a schematic, and locate the defective component with the proper test equipment.*

take the symptom and break it down to the defective part, the more efficient technician you become. Remember, time marches on.

The different symptoms

The most common symptoms found in the electronic products are a dead chassis, weak audio or picture, distorted sound, no high voltage, no raster, no audio, insufficient height or width, and chassis shutdown. Of course, these are only a few of the common symptoms because each product has its own related problems.

Before attempting to isolate or locate a defective part, you must analyze the information from the symptoms. By taking the symptoms, analyzing the trouble, and using a little knowledge, you can service most any electronic product. Of course, you must have the correct test equipment and tools (Fig. 11-4). One must-have tool is the schematic of the product you are servicing. You can save a lot of service time with a correct schematic.

■ **11-4** *The technician is injecting a signal into the CD player output circuits to determine where the audio signal is lost.*

Here are a few actual symptoms; see how they apply and how the defective component was located:

Symptom: Dead AM/FM/MPX receiver

By placing your ear next to the speakers, no sound or even hum is heard, but the pilot lights are on. The radio receiver can be broken down into three sections: the front-end circuits, audio circuits, and the low-voltage power supply (Fig. 11-5). Most electronics technicians go directly to the low voltage power supply with the dead electronic product operating from the ac power line.

You know power is getting to the receiver if the pilot light is on. The fuse is good. This eliminates parts up to the secondary winding of the transformer. A quick voltage test across the largest electrolytic capacitor indicates either a low voltage or no voltage. No dc voltage might indicate a leaky silicon diode or bridge rectifier. If the voltage is normal, trouble must lie in the audio output circuits. Try another symptom.

Symptom: No sound/no picture/no raster in TV chassis

By analyzing the circuits in the TV chassis, no sound can be caused by no high voltage or sweep in the horizontal circuits. You cannot see the picture because there is no raster. No raster might be caused by defective components in the horizontal or high-voltage circuits (Fig. 11-6). If the sound was normal, you would go di-

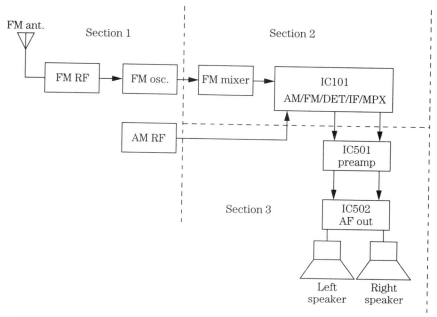

■ **11-5** Break down the block diagram of the AM/FM/MPX receiver circuits into three sections for quick servicing methods.

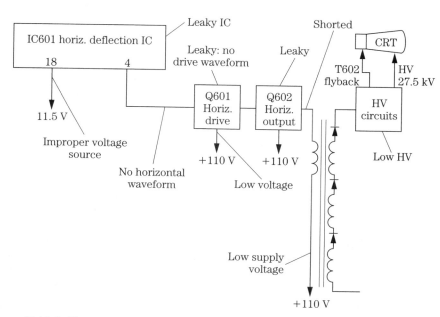

■ **11-6** The no raster/no picture/no sound symptom might be caused by a problem in the horizontal and high-voltage circuits.

rectly to the picture tube circuits with no raster or high voltage. The fuse is normal.

Because most raster and picture problems are caused by the horizontal section, go directly to the horizontal output transistor. Check the dc voltage applied to the horizontal output transistor. By testing the voltage at the horizontal output transistor, you can eliminate a possible problem or determine what the problem is. No voltage indicates that the low-voltage power supply or source is defective. Test the transistor for shorted or leaky conditions.

Symptom: Tape rotates in cassette player—no audio

You might assume that the dc power supply is working with the dc tape motor turning the tape. A low hum in the speaker might indicate that the audio section is normal. By rotating the volume control up and down rapidly, you might hear a noise, indicating that audio is available. By touching the volume control center terminal with a screwdriver blade, a loud audio hum can be heard with the volume control wide open (Fig. 11-7).

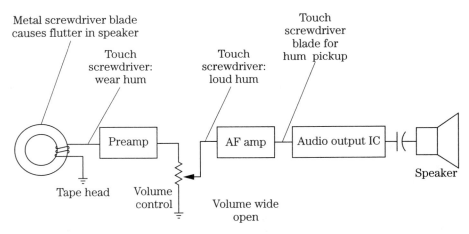

■ 11-7 *A quick screwdriver blade test in the cassette player circuits might indicate that a tape head or audio stage is defective.*

The trouble must lie from the tape head through the preamp audio circuits. A flick of a screwdriver blade across the tape head will produce a flutter sound as the metal blade passes over the tape head. If there is no sound, check the voltage applied to the preamp audio circuits. Suspect that the tape head is open if no sound can be heard when the ungrounded side of the tape head is touched with a small screwdriver.

Symptom: Dead CD player smokes

When the CD player was turned on, the chassis groaned and an acrid smell arose from the case. After removing the cabinet, the symptom pointed to the low-voltage power supply and power transformer. After a close inspection, the transformer shows signs of overheating with wax oozing at the bottom (Fig. 11-8). This transformer was not fused. Two leaky silicon diodes were found in the low-voltage bridge circuit, which destroyed the power transformer. With the dead symptom and smoke or signs of overheating, the defective part could be located.

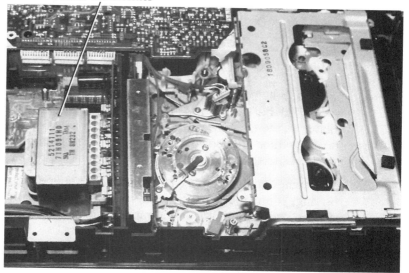

■ **11-8** *Check the power transformer for burns and oozing wax on the chassis of a VCR or TV.*

How to isolate the different circuits

By knowing how each stage operates in a given electronic product with a circuit diagram, you can isolate the problems and make most repairs. After isolating the suspected circuit, check the voltages, resistance, waveforms, and components to determine if the part is defective. Sometimes the defective part might be outside the circuit that you suspect.

Isolate the defective circuit with the symptoms given by the non-operating product. For instance, if there is no AM with the AM/FM/MPX receiver, but it has normal FM reception, go directly to the AM RF stages (Fig. 11-9). The IF, audio, and low-voltage

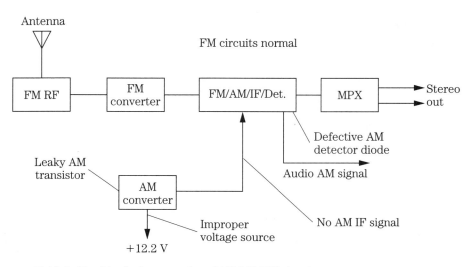

■ **11-9** *The block diagram of an AM/FM/MPX circuit.*

power supply is normal with FM reception. In case the symptom was no FM, but normal AM reception, check the RF, mixer, and oscillator sections of the FM front end. Because the AM is normal, the IF, audio, and power supply are functioning. Compare the symptoms to the schematic or block diagram.

If the symptom in a cassette tape player is no sound, but the tape is rotating, you can assume the low-voltage power supply is normal because the tape motor is operating. Most tape motors operate from a dc source of the power supply circuits. Use the touch-hum system to locate the defective stage with a screwdriver blade or inject an audio signal at the volume control and tape head circuits to determine if the preamp or audio circuits are malfunctioning.

The TV symptom is a narrow vertical picture with good sound (Fig. 11-10). Go directly to the vertical output stage to check for insufficient vertical sweep. Because some vertical sweep is found, assume that the vertical driver and oscillator circuits are normal. Proceed to the two vertical output transistors or ICs for insufficient vertical sweep.

Pinpointing the defective circuit

With the symptoms and suspected circuit located on the schematic, pinpoint the defective circuit on the TV chassis. In this particular schematic, two output transistors form the vertical output (Fig. 11-11). Check the chassis outline schematic for the loca-

■ **11-10** *Improper vertical sweep on the TV raster might be caused by a leaky transistor or leaky diodes in the vertical output circuits.*

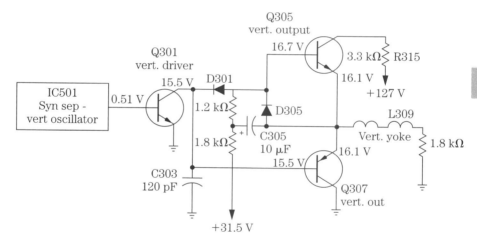

■ **11-11** *From the symptom to schematic, pinpoint the defective component in the vertical output circuits.*

tion of two vertical transistors on the chassis. Because the vertical transistors have a heatsink, they are easily located (Fig. 11-12).

If a chassis layout parts diagram is not given, look on the PC board for the transistor numbers. Vertical output transistors, Q305 and Q307 might be found marked on the bottom side of the PC board or on top. Some PC boards have each terminal labeled. Often, vertical output transistors are located on a separate heatsink within the PC chassis. Each transistor might have its own heatsink.

Pinpointing the defective circuit

■ 11-12 *Most vertical output transistors are mounted on a separate heatsink on the PC board.*

Suspect a defective vertical output IC or component tied to the IC for insufficient sweep, if it uses an output IC instead of transistors. This vertical output IC might be mounted on a heatsink on the PC board. Sometimes, the output IC might use the metal chassis as a heatsink (Fig. 11-13).

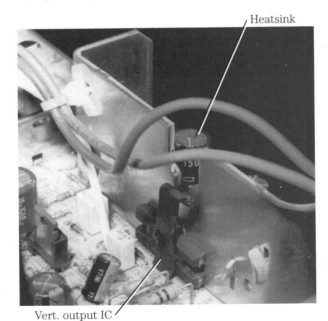

■ 11-13 *The vertical output IC might be mounted on a separate heatsink or on the metal chassis.*

Isolate, locate, and repair

When using the manufacturer's service literature, you will find a chassis component layout of all parts. Check for the transistor or IC symbol number and letter on the layout. In this case, Q305 and Q307 might be found on separate heatsinks. Now locate the suspected parts on the chassis. Rotate the parts layout in line with the TV chassis. You can use larger components on the chart to pinpoint the suspected parts on the TV chassis (Fig. 11-14).

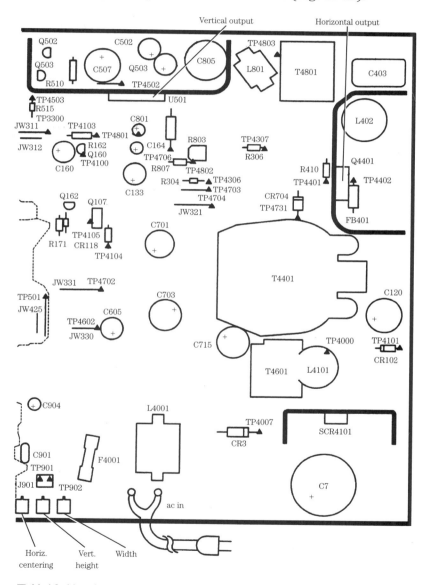

■ **11-14** *Use the parts layout within service literature or identify larger parts to locate suspected components.*

Pinpointing the defective circuit

Insufficient vertical height might be caused by an improper voltage source, leaky or open output transistors, open or a change in bias resistors, and improper vertical drive signal. The drive signal can be checked with the scope waveform. Each transistor can be checked in the circuit and critical voltage measurements taken on each output transistor terminal (Fig. 11-15). Then critical resistance measurements can be taken on terminals and components. Do not overlook the possibility that the defective component might lie outside of the actual vertical output circuits.

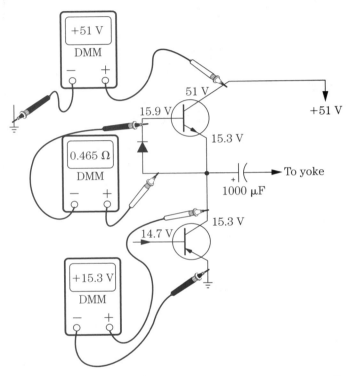

■ **11-15** *Take critical vertical voltage and resistance tests on the suspected vertical output transistor.*

Outside of the suspected circuit

After taking critical voltages and testing various components within the vertical output circuits and insufficient height still exists, look outside the intermediate vertical output circuits. A dried yoke coupling capacitor might cause insufficient vertical sweep. The vertical yoke winding or return resistor might have increased in resistance. Check the return path of components in the vertical circuit.

Another example is poor width in the raster. All horizontal output parts, voltages, and waveforms are normal, but still the sides are pulled in. Look outside of the horizontal output circuits to the yoke and pincushion circuits (Fig. 11-16). A shorted winding in the pincushion transformer or poor transformer connections can produce insufficient width problems. Leaky bypass capacitors within the pincushion circuits might cause improper width. Always look outside the suspected circuit for a defective component in another circuit that is tied to the one that you suspected earlier.

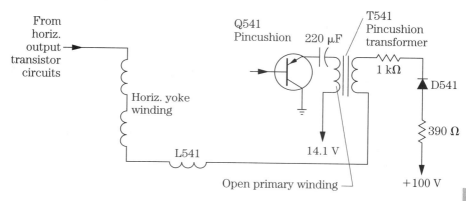

■ 11-16 *Poor width in a TV chassis might be caused by a defective pincushion circuit outside of the horizontal circuits.*

Locating the defective component

After locating the defective component, take voltage, waveform, and resistance measurements, and component tests. Sound injection tests can locate the weak or intermittent component. Which test do you make first? Some electronics technicians make a quick voltage test and then check the suspected component. Other technicians might take a quick scope test, then take critical voltage measurements. Make the test that you feel will locate the defective part the fastest.

For example, if you suspect that a certain transistor is defective in a given circuit, quickly take critical voltage measurements. Voltage measurements on all three terminals might indicate that a transistor is defective. Testing the transistor in the circuit with the transistor test might locate an open or leaky transistor (Fig. 11-17). Then again, when the test probes are placed on the transistor terminals, the open transistor might return to normal. Likewise, when a transistor tests leaky or open in the circuit, but tests normal once it has been removed, it still might be defective. Make all

■ **11-17** Check the suspected transistor in the circuit with a transistor tester.

tests before removing a suspected transistor because removing it takes time.

Likewise, if the IC is suspected to be faulty, take signal in and out tests, and critical voltage and resistance measurements on each terminal. In case of a defective audio output IC, the symptom is no sound and the audio output circuits are suspected. With the unit playing, signaltrace the audio in and out of the IC terminal with an external audio amp (Fig. 11-18). Take sound waveforms with the scope at input and output terminals.

With no signal or audio at the output terminal, suspect a defective IC or components tied to each terminal. Next take critical voltage measurements and compare these figures with the schematic. Any low or high voltages might indicate a leaky or open IC and capacitors tied to the IC terminals. A change in resistance of bias resistors might indicate distortion. Open bias resistors might cause a low-audio or a dead audio channel.

When the voltages are low on certain terminals and do not compare with those shown on the schematic, take resistance measurements to ground. Low resistance measurements might indicate that an IC or component is leaky. Unsolder one lead from the suspected component and take another resistance measurement. Be-

Isolate, locate, and repair

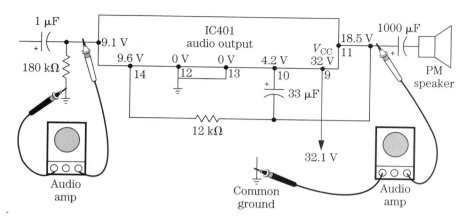

■ 11-18 *Signaltrace the audio in and out of the suspected IC with an external audio amp.*

fore removing and replacing any IC, LSI, or microprocessor, make sure that the component tied to each terminal is normal. Sometimes voltage and resistance measurements might be normal on the IC and the IC will still be defective. Replace the IC if you suspect that it is faulty.

Critical voltage measurements

Voltage measurements are usually made from a component terminal to common ground. Common ground might be a larger copper area on the PC board or metal chassis. Shielded coils and transformer cases are at ground potential (Fig. 11-19). A metal shield provides a common ground. The outside shield of a cable of wires is common ground. Doublecheck the schematics for common ground. Remember, some TV chassis have IC or transistor heatsinks on a hot ground. The hot ground will provide improper voltage measurements if it is used as a common ground.

A quick voltage measurement on the largest electrolytic capacitor of the power supply will indicate if the supply is normal. Then take a voltage source check across the source electrolytic filter capacitor or at the suspected component (Fig. 11-20).

The critical voltage measurement on the collector terminal of the suspected transistor might determine if the transistor is open or leaky. A low voltage at the collector terminal might indicate a leaky transistor or improper voltage source (Fig. 11-21). A higher-than-normal voltage at the transistor collector terminal might indicate an open transistor. Lower voltages on all three terminals might in-

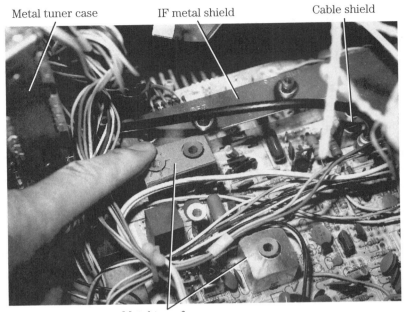

■ **11-19** *The common ground on the TV chassis might be the shielded areas, the metal tuner case, the transformers, and the shielded cables.*

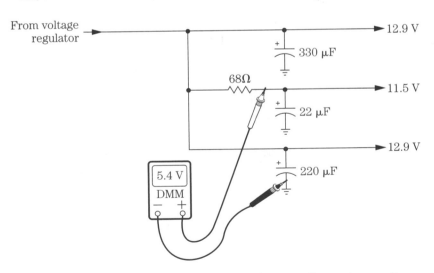

■ **11-20** *A voltage test across the capacitors on the various voltage sources might indicate that power-supply circuit is defective.*

Isolate, locate, and repair

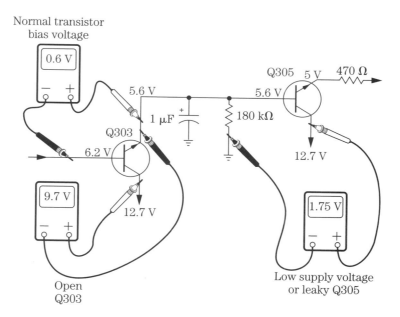

11-21 *Take critical voltage tests on the transistor terminals to determine if they are leaky, shorted, or open.*

dicate that a transistor is leaky or shorted. A forward bias voltage reading of 0.06 V between the base and emitter terminal on an NPN transistor indicates that the transistor is normal. Look for a different voltage measurement on each transistor terminal compared to the schematic.

Critical voltage measurements can determine if the IC is open or leaky. A leaky IC might have a very low voltage at the supply pin (V_{CC}). Remove the pin from the circuit and take another test. Low comparable voltages on any IC terminal might indicate a leaky IC (Fig. 11-22). If the voltage is higher than that which is listed on the schematic, the IC might be open.

Critical resistance tests

A critical resistance measurement from each transistor terminal to common ground might indicate a leaky transistor. Resistance measurements between the transistor elements might show that the transistor is shorted. No emitter voltage might indicate an open transistor or resistor. Take a quick resistance measurement across the emitter resistor. Make sure that no low resistors, coils, or diodes are found in the circuit when taking resistance of transistor

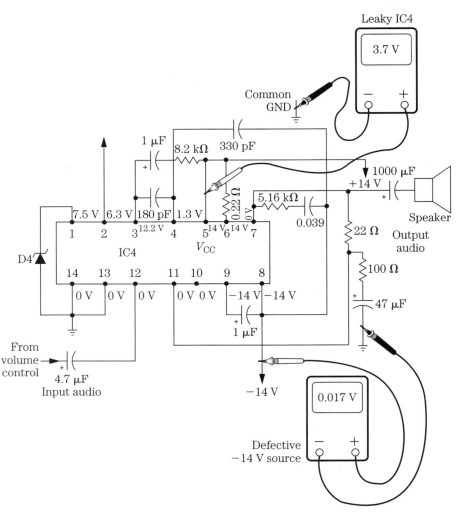

■ **11-22** *Take critical voltage tests on the IC and compare them with the circuit diagram.*

terminals. Doublecheck the schematic for a low-resistance comparison measurement.

Critical resistance measurements on the supply terminal and all other IC terminals might indicate that an IC or component is leaky. Check the resistance from each terminal pin to common ground. If a resistor or capacitor in these circuits is suspected, remove one lead of the component and take another test. Look for diodes and coils at the IC terminals when a low-resistance measurement is made (Fig. 11-23). Remember, the IC might be defective, even if

Isolate, locate, and repair

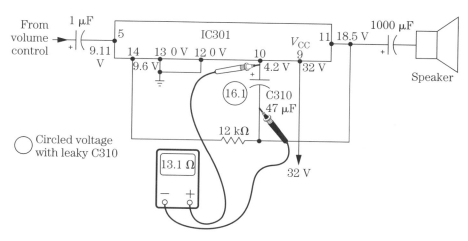

11-23 Critical resistance measurements on the IC terminals might indicate a leaky IC or a leaky component in the circuit.

the resistance and voltage measurements compare with the wiring schematic.

Critical voltage and resistance measurements across a zener or silicon diode might determine if the component is leaky or open. When the zener diode voltage is lower than the voltage shown on the diode or the schematic, suspect that the diode is leaky (Fig. 11-24). Often, the zener diode overheats and becomes lower in resistance with lower regulated voltage. Sometimes the zener diode will overheat and leave the diode open, which increases the dc voltage source.

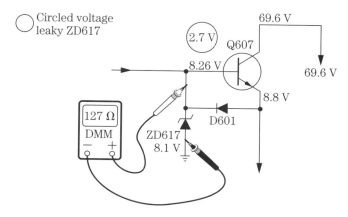

11-24 The leaky zener diode might overheat and lower the voltage in the dc source.

Critical resistance tests

Critical waveforms

The waveform is the shape of a signal and is seen as a sawtooth, square, rectangular, or triangular waveform. Waveforms are taken with the oscilloscope at different points in the circuit to determine if the circuit is performing. The amplitude of the waveform is the gain of a special circuit. Critical waveforms taken on the TV, VCR, and CD chassis quickly analyze the various circuits.

The waveform taken from the horizontal deflection IC might determine if that circuit is functioning (Fig. 11-25). Critical waveforms taken throughout the horizontal circuits might determine what stage is defective or working. The height of the waveform might determine a weak horizontal circuit.

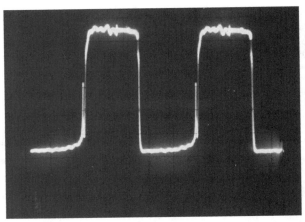

■ 11-25 *The horizontal output waveform from the deflection IC indicates that the drive is applied to the horizontal transistor.*

Waveforms taken in the vertical circuit are not as critical as the horizontal circuits in the TV chassis. A critical waveform taken on the vertical countdown or deflection IC indicates if the circuit is normal or defective (Fig. 11-26). A vertical output pulse at the vertical output transistor or IC drive signal determines if the vertical circuits are normal or problems lie in the yoke circuits. Waveforms taken inside the defective vertical output circuit might not be effective for troubleshooting.

Replacing the defective part

After all tests indicate that the part is defective, remove it from the chassis and perform another test. Notice how the part is mounted.

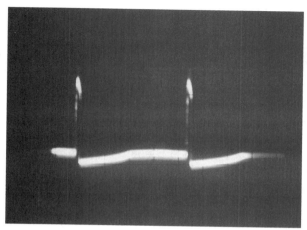

■ **11-26** *The vertical drive pulse from the deflective vertical IC terminal will indicate if the deflection IC is normal.*

Be very careful when unsoldering a defective component from the PC board. Accidentally touching an IC or transistor with a soldering iron can destroy it in seconds (Fig. 11-27). Be careful not to drop solder on the PC board wiring. Use solder wick or solder-sucking tools to remove the excess solder.

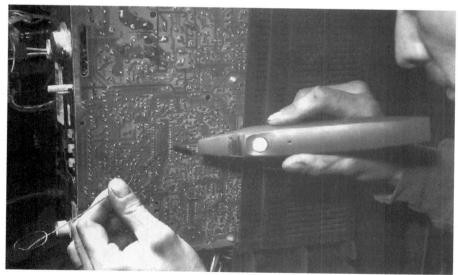

■ **11-27** *Be careful when soldering transistors or ICs on the PC wiring. Use a battery or a low-wattage soldering iron.*

Check the position of the part and how it mounts alongside of other components. The replacement part will mount exactly in the same spot. Mark all terminals of a transistor on top and bottom of the PC board before removing. If this is not possible, draw an outline of the transistor and the leads on a piece of paper with the terminals labeled. Of course, some manufacturers have terminal leads stamped on the bottom of the PC board.

Make sure that the correct transistor terminal is pushed through the right hole in the board. Remember that the transistor is right side up when mounted, but the terminals should be upside down when looking at the PC board wiring. Carefully solder the transistor terminals with a low-wattage iron (Fig. 11-28). Use a pair of long-nose pliers or a clamp on the heatsinks to prevent heat from damaging the transistor.

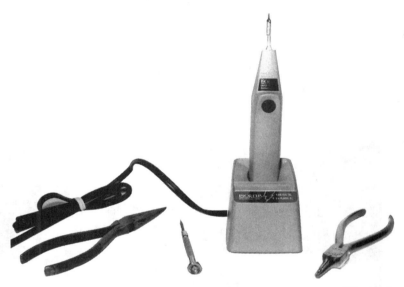

■ **11-28** *Use long-nose pliers on a heatsink and use a small soldering iron when soldering the transistor terminals.*

When removing a defective IC, LSI, or surface-mounted microprocessor, be careful not to unsolder or suck-up a connection from a nearby component. Clean off all solder from the area after the IC has been removed. Mark terminal 1 on the board if it has not been stamped on the chassis. Sometimes the manufacturer has a "U" shape printed right on the top side of the board for easy IC replacement. Be careful when soldering each terminal so that the excess solder does not overlap into another terminal. Again, use a low-wattage soldering iron.

Universal replacements

Always try to obtain the original part number when replacing a transistor or IC. Replace with original parts in critical circuits. Universal replacements are made every day in servicing electronic products. Check the different semiconductor manuals for exact transistor and IC replacements. Most radio receivers and audio equipment can be repaired with universal parts. Use the original power output solid-state devices in high-wattage amplifiers. Critical circuit parts in VCR, CD player, and camcorder circuits should be replaced with the original part numbers.

Replace all faulty surface-mounted ICs, LSIs, or microprocessors in any chassis with exact replacements. Surface-mounted resistors and capacitors might be universal types. Make sure they fit in the required space. Round surface-mounted resistors can be replaced with some flat-mounted resistors.

Special volume, tone, and miniature controls should be replaced with the exact part. An ordinary volume control can be replaced with a universal type by cutting the shaft to the correct length. Small color, tint, brightness, and contrast controls can easily be replaced with universal types. When replacing volume controls with universal parts, make sure that they have an audio taper. Most resistance controls are linear types. Several special controls found in one assembly should be replaced with the original part.

When a special bridge rectifier is not available, use single diodes in the circuit. Most 2.5-A silicon diodes will serve current requirements in the TV chassis. Connect four single diodes in a bridge circuit (Fig. 11-29). Make up the bridge unit with diodes mounted and soldered close together. Then place the correct polarity leads into the bridge rectifier mounting holes.

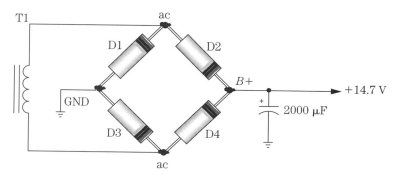

■ **11-29** *Connect four single silicon diodes in the bridge circuit if a regular bridge component is not available.*

Universal electrolytic capacitors can be replaced for the defective capacitor as long as the capacitance is the same or higher and the working voltage is the same or higher. For instance, you can replace a 1000-µF 25-V electrolytic capacitor with a 2000-µF 35-V capacitor, if space is available (Fig. 11-30). But you cannot replace a 1000-µF 25-V electrolytic capacitor with a 1000-µF 15-V type. The working voltage is too low. If a can capacitor contains several capacitors and only one is bad, still replace the entire can.

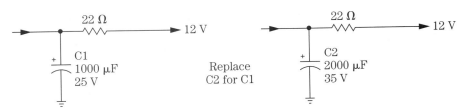

■ **11-30** *Replace the filter capacitor with one that has higher ratings if an exact replacement is not available.*

Multiple filter capacitors can be added or paralleled when the original is not available. For example, if you need a can-type filter of 250 µF, 200 µF, and 10 µF at 180 V and you have a universal replacement 200–300 µF, 200–400 µF, 20–50 µF at 250 V, use the replacement (Fig. 11-31). Often in filter circuits, the added capacitance will improve the filtering action. Be sure that the operating voltage is the same or higher with universal replacements. The TV or radio chassis should not be tied up when electrolytic capacitors can be connected together to safely place the chassis back into service.

Always mark each terminal lead of a power or flyback transformer before removing the component. This also applies to any component with several color-coded leads. You might be called away from the bench and not return until a few days later. It's difficult to remember where each lead connects without a schematic. This means tracing out each wire to see where it is connected. Test all new parts before installing them in any electronic chassis, if possible.

Heatsinks

Large power output transistors and ICs in TVs, receivers, audio amplifiers, and VCRs must have a heatsink to help dissipate the heat generated by the power output component (Fig. 11-32). These power transistors and ICs are bolted to the chassis or on heatsinks.

■ **11-31** *Large can filter capacitors can be added in parallel if the original part is not available.*

Silicone grease is applied between component and heatsink. If the transistor is to be isolated away from the metal heatsink, a piece of insulation is placed between transistor and metal heatsink. Silicone grease is applied on both sides of the insulation so that the heatsink will absorb the heat from the transistor or IC.

Apply silicone grease between the transistor or IC that mounts directly on the metal chassis or heatsink. Silicone grease is available in white or clear grease. The white silicone grease is difficult to remove from clothes and hands. Always apply silicone grease and mount the power output component before soldering terminals and trying out the repaired unit. You might quickly damage the power output replacement if it is not fastened to the heatsink.

Tests after repairs

Do not fail to make correct tests of alignment procedures after repairing the electronic project. This not only makes the product operate efficiently, but it might save a call back. Besides making

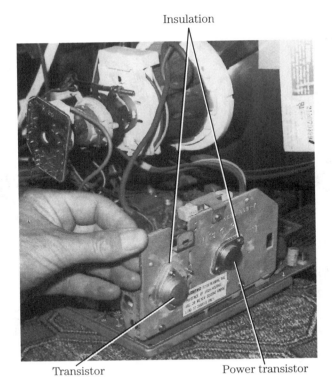

■ **11-32** *Large heatsinks or the metal chassis might serve as a heatsink in the TV.*

the product operate, you avoid problems that might be dangerous for the customer or operator. Sometimes, the simplest adjustments might make the product last longer. Isolating, locating, and replacing defective components occur every day in the life of the electronics technician.

12

Servicing low-voltage power supplies

MOST ELECTRONIC TECHNICIANS CHECK THE LOW-VOLTAGE power supply source of any electronic repair before digging into the defective circuits. These electronic circuits will not function without a voltage source. The voltage source in a portable cassette player might be four "C" batteries. If it operates from the power line, a small plug in the power supply is inserted into the player to operate it from the ac power line. This low-voltage power supply provides dc voltage to the tape player (Fig. 12-1).

■ **12-1** *The low-voltage power supply is used in the table radio, cassette, and tape player, VCR, TV, and CD player.*

The power supply circuits might provide several dc and ac voltages. A separate positive and negative voltage source might be found in some power supplies. Most low-voltage power sources provide ac voltage to rectifier components from a step-down power transformer. Usually radio receivers operate under a 25-Vdc source. The large amplifier might have dc voltages up to 100 V. The TV voltage source might operate at 165 Vdc with lower regulated dc voltages, from a transformerless power supply or directly from the power line.

The low-voltage power supply might furnish low voltage to the horizontal circuits and in turn develop over 35 kV in the flyback and HV circuits of the TV chassis. The flyback or horizontal output transformer circuits might produce different voltage sources in the secondary winding to power other TV circuits. Voltage-regulation circuits are found throughout electronic products for critical electronic stages. These voltage regulation circuits might have transistors, ICs, zener diodes, or a combination of these. Radio amateurs and broadcast stations use heavy-duty transformers in their equipment for higher voltages and current.

Play it safe

To protect yourself, test equipment, and the electronic product you are servicing, plug the unit into a power line isolation transformer. The isolation transformer isolates the electronic product from the power line (Fig. 12-2). Be careful while working in low- and high-voltage circuits. Don't drop tools (test probes, pliers, screwdrivers, etc.) or anything else into live circuits. You can get shocked while working with the ac power line voltages. Remember, dc voltages above 50 V can produce a shock hazard.

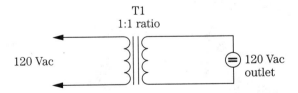

■ **12-2** *The isolation transformer isolates the power line from the consumer electronic product for safety.*

Accidentally shorting a test probe between a filter capacitor and ground might knock out several silicon diodes and destroy semiconductors tied to that circuit. Do not try to sub a filter capacitor or shunt parts in the power-supply circuits. Discharge the filter capacitor. Clip the component into the circuit with alligator clip leads. Do not shunt a filter capacitor across a suspected one to eliminate hum when power is on. The voltage surge and charge of the capacitor might raise the voltage source and destroy transistors and ICs that are tied to the low-voltage sources. Clip the capacitor into the circuit with the power off.

Power-supply components

The low-voltage power supply in a radio or cassette player might consist of a step-down power transformer, rectifiers, a filter capacitor, and regulators (Fig. 12-3). The step-down transformer provides low ac voltage to a half-wave, full-wave or bridge circuit. These small transformers might be fused at the primary or secondary windings and sometimes no fuse is used in the ac circuits. Often the secondary voltage is switched instead of the primary to prevent an ac power line voltage at the switch terminals. This leaves the primary winding of the transformer in the ac circuit at all times. In a full-wave circuit, the transformer might be centered tapped. The center tap is not found in the bridge rectifier circuits (Fig. 12-4).

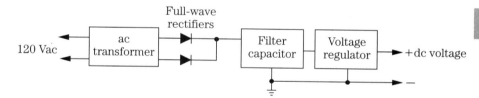

■ **12-3** *Block diagram of a simple full-wave low-voltage power supply, which includes the power transformer, silicon diode, and the filter and voltage regulators.*

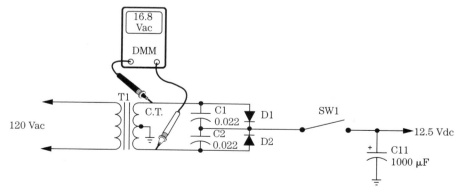

■ **12-4** *A full-wave circuit with D1 and D2 and a center tap of the secondary winding of power transformer.*

You might find a half-wave (one diode), full-wave (two diodes), bridge rectifier (four diodes) in the commercial electronic product. The half-wave rectifier cuts off half of the ac input cycle. The step-down transformer winding connects to a half-wave silicon diode and the other lead goes to common ground. Half-wave rectification is most difficult to filter out the ac ripples unless a good regulator is found at the voltage source. The half-wave rectifier might be found in the TV chassis and in inexpensive low-power power-supply circuits.

The full-wave rectifier removes both halves of the ac input cycle providing better regulation. Most full-wave rectifiers use a center-tapped transformer with two silicon diodes. The center tap of the transformer connects to the chassis ground. Each diode works in a half-wave circuit. A lower voltage drop is noted with a load placed in the circuit than with the half-wave rectifier (Fig. 12-5). The full-wave rectifier circuits are found in radios, tape players, cassette players, and battery ac-operated portable TVs.

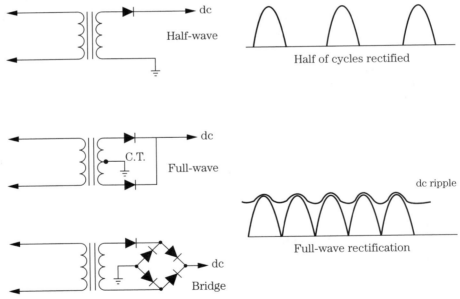

■ 12-5 *The half-wave, full-wave, and bridge rectifier circuits have corresponding rectifier cycles.*

The bridge rectifier is a full-wave rectifier circuit using four silicon diodes in a bridge circuit. The main advantage of the bridge circuit is that the transformer is not center tapped and it provides good voltage regulation. The bridge circuit uses the entire secondary on both halves of the input cycle. These bridge rectifier circuits will deliver high current demands. The bridge circuit is found in receivers, amplifiers, CDs, VCRs, and in special power supplies.

The rectifier

In the days of radio tubes, a cathode and plate elements provided dc rectification. Then along came the selenium rectifiers with large plates and spaced washers. Although these rectifiers worked well, they had a large voltage drop between elements and plates. The large selenium rectifiers have large fin heatsinks to dissipate heat; when overloaded, specks of selenium popped off the plates.

Along came the silicon diode, which was small in size and could deliver a lot of current. These silicon diodes are found today in most low-voltage power supplies. Although you might find 3-A diodes in the TV chassis, most electronic technicians replace the smaller diodes (1 A) with 2.5-A types. Always observe correct diode polarity. The white or dark line at the end of the diode is the cathode or positive terminal.

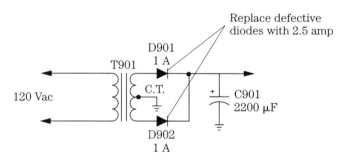

■ **12-6** *Replacing the 2.5-A silicon diodes in the full-wave and bridge circuits.*

Voltage-doubler circuits

In the early radio and TV chassis, the voltage-doubler circuits was used to increase the output voltage with a lower-priced transformer. The vacuum tube required higher dc voltages than solid-state components. By connecting electrolytic capacitors and silicon diodes in a doubling circuit, the input voltage can almost be doubled. Although the voltage doubler circuits are practical in low-current power supplies, very seldom do you see one in today's electronic products. The voltage doubler circuit operates in the whole ac input cycle and is a full-wave voltage doubler (Fig. 12-7).

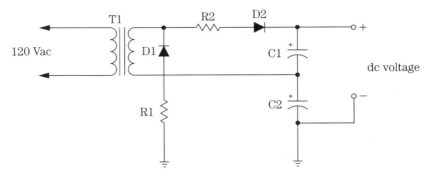

■ **12-7** *The voltage doubler circuit uses two silicon diodes and two electrolytic capacitors.*

In some voltage doubler circuits, low-value isolation resistors (R1 and R2) were wired in series with the silicon diodes. Notice that no center-tapped transformer is found in the doubler circuit. Both electrolytic filter capacitors (C1 and C2) must have a working voltage higher than the output dc voltage. C1 and C2 are used to boost the voltage and provide adequate filtering.

Filter capacitors

The filter capacitor filters out the dc ripple voltage from the rectifiers (Fig. 12-8). Electrolytic filter capacitors are used in the low-voltage power supplies. When replacing the electrolytic capacitor, observe correct polarity. The black or white line down the side of the filter capacitor indicates common ground terminal. If the electrolytic capacitors are placed in backwards, they become hot and they could blow up in your face. Check the polarity of any type of capacitor.

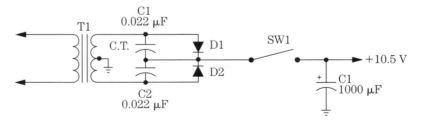

■ **12-8** *C1 filters out the dc ripple from the full-wave rectifier circuit.*

Larger filter capacitors with high capacitance are used in products that contain certain audio frequencies. Hum in the sound might indicate a dried-up filter capacitor or a capacitor that has lost its capacity. Electrolytic capacitors from 1000 to 3000 µF are found in receivers, cassette players, and other audio products (Fig. 12-9). Each different voltage source in the low-voltage power supply might have a smaller electrolytic filter capacitor. Besides filter capacitors, power transistors, and chokes, IC and zener diodes provide better ripple filters in low-voltage power supplies.

2000 μF filter capacitor

■ **12-9** *A 2000-μF electrolytic filter capacitor in a Sharp combination radio, cassette, and CD player. Notice that the white line is the negative terminal.*

Filter chokes

Back in the early days, the electrolytic capacitor and filter choke was known as a *pie filter*. The large filter chokes help smooth out the dc ripple in power supplies. Although the filter choke is not used in the present-day TV chassis, it is still found in auto receivers, amplifiers, and industrial and test equipment. The large filter choke has metal laminations, like the power transformer.

The capacitor input filter consists of an electrolytic capacitor, choke, or resistor. The output voltage is higher with a choke input filter and will deliver higher current. Greater filtering action can be accomplished with two filter capacitors, one on each side of the filter choke (Fig. 12-10). When a heavy leakage is found in the audio output stages or component after the filter choke, the choke coil might run red hot. The overheating shorts out turns in choke transformer, which results in a poor filtering network. Also, hum can now be heard in the speakers.

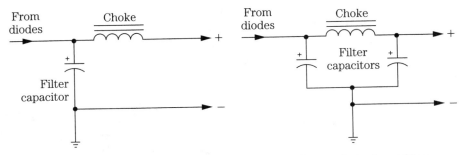

■ **12-10** *The filter choke might operate in capacitor or choke-input filtering circuit for auto receivers, amplifiers, and test equipment.*

How the power supply operates

Ac voltage is picked up from the power line by the ac cord and wired directly to the primary winding of the power transformer. Ac power is applied to the secondary winding with transformer action. You might or might not find a switch in the ac line. Step-down transformers are used in most small electronic products, tape and cassette players, radios, VCRs, and CD players. The secondary winding has a larger gauge of wire and fewer turns than the primary winding. Often, the ac step-down transformer secondary voltage is below 35 Vac. Most electronic products have a 12, 18, or 25 Vac secondary winding (Fig. 12-11)

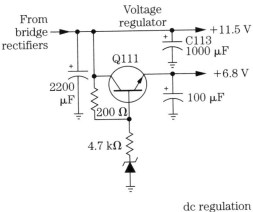

■ **12-11** *The ac secondary winding of a cassette player might be in the 12- or 18-Vac range. The dc ripple voltage is smoothed out with Q11 and C113.*

The step-down transformer is still used in portable TV sets, amplifiers, and other electronic products. Of course, today, the TV chassis operates directly from the ac power line without a transformer. The flyback transformer in the TV horizontal output and high-voltage circuits is considered a step-up for the high voltage and a step-down for secondary derived voltage sources. The power transformer core fits around the center of metal laminations to transfer the energy.

The secondary ac voltage is applied to the anode terminals of the diode rectifiers with dc voltage at the cathode terminals. Small bypass capacitors are used across the diodes to prevent radiation into radios and TV receivers. A large filter capacitor (2200 µF) smooths out the dc ripple. Notice that the dc voltage is always higher at the filter capacitor than at the input ac voltage. In critical and stabilized electronic circuits, additional filtering is accomplished with resistor and voltage regulators.

Voltage regulators

To provide greater and smoother filtering for critical circuits, additional filtering is done with transistor, IC, and zener diode regulators. You might find a combination of transistor and zener diode regulation in a given voltage source. The zener diode is a silicon diode that is especially designed to break down at certain voltages. A 8.1-V zener diode will maintain regulation or voltage around 8 V. Often, the actual measured voltage is a little higher than the zener diode rating. Zener diodes come in several different voltages with 1- or 5-W varieties.

The zener diode regulator circuit is very simple. A voltage-dropping resistor is ahead of the diode and the anode terminal of the diode is at ground potential (Fig. 12-12). Usually the cathode or positive terminal of zener diode is above ground. These diodes

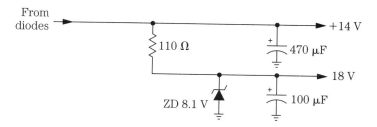

■ **12-12** *The zener diode forms a simple voltage-regulation circuit that keeps the output voltage constant.*

have a tendency to overheat and eventually break down, which results in lower dc voltages. The leaky zener diode can often be spotted because it has burned or overheated marks on its body.

Gas-filled regulator tubes were found in critical voltage sources in early TVs and electronic equipment. Today, the transistor is used to regulate high-current circuits. The positive transistor regulator has a positive voltage from the emitter terminal with rectified voltage at the collector terminal (Fig. 12-13). A negative regulated power source might be the reverse with a negative voltage at the collector. These positive and negative transistor regulator circuits are found in CD players, TVs, and VCRs.

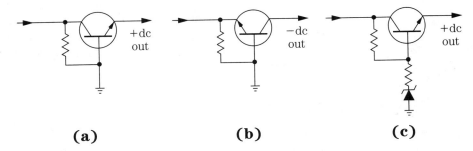

■ **12-13** *The transistor regulator might be in a positive (A) and negative (B) dc low-voltage source. Both transistor and zener diode regulation is used in circuit C.*

Regulator circuits that require greater current capacity might use a transistor, power resistor, and zener diode as regulators. The zener diode is found in the base circuit to ground. This combination of transistor and zener regulation is an inexpensive filter network compared to the filter choke. Often the transistor might become leaky and cause the zener diode to overheat. Replace both components when one is found leaky.

IC regulators

The IC regulator operates like the transistor and diode and looks like a transistor (TO-220 case), flat, in-line IC (Fig. 12-14). This IC regulator connects directly to the main filter capacitor and might supply voltages of 5, 6, 8, 12, 15, 18, and 24 Vdc. The large IC regulators found in the TV low-voltage supply circuits operate directly from a higher voltage (150 to 165 V) and drops the regulated voltage from 100 to 128 Vdc. These large regulators will automatically shut down when a short or leakage overload exists. Usually, IC reg-

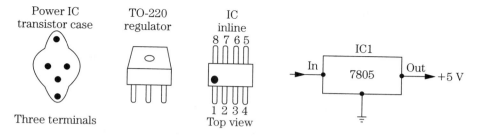

12-14 *The IC voltage regulator might appear as a power transistor with three terminals, TO-220 transistor, and in an in-line IC.*

ulators will be damaged when power line surges or lightning strikes the TV chassis.

Fuses

Very few fuses are found in the small electronic product. A fuse is not found on low priced equipment. You might find a fuse in the secondary winding instead of the primary winding (Fig. 12-15). In early tape players and TV sets, the fuse was found in the primary winding with the on/off switch. To prevent shock to the operator, both the fuse and switch are now eliminated from the ac power side. Now, the transformer is on all the time, but it does not pull any current until the secondary switch is turned on as circuits began to operate in the secondary voltage sources.

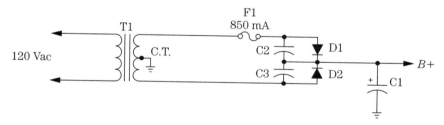

12-15 *The secondary of the power transformer might be fused instead of the primary winding. A shorted diode will blow the fuse instead of opening the primary winding of the transformer.*

Fuses consist of a soft metal that will melt or open when current exceeds the fuse rating. The fuse might slip into a fuse holder or soldered into the PC board with pigtail leads. The house fuse screws into a type of a lightbulb socket. Fuses found in small receivers might range from 650 mA to 1 A. Larger amplifiers and receivers might have 1- to 3-A fuses.

The TV set has 3- to 5-A fuses in the ac power line. You might find two different fuses in the TV chassis. Besides the ac line fuse, a $B+$ fuse feeds voltage to the horizontal output circuits. This $B+$ fuse protects the horizontal and high voltage circuits. When both the line and $B+$ fuse are open, suspect that a horizontal output transistor, damper diode, and flyback transformer (Fig.12-16) is shorted or leaky.

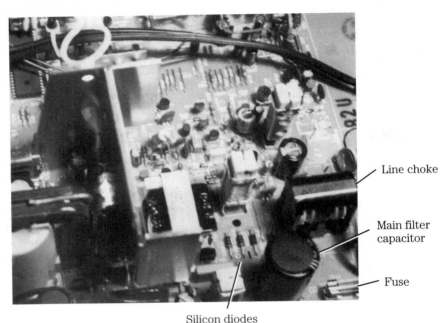

■ **12-16** *The power line fuse is used close to the line cord and to the* $B+$ *fuse in the horizontal output circuits of an RCA 13-inch color TV.*

Fuses in the speaker circuits are to prevent damage to the speaker if an output transistor or IC becomes leaky. Sometimes these speaker fuses open when too much volume or power is applied. Check the speaker circuits for dc voltage when one of these fuses blows before inserting a new one.

You might find slow-blow fuses in consumer electronic products. Slow-blow fuses might have a small spring inside and take a few seconds to open. These are usually found in power circuits where a heavy load exists at turn on. The foil or metal fuse blows almost immediately. Always replace the fuse with the same voltage and amperage as the original one. Look at the end of the fuse for the correct voltage and current rating. You might find the fuse data near the fuse holder. Never wrap tin foil around the blown fuse to make the product operate. Be safe, replace it with a fuse of the correct value.

Circuit breakers

The circuit breaker can be found in the early TV chassis, house fuse box, amplifier, and test equipment. When the circuit breaker opens, all that is needed is to push the small red button to restore operation. Never push the red button while there is an overload or the circuit breaker might be damaged. Pushing the circuit breaker too many times might cause it to kick out. Replace the circuit breaker with the correct amperage when it will not reset or if it keeps kicking out after resetting. Circuit breakers are usually found in the ac power line (Fig. 12-17).

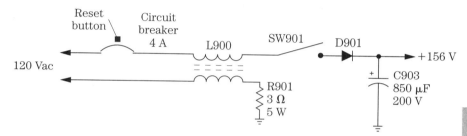

■ **12-17** *A circuit breaker was used in the ac power line of early TV sets. Just push the red button to reset the circuit breaker.*

Power-supply systems

The TV chassis might be dead with a blown power fuse. Excessive hum might be heard in the speakers with a dried-up filter capacitor. A low-level hum might result from a defective decoupling capacitor. Horizontal hum bars might appear in the TV picture with a defective main filter capacitor (Fig. 12-18). The CD player might be dead with an open winding in the power transformer. A left channel speaker might be dead with a blown speaker fuse. The width of the TV picture might pull in from the sides of the CRT. Improper voltage sources might cripple any circuit. All of these symptoms might be caused by a defective component in the power-supply circuits.

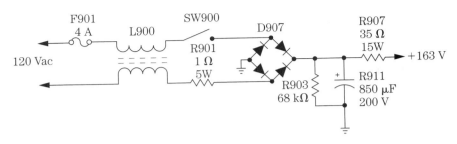

■ **12-18** *The various trouble symptoms that might occur in the low-voltage power supply.*

Critical voltage and resistance tests

Take a quick voltage test across the main filter capacitor to ground or across both capacitor terminals. If the voltage is low, high, or if there is no voltage measurement, check out the low-voltage circuits (Fig. 12-19). Next, proceed to the voltage source that supplies voltage to the section or circuits you suspect are defective. Check the voltage that feeds the circuits that are not operating. Critical voltage tests at these three different areas might indicate if the low-voltage power supply is normal.

■ **12-19** *Low- or no-voltage measurements across the main filter capacitor might indicate that the low-voltage power supply is defective.*

For example, if no voltage is found on the main filter capacitor, suspect that a fuse is open, or that diodes, resistors or the power transformer are defective. Shorted or leaky silicon diodes will open the fuse or damage the power transformer. In products that have no fuses and the primary winding connects directly to the power line, a leaky diode might open the primary winding. Pull the power cord and take critical resistance measurements across the prongs of the ac plug. No continuity indicates that a transformer is open.

The leaky filter capacitor might open the fuse or damage one or two silicon diodes. The voltage might become low if an electrolytic filter capacitor is losing its capacity. The width pulled in both sides of the TV raster might be caused by a defective main filter capacitor or by improper low-voltage adjustments. Failure of the main filter capacitor in the receiver or amplifier produces hum in the speakers.

After locating a defective low-voltage source with voltage tests, take resistance, continuity, and diode tests. Leaky or shorted silicon diodes will measure low resistance in both directions. No matter which way the test probes are applied to the diode, the low resistance is the same. A low-resistance measurement across the filter capacitor might indicate that a filter capacitor is leaky. To make sure that your test was accurate, desolder one terminal wire or lead. Now take another resistance reading.

Low-resistance measurements on separate voltage sources might be caused by leaky voltage transistor and zener diode regulators (Fig. 12-20). Burned resistors in the regulator circuits point to a leaky zener diode or transistor. Check the low-voltage source for a

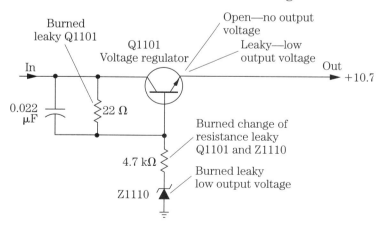

■ **12-20** *Burned resistors and zener diodes might indicate that the transistor regulator is leaky or shorted.*

leaky rectifier and resistor. If the voltage at the low-voltage source is too low, the condition might be caused by a dried-up decoupling electrolytic capacitor.

Low voltage at a separate voltage source might result from an overloaded circuit tied to it. Take a low-resistance measurement at the voltage source. If the resistance is below 150 Ω, remove one lead of the isolation resistor or diode, which provides voltage to that source. If the voltage source resistance is still low, check the circuits that the low-voltage source feeds to. For instance, if the symptom in a TV is weak sound with normal picture, check for leaky components in the audio circuits.

To separate the sound circuits from the low-voltage source, remove the audio circuits at the power source with a pocketknife or sharp tool. Simply remove a small piece or gouge out the PC wiring with the tip of a pocketknife. You are removing the audio voltage from the low-voltage source. After locating the defective component in the audio circuits, place a piece of hookup wire across the broken PC wiring and solder both ends to the PC board wiring. You can remove any overloaded circuit from the low-voltage sources with the same method.

Very low voltages at the voltage source might result from a leaky regulator transistor or zener diode. No voltage at the voltage source might be caused with an open regulator transistor. The open transistor will have no voltage at the emitter terminal of a positive voltage source. A leaky regulator transistor might increase the voltage and produce hum and bars in the radio or TV set with emitter-to-collector leakage. When the transistor regulator becomes leaky from emitter or collector to the base terminal, the voltage is always lower. Check the suspected transistor and zener diode with in-circuit tests. Remove one terminal for correct leakage tests. Sometimes the various resistance measurements are found at the different voltage sources on the service literature.

Besides overloading in the low-voltage sources, a low voltage might result from defective filter capacitors (Fig. 12-21). Small electrolytic filter capacitors of 1 to 22 μF have a tendency to go open and lose capacitance. This might create havoc in a TV circuit with low voltage. After checking the source for low voltage and with no regulator in the circuit, shunt small filter capacitors across those that you suspect. Pull the power plug. Never shunt an electrolytic capacitor with the power on. Discharge the main filter capacitor. Clip or tentatively tack the small filter capacitor across the

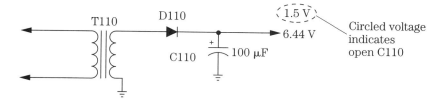

■ 12-21 *A dried up voltage-source filter capacitor might provide low-voltage.*

one that you suspect. Observe correct polarity. Install a new exact replacement for the defective capacitor.

Very low voltage sources in the audio circuits of auto radios might indicate that a power output transistor or IC is shorted or leaky. If the resistance of the voltage source to chassis ground is below 50 Ω, go directly to the audio output circuits. Sometimes where power output transistors or ICs become shorted or leaky, PC wiring and cables might be burned (Fig. 12-22).

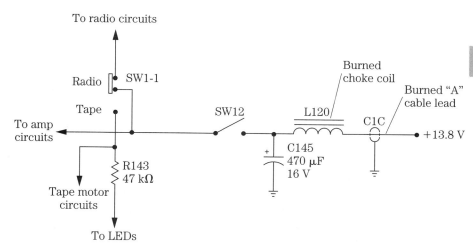

■ 12-22 *A burned cable and fuse socket might point out a leaky output transistor or IC in the power audio output circuits of a car receiver.*

Shunt the main electrolytic capacitor when hum bars are found on the TV screen or if a loud hum is in the sound. Again, turn off the power or pull the ac plug. Clip the same value of capacitor (or one with a higher capacitance and voltage) across the main filter capacitor and then turn the set on. If the hum bars and hum disappears in the audio and picture, replace the main filter capacitor. Do not overlook improper or low-voltage adjustments in the TV chassis; leaky transistors and zener diode regulators in the low-voltage power supply might produce hum bars in the raster.

Power wall adapters

The line voltage adapter operates the power line and plugs into a small tape and cassette players, radios, and battery-operated power tools. The adapter might be nothing more than a small transformer, full-wave silicon diodes, and an electrolytic capacitor in a flat plastic case that plugs directly into the ac receptacle (Fig. 12-23). Some ac adapters use a bridge rectifier circuit instead of the two diodes. The power adapter converts ac to dc voltage. These adapters might come in 3-, 4-, 5-, 6-, 9-, and 12-V types. The total current output is from 450 mA to 1.5 A.

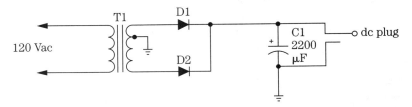

■ **12-23** *The wall adapter might consist of four active components: the transformer, diodes, and a filter capacitor.*

Wall adapter cords sometimes break. They can be checked with pins or needles stuck in the cord up to the plastic box for continuity tests. The cord usually breaks where it enters the box or at the molded plug. New female plugs (2.1 mm) can be picked up at Radio Shack or most any electronic store. If one of the diodes becomes leaky, the primary winding opens, which destroys the unit. Most parts can be replaced in the wall adapter, except for the power transformer, which might cost more than a new wall adapter.

Portable radio and cassette player power-supply circuits

The step-down transformer provides an ac voltage to the anode terminals of the two silicon diodes or to a bridge circuit in the radio-cassette portable. When the power cord is plugged into the back of the portable, switch (S901) provides ac operation. C901 is the main filter capacitor. The 10.1-V source feeds to the audio output ICs or transistors (Fig. 12-24).

Q902 provides ripple dc filtering to other dc sources, 6.4 V, 4.8 V, 4.85 V, and 4.6 V. The 6.4-V source feeds the tuning LED indicator,

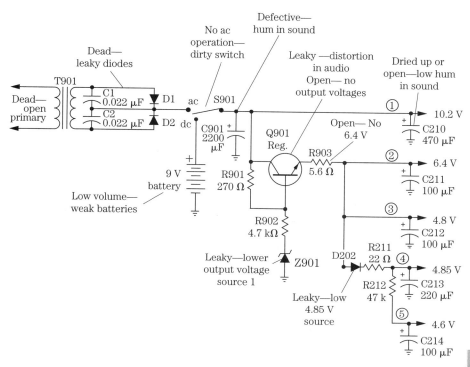

■ **12-24** *The portable radio and cassette player low-voltage circuits might provide five or more low-voltage circuits.*

AM and FM, preamp audio, and the tape motor circuits. The voltage source (4.8 V) feeds to the AM and FM multiplex circuits. The 4.6-V source supplies voltage to the AM/FM/IF and detector circuits. Some radio-cassette power-supply circuits are quite simple, but in the boom-box units, you might find transistor and zener diode regulators.

Auto radio power-supply circuits

Although the car radio operates from the 13.8-V auto battery, additional filtering and regulation might be found in the power-supply circuits. The "A" lead (red) brings in 13.8 Vdc from the battery to a socket at the radio and into a line fuse, choke coil, and electrolytic filter capacitors to form the power-supply circuits (Fig. 12-25). Choke input and capacitor filtering provides a high current drain power supply. D201 automatically blows the fuse if the red lead is connected to a negative voltage or if the auto battery is charged up backwards. This diode prevents semiconductor damage within the auto radio. Capacitor C201 provides extra filtering in the low-volt-

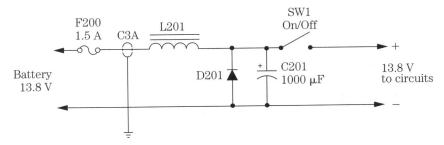

■ **12-25** *The dc auto radio low-voltage circuit consists of a fuse, a line choke, a diode, and filter capacitor.*

age source. You might find IC, transistor, and zener diode regulation in more expensive or deluxe auto radios and CD players.

Besides blowing the fuse, problems in the low-voltage power supply, might result from a burned cable harness, hum in the sound, and lower output voltages. A defective filter capacitor might cause "motor-boating" sounds in the speaker. Shunt C201 for hum heard in the audio. Notice if L201 winding might be burned with shorted or burned windings, resulting in hum and lower voltages. If D201 is shorted, it will blow the fuse repeatedly until it is replaced.

Boom-box power-supply circuits

The boom-box low-voltage power supply circuits are somewhat like the AM/FM/MPX cassette player. You might find additional voltage regulator circuits in the various voltage sources. A bridge rectifier provides dc voltage to the tape circuits and regulated voltage applied to the AM and FM stereo circuits (Fig. 12-26). SW1 switches in or out the battery and ac power circuits. The 10.8-V source feeds to the tape motor, LED, and power IC output amplifier circuits.

The 5-V regulated source feeds to the FM/MPX circuits. The 4.9-V source feeds the FM/AM IF and AM circuits. The 5.6-V source supplies voltage to the FM/MPX IC circuits. The regulated 6.3-V source supplies voltage to the recording tape circuits. The regulated 5.63 V feeds to the FM input circuits. The preamp tape circuits are supplied from the 4-V source

Check Q101, Q102, and D101 for regulation problems. The regulator transistors might go open with no voltage source or become leaky with hum and distortion in the radio and cassette circuits. Small dried-up filter capacitors in the voltage sources might cause

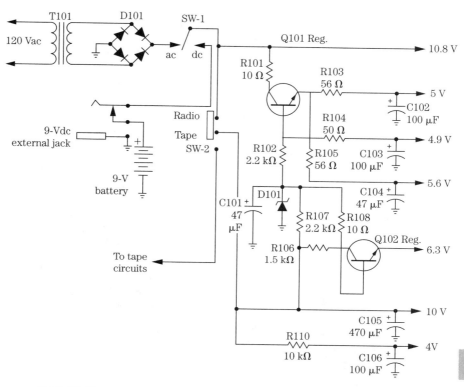

12-26 *The boom-box cassette player might have transistor and zener-diode regulation in the low-voltage circuits.*

low hum, distortion, and low voltages in the radio and tape circuits. Clip in a new capacitor with the same voltage and capacity across the suspected electrolytic capacitor to determine if the capacitor is defective.

Conclusion

Always check the low-voltage power supply with a dead symptom. The voltage sources connected to the various circuits must operate before these circuits can function. When lower or improper voltages are found in a circuit go directly to that voltage source in the low-voltage power supply. Most problems found in the low-voltage circuits are leaky diodes, dried-up filter capacitors, open or leaky voltage regulators, and open power transformer primaries (Fig. 12-27). The transformer might run hot, smoke, and burn when diodes become leaky in the power-supply circuits. When the power transformer is defective, the cost of servicing might be greater than the cost to replace the entire unit.

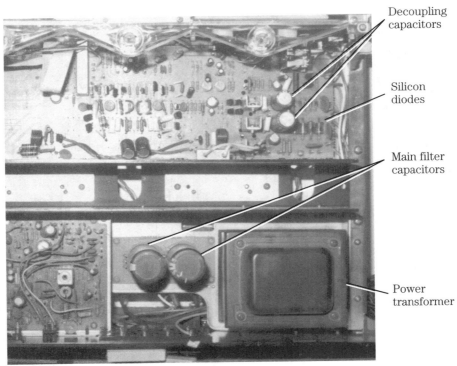

■ **12-27** *Most problems in the low-voltage circuits are leaky diodes, dried-up filter capacitors, dirty on/off switches, and open or shorted transformers.*

Servicing low-voltage power supplies

Servicing simple audio circuits

THE AUDIO AMPLIFIER RECEIVES A WEAK INPUT SIGNAL and delivers a larger power output signal. *Amplification* is the process of increasing the amplitude of a signal that might be current, voltage, audio, or power. The input signal of the amplifier might be weak, but then is amplified by a preamp or AF audio stage. The audio frequency (AF) amp might be the second stage with the final stage of the power output amplifier (Fig. 13-1).

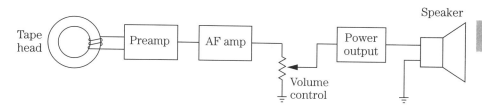

■ **13-1** *A block diagram of the preamp, AF amp, and power output stages in a cassette player's amplifier.*

The small amplifier in the cassette player might have a preamp, AF amp, and power output amplifier. The weak tape head signal is switched to the preamp circuits with the volume controlled in the AF amp and the power output circuits directly coupled to the speakers. You might find one large IC that contains all of the audio circuits.

In early tape players, PNP transistors were often used throughout the audio circuits. The tape head or microphone is switched into the preamp transistors (Q1) and audio is directly coupled to the preamp (Q2). The PNP AF amp is capacity coupled to the preamp stages with a volume control in between to control the audio. The AF amp is transformer coupled to the output transistor in a push-pull amplifier circuit.

One transistor might be classified as an audio amplifier (Fig. 13-2). The input signal is amplified by the PNP or NPN transistor and is either directly or capacity coupled to the next audio amplifier. Two or three transistors might be used for headphone operation. To drive a small speaker, three or four transistors were often used in the audio circuits.

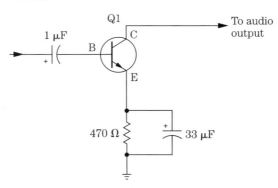

■ **13-2** *One transistor or IC might be considered an amplifier. Here, the AF amp is a class-A amplifier.*

Most of the radio and tape player amplifiers consisted of several transistors. To achieve greater volume, two transistors were wired in a push-pull circuit with a coupling transformer in the power audio circuits (Fig. 13-3).

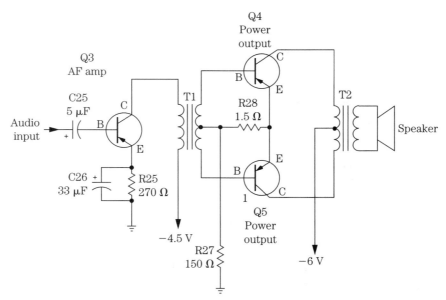

■ **13-3** *Two transformers are used to couple the AF amp to the push-pull output transistor and from the output transistors to the speaker voice call.*

Class-A amplifier

The AF transistor might have capacitor-resistance coupling with transformer output coupling to the speaker. Most input or preamp circuits operate in a weak class-A amplifier circuit. The class-A amplifier operates in the lower portion of the current-voltage curve. The output signal or waveform is the same shape as the input with greater amplitude (Fig. 13-4). The class-A amplifier has very little distortion.

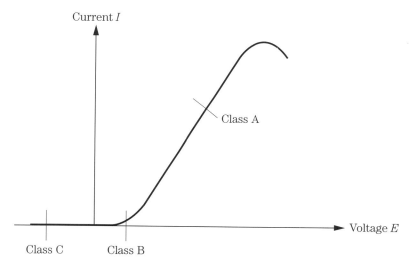

■ **13-4** *The class-A amplifier is used in the linear portion of the current and voltage curve.*

Class-B amplifier

Q4 and Q5 are transformers that are coupled to the AF amplifier. These two transistors operate in a push-pull operation. The power output circuits have the greatest volume or signal in the amplifier circuits. Resistors R27 and R28 provide emitter and base bias for Q4 and Q5. The class-B amplifier operates at zero current with no input signal. In the push-pull circuits, one transistor operates on the positive half of the cycle and the other transistor on the negative half, which provides greater amplification. Notice that the PNP transistors have a negative voltage applied to the collector terminals. In other words, the collector is more negative than the emitter or base terminals.

How the amplifiers perform

The audio amplifier might amplify a radio, tape, cassette, microphone and stereo channels. In the small cassette player or portable radio, the tape head and radio signals are switched into the audio circuits with a function switch. Usually the radio detected signal and tape head circuits are switched in separately. At the same time, the voltage source might be switched to that certain circuit. Often audio switching is done in a cassette player after the preamps and in the radio after the detector stage (Fig. 13-5).

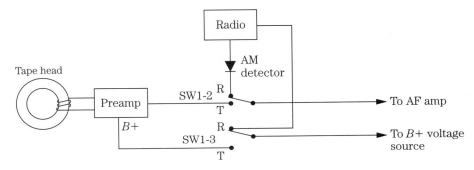

■ **13-5** *Audio switching of the radio and cassette player is performed after the preamp cassette stages, and it switches into the AF amplifier.*

The weak radio signal from the detector circuits are switched into the audio circuits with the function switch. This weak signal is then amplified by the first AF stage (Q1). From S1A, the radio audio signal is coupled to the base terminal with capacitor C100. This is known as *capacitor-resistance coupling*. Q1 amplifies the weak radio signal and it is fed to a push-pull transistor or to IC power output circuits (Fig. 13-6). Notice in this circuit an NPN transistor has a positive voltage fed to the collector terminal.

IC audio amps

Several years ago, transistors might have been found in the preamp and AF stages with a single IC as the power amplifier. In fact, the whole audio circuit might have been contained in one IC of the small amplifier (Fig. 13-7). Here, one IC can pick up the audio signal, control the volume with a potentiometer or variable resistor, and drive a speaker. Notice that this IC amp operates from the power line.

The simple IC audio amplifier found in many hobby and Ham magazines is an LM386 IC. This IC can pick up a phono, tape, or radio

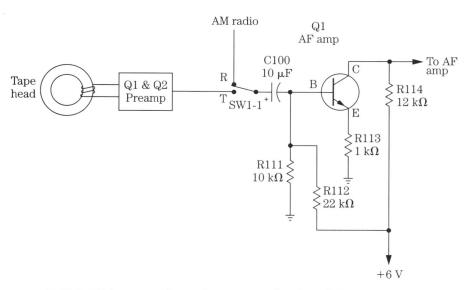

■ **13-6** *Q1 has capacitor-resistance coupling from SW1-1 to the base terminal.*

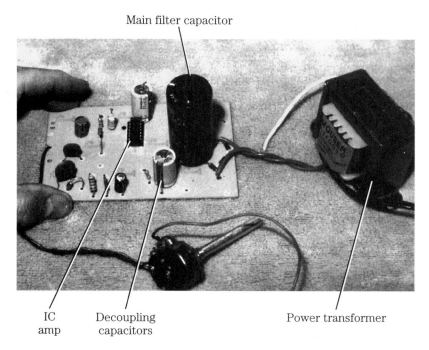

■ **13-7** *Here, one small IC provides audio amplification in a self-contained transformer power supply.*

IC audio amps

signal and amplify it for earphone or speaker reception. The audio signal is controlled by volume control R3 and is capacity coupled with a 10-µF electrolytic capacitor to pin 3 of IC1 (Fig. 13-8). The output signal is found at pin 5 and is capacity coupled with C7 to a headphone jack or small speaker. Pin 6 is the voltage supply pin and can operate from 3 to 15 V. Notice that this small signal amplifier is an 8-pin IC. Although this small IC amp will not drive large speakers, it does supply enough audio for headphone operation or to drive a 4-or 5-inch speaker.

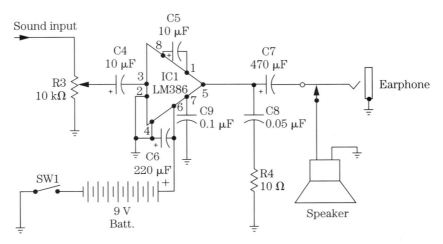

■ **13-8** *This small IC amplifier has capacitively coupled audio signal from the volume control to pin 3 of IC1 (LM386) and a capacitively coupled output to speaker at pin 5.*

Today, IC amplifiers are found in all monaural and stereo circuits. Separate amplifiers can be used for the input audio circuits and power IC amps in the power output circuits.

Stereo audio circuits

The stereo amplifier system is a multi-sound system with two different audio channels. The left and right channels are reproduced at the same time. Stereo tape deck circuits contain two tape heads with a preamp, AF amp, and a power output that feeds audio to a pair of stereo headphones or to several speakers (Fig. 13-9). Although each channel is separated from the other, the supply voltage is applied to both channels from the same low-voltage power supply. In the FM radio stereo circuits, the audio is fed from the multiplex (MPX) system and is switched into the AF audio stereo channels.

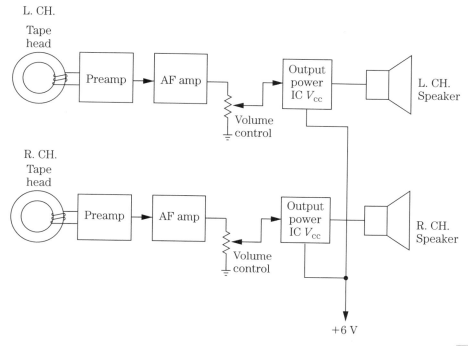

■ **13-9** *The block diagram of a stereo cassette player.*

You might find all stereo sound circuits in a portable stereo radio and cassette player in one IC. These portable cassette players can be strapped to the body and used via headphones. The stereo tape head circuit is wired directly to the IC and the output is capacity coupled to the headphone jack (Fig. 13-10). The supply voltage pin might operate from batteries (3 to 6 V). Very few components are located in the audio circuits.

Larger boom-box cassette and AM/FM/MPX players might have two ICs in the stereo circuits. A dual IC might serve as preamp and the other as dual-audio output. The IC preamp receives the radio or tape signal from a function switch and is capacity coupled to the input terminals. The output terminals are coupled with electrolytic capacitors to the volume and tone controls (Fig. 13-11). The preamp supply voltage is very low compared to the audio output circuits. This prevents less pickup noises and distortion.

IC401 is a dual preamp IC that serves both the right and left audio channels. Both channels are identical. When one stereo channel is defective, the other channel can be used as a signal source to make comparison tests. The voltage and signal in the normal chan-

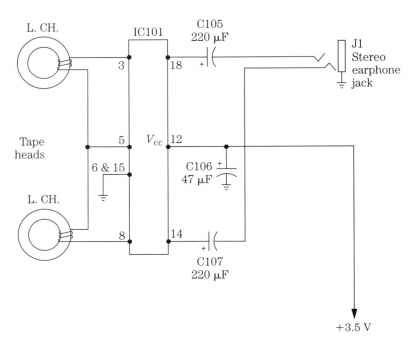

■ **13-10** *A small portable cassette player that uses one IC for both left and right channels with capacitive coupling to the stereo headphone jack (J1).*

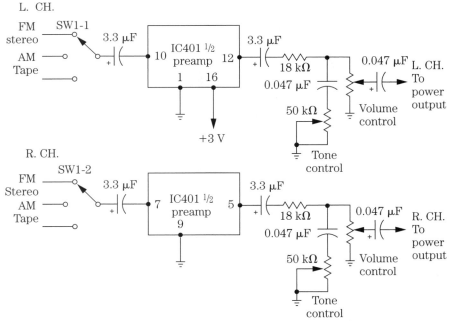

■ **13-11** *Both left and right input switching in capacitive coupling to one large IC preamp with a tone and volume control in the output circuits.*

Servicing simple audio circuits

nel can be compared to the defective channel. This makes the audio stereo channels much easier to service without a schematic.

The input signal of the power amp is capacity coupled (0.47 µF) to each input pin. Often the audio amp has a higher dc voltage source. Again, the power IC output signal is capacity coupled with a larger electrolytic capacitor (470 to 1000 µF) to the speakers. The boom-box portable might have a 6½-inch woofer and a 1-inch tweeter speaker. The dual output IC might be bolted to a separate heatsink on the PC chassis (Fig. 13-12).

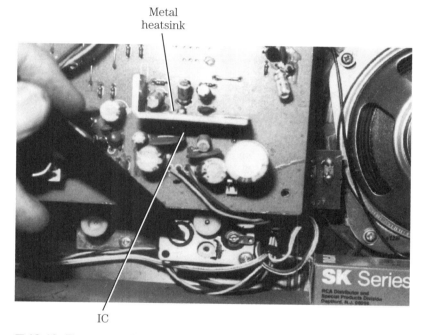

■ **13-12** *The power IC is bolted to a separate heatsink for the left-channel output amplifier and drives a 4-inch pin speaker.*

Different sound problems

Like any other circuit, the audio circuits might appear dead, have weak sound, or have intermittent and distorted audio. The dead audio circuit might be caused by a defective transistor, IC, no supply voltage, or by a dirty headphone jack. Weak sound might result from dried-up or open electrolytic coupling capacitors, or from leaky ICs and transistors. Intermittent sound can be caused by bad head connections, intermittent components, and poor PC wiring connections. Often excessive distortion is found in the audio output transistor or IC circuits (Fig. 13-13)

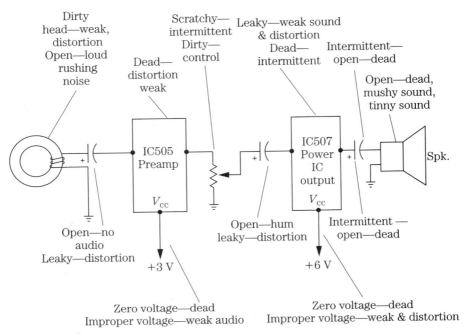

■ **13-13** *Check the stages that might produce distortion in the amplifier and speaker.*

In the stereo circuits, you might have weak sound and intermittent reception in the right channel with a normal left channel, for example. Distortion might occur in either or both channels, except when both channels have weak and distorted symptoms. Suspect the dual output IC.

Weak sound

Although, a leaky transistor might produce a weak audio signal, the transistor itself is not weak. Transistors will either go open, or become leaky and intermittent. Leakage might occur between any two terminals, but in most cases, leakage is found between collector and emitter terminals. When a transistor has an open junction, no sound is heard at the collector terminal. The defective transistor might have a high-resistance junction, which results in a weak signal or in no audio signal.

Defective audio ICs might cause a weak signal. Usually a defective IC with weak signal might also cause audio distortion. The open IC might have a weak signal or no signal at the output terminals, but hum in the speaker. An open bias resistor or capacitor might cause

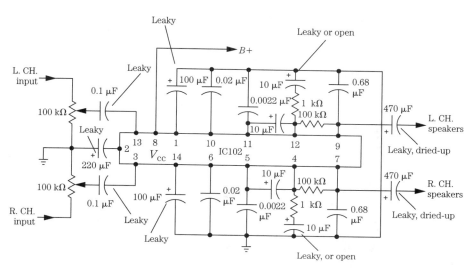

■ **13-14** Check the electrolytic and coupling capacitor if there is weak sound in the stereo channels.

a weak or distorted audio channel. If there is no supply voltage at the supply pin (V_{cc}), there will be no sound in the speaker.

Most weak audio signals are caused by electrolytic capacitors. These small coupling capacitors have a tendency to dry up or become open after years of use or disuse. A poor foil or terminal connection of the coupling capacitor might cause weak audio (Fig. 13-15). Do not overlook a change in bias resistors to produce a weak audio channel.

Dead

The dead audio channel is a lot easier to locate than the weak or intermittent channel. Simply signaltrace the weak or intermittent channel. Most dead audio circuits are caused by a broken board connection, no supply voltage, an open component (coupling capacitor, transistor, or IC), speaker coupling capacitors, and speakers. The open headphone jack might cause a dead audio symptom. If the speaker is dead and earphone reception is normal, suspect that a voice coil is open or that a connection to the speaker is bad. An open tape head or broken connection might cause a dead channel in the cassette player (Fig. 13-15). Check the supply voltage first for a dead audio channel.

■ **13-15** *The defective tape head might be open and cause a loud rushing noise with no music. Check for packed oxide on the tape head, which would produce distortion and weak volume.*

Intermittent sound

Intermittent audio might be caused with a poor junction in a transistor or IC. Poor internal foil connection inside a coupling capacitor might cause intermittent audio. Intermittent sound might result from broken terminal board connections and PC wiring (Fig. 13-16). Noisy or intermittent headphone reception can be caused by a dirty headphone jack and terminals. The broken speaker voice coil might produce intermittent audio. Do not overlook electrolytic coupling capacitors. The output coupling capacitor to the speaker might become intermittent and have weak audio symptoms.

Distorted sound

Distorted and intermittent sound are the most difficult symptoms to find in the audio circuits. It's possible to have both intermittent and distorted sound in one dual-output IC. Within the transistor

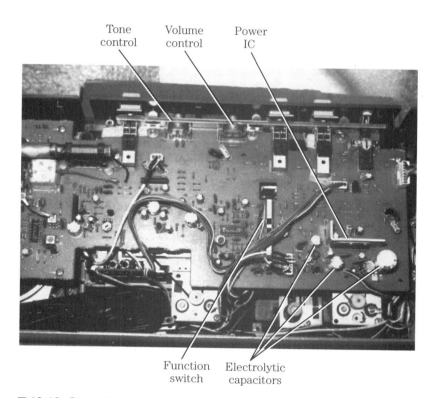

■ **13-16** *Check the components in the audio circuits that might cause distortion.*

output circuits, distortion might be caused with a leaky or open push-pull output transistor. Burned bias resistors or a change in resistance might produce distorted audio (Fig. 13-17), A leaky coupling capacitor to the base terminal of the audio transistor might cause distortion. Also, an improper supply voltage might result in distorted sound.

The leaky IC might cause weak and distorted sound in the speaker. Open or leaky electrolytic capacitors and a dropped voice coil or warped speaker cone can produce distortion. Do not overlook the possibility that packed oxide on the tape head is causing weak and distorted audio.

How to signaltrace audio signals

Missing, weak, or distorted audio can be signaltraced with an external audio amp. Audio signal injected into each stage determine what circuit is dead or distorted. The audio signal can be traced

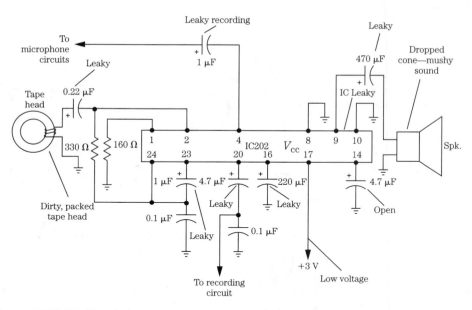

■ **13-17** Check the components in the schematic that might produce distortion and weak audio.

with the oscilloscope and input sine- or square-wave generator to locate the weak or distorted component. Critical voltage and resistance measurements can locate dead, weak, and distorted circuits.

One of the most common methods is to signaltrace the audio signal from a radio or tape player source. Insert a cassette or turn the radio on to a local FM broadcast station and check for distorted music or no music with the external amp (Fig. 13-18).

In stereo circuits, you can compare the signal at the volume control in each channel to determine where the weak or distorted sig-

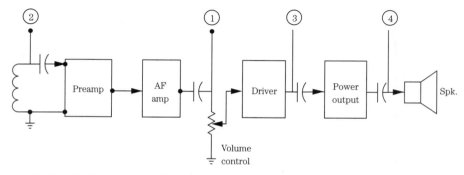

■ **13-18** Signaltrace with an external audio amplifier. Keep the volume as low as possible so that you can hear if the music is distorted.

nal is located. If the left channel is weak at the volume control and normal at the right channel, suspect a component in the preamp stage or tape head. Compare the input and output signal of the preamp transistor or IC to determine where the weak left channel exists. Signaltracing methods are found in Chapter 16.

Servicing audio circuits with a DMM

The portable cassette or radio can be operated from batteries or an external power adapter. If the unit will not play either in the radio or tape player mode, suspect that the batteries are weak or dead. Check the total battery voltage with the unit turned on (Fig 13-19). If voltage is normal, check the voltage across the largest electrolytic capacitor with a DMM. This will indicate if the high voltage is present. Go next to the output transistor or IC and see if the sound is dead, distorted or weak.

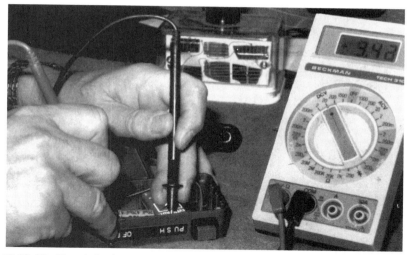

■ **13-19** *Check the battery voltage in the small tape player to see if the batteries are weak or dead with the switch on and battery connected.*

Suspect that the audio output IC is leaky if both channels are dead. Notice if the IC amplifiers of both channels if one channel is dead, weak, or distorted. Check the collector voltage of each transistor in that channel. Compare the voltages of all power output transistors with the normal channel. Now quickly test each transistor with the diode-junction test of the DMM. When both channels are dead, suspect that an improper voltage is coming from the power supply.

When the left channel is dead or distorted, check the voltages on the audio output IC. Carefully measure all voltages on the IC and compare them with those of the schematic. If no schematic is available, compare the voltages with the normal channel. Usually, in a dual-IC with proper supply voltage and leaky conditions, the symptom will occur in both channels. If one of the voltages is off compared with schematic or the normal channel, suspect that the output IC is leaky. Check each component tied to each IC terminal for correct resistance and capacity.

Servicing table radio audio circuits

If the audio of a table radio is very weak and distorted, suspect a leaky audio transistor. Really weak audio might be caused by a small electrolytic capacitor (1 µF to 10 µF). The pocket DMM is very handy when testing voltages, but you can't use it to check electrolytic capacitors.

First, measure all voltages on each transistor. Check each transistor for open or leaky conditions with the diode-transistor test of DMM or transistor tester. Accurately measure the resistance of each bias and emitter resistor. After all of these tests, if the audio is still weak, suspect that a coupling capacitor or emitter-bias capacitor is defective.

These small capacitors might be located by shunting another capacitor across them. Always shunt the capacitor with one that has the same or a higher voltage. Tack the capacitor leads or clip across the suspected one with the power off. Apply this method until the weak one is located. Then, solder the correct replacement on the PC board. Check the defective capacitor in a capacitor tester, if handy.

In early transistorized radio audio circuits, a single-ended transistor is used in the audio output circuits. The AF transistor drives the output transistor with direct coupling or through a resistor. The audio signal is amplified by Q105 and is directly coupled to Q107 (Fig. 13-20). Transformer coupling is connected to the collector terminal of output transistor and stepped-down to a 3.2- or 8-Ω speaker. Notice the high transistor collector voltage in the audio output circuits.

Servicing stereo audio circuits

The stereo audio circuits in a cassette or CD player have identical audio circuits in each channel. The same components, voltage,

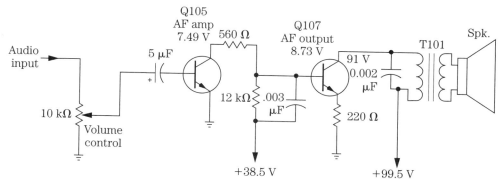

■ **13-20** *In older table radios, single-ended transistors were used with a power output transistor transformer-coupled to the speaker.*

and resistance should be the same in the left and right channels. The great advantage of servicing stereo circuits, is that one channel can be used as a reference point while locating and repairing the defective channel. Often when a dual IC is used as a preamp and another IC is used as a power output, the defective component affects both channels.

Locate the defective channel on the schematic and audio chassis. Check for output components marked on the PC board that are found in the schematic. Another method is to use the external amp and locate the dead stage by starting at the speaker and preceding to the power output IC (Fig. 13-21). Check the audio signal in and out of the IC. If the signal is coming in and not out, suspect that an IC or parts in the same circuit are defective. Take critical voltage and resistance measurements on each terminal pin. Check the supply voltage terminal of the defective IC.

Distortion and weak audio signals can be located with the external amp. Check the audio at the input and output terminals of the suspected IC. Keep the volume as low as possible because the IC will amplify the input signal. If the audio input is normal, but is distorted at the output, suspect that an IC is leaky. Check the signal on the good or normal channel and compare. Test every component tied to the IC before replacing it.

Checking power audio circuits

The small phono amplifier might contain one dual IC or transistorized audio channel. The monaural phono amp might have only three small transistors. If the phono is stereo operated, look for

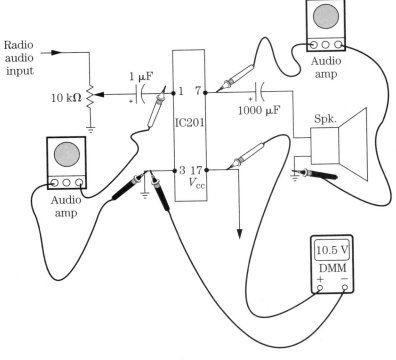

■ **13-21** *Locate the defective IC with external amplifier checking the audio at the input and output terminals. Check the supply voltage terminal for the correct dc voltage.*

one IC in the audio circuits. Suspect that the dual IC is defective if both channels are weak or distorted. Check the voltage at the supply pin terminal. The phono input leads can be switched to determine if the crystal cartridge is defective. Likewise, the two speaker leads can be reversed to locate a defective speaker.

If the right channel is weak compared to the left channel, suspect that coupling and speaker output capacitors are leaky or open. These electrolytic capacitors have a tendency to dry out or contain poor internal terminal connections. Simply clip another electrolytic capacitor across each one (with the power off) to determine if it is defective. If both input and output capacitors sound ok, but still have weak sound, suspect the IC or terminal components. Check C405, C407, and R402 tied to IC401 terminals (Fig. 13-22).

The phono cartridge might have a crystal or magnetic pickup. Lower-priced phonographs have a crystal cartridge. The needle holder is fastened to the end of the crystal wafer. When twisted, the crystal produces a weak audio voltage and signal that is ampli-

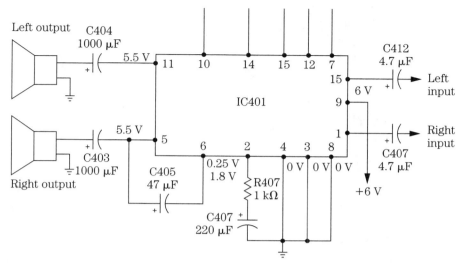

■ **13-22** *With the audio signal at pin 1 and 15 and no sound at the right speaker, suspect that IC401 is defective. Before replacing IC401, check the voltage on each terminal pin, and compare it to the schematic and to the normal left channel.*

fied by the audio circuits. The crystal cartridge is a high-impedance device and is coupled into a volume control or electrolytic capacitor. A defective crystal cartridge might produce distorted and weak audio. A poor crystal cable might cause intermittent sound. Lightly press down on the pickup arm; if the music cuts in and out, replace the cartridge. Excessive dust and dirt collected around the stylus can also cause weak and distorted music.

The magnetic pickup contains a coil with moving vanes connected to the needle or stylus. Because the signal or voltage is very low compared to the crystal cartridge, several stages of amplification are needed before coupling to the AF circuits. Very few problems are found in the magnetic pickup. Measure the continuity or resistance of the magnetic head and compare the resistance of each stereo head connection.

Speaker problems

Speakers come in many different sizes and shapes. The impedance range of speakers might vary from 3.2, 4, 8, 10, 20, 32, to 40 Ω. Some TV speakers might have a resistance of 32 and 45 Ω. Always, replace the defective speaker with one that has the correct circumference, impedance, size and weight of the magnet (Fig.

■ **13-23** *This Sylvania clock table radio has a 5 1/4-inch speaker.*

13-23). If a 3.2-Ω speaker is replaced with an 8-Ω speaker, the volume will decrease by half.

The defective speaker might have an open voice coil, frozen, dropped cone, loose cone, or a loose spider assembly. Lightly feel the cone and push it in and out to see if the cone is rubbing against the pm magnet. Inspect the speaker cone edges for loose or warped areas. The speaker glue might come loose and cause a blatting or buzzing sound. If the speaker has a tinny or raspy sound, the cone is either frozen or has dropped down against the center pole piece. The cone might be warped and rubbing against the center pole piece. Voice coil damage might result from excessive volume. Push lightly in on both sides of the speaker cone with your thumbs. You can quickly determine if the cone is dragging. Replace the speaker if it is defective.

Check the speaker voice coil for open or torn leads on the low-Ω scale of an ohmmeter. Remove one speaker lead from the amplifier for accurate tests. The resistance across the speaker terminals is always slightly lower than the actual impedance. For instance, an 8-Ω speaker voice coil impedance might measure 7.5 Ω of resistance. The intermittent voice coil will show an erratic reading on the ohmmeter, while you push up and down on the cone area. Of course, the open speaker will have no resistance. Replace any speaker that has an open voice coil, torn voice coil leads, or a damaged cone.

14
Checking and repairing PC boards

A PRINTED CIRCUIT (PC) IS A PATTERN FORMED ON A phenolic board with a copper-clad surface and can be etched by several different methods. PC boards are made of a thin sheet of copper cemented to a phenolic base. The foil circuit or pattern remains after the rest is etched away. The circuit might consist of pads and traces of wiring. Usually the pads are used for soldering terminals of components mounted on the board (Fig. 14-1).

■ **14-1** *In this Audiovox car radio, large and small PC wiring is used with pads and socket connections. Notice the many terminal wires that are used on the surface-mounted microprocessor.*

Single- and double-sided boards are both common. Single-sided boards have copper only on one side and double-sided boards have copper cemented on both sides. The double-sided boards are designed for more complex circuits so components can be mounted on both sides. In the latest cassette player or TV chassis, you might see a double-sided board with large standard parts mounted on top and surface-mounted components (SMD) soldered underneath. However, double-sided boards are prone to more problems if the board warps. Small PC boards can be cut from larger pieces for small electronic products and projects.

Small holes can be drilled into the pad terminals where the part terminals are fitted into. Each pad is soldered to tie the part into the circuit. PC wiring connects the pads and parts together and forms a circuit. The PC wiring and layout of the pads might be called a *printed circuit* or *foil pattern* (Fig. 14-2). Soldered holes sometimes tie the bottom and top circuits together.

■ **14-2** *In this VCR, the PC board has large grounded sections of PC wiring and thin PC wiring to the different pads and terminals.*

Making your own PC boards

There are many different printed circuit kits on the market to make circuit boards. The oldest and inexpensive PC board method is with dry transfer labels and etching solution. The circuit is laid out on the copper side with dry transfer donut pads, tape, circles, wiring lines, different pattern materials, and a special ink pen. After the circuit is laid out, the wiring is etched with etchant. The etchant is available in dry or liquid form.

The complete kit might include a couple of small PC boards, a plastic pan tank or cover that contains all components, polishing pad, and $\frac{1}{16}$-inch drill bit. Be careful when working around the etchant because it can stain clothes, hands, and your working surface. Lay down several pieces of old newspaper in case the solution is spilled. Use a wooden pencil or tongs to handle the board for inspection and to remove it from the etching tray. Try to avoid any spills. These kits can be found at most electronics stores and Radio Shack (Fig. 14-3).

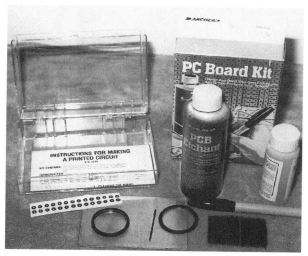

■ **14-3** *You can etch your own PC boards with a simple printed board kit from most electronic stores.*

Draw and lay out the circuit on a piece of paper. Check with other project boards found in electronic hobby and project magazines. Doublecheck the layout to make sure that the whole circuit is correctly connected together. Keep the donut pads and tape press-on printed circuit tape from running circuits together (Fig. 14-4). Always wash off the copper side with soapy water, rinse it clean, and dry it before laying out the circuit or applying PC wiring transfers to the copper side.

After all pads and circuit tape have been applied, begin the etching process. Carefully slide the board into the solution or place it in the etching glass or plastic container and then pour enough etchant to

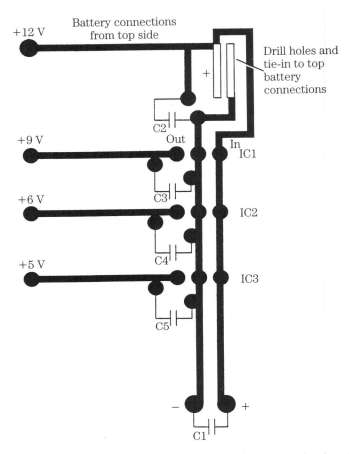

■ **14-4** *Use a drawing or PC wiring layout chart and place the donut pads and trace wiring on the upper side of the board.*

cover the board from ⅛ to ¼ inch. Do not put too much liquid into the container or it will spill out when rocked back and forth.

The etchant might come in a dry form as powder or in a plastic bottle. The powdered etchant must be dissolved in water according to the directions on the package. Ferric chloride and ammonia sulfate are often used as etchant. Keep the lid tight on the plastic bottle so that if it is accidentally tipped, it will not pour out and damage the floor or table.

Rock the plastic tray back and forth to keep the liquid moving over the board area, which speeds up the etching process. Place a wooden pencil under the center area of the tray and carefully rock the tray back and forth. Handle the PC board with the same old pencil because the etching solution will mar and corrode metal tools. A small PC board can be etched within 15 or 30 minutes. Larger boards take a little longer. You can warm the solution by placing the etching tray in a pan of warm water to reduce time for the etching process. If you decide to use the etching solution over again, place it in another plastic bottle and mark the date. Remember that it will take longer to etch the next board with old solution.

Check the board every five minutes to see how the etching process is working. Do not leave it in the solution too long or it will destroy some of the wiring. Pour off the etchant and wash off the board. Remove the etchant-resistant labels with a kitchen-type scouring pad. Soapy water will clean up the board and the solvent. Let the board lay under running water for 15 or 20 minutes. Clean up the tray and wash out all etchant. Be careful not to splash it on your clothes or skin.

Drill out each donut pad where each component terminal lead goes. Place a soft pine board underneath the PC board when you drill small holes. Use the smallest bit possible. Large bit holes are difficult to fill with solder and you might make a poor connection. Drill the hole from the copper side out to prevent chipping the PC board and wiring.

PC board wiring problems

Broken PC wiring and poorly soldered connections on a PC board might cause intermittent problems. Large boards have a tendency to warp and might snap the PC wiring at the pad or terminal connection (Fig. 14-5). PC wiring foil can crack in a way that the naked eye cannot see. Poorly soldered eyelets on double-sided boards might break or open the circuit. The chassis might appear dead with a piece of broken wiring at the pad, terminals, or at the component. Check the PC wiring with a lighted magnifying glass. Sometimes by showing a strong light behind the board, you can locate PC wiring foil breaks.

The intermittent board

When the PC board is moved or flexed in a certain section, look for the broken wiring or intermittent component in that area. Take the symptom and try to locate the section on the board where these parts are mounted. For instance, in the TV chassis, if the picture collapses down to a bright horizontal line, you know that no vertical sweep exists. Go directly to the vertical circuits and prod around on parts with a plastic pen or with an insulated

■ **14-5** *A large, one-piece PC board is used for the entire chassis in this RCA TV set.*

tool to try and locate the defective component or broken wiring board connections (Fig. 14-6). Isolate the component to a given section by flexing the PC board.

Intermittent board connections might exist where the PC eyelet feeds the circuit from one side of the board to the other. Often solder lays in these holes and ties the top and bottom board circuits together. Repair the eyelet connection by pushing a bare hookup wire through the eyelet with soldering iron on the eyelet area. Leave ¼ inch of wire on each side and solder the ends to the PC wiring on both sides of the board. This will cure most broken eyelet connections. Cold soldered joints or blobs of solder on the PC wiring might also cause a poor connection; heat the joint with the soldering iron (Fig. 14-7). Sometimes a poorly tinned lead of a component will make a poor connection inside the blob of solder.

Usually a broken board will crack a section or break where a heavy part is mounted. If the chassis is dropped in shipment or in the

■ **14-6** *Probe around the different parts and sections of the chassis to isolate a defective component or an intermittent PC board.*

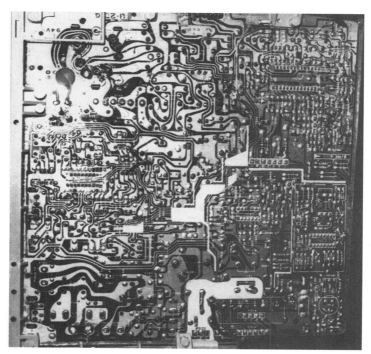

■ **14-7** *Check for poor solder joints. Correctly resolder these joints.*

handling process, the weight of a transformer or standoffs might crack the board. Repair the broken wiring by shunting a bare piece of hookup wire across each PC wire break. Do not rely on just solder. Doublecheck the connection. Make sure that the shunted wire is on the same PC wiring foil. Most service literature shows the correct PC wiring layout and the layout of mounted components.

Check around the terminal sockets and standoffs (or insulators) for breaks. The PC board is fastened tight in these areas and the wiring might break on either side of the standoff. Inspect the areas where a female socket is mounted on the PC board. Often, resoldering the socket connections solves poor socket connections (Fig. 14-8). Poorly soldered IC, LSI, or microprocessor pins can cause intermittent problems.

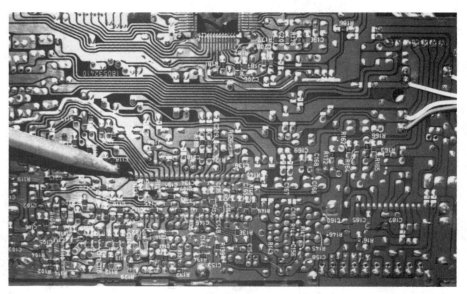

■ **14-8** *You might find poor socket terminals in the PC wiring. Slightly move the socket on top of the board to see if the chassis cuts in and out. Then, resolder all socket PC connections.*

Cracked and broken boards

As was mentioned earlier, you sometimes can see a crack or break in the PC wiring if you place a bright light behind the circuit board (Fig. 14-9). Notice if the board is broken over several areas of the PC wiring in a TV chassis. If all of the cracks are around the flyback transformer, the TV was probably dropped or knocked off of the TV stand. Several separate PC wiring areas can be repaired with insulated and bare hookup wire. Notice the different component

■ **14-9** *Notice that bare and covered hookup wire was used to repair this broken section of the PC board around the leaky flyback transformer.*

part symbols and outlined areas. If the board is broken in several different places, it might not be repairable.

If the PC board has a portion broken off entirely, it is difficult to repair. On the other hand, holes that are burned into the board with overheated resistors or components can be repaired. Simply cut out the burned areas. Check the schematic, PC board, and board parts layout found with the service literature. If there is room for an insulated standoff to be mounted, place it in the circuit and tie parts and hookup wire to join the various circuits. Another method is to design a small PC wiring on a blank copper PC board and tie it into the main board with hookup wire.

Circuit tie wires

You might see several pieces of bare wire bent over like staples that are fed through the PC board. Often, these wires tie another circuit or parts together. Notice that this is done on the top side of the chassis, instead of on the PC wiring side. This can be a source of trouble. Check the solid tie wire for poorly soldered connections on the PC wiring that might produce a dead or intermittent symptom. Surface-mounted tie wires or ends are solid flat pieces that look like a resistor or capacitor and do nothing more than tie the circuits together. Do not mistake a shorted component for a surface-mounted tie-end.

Surface-mounted PC components

Surface-mounted devices or components are mounted directly on the PC wiring. These miniature parts are difficult to see without a magnifying glass. Today, you will see SMD parts in practically all consumer electronic products (Fig. 14-10). The larger components are found mounted on the top side of PC chassis.

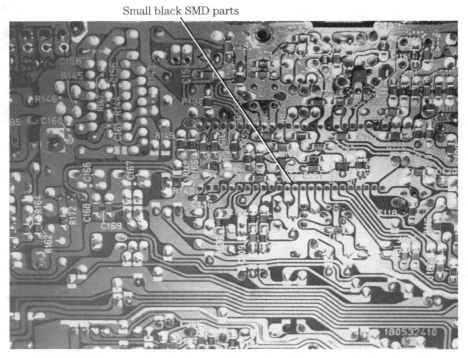

■ **14-10** *The many terminal connections of an IC are mounted on top with terminals soldered to the PC wiring pads. Notice that several SMD parts are used between the rows of IC terminals on the PC wiring.*

Poorly soldered end-contacts or breaks in wiring made at the SMD terminals might cause intermittent operation. After isolating the area that might be producing the intermittent problem, touch up the soldering pads, connections, and SMD soldered connections with the soldering iron. Be careful not to destroy small parts or nearby PC wiring. Inspect the foil wiring where the SMD part mounts for possible cracked wiring. Take continuity tests from one component to the next to locate a possible break in the wiring.

Locating parts on PC wiring

Check the service literature for SMD parts layout on the PC board. Large components found on the top side might tie into these same circuits. Surface-mounted components might be located with a parts layout chart. Notice on the top side of the board that white and black line symbols identify each part (Fig. 14-11). Underneath the chassis, white dotted lines might outline the various circuits. Large LSI or microprocessors that have many terminals might have the part outline printed in white on the top of the chassis, but are mounted underneath the chassis.

■ **14-11** *Check on the top and bottom of the PC wiring for part symbols and white outlines of the various circuits to locate the defective component.*

Board replacement

Complete board replacement of the small consumer electronic products might be too expensive and not warrant repair. Separate board sections might be replaced at factory depots and manufacturer service stations. Board replacement on the TV chassis might be sent into various electronic repair firms that specialize in board repairs and replacements. For instance, board plug-in modules might be repaired in this manner. First, try to repair the board before condemning or attempting to receive board replacement. Removing and replacing components on the PC board occurs every day in the life of the electronics technician.

Required test equipment and how to use it

15

SERVICING ELECTRONIC PRODUCTS WITHIN THE CONSUMER electronic field can be most difficult without good test equipment. A combination of electronic know-how, experience, and correct test equipment will solve most electronic problems found in consumer electronics. Of course, you are not required to have several shelves full of different test equipment to collect dust. Only three or four test instruments are used daily by the electronics technician (Fig. 15-1).

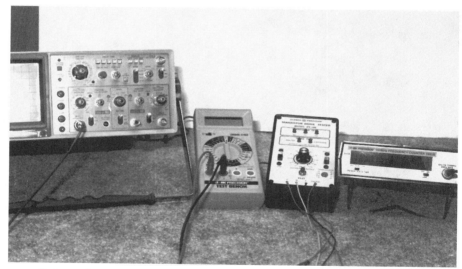

■ **15-1** *Only a few test instruments are needed for the intermediate electronic technician.*

Test equipment is quite expensive. At first, only a few test instruments are required to do the job. You can add a new test instrument to the bench each year to suit your electronics budget.

Read the test instrument manual over and over. Know how the equipment works and know what tests are made. Remember, the test instrument performs no better or worse than the electronics technician operating it. The more you know how the instrument works, the more products you can service in shorter periods of time.

The most-used test equipment

The VOM

You can purchase a volt-ohmmeter (VOM) for under $20.00. These instruments are fine for continuity and rough voltage measurements (Fig. 15-2). Choose a VOM with high resistance, voltage, and current ranges. Some VOMs have a reset pushbutton release when overload occurs and will not reset until the overload is eliminated. Choose a more expensive VOM ($175–250) with overload protection. The Simpson model 260 VOM series has provided rugged electronic servicing for many years.

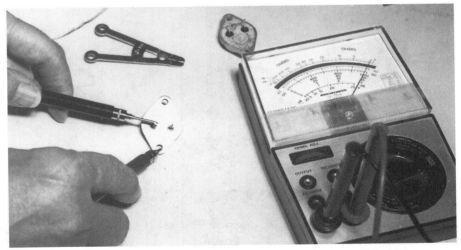

■ **15-2** *The volt-ohmmeter (VOM) was the first test instrument used to check voltage, resistance, and current in a circuit.*

The VOM is ideal when connected as an output meter in sound adjustment procedures. The analog meter will track the volume and audio signal of an audio oscillator for gain, level, and head alignment. The VOM outshines the DMM in making these adjustments because the numbers that move up and down the scale are difficult to follow upon the DMM display.

The VOM was followed by a vacuum tube voltmeter (VTVM). Because tubes are a scarce item, the VTVM has been replaced with a transistor or FET voltmeter. Because the FET transistor input has high impedance, like the vacuum tube (VTVM), the circuits are not loaded down when connected into the circuit to be measured. The electronic voltmeter, VOM, and VTVM are analog meters.

The FET analog meter is still available and might have a high impedance (up to 10 MΩ), making it ideal for many electronic applications (Fig. 15-3). Choose an FET meter with a large 4½-inch mirrored scale, jeweled meter movement, polarity reverse switch, zero scale movement, and overload protection for both meter and internal circuitry. Select the FET analog meter instead of a low-priced VOM meter for most audio and electronic measurements (Fig. 15-4). Tests are made with the VOM and digital multimeter (DMM).

■ **15-3** *The FET analog meter is a high-impedance meter that will not load down the electronic circuits when making critical electronic tests.*

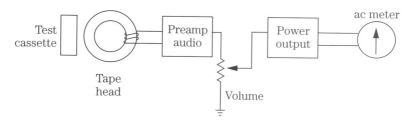

■ **15-4** *The FET meter can be used as an alignment indicator in radio and cassette player alignment procedures.*

The digital multimeter (DMM)

The digital multimeter (Fig. 15-5) is covered in Chapter 3.

■ **15-5** *The digital multimeter (DMM) is ideal for electronic servicing. It can be used to check voltages, resistance, current, diodes, transistors, and capacitance.*

Transistor-semiconductor tester

A low-priced semiconductor tester provides in- and out-of-circuit testing on transistors, FETs, SCRs, and diodes (Fig. 15-6). Several low-priced transistor and diode testers on the market will not test SCRs or triacs. These testers will identify if the transistor is an NPN or PNP type, if the FET is an N-channel or P-channel type, and will identify silicon-germanium diodes. Most semiconductor testers are priced under $100.00. The deluxe meter might have an audio tone or meter movement. The semiconductor tester can be found in kit form (see Chapter 8).

The oscilloscope

The cathode-ray tube or scope provides visual pictures of waveforms, time, frequency, phase, and voltage found on a small screen. When signal or input voltage is applied to the input terminal (y-axis), a waveform might be formed with the internal horizontal (x-axis) providing linear range voltage. The input signal or voltage

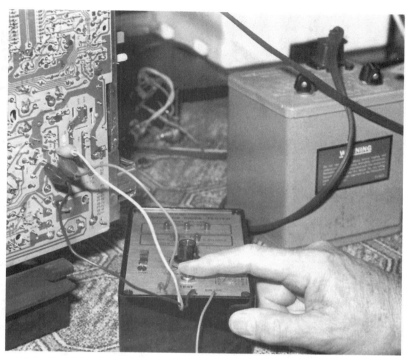

■ **15-6** *The transistor-diode checker will test transistors or diodes in and out of the electronic circuit.*

can be controlled with a gain control to enlarge or make the waveform smaller on a phosphor (fluorescent) screen. The scope is the ideal instrument when taking waveforms in the TV chassis. Choose a scope with a 5- or 6-inch rectangular screen (Fig. 15-7).

If you do not have an oscilloscope or one that is fairly old and will not respond to the various waveforms, select a 40- to 100-MHz dual-trace model. Today, the electronic circuitry demands higher bandwidth than ever before. Besides troubleshooting the TV chassis, higher bandwidth, delayed sweep and dual time base is needed for servicing VCRs, computers and other complex electronic products.

The scope can determine if a signal is present; the wave slope of square, sine, triangle, and pulse waveforms; basic voltage and peak-to-peak measurements; and frequency measurements. Two different audio signals might be observed at the same time with dual-trace operation. A loss of gain in one audio channel can be compared with the normal stereo channel. The amplitude of the waveform might be controlled by the vertical gain control.

■ **15-7** *The oscilloscope checks the various waveforms used in the TV, VCR, camcorder, CD player, and audio circuits.*

Always read the oscilloscope instruction manual before setting or making adjustments on the instrument. Keep the spot or base line at the lowest intensity when a signal is not moving on the screen. Do not keep the brightness too high when taking waveform measurements (Fig. 15-8). Keep the trace focused at all times and keep the waveforms in the center of the screen. Avoid bright sunlight, bench lights, and strong magnetic fields.

Learn how to use the scope and prevent it from setting on a shelf collecting dust. The oscilloscope can save valuable service time in troubleshooting the various circuits in consumer electronics. Besides the DMM, the scope is the electronics technician's best friend. Of course, service time can be wasted if the scope is used in servicing low-cost items, when other test instruments will fill the bill.

Plug in the scope, push in the power on/off switch, turn up the intensity control to bring up the brightness. Adjust the focus control

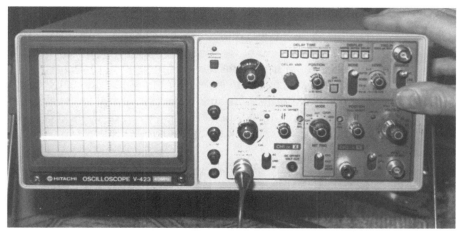

■ **15-8** *Lower the brightness on the scope screen when it is not in use or taking critical waveforms.*

until the display is the clearest. CH1 and CH2 are the BNC connector probes for signal input of the vertical x-axis. The variable gain control amplifies the picked up signal of the probe and can be varied for normal signals. The time/div. select switch and sweep variable control selects more than one square-wave and waveform horizontally on the screen. Follow the manufacturer's operation manual for typical scope operation (Table 15-1).

■ **Table 15-1**
A typical set-up chart to obtain a display on the oscilloscope.

Power	Off
Intensity	Counterclockwise or full on
Focus	Midrange
ac and dc	Gnd
Up and down position	Knob at midrange
V.mode	Auto
Trigger source	Int.
Display	Normal
Time/division	0.5 ms/Div
Horizontal position	Midrange

The most-used test equipment

Audio generator

The precision audio generator might have a frequency range of 10 Hz to 1 MHz in four or five different ranges, while the hand-held or portable generator might have a range of 20 to 220 Hz in four different bands. The audio generator might provide sine, square, and triangle waveforms. The sine-wave output voltage might vary from 1.2 V (no load) to 20 V peak to peak (Fig. 15-9). Square-wave output might vary from 5 V to 20 V peak to peak.

■ **15-9** *The audio signal generator provides a sine- or square-wave signal to troubleshoot the audio amplifier circuits.*

A function audio generator might contain an audio generator and digital frequency counter. Usually, both instruments can be purchased in one unit rather than found in single test instruments. Hand-held portable generators start at $60.00, audio generators sell for under $125.00, and combination audio and frequency counter under $250.00.

The audio signal generator can be used for signaltracing, signal injection, and audio alignment in the audio amplifier, stereo audio circuits, CD players, VCRs, camcorders, car stereos, and cassette players. A sine-wave waveform might be used for signaltracing and signal injection in any audio circuit. Locating distortion in the audio circuits can be located with a square-wave signal and oscilloscope. Always keep the signal generator output signal as low as possible to prevent clipping and distorting the waveform.

Inject a 1-kHz audio sine-wave signal into the input of an audio amplifier or cassette player and check each stage with the scope for loss of signal, or dead or weak reception. Connect the 1-kHz sine-wave signal to the audio input of a cassette player and check the output at the speaker with a frequency counter to determine the amplifier's frequency response and level adjustments. A test cassette with a 1-kHz or 3-kHz audio tone can be inserted into the cassette player and the frequency counter can be attached to the speaker output terminal to indicate the correct speed of the tape player.

By injecting a square-wave signal into the input or output stages of a stereo amplifier, you might determine what stage has distortion with clipped or rounded corners of the square wave on the oscilloscope. When a square-wave audio signal is injected into both the right and left channels of the stereo amplifier, both channels can be seen on a dual-trace oscilloscope (Fig. 15-10). Each stage can be checked within the stereo stages by connecting both scope inputs into the audio circuits at the same point in each audio channel, to show gain, and weak or distorted square waveforms.

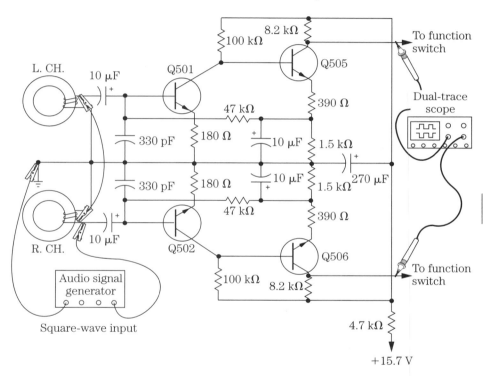

■ **15-10** *Both stereo channels can be checked at the same time with a square-wave signal at the input and a dual-trace scope in each audio circuit of audio amplifiers.*

Sweep generators

The sweep generator can be used as a standard function or sweep generator. This generator produces square, sine, triangle, ramp, and pulse waveforms. The sweep generator can be used for audio, stereo, TV, VCR, and camcorder repairs. You can purchase a digital function generator with the same features and a frequency counter in one test instrument for under $300.00.

The most-used test equipment

Besides injecting the sine, square, or triangle waveform into audio electronic products for signaltracing and alignment, the ramp and pulse sweep waveforms can be used in the TV and VCR circuits (Fig. 15-11).

■ **15-11** *The sweep generator sends out several different pulses to service TV, VCR, CD, and audio circuits.*

The sweep-marker generator, found in older TV shops, is collecting dust on the shelf because the present-day RF timer circuits have frequency-synthesis tuning (FS). The sweep-marker generator was used for front-end and IF alignment in the TV chassis. Today, the IF circuits have ceramic filter networks with fixed frequencies that seldom change frequency. The sweep-marker generator can be used as an FM-stereo generator for troubleshooting FM circuits in the AM/FM/MPX chassis.

RF signal generator

The RF signal generator provides stable RF and IF signals in servicing AM/FM receivers, signal injection, and alignment. A wide range of frequencies from 100 kHz to 450 MHz covers most RF circuits (Fig. 15-12). Some RF generators might also have a 1-kHz oscillator built-in modulated signal. Five or six individual bands might be switched in to cover the wide range of frequencies. The RF signal generator might be purchased separately or with a combination frequency counter and generator.

The RF signal generator can be used to provide complete RF and IF alignment for the AM and FM bands. Although the new AM/FM receivers have fixed IF frequencies (455 kHz and 10.7 MHz) with

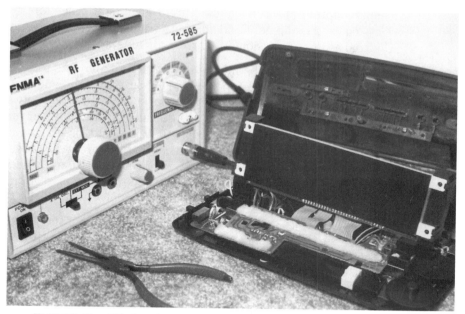

■ **15-12** *The RF signal generator is used to align RF IF circuits in AM/FM /MPX receivers.*

ceramic filter networks, the RF stages might need a touch-up in alignment. The IF signal can be injected at the mixer or at the IF stages for IF alignment (Fig. 15-13). Connect the signal generator output to the RF circuit of the AM and FM circuits for correct RF

■ **15-13** *Do not touch the AM or FM IF coils to improve either reception in the AM/FM/MPX radio. Use an RF signal generator for correct alignment.*

The most-used test equipment

alignment. Follow the manufacturer's alignment procedures when specialized equipment and adapters are used. Keep the generator signal as low as possible for correct alignment.

An RF signal might be injected into the radio circuits to isolate a defective stage and component. Make sure that the audio stages are normal and inject a 455-kHz signal at the base of the converter transistor in the AM circuits to determine if the detector and IF stages are normal. Likewise, inject 600-kHz and 1400-kHz RF signals at the antenna terminal to check the RF circuits. If a weaker signal is heard at the antenna circuit or at the base of RF transistor, inject an RF signal at the base of the mixer transistor or converter to determine if the RF transistor is defective.

Likewise, inject an RF signal into the FM circuits if the FM signal appears to be weak or dead. Inject a 10.7-MHz signal at the base of the mixer transistor or IC to determine if FM IF circuits are normal. Check the high end of the FM band with the tuning dial at 108 MHz and inject 108-MHz RF signal at the base of the RF transistor or IC. The 1-kHz modulated RF signal can be heard in the speaker with normal RF and IF circuits.

The RF signal generator is ideal when servicing RF IF circuits in the radio circuits, especially when someone has tried to adjust the IF transformer adjustments for weak reception. Do not turn any RF or IF adjustments unless correct alignment test instruments are connected (Fig. 15-14). Usually RF or IF alignment does not change frequency by itself. Someone diddles with the RF and IF core adjustments to improve the reception and throws the whole set out of alignment. You can purchase a reliable RF signal generator for under $150.00.

The RF signal generator might use the radio speaker, scope frequency counter, and output meter as indicators. The radio speaker and output meter are connected to the speaker output connections. Most RF and IF alignments are adjusted for maximum output of sound in the speaker or on the output meter. You might find a combined frequency counter and generator in one test instrument.

The frequency counter

The basic frequency counter test instrument counts the range of frequency in most consumer electronic products. A digital frequency counter might be purchased under $200.00 and in a multifunction generator and counter for under $300.00. A normal counting range might result from 10 Hz to 100 MHz. The multi-

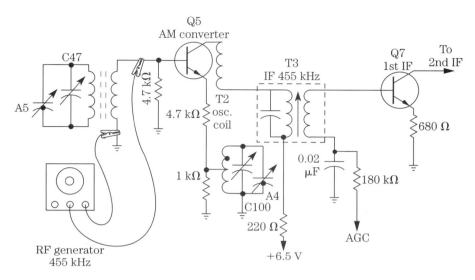

■ **15-14** *How to connect the 455-kHz IF signal from the generator for alignment or troubleshooting.*

function frequency counter might have a precision counting range of 10 Hz to 1 GHz. The digital frequency counter has a digital readout display (Fig. 15-15).

The frequency counter can be used to check or align the frequency in RF or audio circuits. Besides checking frequencies, the counter can be used to check the speed of tapes, VCRs, camcorders, and cassette player motors. The frequency counter can

■ **15-15** *The frequency counter can check for correct frequency in the TV, VCR, CD player, and cassette player. It can also be used to align equipment.*

The most-used test equipment

be used to determine the correct output frequency of an audio or RF signal generator.

Frequency counter accuracy should be checked every two or three years, by sending to the wholesaler or factory for correct calibration. You can check the frequency counter on a shortwave receiver using signals from government broadcast station WWV. These signals are very accurate at 2.5, 5, 10, 15, and 20.

Besides a frequency counter, an RF signal generator must be used with a receiver's S-meter and ac output meter. The RF signal generator and frequency counter are loosely coupled to the receiver antenna input terminal (Fig. 15-16). For instance, tune in a 5-MHz WWV station on the shortwave receiver with the S-meter at maximum. Also, the speaker or ac output meter (VOM) should be set for maximum audio and signal. Let all instruments warm up for at least 30 minutes before taking measurements.

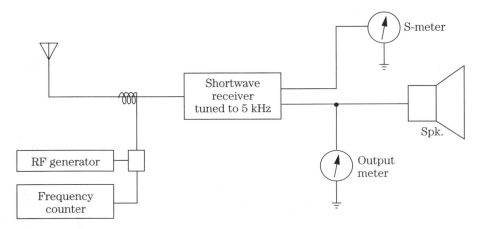

■ **15-16** *Frequency counter accuracy can be checked with a shortwave receiver, RF generator, and null indicator.*

Tune in WWV at 5 MHz with the maximum reading. Turn the signal generator to the 5-MHz frequency and adjust it for a whistle or beat tone in the speaker. If the generator and receiver signal (5 MHz) is exactly the same, a null or no signal can be heard or zero beat. Check the frequency on the digital frequency counter.

Another method is to check the frequency counter on a crystal-controlled oscillation kit or project. Choose a crystal at 5 kHz, found in most wholesale catalogs, for the oscillator frequency. You can add several crystals with a switching arrangement to check the various counter frequencies (Fig. 15-17). These frequency

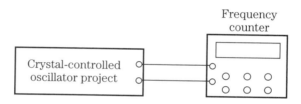

■ **15-17** *A crystal-controlled oscillator can also be used to check the frequency of the frequency counter.*

checks are accurate enough for most consumer electronic frequency checks.

Capacitance meter

Most capacity meters found inside the digital multimeter (DMM) might have a capacity range of 1 pF to 20 µF. Of course, this meter will not measure large filter capacitors found in today's electronic consumer electronic products (Fig. 15-18). Choose a capacitance meter or one with extended range to accurately measure most capacitors (0.1 pF to 20,000 µF). These meters can be purchased under $100.00. A battery-operated capacitor tester with a crystal-oscillator time base is best.

■ **15-18** *Choose a capacitance meter that will test the large capacitances that are used in the TV chassis.*

Before measuring capacitors in or out of the circuit, make sure that the voltage is discharged across the capacitor terminals. Keep ac products turned off with correct capacitor measurements. Select the correct range of the capacitor to be tested, found on the body of the capacitor. Connect the test leads across capacitor terminals and read the capacity value on the digital display.

An open or dried-up electrolytic capacitor will have a very low reading compared to the capacitor markings. The shorted or leaky capacitor might not indicate properly. You might assume that the electrolytic capacitor is normal if it is within a few microfarads of the original capacitor markings. Unsolder one capacitor terminal to accurately measure small bypass capacitors or if the previous measurement seems to be way off. Capacitance measurements out of the circuit are more accurate than those within-circuit tests (Fig. 15-19).

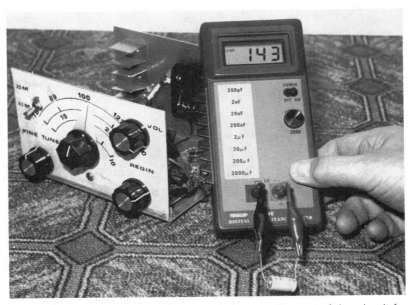

■ **15-19** *Check the suspected electrolytic capacitor out of the circuit for an accurate measurement. Here, a 150-µF capacitor has a reading of 143 µF.*

Color pattern generator

The color pattern generator provides a variety of test signals and patterns for TV monitors, camcorders, video disc players, and VCRs. The generator might provide a dot, crosshatch, color bars, a full-color raster, a blank raster, purity, and a wide white bar or a half screen white bar (Fig. 15-20). Originally, the color pattern

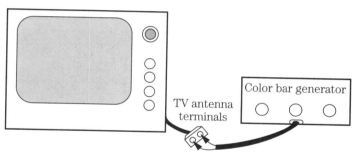

■ **15-20** *Connect the color pattern generator to the antenna terminal of the TV receiver and check for yoke level and pincushion effects (crosshatch), red color for the purity screen check, and the color base for the missing color.*

generator was designed to provide patterns for the TV chassis. The dot, crosshatch, and color bars were used to set up the TV color tube. The crosshatch or lines are used to level the picture while making adjustments on the yoke assembly.

The blank raster, purity, wide white bar, and half screen bars are used for purity, static, and dynamic convergence and screen adjustments. Correct adjustment of the yoke and position on the neck of the picture tube are used for purity adjustments. The color bars indicate what color is missing on the TV screen. Chroma adjustments can be made from red to green or from only a color raster.

Although the new TV receivers have a fixed convergence system, the dot and crosshatch patterns were used to adjust each color dot on each other to receive a white dot. The crosshatch pattern was used with convergence adjustment controls (Fig. 15-21). When all three colors are properly adjusted, the dot pattern and lines are white. The dot pattern is used for static convergence. The crosshatch pattern is used for dynamic convergence, and vertical and horizontal linearity. The crosshatch pattern is used to check pincushion circuits, with the same size squares, and bend-in lines at the outside edges of today's large picture tubes.

The color pattern generator can be used to service the various IF, video, and color circuits in the TV chassis. Scope waveforms taken within these circuits will quickly locate the defective stage, with the color generator connected to the TV antenna terminals. Like the Howard Sams Photofact waveforms of these circuits, these tests are performed with the color pattern generator. The color pattern signal generator has a horizontal sync output of 15,750 Hz with a chroma sub-carrier of 3.563795 MHz (colorburst frequency).

The most-used test equipment

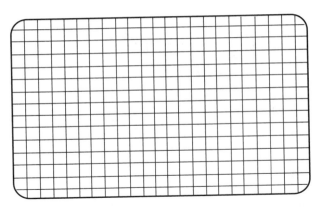

■ **15-21** *The crosshatch pattern will indicate if poor linearity is used at the middle and toward the outside ends of the picture tube.*

Although most color pattern test instruments are quite expensive, you might find portable units under $150.00. The overall performance of the TV chassis can be checked with the color pattern generator. The portable color pattern generator is ideal to take on service calls for TV adjustments in the home. The color pattern generator or NTSC generator is a must for the TV service bench.

NTSC color generator

Like the color pattern generator, the NTSC generator is used to service TVs, VCRs, and computer monitors. The NTSC generator produces sync pulses as those produced by a TV broadcast station. Besides TV patterns, the NTSC sync pulses allow proper adjustment of sound and switching circuits in VCRs.

The NTSC color generator produces NTSC color bar patterns with –1WQ, and gray, yellow, cyan, green, magenta, red, blue, and black with a split field –1WQ. The black-and-white reference signals can be used to adjust the black-and-white clip level of VCRs, video cameras, and camcorders. Besides the NTSC color patterns, the NTSC generator can provide a raster of red, green, blue, black, burst patterns and with a crosshatch pattern.

The RF IF outputs are crystal controlled and can be tuned in on channel 3 (61.25 MHz) and channel 4 (67.25 MHz). Most signals are modulated with a 1-kHz audio tone. The NTSC color generator is quite expensive compared to the color-pattern rainbow generator. The NTSC generator is ideal when servicing both VCR and TV units. The NTSC color generator can be added to the list of future test equipment purchases—especially in extensive servicing of VCRs.

Additional test equipment

CRT tester and restorer

The combination picture tube tester and analyzer that tests the condition of the CRT also restores the picture tube. Most CRT testers will check picture tube emission, leakage, and shorted gun assembly. Intermittent tests can be made with the tester connected and tapping lightly on the end of the CRT. The new CRT restorer-analyzers will test most picture tubes with several socket adapters.

A restoring function consists of removing shorts in the gun assembly, gun cleaning, balancing, and cathode rejuvenation. One color might have low emission or weak color and can be restored by removing electrons that pile on the cathode element. By stripping the cathode element, the tube might be restored for several months or years of service. You can see the firing action inside the gun assembly while it is in the restoring process (Fig. 15-22).

■ **15-22** *The CRT tester and restorer checks the condition of the picture tube and it also cleans the cathode elements in each color gun assembly for balance and weak emission.*

Restoring the picture tube might save a TV repair job. You might repair the chassis and find later that the picture tube is defective. Removing internal gun shorts and restoring the emission within seconds might place the TV chassis back into operation. This method is especially profitable when servicing used TV portables for resale purposes. It's best to purchase a CRT tester with restorer features for TV servicing.

Tube tester

Although tube testers are seldom used any more, they can be picked up at surplus stores, ham conventions, and from used electronic test instrument suppliers. The tube tester is needed to test emission and weak conditions of tubes that are still being used in audio amplifiers, intercoms, and industrial manufacturing (Fig. 15-23). Today, some new audio amplifiers contain many vacuum tubes.

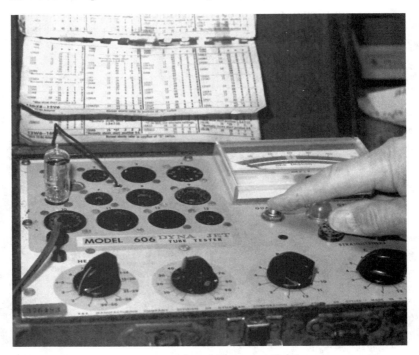

■ **15-23** *Testing tubes in a small emission tube checker.*

Years ago, you could substitute another new tube for the suspected one instead of having the tube tested. Of course, now tubes are hard to find and are quite expensive, so this method cannot be used. Check with your local TV technician when a tube tester is required because he or she might have an extra one collecting dust on the top shelf.

Isolation transformer

The isolation transformer provides power-line insulation while servicing the "hot" chassis. This isolation transformer should be used while servicing any electronic product that operates from the ac power line. The TV chassis is plugged into the isolation transformer to prevent shock, damaged test instruments, and TV circuits. These transformers should have a rating of 5 to 10 A (Fig. 15-24).

■ **15-24** *Always plug the electronic product to be serviced in the isolation transformer to prevent shock and damage to the equipment.*

A variable isolation transformer is ideal to service the chassis that might shut down from excessive HV, defective shutdown circuits, and intermittent symptoms. Some isolation transformers have a variable ac voltage with a switchable transformer winding. Others might use a separate isolation transformer with a variable power transformer plugged into the isolation transformer to provide isolation and also a variable line voltage. A built-in current and voltage meter might be found in several variable ac supply test instruments.

Variable power supply

The variable power supply should include at least two different variable power sources of 3 A. Both power sources should be capable of variable voltage of 0 to 30 V. Line and load regulation should be under 0.01 to 0.2% (Fig. 15-25). These power supplies can be purchased for under $200.00.

■ **15-25** *A variable power supply can furnish dc voltages to the various electronic circuits to be checked with the scope and DMM.*

Use the variable power supply to provide voltage for different power sources, motors, camcorders, and low-powered auto radio repairs. Two different low-voltage sources might be needed in servicing vertical and horizontal circuits in the TV chassis. Providing a dc source to the vertical or horizontal sweep IC might determine if the sweep IC, circuit, or stage is defective when the regular voltage source is taken from a derived or secondary winding of the flyback (Fig. 15-26). Variable dc voltage injection can help you to isolate and determine if a certain circuit is defective or normal.

Signal injector/tracer

The signal injector/tracer can troubleshoot audio circuits from input to the speaker. Some injection testers have a switchable detector for troubleshooting AM circuits (Fig. 15-27). Usually a 1-kHz tone provides injected signal with a VU meter and speaker as indicator. A continuously variable 1-kHz square-wave signal up to 5 V p-p can be injected. Speaker output and oscilloscope jacks might be provided.

The signal injector/tracer might be constructed of two different, separate, test instruments. The signal injection can be built for less money or from a kit to provide a 1-kHz tone for signal injection. An external monaural or stereo amplifier can be constructed for the signaltracer. Of course, the VU meter can be used for audio level and sound adjustments. The cost of a factory-assembled signaltracer might be less than $150.00.

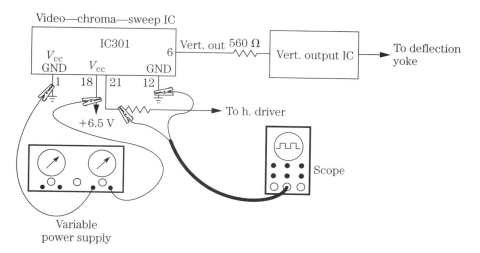

■ **15-26** *The variable power supply provides a supply voltage to the horizontal and sweep IC.*

■ **15-27** *The signal tracer/injector provides quick IF and AF signals for servicing the radio and audio circuits.*

Additional test equipment

Special test equipment

Tuner-subber

The tuner-subber is a transistorized tuner that operates from batteries and subs for a possible defective tuner in the TV chassis. With the tuner-subber, you can eliminate the tuner in a no-picture or snowy-raster situation. The subber determines if the original tuner, IF stages or AGC circuits are defective. Although the tuner-subber is no longer sold at electronics stores, they might be picked up as surplus or constructed with another transistorized tuner (Fig. 15-28).

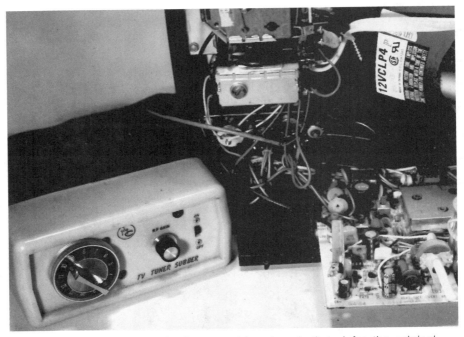

■ **15-28** *The transistorized tuner-subber is substituted for the original tuner. It indicates if the tuner, IF, or AGC circuits are defective.*

Simply clip the output of the tuner-subber to the IF cable or plug the IF cable into the subber. Most IF cables can be unplugged from the tuner, although some plugs might be soldered down. Connect the external antenna leads to the subber and turn the tester on. Rotate the channel selector to a local station and notice if the picture appears on the TV. If the picture is clear and clean, suspect that the original tuner or AGC circuits are faulty. A new tuner can now be ordered out or sent in for repair.

Laser power meter

The laser power meter measures the output of laser and infrared sources used in servicing CD players, video disc players, VCRs, and remote-control units (Fig. 15-29). A test on the laser head assembly might determine if the compact disc player is worth repairing. Most CD laser optical assemblies are so costly that many owners will dump the CD player instead of having it serviced. The electronics technician can save a lot of valuable service time and trouble by making certain that the CD head and beam assemblies are functioning with a laser diode head test.

■ **15-29** *The laser power meter will indicate if the optical head assembly of the CD player is emitting the correct infrared signal.*

Some technicians might use an infrared indicator card to check the laser diode. The infrared card might show that the infrared signal is present, but not the amount of infrared radiation. The laser power meter is ideal to measure the infrared signal and compare this measurement with the CD manufacturer's power requirements. The laser power meter will provide accurate infrared power measurements.

Several low-priced laser power meters are on the market for under $200.00. Most CD and laser manufacturers recommend and distribute laser power meters. The laser diode frequency wavelength is 750 to 820 nM in the compact disc player. The power-range settings should be at 0.3, 1, and 3 mW. The CD laser output might vary from 0.15 to 0.7 mW.

Insert the laser power meter probe between flapper or keeper and optical line assembly. A safety interlock must be defeated in the CD player before the table or portable CD player will emit infrared light. Move the meter probe around until the highest reading is obtained. Compare this laser power measurement with those found in the manufacturer's CD service manual. A power laser adjustment might be needed to compare with the service literature. A low laser measurement might indicate that the laser diode assembly is defective.

The laser power meter can be used to service TV remote-control units. Place the black probe against the end of the remote transmitter (Fig. 15-30). Press the remote control functions and measure it on the laser power meter. The 1- and 3-mW ranges might be used in remote-control tests. A weak remote infrared signal might only indicate on the lower power range, although a dead remote will not emit an infrared signal. Comparable remote transmitter tests can be marked on the scale for good and bad measurements.

■ **15-30** *The laser power meter might also be used to check the TV, VCR, CD player, and receiver infrared remote-control transmitter.*

Magnameter

The magnameter is a specialized test instrument to speed up and safely test microwave ovens. This instrument will test the oven high voltage (1.8 to 4.5 kV) and plate current with one setup. Once the test leads are connected, both current and high-voltage tests can be made (Fig. 15-31).

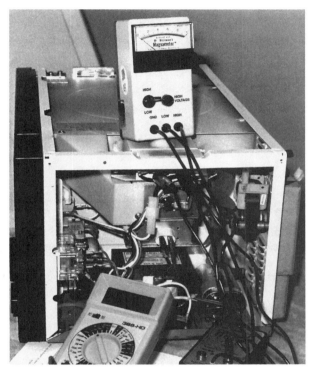

■ **15-31** *The magnameter is a specialized test instrument that is used to safely test high voltage and current in the microwave oven circuits.*

The magnameter can locate a leaky, low-emission open filament and overheated magnetron tube. Besides the magnetron, a leaky diode, and HV capacitor, shorted or open HV transformer, and fuse can be located with the magnameter. The biggest advantage of this meter is the safety factor in testing the high voltage applied to the magnetron.

Testing

Before making any oven tests, pull the ac cord, remove the outside wrap on the oven, and discharge the HV capacitor. Connect the black lead to ground (chassis). Clip the red lead to the HV capacitor or rectifier of the high-voltage side. Next, connect the green lead between plate resistor and magnetron tube plate lead wire. If no plate resistor is in the circuit, disconnect the HV diode ground lead and insert a 10-Ω test resistor in series with the diode (Fig. 15-32). Most ovens have a metal screw to hold the diode lead to the chassis ground.

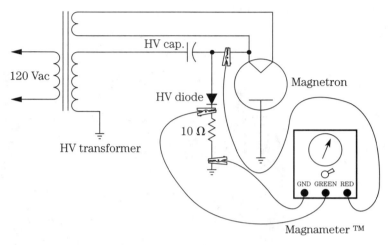

■ **15-32** *Insert a 10-Ω resistor in series with the HV diode and ground to check the current and HV in the voltage-doubler circuit.*

Insert the ac power cord, set the meter toggle switch to "high" and insert the water load into the microwave (usually 16 oz. of water). Turn the oven on the high-power setting and read the meter for correct high voltage. If the meter hand appears in the green, the microwave is operating OK.

A leaky magnetron, diode, or capacitor, or a shorted or open HV transformer, or an open HV fuse might show a measurement in the yellow meter area. If the meter hand appears in the red area, suspect an open magnetron, open filament, bad filament transformer, poor filament lead contacts, and open HV wire.

Turn the meter switch to the low position and read the small voltage or plate current. A home microwave oven should read 1.6 V to 4.5 V (160 to 450 mA). For commercial microwave ovens, the meter should read 2.0 V to 7.0 V (200 to 700 mA).

LCR meter

The LCR portable meter is a convenient way to accurately measure inductance, capacitance, and resistance. Inductance can be measured from 10 µH to 200 H. Measure capacitance from 10 mH to 2000 µH in 7 or 8 different ranges. Resistance might be checked from 20 Ω to 20 MΩ. The dual-display LCR meter provides simultaneous measurement of inductance and Q or capacitance and dissipation factor (Fig. 15-33).

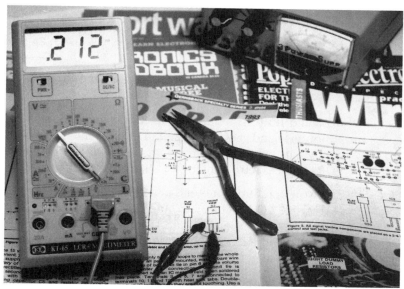

■ **15-33** *The LCR checks capacity, inductance, and resistance in or out of the circuit.*

Wow and flutter meter

Wow and flutter sounds might occur in the phonograph when the turntable changes speed. A simple cardboard strobe disc with lines indicates the slowing down, uneven, or fast speed of the turntable with a neon or fluorescent light above the indicator. The wow and flutter meter is ideal when servicing any record/playback equipment, such as VCRs, VTRs, reel to reels, and home or auto cassette players. The wow and flutter meter will indicate improper speed in motor-controlled audio and video equipment.

The wow sound might be a slow flutter, or an uneven or wavering rotation of audio. Flutter and wow are changes in pitch of an audio recording or reproduction system. Wow fluctuations in sound occur in a longer duration, while the flutter sounds are shorter in duration. These sounds are very annoying and they can be indicated with a wow and flutter meter (Fig. 15-34).

A 3-kHz test disc or tape with frequency counter at the output stages can locate most wow and flutter conditions. Usually the

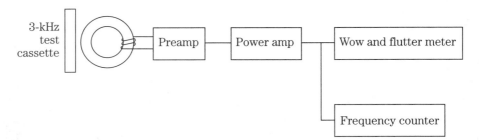

■ **15-34** *The wow and flutter meter might determine the pitch of an audio recording or reproduction system.*

wow and flutter meter is an accurate measurement in checking out precise and specialized sound equipment. The wow and flutter meter can be added a few years later to the service centers of VCRs and cassette players.

The wow and flutter meter should be compatible with JLS (Japan), NAB (USA), CCIR (France), and DIN (Germany). Japan, USA, and France use a 3-kHz tone and Germany has a 3.15-kHz test tone. The input impedance should be at least 330 kΩ at 3 and 3.15 kHz. This meter might have an analog meter or digital display.

The latest wow and flutter meter might have a signal oscillator source of 3 and 3.15 kHz. The enclosed frequency meter might have a frequency range of 10 Hz to 999.9 kHz. Accurate measurements of 0.005% to 3% are found in the wow and flutter meter. External outputs are needed for oscilloscopes and recorders. A good wow and flutter meter might cost above $1000.00.

The distortion meter

Usually, the distortion meter is found in specialized shops that service audio and sound equipment. The distortion meter might also be found in factory and audio servicing depots. The radio and TV establishment has no use for a distortion meter as long as clear sound can be heard in the audio circuits. In most shops, the distortion meter might sit on the top shelf and collect dust. Besides, a good distortion meter might start at $800.

The distortion meter measures distortion found in the audio amplifier circuits. Distortion is a change in the shape of a waveform and results in a form of harmonics. Extreme distortion might be found in the audio output circuits. A distortion meter is designed to measure the total distortion at any frequency between 20 Hz and 20 kHz. The enclosed dual-analog meter might act as a level meter, allowing simultaneous measurements of signal level and distortion.

The distortion instrument should measure distortion from 0.1% to 100% and signal levels from 1 mV RMS to 300 V RMS. The instrument should have automatic and hold features. Choose a distortion meter with x and y outputs to observe waveforms of the input signal and total harmonics with the scope.

Video head tester

The VCR and camcorder video head tester for VHS video heads provides a bridge measuring circuit to determine if the head is worn or defective. The analog meter might have a green and red scale to indicate if the head is good or bad. The measuring frequency might be 1 MHz with three different ranges, A = 0.2 to 3.5 µH, B = 8 to 3.0 µH, and C = 0.5 to 15 µH. A three-head tester can be purchased for $100.00.

Auto heavy-duty power supply

The variable dc power supply of 1 to 15 V up to 25 A is perfect for servicing and powering auto stereo products. The heavy-duty supply should have regulated, dependable voltage and current meters. Overcurrent protection automatically shuts off current flow when the maximum output exceeds 30 A. Some deluxe heavy-duty power supplies have separate current terminals for external current measurements.

The high-current/low-voltage power supply is a must for auto radio servicing. These power supplies can be purchased for under $200.00. Separate 5- and 10-A regulated power supplies with a fixed 13.8-Vdc source for auto receiver repair can be purchased for under $75.00 (Fig. 15-35). The high-current dc power supply can also be used to service high-power car stereos, car stereo accessories, camcorders, and ham radios. The high-current supply can be used to substitute the auto or trickle battery charging.

High-voltage probe

The high-voltage probe will measure the HV found at the anode terminal of the picture tube in the TV chassis. The latest high-voltage multiplier probes measure up to 40 kV with factory calibrated marking at 25 kV. A metal probe point must contact the HV button or socket on the CRT anode terminal (Fig. 15-36).

■ **15-35** *The small 3- to 4-A dc power supply can be used to service auto cassette players and radios.*

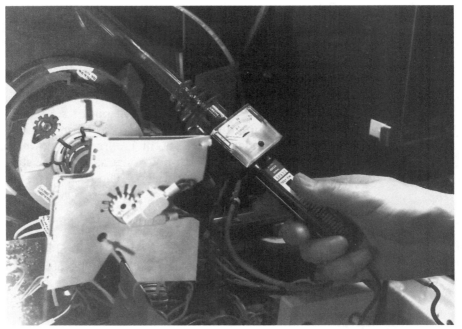

■ **15-36** *The HV probe is inserted under the rubber cover of the TV anode lead to measure the high voltage applied to the CRT.*

The metal probe must be inserted under the rubber CRT anode rubber cover to engage the anode terminal. Sometimes the HV lead and plug will snap off when inserting the test probe. The high voltage on the CRT lead might arcover to the TV chassis.

Always keep the ground wire of the test probe clipped to the metal TV chassis or to the ground strap on the picture tube. If not, you can receive a terrible shock by holding on to the probe tester. You might drop the high-voltage probe if the ground wire comes loose. Keep the ground wire attached at all times while taking HV measurements.

Specialized test equipment

Most specialized professional test equipment is quite expensive and can be added as needed on the service bench. These test instruments are designed for special fields of servicing (Fig. 15-37). For telephone repair, the B & K precision model 1050 telephone product analyzer and B & K telephone product tester model 1045A help solve telephone service problems.

■ **15-37** *Choose the correct test instrument for servicing TVs, VCRs, CD players, and waveform analyzers.*

Choose a B & K (120/SR) television frequency converter/modulator and a Sencore (TVA92) TV video analyzer for tough TV defective chassis. For video work, select a Sencore (CVA94) Video Tracker, VG91 universal video generator, and VA62A universal video analyzer for camcorder and video servicing.

To analyze circuits, select a B & K 545 PC board circuit analyzer or a Sencore 5C3100 Auto/Tracker waveform and circuit analyzer. The Leader NTSC vectorscope (5850C) speeds up in color TV and camcorder repair. Sencore's VC93 makes VCR troubleshooting and repair easy.

16
How to signaltrace various circuits

SIGNALTRACING IS ONE OF THE MOST EFFECTIVE METHODS to locate a defective stage in the electronic chassis. Signal injection or tracing tools consist of a noise generator, harmonic pencil injector, RF signal and audio generator, scope waveforms, and external audio amplifier tests. Usually, after locating the defective stage or circuit, voltage and resistance measurements help locate the defective component (Fig. 16-1).

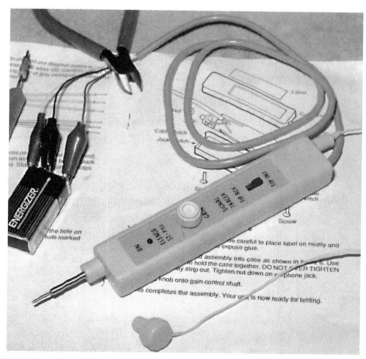

■ 16-1 *The simple harmonic signal generator and IC audio amp circuits with headphone output might help you to signaltrace defective circuits in the audio and RF circuits.*

Indicators

Besides a signaltracing test instrument, some type of indicator must be used to identify the results of loss or gain in a given circuit. When using the pencil or noise generator in the radio circuits, the radio speaker is the indicator. The speaker will indicate when no, low, or loud audio is found in the sound circuits. The ac meter can be used as an indicator in tape or cassette head alignment. The frequency counter can be used as an indicator in radio and audio circuits. The speaker or headphones in the external audio amplifier indicate a loss or gain in the audio circuits (Fig. 16-2).

■ **16-2** *A speaker, analog VOM, FET, and digital frequency counter might be used as a signal indicator.*

The greatest indicator in signaltracing might be the oscilloscope. Troubleshooting the various circuits within the TV chassis is made easy with scope waveforms. Distortion in sound circuits can be located with the oscilloscope. Troubleshooting with the oscilloscope is found in Chapter 17.

Signal injection

The various electronic circuits can be signaltraced by using signal injection methods. A 1-kHz pencil harmonic and noise generator can be injected in the various RF, oscillator, mixer, IF, and audio stages of a radio to locate the defective stage. The separate RF, audio, sine-wave, square-wave, sweep, and triangle generators can be injected in the various radio, TV, audio, and amplifier circuits. Today, most of these types of generators are found in one function generator test instrument.

The audio signal generator signals can be injected in the cassette, preamp, AF amps, and power audio output circuits. Besides checking for a loss of audio, the audio signal might be injected in the various circuits to locate extreme or low distortion. Signal injection can be injected at any RF or audio circuit to determine the condition of a given circuit.

Signaltracing

Signaltracing can be accomplished by using a radio, tape, cassette, or an audio signal as a generator and an audio amp or the enclosed amplifier circuit as an indicator (Fig. 16-3). Extreme distortion or a weak audio signal might be signaltraced through the suspected amplifier circuits with an external audio amp or scope. By tuning in a local radio station, the RF signal can be signaltraced with a demodulator probe and meter or audio amplifier. Likewise, insert a cassette tape and turn on the player to locate a weak, low, or distorted stage in the cassette player, with an external audio amp or scope as the indicator. By connecting an audio signal generator to both input stereo channels, the defective audio channel can be compared to the normal channel at any point in the audio circuits.

■ **16-3** *Signals from the radio and tape player can be used to signaltrace with the enclosed speaker as an indicator.*

Both signal injection and tracing can quickly locate the defective circuit.

Pencil and noise generator tests

The pencil or noise generator has a wide band of frequencies that can be used in signaltracing RF and audio circuits. Often, the pencil injector has no control over the amount of signal that is injected into the circuit. The volume must be controlled by the audio volume control of the radio receiver. The pencil generator is a quick injection method to locate a defective RF or audio circuit.

You might find a pencil generator and headphone amplifier found in one signaltracing test instrument. The signal is injected into the various circuits with the earphone used as amplifier and indicator. Here, a separate volume control is used to control the audio gain found in the various sound stages (Fig. 16-4).

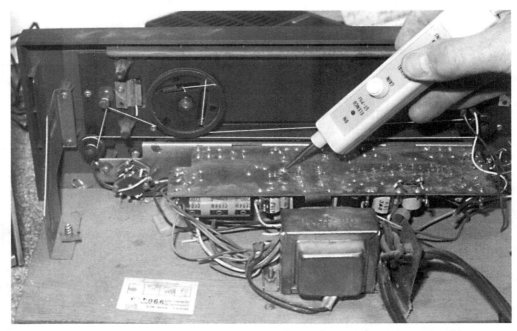

■ **16-4** *The pencil or noise generator can inject a signal into the radio AM or RF sections, with the speaker as indicator.*

The noise generator consists of a wide band of noise frequencies that can be used in signaltracing the RF, IF, and AF circuits of the AM/FM receiver. Always keep the volume control of the noise generator as low as possible when injecting noise into the various radio and audio circuits. Start at the volume control of the radio circuits and inject the noise generator signal in the detector, IF, and AM and FM RF circuits until the noisy signal stops or disappears (Fig. 16-5). After locating the defective circuit, take voltage, transistor, and resistance measurements to locate the defective component.

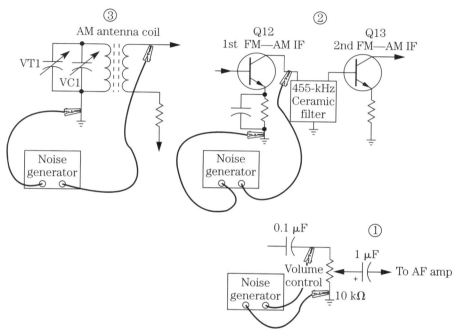

■ **16-5** *Inject the noise generator at the RF antenna coil, fast IF stage, or ceramic filter and volume control to signaltrace the radio and audio circuits.*

You can build your own noise generator with only a few electronic components. Place the noise generator in a metal base or cabinet to prevent external noise radiation. The noise generator can operate from batteries or an ac power supply. The noise generator works best with a 12- or 15-V source (Fig. 16-6).

The AM radio circuits can be signaltraced by injecting the noise generator at the top side of the RF variable capacitor or at the secondary coil of the antenna coil. Inject the noise generator into the base of the RF or FET transistor, then on to the 455-kHz ceramic filter or IF circuit (Fig. 16-7). Inject the signal into the input terminal of the AM/FM IF IC. The volume of the radio should be turned up at least halfway to register the noisy generator signal. Readjust the volume as the noise increases.

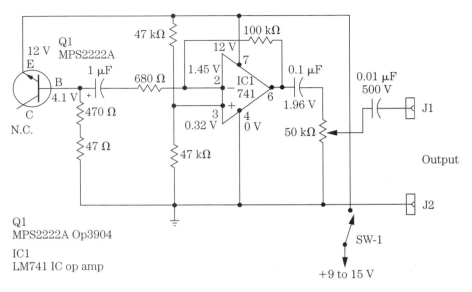

■ **16-6** *Build your own noise generator out of a few electronic parts. The higher the power supply source, the greater the noise generator output.*

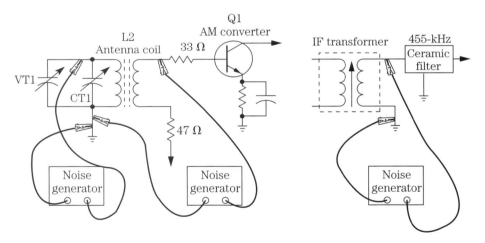

■ **16-7** *To signaltrace the RF stages of an AM radio, inject the noise generator signal at the RF stage, at the base of the converter transistor, and at the 455-kHz ceramic filter.*

When the noise becomes real loud, you have located the defective stage. Back up one stage and take voltage, transistor, and resistance measurements on the defective circuit. Remember, harmonic and a wide band of noise generator signals are not precise test instruments. But they might quickly locate the defective stage without connecting up several larger test instruments.

How to signaltrace various circuits

External audio amp signaltracing

The external audio amplifier can quickly locate a dead, weak audio, or distorted sound stage. The outside amplifier is used as an audio indicator, separate from the regular audio circuits. This external amplifier might be a one IC or it might be followed by stages that have several watts of amplification (Fig. 16-8). If the trouble lies in the audio stage, use the external audio amp to locate the defective stage.

■ **16-8** *The external audio amp for signaltracing might be a small one IC or one that will drive a larger speaker.*

Keep the volume as low as possible on the external amp. Remember, as you proceed through the audio circuits toward the audio output circuits, the audio volume should increase. To troubleshoot the audio stages, tune in a local radio station or insert a cassette and check each audio circuit. Start at the volume control and work both ways. If the volume or audio is normal at the high terminal of the volume control and weak at the speaker, proceed toward the speaker with the external audio amp (Fig. 16-9).

Proceed toward the tape preamp or radio circuits if the audio is low or lost at the volume control. The external audio amp can be connected directly to one side of the tape head circuit and proceed to the base of the preamp transistor or IC. The audio signal should increase as you proceed from tape head toward the volume control (Fig. 16-10).

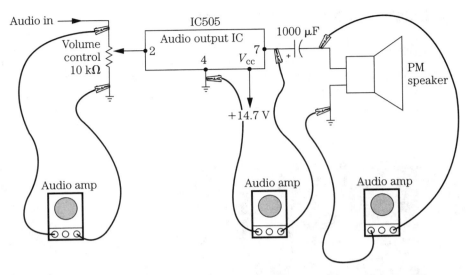

16-9 Check the audio at the volume control and proceed toward the speaker if no sound or low sound can be heard.

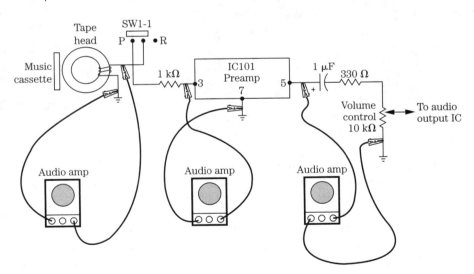

16-10 The external audio amp can be used to check the audio from the tape head through the preamp stages with a cassette playing.

Noisy amp

An external audio amp can be used to locate a weak or noisy stage in the AM/FM receiver, cassette and tape player. Often, transistors or ICs might appear noisy with a low frying noise and they might be difficult to locate. Turn the volume control down to determine if the noise disappears. If the noise can no longer be heard, signaltrace from the volume control back to the radio or cassette circuits. Go from stage to stage until you can no longer hear the noise. Check the preceding circuit for a noisy transistor, IC, or ceramic capacitor.

Replace the component that you suspect. If the noise is found on the collector and not in the base terminal, replace the transistor (Fig. 16-11). Transistor or voltage tests will not indicate a noisy component.

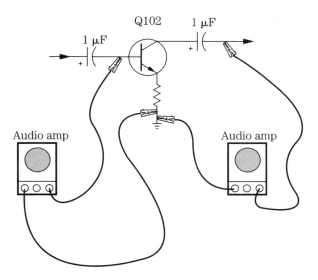

■ **16-11** *Low-frequency noises caused by a defective transistor, IC, or ceramic capacitor can be signaltraced with the external audio amp.*

A small loss of audio is very difficult to locate. Keep the volume control as low as possible on the external amp and notice the gain of volume after each audio stage. Because small electrolytic capacitors have a tendency to dry up, check the audio on both sides of the capacitor. Shunt another electrolytic capacitor across the suspected capacitor and notice if the sound increases. Compare the good channel with the weak channel at different points in the stereo audio circuits.

Audio signaltracing

The audio circuits can easily be signaltraced with an audio signal generator and an external audio amp. The audio generator might be a simple 1-kHz signal with a sine- and square-wave output (Fig.

■ **16-12** *The audio signal generator is ideal for signaltracing or aligning the audio circuits.*

16-12). The precision audio generator should have a frequency range of 10 Hz to 1 MHz. A hand-held audio generator might have a frequency range of 20 Hz to 150 kHz. You can purchase a precision audio oscillator and frequency counter in one instrument.

The audio signal generator sine- or square-wave signal might be injected at any stage in the audio circuits to locate a dead or distorted circuit. Always keep the volume as low as possible on the generator so as not to distort or clip the top and bottom sections of a sine or square wave (Fig. 16-13). Use the speaker as an indicator or connect the VTVM, FET, or analog VOM through a bypass capacitor at the audio output or speaker terminals. Rotate the ac meter to a lower range to indicate the amplified audio generator signal.

Distortion found in the audio stages might be located with the audio signal generator and scope. Clip the output terminals of the audio generator to the cassette tape head phono input or auxiliary input jack. Set the audio frequency at 1 kHz and keep the output attenuation control as low as possible. Rotate the control to the square-wave output. Although, a sine wave can detect audio distortion, the square wave shows many different forms of distortion.

Now scope the various audio stages and notice if the square wave is normal, rounded off, or has sharp edges. Compare the good channel with the distorted channel in the stereo audio amplifier.

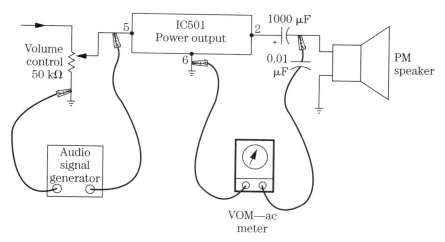

■ **16-13** *Inject the audio signal at the volume control, AF, and IC output circuits if there is no sound at the speaker. An FET or VOM can be used as an indicator across the speaker terminals.*

The square waveform might indicate a poor overall response, low-frequency shift, reduced high-frequency, poor high-frequency, poor-low frequency, reduced low-frequency gain, and narrow-band emphasis. If the volume is too high, you can distort or clip the waveforms by overdriving the audio stages (Fig. 16-14).

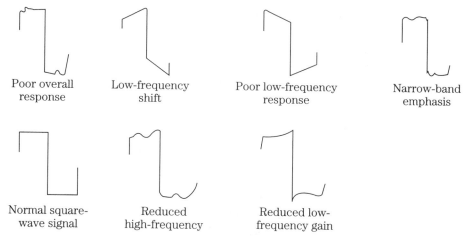

■ **16-14** *A square-wave signal injected at the audio amp input terminals and scoped at each stage might show the different signs of distortion.*

The oscilloscope is the ideal test instrument for audio troubleshooting and alignment. Inject the audio signal at the input audio circuit and check each stage with a scope waveform. Besides locating a dead or weak audio circuit, you can locate components that are causing distortion. Use the sine- or square-wave audio signal for signaltracing and the square wave for locating a distorted stage (Fig. 16-15). Usually extreme distortion occurs in the audio output stages. You can check both audio stereo circuits at the same time with a dual-trace oscilloscope.

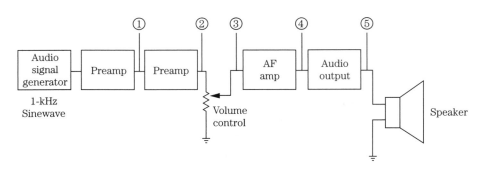

■ **16-15** *Inject the audio signal generator at the input of the cassette player preamp circuit. Check the waveform and amplitude of each section of the audio circuits with the scope.*

RF signal generator tracing

The RF signal generator can be used to signaltrace RF circuits in AM/FM/MPX receivers, TVs, CD players, and VCRs. The regular RF signal generator can provide RF signals from 100 kHz to 450 MHz. You might find an RF, sweep, and frequency counter in one digital test instrument.

The RF frequencies used in the AM radio band are 455 kHz (IF), 600 kHz (low), and 1400 kHz (high) band. The FM IF signal is 10.7 MHz, with 88 MHz (low), and 108 MHz at the high end of the FM band.

If the FM stations are not heard and the AM stations sound normal, suspect that a circuit in the FM stages is defective before both AM and FM signals are common to the IF stages. Inject a 10.7-MHz signal at the first 10.7-MHz IF or ceramic filter network, through a 0.001-µF ceramic capacitor. Make sure that the FM or AM signal is modulated so that you can hear a sound in the speaker. If the IF signal is audible, proceed to the base of the FM mixer transistor or IC with the 10.7-MHz IF signal (Fig. 16-16).

■ 16-16 *Locate the FM IF transformers or ceramic filter network to inject a 10.7-MHz RF signal to determine if the remaining FM circuits are normal.*

Start at the FM antenna terminal and clip the FM signal generator (set at 88 MHz) to the antenna coil and to common ground. Make sure that the receiver is tuned to 88 MHz on the dial. Rock the signal generator knob back and forth so that you can hear the RF FM signal. Suspect that an RF transistor or IC is leaky or defective if the signal is weak or missing.

The AM section of the receiver can be signaltraced in the very same manner. Inject a 455-kHz signal at the first IF transformer or ceramic filter network (Fig. 16-17). If a signal generator tone is audible, proceed to the base of the RF transistor or IC. Make sure the tuning gang-capacitor is fully open or tuned above 600 MHz. Suspect that an RF or converter transistor is defective if the IF signal cannot be heard in the speaker. Likewise, check the RF frequency with radio tuned to 1400 kHz and the generator set at the same frequency. Suspect that an RF transistor or IC is open or leaky if the AM signal is weak or missing. Take critical voltage and resistance measurements on the RF and converter stages. Do not over-

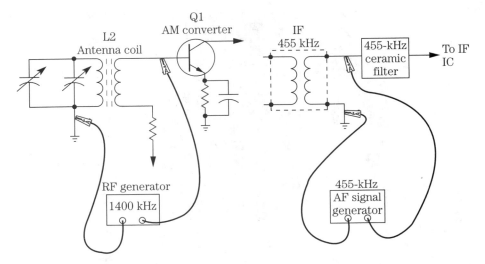

■ **16-17** *Inject a 455-kHz AM IF signal at the IF or ceramic filter to determine if the remaining AM stages are normal. Clip the RF signal generator to the base of the RF amp or converter with a 1400-kHz signal to determine if the RF or converter stages are normal.*

load the improper dc source feeding the RF circuits from the low-voltage power supply.

Scope waveform tests

Signaltracing with the oscilloscope is one of the best methods to troubleshoot circuits on the electronic chassis. You can actually see the results on the scope screen. Your eyes see it as it is. Sometimes your ears might mislead you. Scope waveforms can be taken on the radio, cassette player, CD, VCR, and TV chassis (Fig. 16-18).

Besides signaltracing, the scope can be used as an indicator to align and troubleshoot with audio and RF signals. The scope can be used to FM align the AM/FM/MPX chassis. Critical alignment of the tape azimuth and signaltracing the audio signals in the audio circuits can be taken with the oscilloscope. The scope can be used to check the various EMF, eye pattern, oscillator, servo, and control waveforms in the compact disc players. The servo, signal, and motor waveforms can easily be checked with the scope in the VCR. The oscilloscope is worth its weight in gold while servicing the horizontal, vertical, and color circuits in the TV chassis.

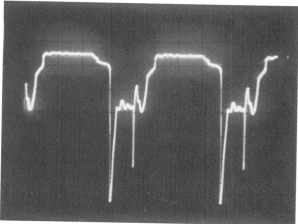

■ **16-18** *A normal scope waveform taken on the base terminal of the horizontal output transistor in the TV chassis.*

Power-supply voltage injection

Although the variable dc power supply can't be used as an indicator in troubleshooting, it can be used effectively by injecting a dc voltage in a given circuit (Fig. 16-19). Certain circuits in the TV chassis, such as the horizontal and vertical circuits, depend on secondary-derived circuits. These derived voltage sources appear at the secondary winding of the flyback transformer. When the

■ **16-19** *With the universal dc power supply, it might be difficult to inject the voltage into the various electronic circuits that might not be functioning.*

horizontal circuits are not functioning, several circuits that depend on the flyback voltage source are also not performing. Of course, any circuit can be energized by the external dc power supply source.

When other circuits depend on another power voltage source, the external power supply can be adjusted and injected into that circuit to see if it is functioning. Likewise, if the TV chassis shuts down right away, with no voltages from the flyback secondary circuits, inject a dc voltage into the horizontal oscillator, countdown, or deflection IC to see if the horizontal drive circuits are normal (Fig. 16-20). Scope the output terminal of the deflection IC for a horizontal drive pulse or waveform. Check the schematic for a correct voltage supplied to the supply voltage (V_{CC}) terminal of the deflection IC. Adjust the variable dc supply voltage to feed this voltage source.

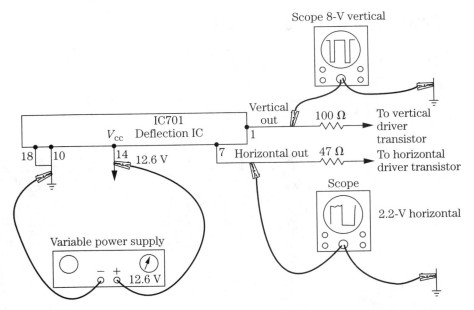

■ **16-20** *Inject 12.6 V at the supply pin (V_{CC}) of a horizontal deflection IC in the TV chassis to determine if the IC circuits are normal. Take both a horizontal and vertical drive waveform with the scope of pin 7 and 1.*

Inject or connect the dc voltage tied to the supply voltage source pin of the horizontal deflection IC and the common chassis ground. Take a scope waveform at the horizontal deflection drive pin. If the waveform is normal, you know the deflection IC was not causing the chassis shutdown symptom. At the same time, check the vertical drive waveform. By injecting dc voltage from an external power supply to a given circuit, you can determine if that circuit is functioning properly.

Random voltage measurements

Signaltracing the various electronic circuits can be accomplished by taking quick voltage measurements on transistors and ICs. A random collector and bias voltage test on each transistor can quickly determine if improper voltage is being applied to the transistor or if the transistor is defective. Take a quick voltage test on the collector terminal and it might indicate very little or no voltage from the voltage source. Low dc collector voltage might indicate that a transistor is leaky. Check the bias voltage between the base and emitter terminals and it might indicate that the transistor is leaky or shorted (Fig. 16-21). Also, this voltage test will indicate if the transistor is an NPN or PNP transistor.

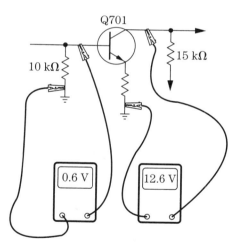

■ **16-21** *A random-voltage test on each transistor collector terminal might indicate that there is an improper voltage source or a defective transistor. A quick bias voltage test between the base and the emitter indicates that there is a normal, or open, or leaky transistor.*

A quick and critical voltage measurement on the supply voltage terminal of an IC will help you to determine if the IC is leaky or shorted. An improper or low voltage on the supply pins (V_{CC}) might indicate that the voltage source is defective. A random voltage test on the low-voltage power supply might indicate if power supply is defective or overloaded.

Usually, critical voltage and resistance measurements are taken after isolating and signaltracing the suspected circuit. When several transistors are found in an electronic circuit, random voltage measurements might quickly indicate a defective transistor or circuit. When only a couple of transistors or ICs are found in the suspected circuit, it might be best to take both a transistor test and a voltage test.

Random transistor tests

By quickly checking each transistor in the circuit, you can locate the defective circuit and component. Of course, randomly checking each transistor might take a little time by placing a probe or clip to each transistor terminal. You can speed this process up by taking transistor-junction tests with the diode test of the DMM. A quick measurement of the two probe tips from the base terminal to the collector and then to the emitter terminal might indicate that a transistor is open or leaky. You can quickly locate a defective transistor with the in-circuit transistor diode test of the digital multimeter (DMM) because only two prods are used with "touch-and-go" methods (Fig. 16-22).

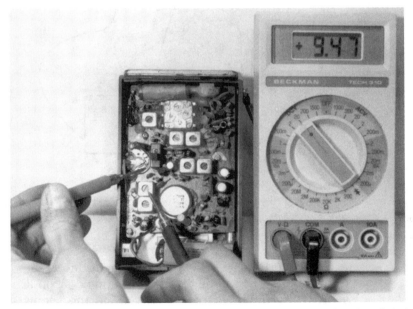

■ **16-22** *Taking a quick in-circuit junction transistor test with the digital multimeter (DMM).*

Do not try the transistor in-circuit tests if the symptom is intermittent conditions. The intermittent transistor might snap on when a probe tip touches one of the transistor terminals. Usually, in-circuit transistor tests on an intermittent transistor are not practical. Likewise, taking critical voltage measurements on the intermittent transistor might cause the transistor to return to normal operation. Intermittent transistors can be located with extreme cold and heat treatments, and then sneaking up on it with a critical voltage measurement.

Tracing signals in radio circuits

Signaltrace the radio circuits by the number. Use the harmonic or noise generator, RF generator, audio generator, and external audio amp (Fig. 16-23). Start at the volume control. If the radio signal can be heard at the volume control on external audio amp, proceed checking each audio circuit toward the speaker (1–5 num-

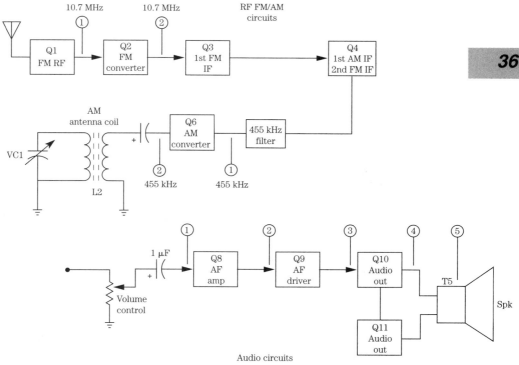

■ **16-23** *Signaltrace the radio circuits by the number with an RF signal generator and an audio generator. Signaltrace the audio section with the audio signal generator or external audio amp.*

bers). When no AM signal can be heard with normal FM reception, suspect the AM front-end stages. Inject a signal at the AM RF or converter transistor or IC and proceed toward the 455-kHz IF circuits (1 & 2 numbers). Likewise, signaltrace the case with no FM reception by injecting a 10.7-MHz signal from RF generator at the base of the mixer transistor or IC (1 & 2 numbers). Weak FM reception might be caused with leaky or open RF FM transistor.

After locating a weak or dead stage in either the audio, AM, or FM circuits, take critical voltage and resistance measurements. Check the suspected transistor in the circuit with a transistor tester or with the diode-junction test of the DMM. Critical resistance measurements might indicate that a transistor or capacitor is leaky, that the bias has changed, or that the resistance of the collector load resistors has changed.

Signaltracing audio stereo circuits

Signaltracing the stereo channels can be accomplished by injecting audio signal, signaltracing with external audio amp, scope, and test cassette. Any portion of a stereo audio channel can be injected with an audio signal and compared with the normal channel. Start at the volume control and signaltrace both ways, if needed.

The great advantage of the stereo circuits is the comparison of the two channels. Not only can an audio signal be injected to locate the defective stage, critical voltage and resistance measurements can be compared. If distortion is the symptom, inject a square-wave signal at the volume control and check each output stage with the scope. Keep the audio signal generator volume as low as possible for a good waveform. Likewise, check both audio channels at the input and output transistors or ICs and notice what stage or circuit develops the distorted waveform (Fig. 16-24).

If the left channel is weak compared to the right channel, insert a 1-kHz or 3-kHz audio test cassette and scope each stage. First, check the signal at both (top side) volume controls. Notice the difference in amplitude or height of waveform (Fig. 16-25). When the signal appears weak in the left channel at the volume control, proceed to the AF and preamp circuits. A quick comparison scope test on each stage should locate the defective circuit. When a test cassette is not handy, inject a 1-kHz sine- or square-wave signal from the audio signal generator at the tape heads.

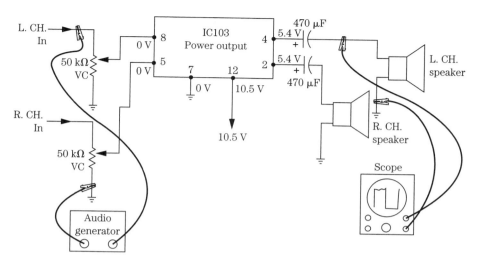

16-24 Signaltrace the audio output circuits with the audio signal generator signal injected at the volume control and scope each stage until you locate the distortion.

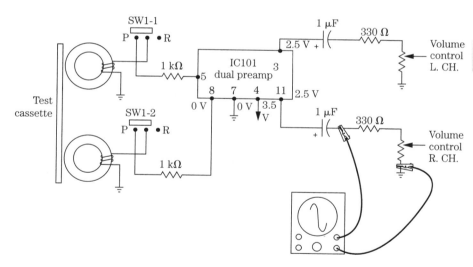

16-25 Check the height or amplitude of the scope waveform to locate a weak channel in a stereo amplifier within the preamp or AF amp circuits.

Conclusion

Signaltracing electronic circuits can be made quickly and very easily after making a few signaltracing tests. With a given symptom, go directly to the circuits involved and proceed with signaltracing to locate the defective circuit. Then take critical voltage and resistance measurements to locate the defective component. Locate the defective transistor with in-circuit transistor tests. Take critical voltage measurements on any IC that you suspect before you replace it.

Waveforms to the rescue 17

SCOPE WAVEFORMS MIGHT INDICATE IF THE STAGE OR circuit is normal or defective within the electronic circuits. The oscilloscope is an instrument that presents waveform inspection patterns on a screen representing variations within the electronic circuits. The dual-trace scope has two separate input cables. The cable input signal is applied to the y-axis (vertical), while the horizontal x-axis provides sweep inside the scope (Fig. 17-1).

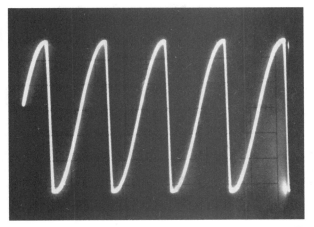

■ **17-1** *The vertical Y-axis provides the input signal of the oscilloscope.*

Always use the correct probe when taking electronic waveforms. Do not exceed the input voltage that might come in contact with the scope probe or arcover might occur and damage some scope components. The demodulator probe is used to check RF, IF, and detector circuits in the TV or radio chassis. A direct X1 or X10 probe (low signal level) is used to take waveforms in TV, CD, VCR, and audio circuits. Never attach the direct probe tip to a voltage higher than 300 Vdc or Vac. Some probes might have an input X10 probe rating of 400 Vdc or Vac.

Keep the brightness or intensity control as low as possible, to prevent eye strain or burn the fluorescent surface of the small CRT. When taking measurements within the circuits, keep the ground clip close to the probe tip area. This prevents distorting, ringing, and overshooting waveforms in rapidly rising signals or in a high-frequency signal. Read the scope instruction manual several times before taking critical waveforms.

Waveform tests in audio circuits

Use the oscilloscope to troubleshoot, check gain and distortion in the audio circuits. Stereo circuits are easy to service with a normal audio channel to compare to the defective one at any point in the circuits. Weak audio and low distortion are difficult to locate in the audio stages. These audio circuits can easily be signaltraced or waveforms can be taken throughout the sound circuits (Fig. 17-2).

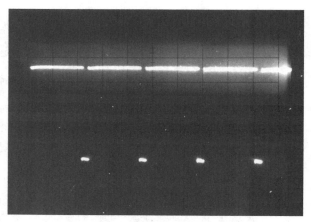

■ **17-2** *The square-wave generator is connected to the input phono auxiliary or to the tape head jack of the audio amplifier and the signal is tuned with the scope.*

To check out the whole audio circuit clip a square-wave generator to the tape, phono, auxiliary, or radio input jack. Clip the high side or the ungrounded wire from the generator to the input jack and ground clip to common ground. A short phone plug cable can be used to insert the signal into the input jack. Make sure that the input jack does not have a shorting-type contact if the generator probe is to be clipped directly to the amplifier input. Clip the generator wires to the input circuit when no jacks are provided.

For instance, if the symptom is distortion in the right channel with a normal left channel, check for distortion at the R CH volume control (Fig. 17-3). Likewise, check the left channel at the volume control for distortion. When both channels are normal at the volume control, suspect the right audio output channel. Notice the gain of both sets of square-wave signals on the scope screen. Distortion found in the right channel at the volume control indicates a defective preamp circuit.

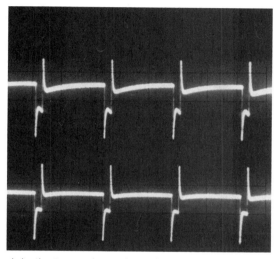

■ **17-3** *Check both stereo channels at the volume control and note which channel is distorted.*

Check the scope waveforms and notice if corners are rounded off or sharp peaks are added in the defective channel. If both channel waveforms are somewhat similar at the volume control, go directly to the audio output circuits (Fig. 17-4). Keep the signal input lower because greater amplification is found in output circuits. Check the output signal at the right speaker terminals. When the right output square waves are distorted, signaltrace each stage with the scope probe and compare them with the normal left channel.

Defective transistors, ICs, bias resistors, and electrolytic coupling capacitors produce distortion in the audio output circuits. In directly coupled transistor output circuits, a leaky Darlington or audio transistor can change the voltage source on each directly coupled transistor (Fig. 17-5). Open or poor component grounds can cause low distortion. Simply scope the input and output terminals of a power output IC for possible distortion.

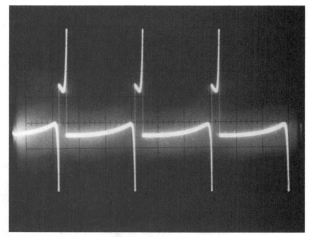

■ **17-4** *The waveform taken at the output of the amplifier with a speaker load attached.*

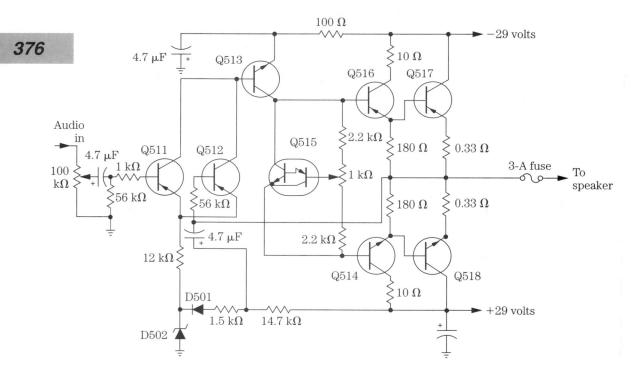

■ **17-5** *Suspect transistors, bias resistors, and improper voltages on the transistor element in directly coupled transistor audio output circuits.*

Waveforms to the rescue

When one of the channels has a small gain or height of a square wave compared to the other channel, at the volume control, suspect a defective preamp circuit. Again, take the probe tip and check the square-wave signal at the base and collector terminals of each preceding transistor. Likewise, check the input and output terminals of a preamp IC. When the square wave appears normal the defective stage is next in line. Take critical voltage, transistor, and resistance measurements of the defective stage.

Troubleshooting the radio circuits with the scope

The AM/FM/MPX receiver circuit can be signaltraced with the scope as an indicator. Connect the signal generator to the RF or IF circuits and to the oscilloscope after the detector diode. You might find a test point in some chassis. Clip the scope probe to a resistor or detector diode and ground clip it to the metal chassis or to the IF shielded transformer (Fig. 17-6).

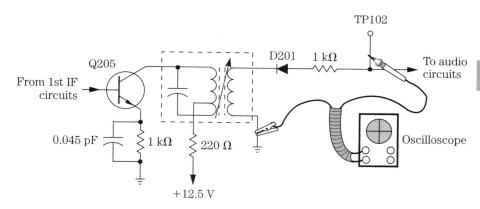

■ **17-6** *The oscilloscope can be used as an indicator while signaltracing or receiver aligning.*

Clip a 0.01-μF bypass capacitor to the RF signal generator and use it to inject a signal into the IF circuits. Set the signal generator at 455 kHz and turn the modulation signal on. You should be able to hear a tone in the speaker. Start at the base of the last IF transistor and proceed toward the front end (Fig. 17-7). If the AM receiver has only two IF stages, start at the second IF transistor. If the modulated 455-kHz signal is normal on the scope screen, proceed to the base of the first IF transistor.

Keep the signal generator signal as low as possible for a normal signal. When the signal cannot be seen, check that circuit for a loss

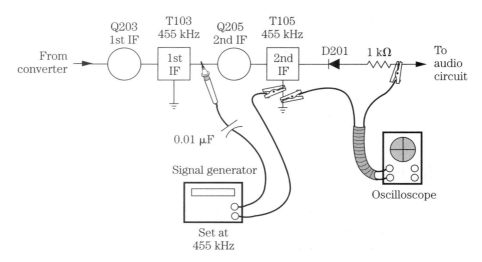

■ **17-7** *Injecting a 455-kHz signal at each base terminal of the IF and converter transistor can quickly help you to troubleshoot the receiver circuits.*

of radio signal. Take voltage, transistor, and resistance measurements. This same scope troubleshooting arrangement can be used for complete IF alignment. Adjust each IF coil for maximum signal on the oscilloscope (Fig. 17-8). Connect the signal generator at the base of the RF amp or converter transistor for alignment.

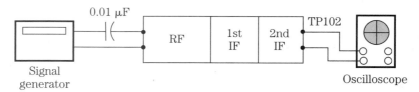

■ **17-8** *The oscilloscope and signal generator are used to align the RF and IF stages in the AM/FM/MPX receiver.*

Important TV sweep waveforms

Be careful when taking waveforms not to short the scope probe against two different terminals and cause damage in the sweep circuits. Dead chassis, shutdown, no picture, and no raster can result in no horizontal sweep. Also, a vertical straight line might indicate that the horizontal sweep circuits and high voltage are normal with an open horizontal yoke winding (Fig. 17-9). The horizontal circuits are working right up to the yoke assembly.

A horizontal white line on the raster indicates no vertical sweep. The three- or four-inch raster might provide insufficient vertical sweep.

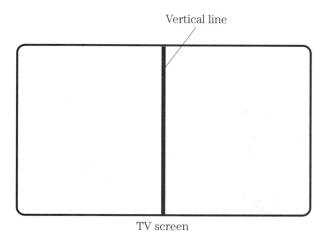

■ **17-9** *An open horizontal yoke winding might produce a white vertical line, indicating that the high-voltage and horizontal circuits are normal up to the yoke assembly.*

Horizontal bright lines at the top of the raster might be caused by an open resistor or by defective vertical output transistors.

Critical horizontal sweep waveforms

In the early 1950s, the horizontal oscillator circuits were developed with tubes. Later, transistors were used in the horizontal oscillator circuit. Today, both horizontal and vertical sweep might be found in one large IC with many other TV circuit functions (Fig. 17-10). Go directly to the horizontal sweep IC terminal when no horizontal sweep or high voltage is found in the TV circuits. Scope the horizontal and vertical waveforms to determine if the sweep circuits are functioning.

Check the waveform pin terminal at the sweep IC to determine if the horizontal sweep circuits are normal. The horizontal oscillator or sweep must operate before the waveform can be amplified and shaped for the driver or horizontal output circuits. If the square wave is normal at this terminal, proceed to the base terminal of the horizontal driver transistor (Fig. 17-11).

Suspect that an IC is defective or that no supply voltage is there if the horizontal drive pulse is missing from the sweep IC. Doublecheck the vertical oscillator waveform at the vertical drive pin terminal. You might assume the IC is defective or that insufficient supply voltage is at the sweep IC if both waveforms are missing. Check the supply voltage at terminal pin 15 of IC402 (Fig. 17-12).

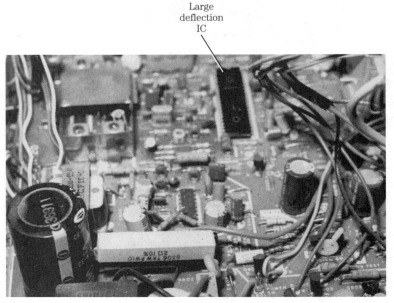

■ **17-10** *One large IC might contain the horizontal and vertical-deflection circuits, and many other TV circuits.*

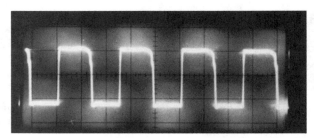

■ **17-11** *The horizontal drive square waveform taken with the scope at the horizontal output pin.*

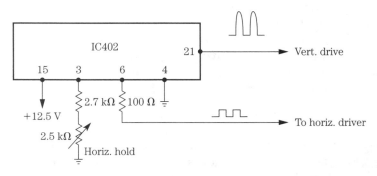

■ **17-12** *With no deflection, vertical or horizontal signal at pins 21 and 6, suspect a defective IC or an improper supply voltage at pin 15.*

With a normal horizontal oscillator waveform, touch the base terminal of horizontal driver transistor (Fig. 17-13). The horizontal driver stage amplifies or drives the horizontal signal through a transformer to the base of horizontal output transistor. The weak 0.5- or 0.8-V horizontal waveform on the base terminal can provide an 18- to 30-V drive voltage at the secondary winding of the driver transformer. An improper or insufficient driver waveform at the collector terminal of the driver transistor might be caused by a leaky transistor or by an improper collector voltage.

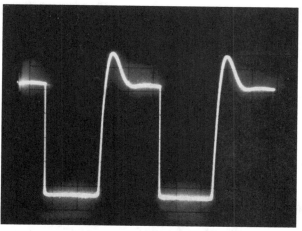

■ 17-13 *A normal horizontal waveform at the base terminal of the horizontal driver transistor.*

Next, touch the base terminal of the horizontal output transistor. This might be difficult in some TV chassis because the base and emitter terminals are shielded by the chassis or subchassis. If it is difficult to get at, locate the driver transistor and take the base measurement off of the secondary winding. Notice that one side of the secondary winding is at ground potential (Fig. 17-14).

An improper horizontal output base drive waveform might result from a leaky horizontal output transistor, open the horizontal output transistor, and overloaded circuits in the flyback circuits. Check for poorly soldered connections of the horizontal driver transformer. Suspect that the horizontal output circuits are defective if there is no raster, but a normal base drive waveform (Fig. 17-15).

Be very careful when taking horizontal output waveforms in the horizontal output circuits because of the high peak voltages. A horizontal output waveform can be taken from the collector terminal or metal body of most horizontal output transistors (Fig.

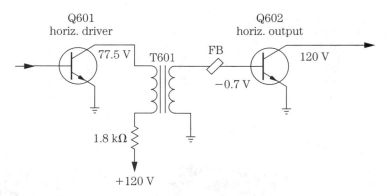

17-14 The horizontal driver transistor and transformer circuits connected to the horizontal output transistor.

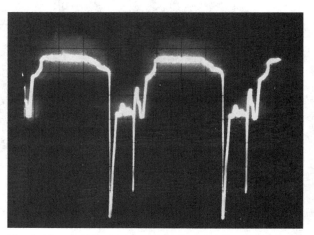

17-15 The horizontal base drive waveform at the base terminal of the horizontal output transistor. Often, when the base waveform is missing, the output transistor is damaged.

17-16). You should have a clean-cut waveform if the circuits are normal. Voltage measurements on the collector terminal should be avoided because most analog multimeters and DMMs will not respond at the collector terminal. For a quick test, place the probe near the flyback winding and it will indicate a large output waveform if the horizontal circuits are normal. Ringing waves might be noted at the bottom of the horizontal output waveform (Fig. 17-17).

Vertical sweep waveforms

The vertical circuits consist of a deflection oscillator or sweep waveform, vertical output transistors or IC, and vertical deflection

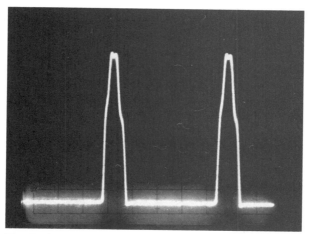

■ **17-16** *The horizontal output waveform at the collector terminal of the horizontal output transistor.*

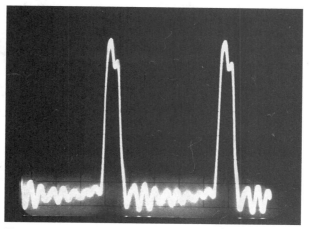

■ **17-17** *The horizontal output waveform taken with probe held next to the plastic flyback transformer.*

yoke winding. The early transistor chassis might consist of two or more transistors in a vertical oscillator circuit. Today, the vertical sweep signal might originate in the deflection IC circuits. The vertical sweep circuits might be included in one IC with the sync, x-ray protector, and video and chroma circuits. Vertical output waveforms are very unstable on the scope.

Check the vertical drive waveform at pin 26 of deflection IC801 (Fig. 17-18). If the signal is not found here, scope the horizontal driver pin to see if the deflection IC is normal. Often, vertical waveforms in the output circuits are not reliable if the vertical cir-

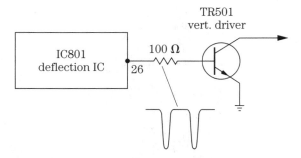

■ **17-18** *The vertical drive waveform taken from the deflection IC801 at pin 26.*

cuits are not functioning. A quick scope probe test at the vertical output emitter terminal or electrolytic coupling capacitor will indicate if the vertical circuits are functioning (Fig. 17-19).

■ **17-19** *The vertical output waveform taken at the vertical output stage or coupling capacitor.*

To quickly check the vertical circuits, take a waveform test at the deflection IC pin and make an output test at the vertical electrolytic coupling capacitor. If only a white line is shown on the raster with normal vertical output waveform, suspect a vertical yoke winding or return resistor is open.

In today's vertical circuits, you might find one IC as the vertical output component or the entire vertical circuits might be found in one complete IC. Simply scope the output IC deflection vertical waveform and the input at the power vertical output IC. Then check the vertical output pin of IC going to the vertical yoke wind-

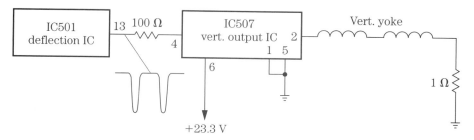

■ **17-20** *Usually, the vertical waveform is shaky and is sometimes difficult to take at pin 2 of IC507.*

ing. The vertical input and output waveforms should be the same as in transistorized vertical circuits (Fig. 17-20).

Important CD waveforms

Nine important waveforms are found in the compact disc player. The RF or EFM, tracking error, focus error, tracking coil, focus coil, PLL output, motor movement, SLED motor, spindle motor, and oscillator waveforms are critical waveforms. Most CD manufacturers have test points where waveforms can be taken (Fig. 17-21).

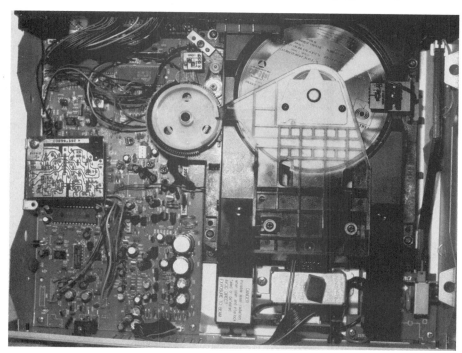

■ **17-21** *The layout of components in this CD player. Locate the correct components from a parts layout chart.*

RF or EFM signal waveforms

A normal RF, HF, or EFM waveform will indicate if the laser diode assembly and RF transistor or IC are functioning. The EFM waveform is taken at pin 20 of RF amp IC110 (Fig. 17-22). If the EFM signal is not present, the CD player might search and shut down. Check the laser diode and RF circuit for constant shutdown of the CD chassis.

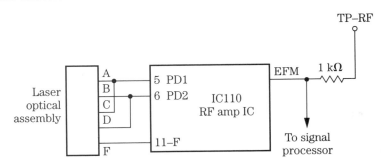

■ **17-22** *The EFM, HF, or RF waveform is taken after the first RF amp IC or transistor. Sometimes a test point is located for this test (TP RF).*

The normal EFM waveform should have a good eye pattern represented by a clear-cut diamond shape or outline. If the eye pattern is vibrating or jumping, poor adjustment or improper EFM signal might be found. Adjust the diamond-shaped area so that it is as clear as possible without excessive jitter (Fig. 17-23).

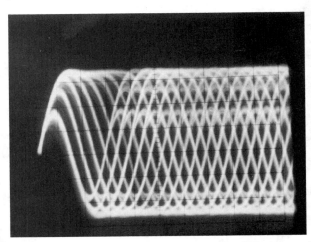

■ **17-23** *The EFM waveform must be present or the CD player might shut down.*

Waveforms to the rescue

Tracing error signal (TE)

Signaltrace the tracking error signal (TE) from the RF amp to the focus-tracking IC. Scope the input waveform (TE) and output waveform (TEO) at the IC (Fig. 17-24). Usually the tracking gain control is located at the TE input terminal of focus-tracking IC. Normal input and output waveforms indicate that the tracking signal is good at this point.

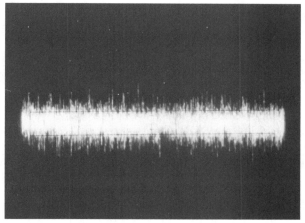

■ **17-24** *The tracking error (TE) waveform taken at the focus-tracking IC.*

When a normal tracking error signal (TE) is found at the input terminal and not at the output terminal of IC (TEO), suspect that there is an improper supply voltage or defective tracking-focus IC. Check the focus (FE) signal input and output before replacing the IC.

Tracking coil waveform

When the CD player is first turned on, the tracking and focus assembly can be seen searching—even if the CD player shuts down and ceases to operate. Often, this indicates that the focus and tracking assemblies are normal. The tracking output (TEO) can be scoped at the TEO terminal of the focus tracking - SLED servo IC. The tracking signal is fed to a tracking driver IC or to dual driver transistors. The driver output signal is applied to the tracking coil winding (Fig. 17-25).

If no signal or waveform is found at the tracking coil, suspect a driver transistor or IC, supply voltage, or improper tracking error output signal. Take critical voltage measurements on the driver transistor or IC.

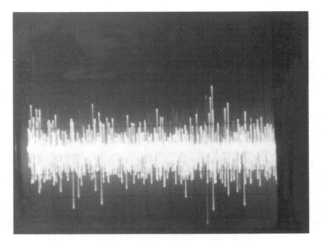

■ **17-25** *The tracking waveform taken from the tracking coil.*

Focus error waveform

The focus error waveform indicates that a correct focus signal is found from the RF amp to the focus error IC. Focus gain and offset adjustments are adjusted for maximum output. The focus offset might contain the RF and EFM adjustments, one of the same or separate adjustments found in different CD players. Check the focus error (FE) waveform at the input pin of focus servo IC (Fig. 17-26).

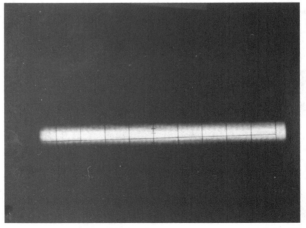

■ **17-26** *The focus error (FE) waveform at the input pin of the focus servo IC.*

Focus coil waveform

The focus coil waveform indicates that the focus signal is applied to the coil itself. The FE signal is sent from the focus-tracking servo to a focus driver IC or to transistors that drive the focus coil, which is located in the optical pickup assembly (Fig. 17-27). Only a white line will be noticed if the focus drive signal is missing at the focus coil. The focus and tracking coils can be checked with the RX1 ohmmeter range, with the power off. The focus error signal can be checked from the FE output terminal of the focus tracking servo IC to the focus driver and on the focus coil leads (Fig. 17-28).

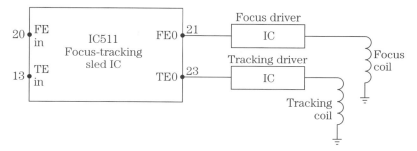

■ **17-27** *Both the focus and tracking coils are driven by IC504 and a separate driver IC or transistors for each coil winding.*

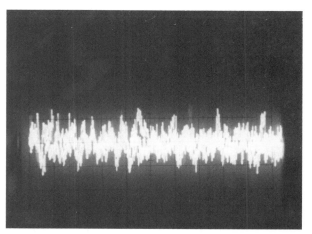

■ **17-28** *The focus coil waveform taken right on the focus coil lead terminals.*

PLL VCO waveforms

The phase locked-loop (PLL) voltage controlled oscillator (VCO) waveform indicates that the EFM signal has reached the digital signal processor and the servo IC. The VCO oscillator output signal is compared to the edge of the EFM signal, which is read from the disc. The PLL VCO oscillator can be adjusted with the oscilloscope and frequency counter. Some manufacturers use a frequency counter to set the frequency of the PLL VCO, and others make the adjustment with the oscilloscope (Fig. 17-29). Proper adjustment of the PLL frequency is needed so that the disc motor follows the optical lens assembly and responds to dropouts caused by scratches or defects on the surface of a CD disc.

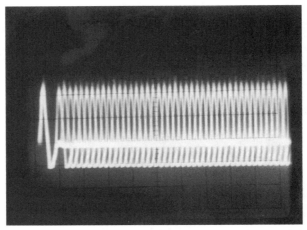

■ **17-29** *The PLL VCO waveform used on the PLL circuits of the digital control servo IC.*

No motor movement

No signal is applied to the motor when a scope waveform is taken at the motor terminals. Although a scope waveform will not indicate too much motor movement, several lines might be seen close together, indicating that the motor is rotating. A straight white line indicates that no signal is applied to the motor terminals (Fig. 17-30). Check the voltage at the motor terminals with a voltmeter.

SLED motor waveform

It is rather difficult to see a slide or SLED motor operating because the slide assembly moves rather slowly and is located under the main chassis. A quick waveform taken on the SLED motor terminals will indicate that the motor is operating with movement of several thin lines on the scope (Fig. 17-31). Trace the motor wires back to the main PC board, often, these wires are connected to a plug that sets in a socket on the PC board.

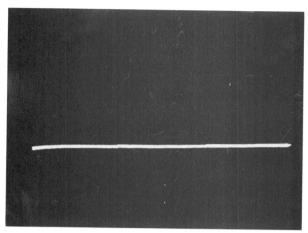

■ **17-30** *Only a white line on the scope screen indicates that there is no motor movement when waveform is taken on the motor terminals.*

■ **17-31** *The SLED or slide motor waveform is taken directly off of the motor winding terminals.*

Check the motor waveform at the motor plug connections. A sharp pointed probe tip is needed to get down inside the plug area. Clip the ground lead of probe to common ground because most motor windings are at ground potential.

The slide or SLED motor is controlled by a signal from the same focus-tracking and SLED servo IC. The slide motor control signal is applied to the slide IC amp, which consists of two transistors or IC drivers (Fig. 17-32). In some complex SLED circuits, push-pull driver transistors might be found in each leg of the SLED motor

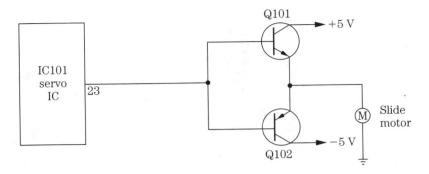

17-32 *The slide motor might be driven by two transistors or the driver IC might be controlled by the servo IC101.*

terminals. In this case, one side of the slide or SLED motor terminals are above ground potential.

Spindle motor waveform

The disc or spindle motor rotates the compact disc laying on the disc platform. Most spindle motors are controlled by a signal from the digital signal processor and the CLV servo IC. This disc or turntable motor operates at a variable speed, starting around 500 rpm and slows down as the laser pickup assembly moves toward the outer rim of the CD (approximately 200 rpm).

The master clock and spindle motor control circuits are fed to a spindle motor drive circuit. The disc motor is locked in with a correction signal from the constant linear velocity (CLV) circuitry. A defective spindle or disc motor might be dead or operate at intermittent speed. The spindle or disc motor waveform in a Sanyo CD player is shown in Fig. 17-33.

Conclusion

Critical waveforms in electronic circuits might indicate what circuit is dead during intermittent and improper operation. Waveforms can point out the defective circuit and component. Use the scope to help locate the defective stages in the radio, TV, CD player, and audio amplifier. Critical waveforms and knowing how to operate the oscilloscope is essential to become a better electronics technician.

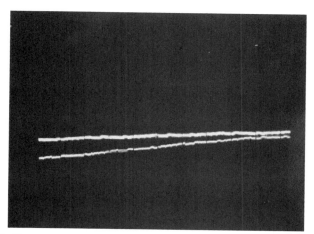

■ **17-33** *The spindle, disc, or turntable motor waveform at the motor terminals.*

Conclusion

18
Identifying and replacing unknown parts

LOCATING THE DEFECTIVE PART IS THE MOST DIFFICULT part of troubleshooting electronic circuits. Removing the component from its PC board is easy compared to finding the part number of an unknown component (Fig. 18-1). Testing the component before replacing it might save hours of frustration. Replacing the component takes time and careful attention by the electronics technician.

After locating the defective component, it must be checked for correct terminal connections, removed from the PC board, and replaced. Before the new component can be replaced, it must be replaced with a component that has the original part number (or is a universal replacement). If the component has a part number with a schematic nearby, locating the part might be easy or quite difficult with older and antique chassis. Simply look up the part in the schematic parts list or with transistorized components within the semiconductor replacement manuals.

Components without part numbers are difficult to locate. The application of unknown semiconductors can be looked up in the semiconductor manuals. Checking parts from one chassis to another might help you to locate the defective part.

Components found in Japanese and other foreign-made products were, at one time, difficult to find. Today, many manufacturers, electronic part wholesalers, manufacturer depots, and mail-order firms have direct foreign replacement parts. Sometimes it might take several months before a replacement can be found. The service technician should exhaust all methods of obtaining a replacement component before totaling out the entire product.

■ **18-1** *Removing the component from the PC board is much easier than finding the unknown part number.*

IC body part numbers

Sometimes you can locate the section that a defective component is in by looking at the part number stamped on top. The large IC (TA7644BP) shown in Fig. 18-2 contains the video, AGC, sync, color, brightness, contrast, horizontal, and vertical. Likewise, the vertical output IC might have the part number stamped on the body area. Not only will the part number show you what circuit it works in, but it will also show you the correct replacement.

Simply look up the part number in a semiconductor manual for the correct universal replacement. You can look at another chassis of the same brand, locate the same IC, and compare the voltage measurements with the other schematics.

Part numbers and letters found on ICs and transistors will help you locate the correct circuit or component, and sometimes the operating voltages. RCA, GE, Sylvania, Workman, and NTE semiconductor replacement manuals cover American, European, and Japanese solid-state devices. Howard Sams has a semiconductor cross reference data book for replacement and substitution of semiconductors. These manuals can be obtained from manufacturers, electronics stores, dealers, and mail-order firms.

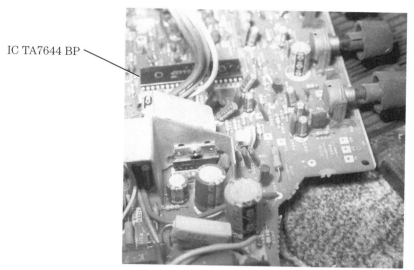

■ **18-2** *The large IC (TA7644BP) has the part number printed on top.*

Unknown transistors

You might need to test transistors without any part number on the body or PC board several times before finding a proper replacement. You must know if the transistor is a PNP or NPN type, what the working voltage is, and what stage or circuit it performs in. There are many different replacements for AF or diode driver transistors (Fig. 18-3). Look up the transistor in a schematic, if available. Check Sams Photofacts for a circuit diagram.

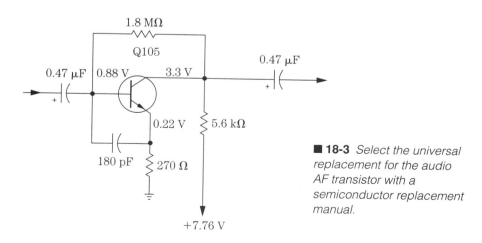

■ **18-3** *Select the universal replacement for the audio AF transistor with a semiconductor replacement manual.*

If you still cannot locate the type or part number of the transistor, check first for an NPN. Today, the majority of transistors are NPNs. Test the transistor with a transistor tester or with the diode-junction test of the digital multimeter (DMM). Exchange the test leads on the transistor until a reading is found. Locate the base terminal with the DMM diode-junction test. Remember, the base terminal is common with the emitter and collector terminals.

Another method to check for the different transistor terminals is with voltage measurements (Fig. 18-4). If the transistor is in the circuit with the highest dc positive voltage, you have located the collector terminal and the transistor is an NPN. Usually, if the highest negative voltage is at the collector, the transistor is a PNP. The audio transistor emitter element is tied to a low-resistance emitter resistor and ground.

The lowest junction DMM test is between base and emitter terminals. For instance, with a horizontal output transistor, the junction reading is 0.471 Ω from base to collector and 0.536 Ω from base to emitter terminal (Fig. 18-5). Now these measurements are not ex-

■ **18-4** *Check the collector voltage on the transistor for the correct universal AF replacement.*

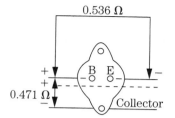

18-5 The normal measurements of a horizontal output transistor taken between the terminals with the DMM.

actly the same on each and every horizontal output transistor, but they will show the big differences between the two elements. At least you can tell which element is the collector and which is the emitter.

You can look up the dimensional outlines and terminal diagrams in the semiconductor manual for the base, emitter, and collector terminals. On the flat side of some power output transistors, the E and B letters are stamped right onto the metal base. This applies to power transistors with T-040, T-041, T-042, and T-043 base diagrams.

Transistor working voltages

After determining if the transistor is either an NPN or PNP and has no markings, determine what circuit and voltage the transistor is found in. Check alongside the cabinet or open the bottom area for a parts layout. Remember how each part was labeled inside the TV or radio, when no schematic is available.

For example, you have a transistorized boom-box or table receiver on the bench with no AM reception. The radio has normal FM and FM/MPX or stereo operation. You pinpoint the possible trouble in the AM converter section because both the AM and FM sections usually have the IF circuits in common. You locate the AM converter transistor that is mounted close by or trace the antenna coil to the nearest transistor base terminal (Fig. 18-6).

Because you have already found the base terminal, check the other two elements to determine if it is an NPN or PNP transistor with the junction-diode test of DMM. You find that the transistor is an NPN. Prove it by measuring the highest voltage on the transistor terminals (collector). Go a step further and check the voltages on the base and emitter terminals. Usually, the base voltage has a higher positive voltage than the emitter terminal. Check for the lowest resistance to ground for the correct emitter terminal.

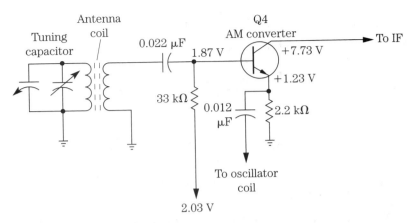

■ **18-6** Locate the antenna ferrite coil and trace the wiring to the base terminal to find the AM converter transistor.

If the transistor is defective and leaky, the voltages are quite close on two elements. If the transistor becomes leaky, it will probably occur between the collector and emitter terminals. If all three elements measure about the same voltage, the transistor might be leaky between all three elements. Measure the resistance to ground between all elements. Often the lowest resistance is the emitter terminals (Fig. 18-7).

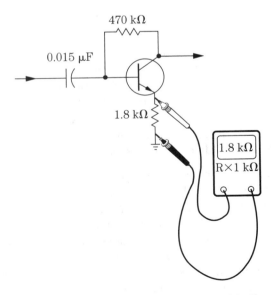

■ **18-7** Often, the lowest resistance measurement to the chassis ground might indicate the emitter terminal.

Identifying and replacing unknown parts

Disconnect one of the other two terminals (base and collector) from the circuit. If the voltage goes quite high where the lead was connected, this is the collector terminal. This voltage might equal the total supply voltage of the power supply or voltage network. Now the other terminal must be the base terminal. Before removing the transistor mark the transistor terminal leads on the chassis, PC board, or on a separate sheet of paper with a few connecting parts (Fig. 18-8). After you have located the unknown transistor replacement terminal, decide on the working voltage. Choose a converter transistor with a voltage that is not lower than the power source or battery. Measure the voltage source that is connected to the collector terminal.

■ **18-8** *Mark or locate the transistor terminals on the PC board before removing it.*

You know that the transistor works in a converter stage, NPN, and working voltage less than 10 V. Check the semiconductor manual for a list of converter or RF transistors that will work in the radio circuit. A ECG123AP, GE123AP, NTE123AP, SK3854, or WEP736 Workman transistor will operate in this converter circuit. This AF/RF small-signal amplifier (NPN), with a collector-to-emitter voltage of less than 40 V, is the correct replacement.

Power output transistors working voltages

Large audio and power output transistors are much easier to locate and replace. Often, they have separate heatsinks or they are mounted on a metal heatsink. Most power output transistors have part or transistor numbers stamped on the body of the metal transistor. If the power transistor is removed from a socket, the replacement is mounted back in the same way (Fig. 18-9). Often, the collector terminal is the metal body, except on some Sanyo and Sears horizontal output transistors.

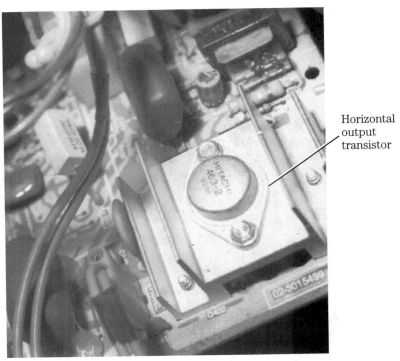

■ **18-9** *Replace the power output transistor that is mounted in the socket. This can only be mounted correctly one way.*

Mark down the color of the leads that are soldered directly to the transistor terminal, which does not mount in a transistor socket. Take the outline of the power transistor and match the universal and transistor terminals on the transistor base drawing. The elements and working voltage can be checked in the same manner as the small signal transistor; except that the power transistors operate at a higher wattage, voltage, and temperature.

For example, the horizontal output transistor in the TV set was found to be leaky between the collector and emitter terminals. The number found on the transistor was 2SD950. Looking on the RCA and Sylvania semiconductor manuals, the SK3710 or ECG369 universal replacements will work. If the number is unknown, look up the application (TV horizontal deflector) and breakdown voltages. The breakdown voltage between the collector and base is 1500 V, and 600 V is between the collector and emitter. It has a power dissipation of 50 W. If the transistor has a damper diode inside, replace it with the same type of transistor (Fig. 18-10).

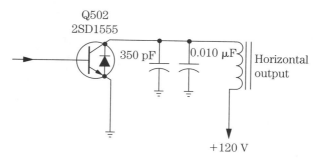

■ **18-10** *Make sure that the damper diode is mounted outside or replace the horizontal output transistor with the damper diode inside of the body of the transistor.*

When the audio output transistor part number is unknown, check in the semiconductor manual for AF power output stages and high-fidelity audio amplifier stages application. If the transistor is a silicon NPN and operates in a 100-W amplifier with 60 V on the collector, choose an output transistor with 150 W with 180 V between collector and emitter (SK3636). Cross reference the part number on the metal case with the semiconductor manual for the correct universal replacement. Check each servicing chapter for universal transistor charts.

Different model, same chassis

The electronics manufacturers might use several different chassis that are the same, year in and year out. The cabinet and outside dressing are changed, but the insides are essentially the same. In fact, the same chassis might be found in other brands of TVs, VCRs, CD players, cassette players, and radios. Compare these schematic diagrams with the one that you are servicing to find the correct part.

Let's take the RCA XL100 chassis, for example, which can be found in many different TVs. The CTC87, CTC97, CTC107, CTC108, and CTC109 have practically the same circuits (Fig. 18-11). Although you might find some modifications, they have the same basic circuits. If you have a schematic for an RCA CTC107 chassis, why not

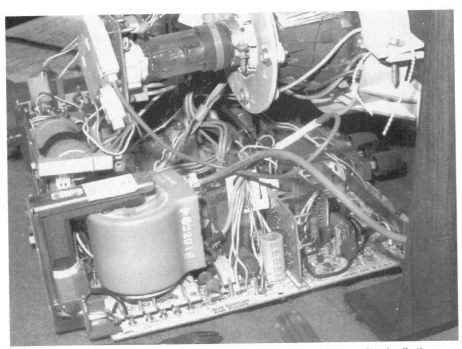

■ **18-11** *The RCA CTC87, CTC97, and CTC107 chassis have basically the same circuitry as the CTC109 chassis.*

use it for the other chassis when the correct schematic is not handy?

Radio circuits are quite common in the same brand receiver with a different model number. Transistors, ICs, and small components can be used in the chassis that has no number on the part or no schematic. Usually large components are not interchangeable, such as transformers, coils, and special types of dc motors.

If the exact schematic is not available, look in another schematic from the same manufacturer that uses the same part number. Compare the voltage measurements to those of the defective chassis. Sometimes the same part in other brands will have about the same voltage measurements.

Unknown IC part number

When no numbers are found on the IC or alongside, it is difficult to find a replacement. Try to locate a schematic for the unit or one that is similar. Locate a diagram with the same type of circuit and stage that the IC performs in. Notice how many pins are at the bottom of the IC. Check the semiconductor manual for a replacement. Make sure that the replacement has the same amount of pins.

For instance, in a Sears 304.21460350 boom-box, the symptom was no AM or FM reception. Because both AM and FM reception were lost, the IF stages are common. The AM or FM LED tuning light was not lit. Both the FM IF and AM IF (10.7 MHz and 455 kHz) generator signals were injected into the AM/FM IC, but no sound could be heard. The supply voltage was fairly normal. The common IF IC needed to be replaced.

The numbers on top of the IC were not clearly visible because only the first two letters were distinct (HA-). Upon checking the RCA series semiconductor manual, under applications, a universal replacement was selected (SK7605). This does take a little time to look up the universal replacement. Another method is to check if this Sears model number is found in Howard Sams Photofacts. If so, the part number will be listed on the schematic.

Replacing the defective transistor

Before removing the transistor or IC from the PC board, mark down the element terminals. Either mark the transistor terminals on top or at the bottom of the board or both. Mark terminal one or the front part of the defective IC before removing it from the board or socket. When looking down at the "U" or white line at the end of the IC, terminal one is the first one on the right (Fig. 18-12). Some ICs have an indentation or white mark for terminal one. You might find that transistor and IC terminals are marked on the top or bottom of the PC board.

Be careful when removing transistor or IC terminals so that you don't apply excessive heat and pull off some of the PC wiring. The PC wiring can easily be pulled off of the transistor or IC terminals, if it is not entirely unsoldered. Suck up all loose solder at the bottom of the PC board and terminals (Fig. 18-13). After the component is removed, clean off all excess solder and brush off the rosin residue.

Test the transistor after it has been removed from the circuit. Make sure where each terminal lead was mounted. If the transistor is special and has to be ordered out, keep the transistor terminals marked on the back of the work ticket or on a piece of paper. You might not remember where it went if it takes several weeks or months to obtain the part.

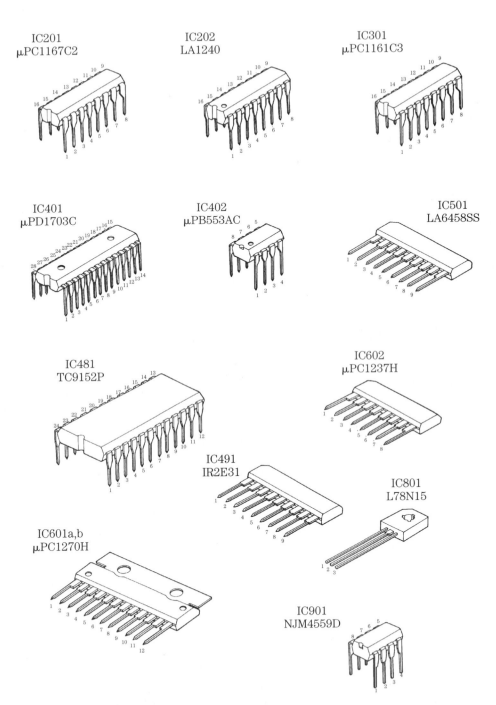

■ **18-12** *Locate terminal 1 (it is marked with a dot, line, or U shape) on the IC.*

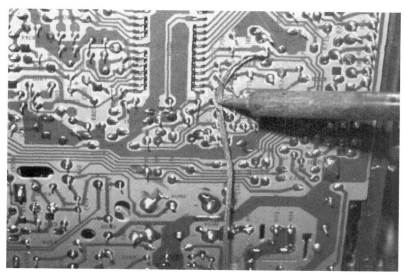

■ **18-13** *Suck up the loose solder around transistor terminals with solder wick.*

When replacing the new transistor, make sure that each wire is in the correct hole. Remember, the transistor terminals are upside down when mounting the transistor from the top. First, select the base terminal and place it in the hole and then place in the other two. Look up the transistor terminals in the semiconductor manual or find them listed on the part box or envelope. Doublecheck each terminal before soldering them.

Clip a soldering heatsink to the transistor lead being soldered to prevent overheating and destroying the new component. Some technicians use a pair of longnose pliers to absorb the heat and also to help in mounting the transistor leads. Make good, clean soldered connections, but do not leave the iron on the terminal too long. Test the transistor in the circuit to make sure that no damage was done while soldering (Fig. 18-14).

Large power transistors are easily removed and replaced. Remove the two screws that hold the transistor in position. Unsolder the base and emitter terminals. Test the defective transistor out of the circuit. Some power output transistors are not mounted in sockets. Mark down the color code of the wires that should be soldered directly to the power transistor terminals. Place silicone grease on the transistor body. If a piece of mica insulation isolates the transistor from the heatsink, apply silicone grease on both sides of the insulator (Fig. 18-15). Do not overtighten the screws or bolts to the transistor and metal heatsink.

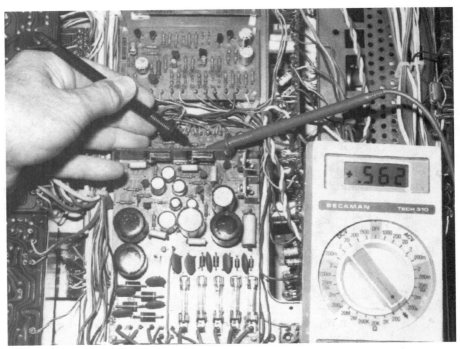

18-14 *After you replace the transistor, test it in the circuit with DMM or transistor tests.*

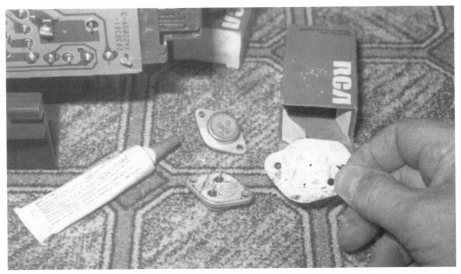

18-15 *Apply silicone grease to the transistor and the heatsink before replacing it.*

Identifying and replacing unknown parts

Replacing defective ICs

Make sure that pin terminal one is in the right corner before pressing the IC into the socket (Fig. 18-16). Likewise, if the IC is to be mounted right on the circuit board, make sure that it's turned the correct way before inserting it into the respective holes. Make sure that each terminal is in the correct hole. Doublecheck each terminal on bottom and top of PC board for correct mounting. Solder each terminal with a low-voltage iron. Do not leave the iron in too long.

■ **18-16** *Make sure that terminal 1 is inserted correctly in the socket or into the PC board.*

Place silicone grease behind power output ICs and regulators that are bolted to the heatsink. Recheck that each IC terminal is in the correct place. Make a resistance check between IC terminals so that no two terminals are shorted. Run the back edge of the screwdriver or pocketknife between each IC terminal to clean out the points of solder or paste.

Large surface-mounted LSIs

The large-scale integration (LSI) circuits contain many transistors and circuits in a single IC chip (Fig. 18-17). The LSI surface-mounted component has many terminals and it solders directly to the PC wiring. These LSI components can be removed by snip-

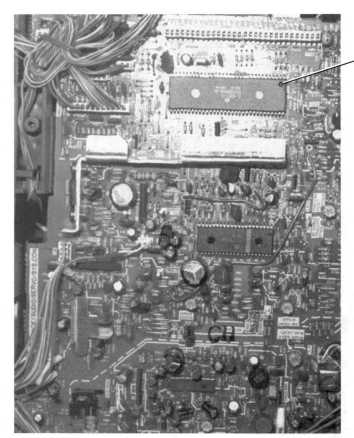

■ **18-17** *The LSI has many terminals. Locate terminal 1 before installing the new replacement.*

ping each lead close to the body of the component. If they cannot be removed in this manner, unsolder each terminal pin and pull upward. Remove each terminal lead with the soldering iron. Clean off the entire area where the new LSI or IC will mount. Check for terminal 1.

Terminal 1 might be marked with a white or indented dot on the body of the LSI. Some manufacturers stamp the numbered terminals at each corner of the component. Match up terminal 1 on the board. Place the new component exactly on each PC wiring and solder it into position. Apply heat from the soldering iron at the ends of the terminal and solder it to the PC wiring.

Unknown diodes

Diodes found in the power supplies are usually 1-A diodes and they can be replaced with universal 2.5-A 1000-V diodes (Fig. 18-18). Select a 3-A diode for larger silicon diode replacement. A signal-rectifying diode in radio circuits can be replaced with a 1N34A or a 1N60. For a fast-recovery circuit, try a 1N914 or an SK3100 diode. Damper diodes, found in the collector circuits of the horizontal output transistor of TVs, should be replaced with a special high-kV voltage rating (Table 18-1). Check the derived voltage for zener diodes.

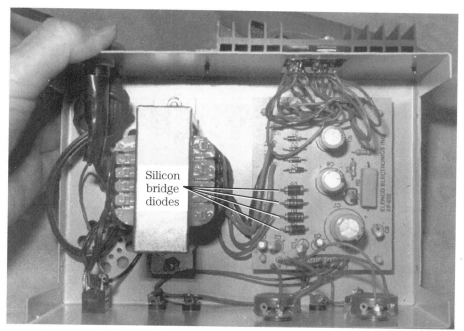

■ **18-18** *Replace all standard diodes with a 2.5-A 1000-V universal replacement.*

■ **Table 18-1 A typical universal diode replacement chart with current and voltage listed in a particular circuit of the TV chassis.**

Circuit	Amperage	Voltage	Universal replacement		
Power supply	1 A	200 V		SK3311	
	1.5 A	400 V	ECG116	SK3312	GE504A
	@.5 A	1000 V	Standard		
Damper	1 A	600 V	ECG552	SK9000	GE511
Boost voltage	1 A	600 V	ECG551	SK9000	GE511
	1 A	200 V	ECG525	SK3925	GE533

■ **Table 18-1 Continued.**

Circuit	Amperage	Voltage	Universal replacement		
Flyback secondary	1 A	600 V	ECG552	SK9000	GE511
Video	0.75 A	75 V	ECG519	SK3100	GE514
Sync	0.75 A	75 V	ECG519	SK3100	GE514
Horizontal	0.75 A	75 V	ECG519	SK3100	GE514
	1 A	600 V	ECG552	SK9000	
Vertical	0.75 A	75 V	ECG519	SK3100	GE514
Switching	0.75 A	75 V	ECG519	SK3100	GE514

Check the schematic and symbol for the correct diode part number. Then look the part number up in the semiconductor manual. In Howard Sams Photofacts, transistor, IC, and diode part numbers and universal replacements are given. Universal diode replacements replace most fixed diodes without any problems. Check the diode polarity before soldering them (Fig. 18-19).

■ **18-19** *Check the diode polarity before removing it and installing the new diode. Notice the white bands in diodes in the VCR power supply.*

Unknown bridge diodes

Just look at the size of the bridge rectifier to discover its wattage. Full-wave rectifiers can be replaced with universal types. Replace the 1-, 1.5-, or 2-A bridge rectifier with a 4-A bridge. Select the correct voltage (50, 100, 200, 400, 600, or 1000 V). Large bridge rectifiers are available in 5-, 8-, 25-, and 35-A ratings. Always observe correct polarity when replacing bridge rectifiers. Half-wave bridge rectifiers can be replaced in the same manner. The half-wave rectifier is one half of the bridge circuit.

The solid-state manual

Most semiconductor manufacturers have their own replacement manual. The different solid-state manuals can be obtained from Sylvania, General Electric, NTE, RCA, Workman, and others (Fig. 18-20). If you know the solid-state part number look it up in the manual. Most semiconductor manuals list the different devices in the contents, tell what it is made of (SK, GE, etc.), polarity, power dissipation, breakdown voltages, current, figure or base number, and what type replaces the defective component. These manuals are very valuable when original parts are not available.

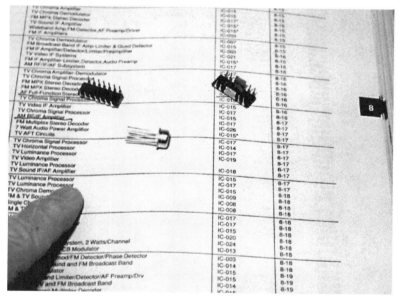

■ **18-20** *Look up the correct transistor for universal replacement in the semiconductor manual.*

Comparison circuit components

You might use another schematic of the same manufacturer to locate the circuit and find the correct part number. If it is not made by the same manufacturer, check what circuit the defective part is from and compare it with another schematic. Use that part number to look up the defective component in the replacement manual. Sometimes in new and special circuits, the original part number should be used.

Manufacturer data

Most manufacturers will furnish service and part data on the products that they manufacture. Naturally, the service company that sells and services warranty products for a certain brand might receive schematic diagrams for free or a fee. Schematics and service data can be obtained from the manufacturer by producing the model number and make of product (Fig. 18-21). Usually, by calling or writing to the manufacturer, they will either supply the schematic or notify what electronics distributor or depot it can be obtained from. Just look up the part number in the service manual and order it.

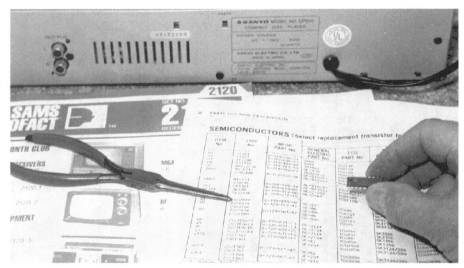

■ **18-21** *Look up the model and part number in the Photofacts for the correct universal replacement.*

Manufacturer parts depot

Many electronics manufacturers have their own or independent electronics wholesalers or part houses that supply the original part. Some manufacturers have several parts and service depots in several different regions of the U.S. Here, the original component can be obtained. Local electronics wholesalers might carry original parts for several different manufacturers. A list of different manufacturers and part suppliers with replacement parts, component directory, product directory, services directory, and mailing addresses appear annually in the March issue of *Electronic Servicing & Technology* Magazine.

Photographing parts

Where critical parts, such as ICs, microprocessors, and LSIs are difficult to obtain or the correct ones are sent with or without a part number, take a photo of it (Fig. 18-22). Sometimes you might order a part several times and they keep sending the wrong one. The chassis or model number might be missing from the cabinet or chassis, but it can be identified by a photo. Take a photo with a Polaroid or a snapshot and send it in with the part number, if it is available. Electronic components can be quickly obtained by faxing the part and model number into the manufacturer or parts depot.

■ **18-22** *Take a photo of the suspected part with the part number on top to ensure the correct replacement.*

Help from local technicians

Do not be embarrassed to call the local technician down the street for help with a problem repair. They might run into the same symptoms and problems. If the service establishment repairs certain brand products, they might have the part you need. Exchanging parts and service data might help everyone around.

Local electronics stores

(Check with the local electronic parts supplier for electronic components.) If they don't have it, they can direct you to another source. Often, the local wholesaler might not sell directly to a student or person, but require a tax permit. Of course, there are many that do. Check for electronics stores in larger cities nearby. Although Radio Shack does not carry special part numbers, you might find the universal part right in your own city.

Conclusion

Sometimes locating the correct component might be difficult. Locating the unknown part might take several attempts before finishing the job. Completing any difficult repair with pride makes electronics servicing worthwhile.

19

How to locate mechanical defects and problems

MECHANICAL AND ELECTRONIC PARTS WORKING TOGETHER in electronic products, produce many service problems. The cracked or loose belt might cause improper speeds or no action (Fig. 19-1). Extremely worn parts can slow down the belts or cause the operation to drag. Squeaky or rattling noises might indicate a dry bearing or lubrication symptom. Noise created in the audio might result from a defective motor.

■ 19-1 *Cracked or loose belts in the camcorder might cause slow or erratic speeds.*

In action

Before the malfunction will show up, the unit must be placed in operation. If the unit refuses to proceed or malfunctions, you must be able to see, hear, or determine what part or components are causing the problem. Knowing the process that the cassette deck, CD player, VCR, and camcorder go through might help solve the problem.

For instance, when the cassette player is pressed into play mode, the cassette is first loaded. When the play button is pressed, voltage is applied to the cassette motor. The motor pulley drives a belt that rotates a flywheel and capstan drive assembly. A pressure or pinch roller applies pressure against the tape and capstan drive, pulling the tape across the tape head. The excess tape is taken up by the take-up reel, which prevents spill out or tape-eating symptoms. Any component in the tape tracking sequence can cause a malfunction of the cassette player.

Cracks and breaks

Cracks or large breaks in the motor drive belt, capstan, or loading belts might slow up or cause the unit not to function. Often, the small cracks in the drive belt will become larger and stop the unit entirely. Cracked or broken metal parts might prevent the tracking sequence in the VCR, CD, camcorder, and cassette player (Fig. 19-2). Bent parts might slow down or prevent the unit from performing.

Broken belts

You might hear the drive motor rotate without any action. Suspect a broken or excessively loose belt when the motor rotates, but there is no further action from the unit. Slippage occurs between belt and pulley when the belt becomes loose, worn, and stretched. Check the motor pulley for bright or worn areas. Intermittent speed might be caused by a defective drive belt, oil on the belt or pulley, dry bearings, or a defective motor. Hold the flywheel with the motor running to determine if the belt is too large or if it is slipping around the motor pulley. The motor should stop when the capstan is stopped. If it doesn't, install a new belt.

The VCR or camcorder might not load with a broken motor drive belt. The motor belt might become worn, loose, or broken in two. Foreign objects pushed into the VCR opening might lodge against the loading mechanism and cause the belt to break. In VCRs,

■ **19-2** *Check for cracked belts, plastic pulleys, and bent arms within the cassette player.*

check for loose drive belts and worn rubber drive wheels and idlers. Clean drive belts and rubber drive wheels and pulleys with isopropyl alcohol.

Squeaky bearings

Grinding or squeaking noises when the unit is operating might be caused by a worn or dry bearing. A scraping noise from the capstan assembly might result from a dry bearing or if the capstan has worked its way down on the drive shaft. Noisy spindles in the VCR or camcorder might be caused by worn or dry spindle bearing surfaces (Fig. 19-3). Sometimes an incorrect motor belt might place more pressure on the motor pulley and capstan and cause noisy rotation.

Usually a good cleanup of all pulleys with a new belt replacement solves most speed problems. Clean all pulleys with alcohol and a

■ **19-3** *Noisy spindles in the VCR or camcorder might show signals of dry or worn bearings.*

cleaning stick. Wipe off the new belt with cleaning fluid and a cloth. Clean out all bearings with alcohol and a cleaning stick before lubricating.

Lubrication

Very little lubrication is needed with today's sealed bearings and motor bearings. Too much lubrication results in oil spilled over critical moving parts and belts, which results in a much-needed cleanup job (Fig. 19-4). Lubricate the capstan bearing at both ends with a drop of light oil or phono-lube. Use light grease or phono-lube on the sliding areas.

Periodically inspect camcorders and VCRs for lubrication after every 1000 hours of use. First remove the old oil and grease. Do not over-oil; only a drop or two will do. Squeaky bearings should be lubricated. It's best to not lubricate at all than to over-oil, where it could drip down onto moving parts. For bearings, use pan oil or Sonic Slidas oil. Apply Hitazal or Froil with a small stick or brush on sliding areas. Many camcorder and VCR manufacturers have their own grease and lubricants. Before replacing a moving part, apply oil or grease that is required.

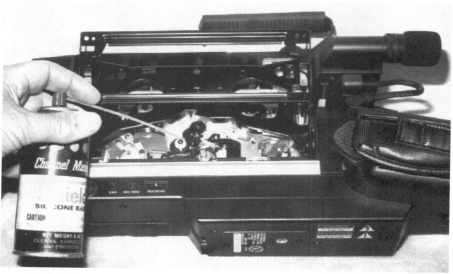

■ **19-4** *By spraying lubricants on the bearings, the liquids might spread over the rubber driving rollers and idlers. Then, you would need to clean the parts thoroughly.*

Within the camcorder clean off the following parts with a cloth dipped in isopropyl alcohol or freon. Also Kim wipes and solvents do the job on the following parts (Fig. 19-5):

- [ ] Capstan shaft
- [ ] All idler wheels
- [ ] Impedance roller
- [ ] Pressure roller
- [ ] All tape guide posts
- [ ] Supply wheel
- [ ] Capstan belt

Apply a light grease, such as Hitazal or Froil, on the following areas for every 1000 and 5000 hours use:

- [ ] Between pressure roller bar and shaft
- [ ] Between pressure roller and gear assembly
- [ ] Gutter of cam gears
- [ ] Loading gear shaft
- [ ] Loading gears
- [ ] Ring idler gears
- [ ] Shaft of loading rings
- [ ] Shaft of idler gears

Lubrication

■ **19-5** *Clean moving components with isopropyl alcohol, cloth, and a cleaning stick or use a camcorder cleaning kit.*

Motor noises

A dry motor end bearing or excessive arcing inside the dc motor might cause a squeaky noise or rotating noise in the audio stages (Fig. 19-6). Today, most motors have sealed bearings and they do not require lubrication. Sometimes a drop of light oil might let the unit play for a few months without motor replacement. If the motor produces noise in the sound stages while operating, replace the defective motor. Large electrolytic capacitors on the motor terminals might help eliminate some motor noise.

Unusual noisy operations

When the VCR, CD player, and camcorder go through an operating sequence with an extreme noisy operation, suspect that a component is dry, worn, or bent. Loud popping or cracking noises might result from extremely worn areas that slow down the unit's operation (Fig. 19-7). Check each moving component for worn or dry pivot or sliding areas. Excessively worn parts might place extra strain on normal components.

■ **19-6** *The noisy motor might cause noise in the audio circuits; it must be replaced.*

■ **19-7** *Check for foreign matter in gears, levers, and idlers, which would produce noisy VCR operation.*

Mechanical problems within the cassette player

Improper and intermittent speeds in the cassette player might result from a loose motor drive belt or motor. The binding pinch roller might cause tape to spill out. Oil on the idler wheel might prevent fast forward or cause uneven speeds. Checking each individual moving component within the cassette player might solve a mechanical problem.

Defective motor

The small dc motor might have a dirty commutator or brushes, which results in poor speed and noise created in the audio circuits of the player. Fast rotation of the capstan might be caused by a defective motor. If the motor speeds up or slows down when the end bell of the motor is tapped, replace the motor. Suspect that the motor is defective if it will not start or can be started with a spin of the motor pulley (Fig. 19-8). When the correct voltage is applied to the motor terminals and it will not start, replace the motor.

■ **19-8** *Spin the small motor pulley to see if the bearings are frozen if the capstan will not rotate.*

Capstan

Often, the capstan is rotated by the motor drive belt. In some models, an idler pulley can be driven from a belt pulley on the capstan. The capstan bearings can become dry and start slowing down the rotation of the tape (Fig. 19-9). The drive shaft of the capstan pulls the tape across the tape head with added pressure from the pinch roller. Sometimes, when tape spills out, it might wrap excessive tape around the capstan drive shaft. Check for excessive tape around the capstan with normal forward operation and high speed in reverse.

■ **19-9** *The capstan bearing might not move or it might have dry bearings if the motor rotates and there is no tape movement.*

Remove a bottom bracket or "C" washer to pull out the dry bearing on the capstan assembly. Clean off any residue with alcohol and cloth. Place light grease or phono-lube on the bearing area. Wipe off excess grease so that it will not touch the small drive belt.

Drive belts

Inspect the motor, pulley, and capstan belts for cracks and slippage (Fig. 19-10). A shiny belt might indicate slippage. Suspect that the speed is improper if the drive belt is enlarged or loose. Clean belts with alcohol and cloth.

Idler pulleys

The idler pulley is shifted from one position to another for reverse, fast forward, and normal operation. Oil on the rubber tire pulley might produce an improper speed. A worn tire or rubber pulley should be replaced. Inspect the pulley for worn areas and clean it with alcohol and a stick. Suspect that the idler wheels are dry if squeaks can be heard in the play mode.

■ **19-10** *Inspect the motor pulley and capstan drive belt for wear, cracks, and oil on the belt.*

Soft-action door

The new type door found on most cassette players uses a slow opening action. When the release button is pressed, the door slowly opens. If the door snaps open at once, suspect that the gear assembly is broken or defective. Do not oil the soft-action door assembly; just replace it.

Pinch roller

The pinch roller presses against the tape and capstan drive shaft. As the tape is pulled through, the pinch roller rotates with the speed of the tape. A dry or worn pinch roller might slow down the speed of the tape (Fig. 19-11). Excessive tape oxide on the pinch roller might cause tape to spill out or pinch up and wind around the pinch roller bearing.

Clean all excess oxide off of the pinch roller. Inspect the roller bearings for tape wrapped down inside of the roller assembly. Pull out and remove all pieces of tape. A drop of oil on the bearings of the pinch roller might keep the assembly from binding or slowing down the tape. Clean off all loose oxide around the pinch roller and capstan area with alcohol and cleaning stick. Wipe off the capstan drive area in the cleaning process.

■ **19-11** *Suspect a dry or worn pinch roller if the tape speed slows down.*

Tape head

Although the tape head does not move, a dirty tape head packed with oxide might cause the tape to slow down. An excessively packed tape head might cause distortion or no sound in the speaker. Check the mounting of the tape head and make sure both mounting screws have not loosened up. Sometimes the tape head will move backward with pressure from the tape and when not contacting the tape, no sound is heard. The tape head azimuth adjustment should be made with a test tape for maximum signal at the speaker terminals.

Take-up and supply reels

The supply reel provides reverse procedures and supply of tape to the tape head. The take-up reel takes the excess tape from the tape head and operates in the reverse and fast forward modes. A worn or dry reel bearing might cause slow speeds or spill out the tape. Suspect that belt slippage is on the reel pulley if it is sluggish in operation. When the tape speed slows up or stops, tape spills out and often wraps around the capstan and pinch roller (Fig. 19-12). Remove the "C" washer which holds the reels and clean the bearing

Mechanical problems within the cassette player

Take-up reel   Supply reel

■ **19-12** *Check the take-up reel for intermittents or stoppage if tape spills out of the cassette.*

area. Only a drop of light oil is needed for lubrication. Wipe off any excess oil from the reel areas.

Mechanical problems in the phonograph

Most problems related to the phonograph are either electronic or mechanical. The electronic problems are found from the phono cartridge through to the speakers. Slow speed, arm cycling, dry motor bearings, and turntable problems might be caused by defective mechanical components. The most common problem is slow speeds caused by slippage between the idler wheel and the turntable rim.

Dead: no turntable movement

Flick the stylus or needle with your finger with the volume wide open to determine if the amplifier section is working. A scratching noise in the speaker indicates that sound is present. Suspect improper voltage, a defective switch or motor when the motor is not rotating. In some models, the motor and amplifier circuits are turned on with the phono function switch. In other models, the turntable begins to rotate when the arm is picked up and placed on the record. A separate switch might turn on the phono motor in low-priced models. Check the "C" washer off of the switch arm assembly and don't strike the on/off switch.

Motor rotates: no turntable motion

Look for a loose or broken motor drive belt when you can hear the motor run and the turntable does not rotate. Replace the motor drive belt if it will not stay on the turntable drive area. Rotate the turntable by hand to determine if bearings are dry or frozen. Remove the turntable, clean the bearings with alcohol, and lubricate them with phono-lube.

Speed problems

Slow speed is quite a common problem with all models of phonographs. Dry turntable and motor bearings might cause slow-speed problems. Slow turntable motion might be caused by a worn or slick idler wheel (Fig. 19-13). The idler wheel drives the turntable on the inside rim area. Check for a worn or loose motor drive belt. Inspect the motor pulley for black rubber marks, which indicates that the belt is slipping.

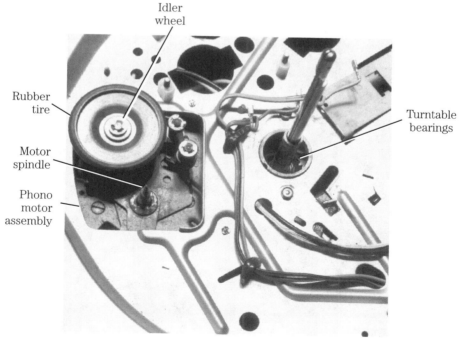

■ **19-13** *Check the parts that might cause speed problems within the phono turntable.*

When the phonograph has accidentally been left on overnight, the motor bearings heat up, become dry, and cause slow speed or frozen bearings. Spin the motor pulley by hand. If the motor is sluggish or will not rotate, disable the motor and wash out the bearings with cleaning fluid. Lubricate the bearings with a light motor oil. Hair and foreign material gathered around the turntable bearing might cause slow or erratic speed. Clean out the bearings and washers with cleaning fluid and lubricate them with phono-lube or light grease.

When all speeds are about the same (78 rpm is a little fast, 45 speed normal, and 33⅓ rpm is about the same as 45 rpm), suspect that the idler wheel is operating on part of the 45-rpm drive area. Jerky rotation might be caused with the 33⅓ speed on part of the 45-rpm spindle. Adjust the speed set screw so that the idler wheel will run on each speed of the motor shaft or spindle.

Will not reject records

If the pickup arm finishes playing the last song and the stylus keeps clicking back and forth or goes into the center of the record spindle, suspect improper rejecting. The pickup arm lever does not trip a reject lever to the next or last record. Check for a dry and gummed up trip lever on older automatic changers (Fig. 19-14).

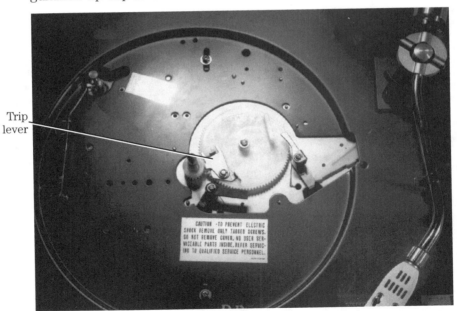

■ **19-14** *The small clip or lever might be dry or frozen and not trip the cam assembly to reject the last record.*

Usually, the trip lever is located in the center of the plate assembly. Clean each piece with alcohol and a cleaning stick. Make sure that the small spring on the cam assembly pushes the lever back into position. Check the large cam bearing for rotation and lubricate it with light oil. Wipe up all excess oil from the cam and turntable.

Will not play the last record

Make sure that the arm keeper is down to play the last record. Check for a bent arm keeper and inspect the shut-off lever assembly. If the record arm keeper is dry and will not fall down, clean it with alcohol and a cloth. Wipe light oil on the arm plunger. The improper weight of the pick-up arm might not keep the needle or stylus in the record groove—especially at the end of the record. Readjust the arm weight screw. Make sure that the tripping lever and cam are not dry and frozen.

Miscellaneous phono problems

When the spindle does not drop or eject records off of the spindle, inspect if the slider lever (inside of the center spindle) is bent or is binding. Also check the small metal lever that holds up the tripping lever. Spray silicon oil on the metal plunger of the arm keeper if it won't slide down under its own weight.

Check for a broken top pin on the metal arm plunger if the arm will not lock into position. The arm keeper will turn completely around if the pin is broken. Most records are tripped with one push of the rejection lever. Sometimes, when a bunch of records are to be played once again, the small record keeper (inside of the center spindle) will come all the way up and might not go down inside of the record hole (Fig. 19-15). Make sure that the keeper falls down before loading the records. Check Table 19-1 for additional phono mechanical problems.

Mechanical problems with the CD player

Although the CD player only has a few moving parts, the mechanical movements are operated by dc motors. In larger CD players, the loading motor pushes out the CD tray and returns the loaded tray. A disc, spindle, or turntable motor turns the CD at a variable speed. The disc motor travels faster at the beginning and slows down toward the outside rim of the CD. The slide or sled motor moves the laser pickup assembly from the center to the outside of the CD on sliding rods. The portable CD player might have two basic motors: disc and sled motors (Fig. 19-16).

■ **19-15** *Make sure that the small record keeper is down to ensure that only one record drops each time.*

■ **Table 19-1 A phonograph troubleshooting chart.**

Symptoms	Cause	Repair
Motor does not rotate	No power to motor	Check leads to power circuit and switch. Check ac voltage.
	Motor coils open, shorted or burned	Check motor continuity with 200-Ω scale of DMM. Inspect burned field coils. Replace motor.
	Check for frozen bearings	Remove motor armature. Wash out bearings. Lubricate and reassemble.
Motor runs hot	Shorted field coil	Replace motor.
Slow rotation	Dry bearings	Clean out bearings and lubricate.
Motor runs fast	Check speed	Replace motor pulley. Replace defective motor.
Turntable runs slow	Oily or loose belt	Clean belt and motor pulley.
	Idler wheel slippage	Replace if worn or cracked.
	Slipping on turntable	Clean rim and idler pulley. Resurface with liquid rosin.
	Turntable slow	Clean out old grease around turntable bearing. Lubricate with phono-lube.
Turntable will not rotate	Check motor	Belt off or broken. If motor rotating check idler.

How to locate mechanical defects and problems

Symptoms	Cause	Repair
	Dry or frozen bearing	Remove turntable. Clean out old bearing. Wash out old grease. Lubricate with phono-lube.
	Center turntable wobbles	Replace turntable.
	Function SW arm loose	Replacing missing "C" washer holding arm level.
Erratic speed	Oil or dirt on belt	Clean with alcohol.
	Oil and dirt on idler wheel	Clean with alcohol.
	Excessive dirt and residue on idler wheel	Brush off dirt—clean with alcohol.
	Jerky music	Improper adjustment of idler. Make sure idler wheel lands on correct speed area of motor spindle.
Will not change speeds	Loose speed lever	Inspect lever arm—check for missing "C" washer.
	Jammed will not move	Inspect arm assembly—check for dry speed lever or bent lever.
	Idler wheel stops	Lubricate idler shaft; readjust height screw.
	Plastic selector will not move	Wash out/clean up/lubricate. Check tension spring.
Pick-up arm will not move	Improper height	Remove keeper or arm up and out of position.
	Arm frozen	Move arm back and forth. Check for dry bearings. Lubricate and make sure that arm is free.
Tone arm sets down ½ inch into record	Improper height adjustment	Reload record-adjust screw slightly before start of music. Several adjustments may be needed.
Tone arm will not raise high enough	Improper adjustment	Check height screw under pivot arm or at bottom of arm pivot post.
Tone arm will not reject records	Check for reject-end of playing cycle-start of record.	Look for bent trip pawl on cam assembly. Check for loose spring or "C" washer.
	Check poor alignment of auto lever	Dry or sluggish trip lever. Wash out and lubricate the dry cam assembly.
	Stylus goes into the center of the record	Replace the needle or stylus. Recheck set down.
Changer will not cycle	Loose cam assembly	Check the cam for dry or gummed up bearings.
	Arm lever out of place	Lubricate all sliding areas. Remove old grease.
Won't play last record	Arm keeper not down	Make sure that the arm keeper is down. Check for a bent arm keeper. Inspect the shut-off assembly.
	Dry or gummed trip lever	Clean the small lever. Lubricate and make sure that the spring is in place.
	Record arm keeper too high	Lubricate the arm keeper. Make sure that it goes completely down. Check the arm adjustment.
Record arm keeper loose	Check for a broken metal pin	Replace the entire record arm assembly.

Mechanical problems with the CD player

■ **Table 19-1 Continued.**

Symptoms	Cause	Repair
	Shaft is bent or binding	Remove the "C" washer. Remove the keeper. Straighten the arm or replace.
Tone arm skips on record	Suspect that the stylus is faulty	Remove it and replace the needle or stylus. Make sure that the turntable is level. Remove the excess dirt from the stylus.
Rumble noise	Check the motor touching the mounting board	Inspect the "C" washer and the rubber mounts. Replace the flattened rubber mounts.
	Dirt on the motor belt	Clean the motor pulley, belt, turntable, and rim.
	Make sure that the turntable is floating	Loosen the turntable screws. Check for missing springs. Make sure that the damping pads are in place.
Howling noise (feedback)	Cartridge pickup vibration	Is the turntable floating? Check the binding arm wire leads.
Two records drop together	Enlarged record hole	Replace the record.
	Control lever not fully down	Lever pulls up when unloading records. Keep it down.
No muting	Check the muting switch	Clean the contacts. Check the switch continuity. Adjust the muting spring. Check the tension.

■ **19-16** *The small portable CD player has only a slide and disc motor.*

Loading motor

The loading motor or tray motor moves the tray in and out to load and unload the CD disc. The loading motor might rotate a large drum with plastic gears that moves the plastic loading platform. Visually inspect the drawer assembly for foreign objects if it will not move in or out. Sometimes small children like to place candy, gum wrappers, and crayons inside of the drawer area. Notice if the tray plastic rails or gears are binding. Clean the area and apply a coat of light lubricant to the sliding surfaces.

If the tray motor is rotating with no tray action, check for a broken drive belt or a worm gear assembly. In larger CD turntables, that automatically change five different discs, the tray motor drives a small belt to a worm gear assembly. Check for a broken or missing belt (Fig. 19-17). Check the voltage at the motor terminals if the disc motor won't rotate.

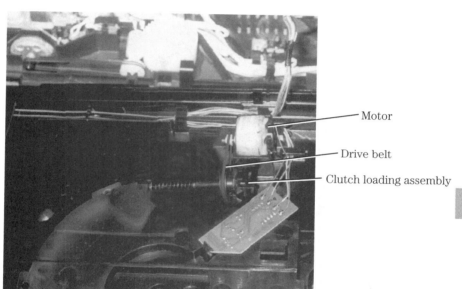

■ **19-17** *The CD turntable-loading motor might be belt or worm-gear driven. Check for a broken or motor belt if the motor operates and there is no CD loading.*

Disc or turntable motor

The spindle, disc, or turntable motor starts to rotate after the disc is loaded. The CD mounts directly on the small platform that is at the end of the disc motor shaft. The disc motor starts out at approximately 500 rpm and slows down as the laser pickup assembly moves toward the outer rim of the CD (around 200 rpm). The turntable motor might be controlled by a CLV (constant linear velocity) motor circuit and be controlled directly from one large IC.

Slide or sled motor

The slide motor might also be called a *feed motor* by some manufacturers. The sled or slide motor moves the optical pickup assembly across the disc from the inside to the outside rim of the CD. Often, the motor is gear driven to a rotating gear that moves the laser beam down two sliding rods or bars (Fig. 19-18). The slide motor might have a fast-forward and rewinding mode operation.

■ **19-18** *The finger points to two separate slide assembly rods, where the optical assembly moves up and down.*

Erratic or intermittent operation of the slide motor might be caused by a gummed-up track or by poor gear meshing. It is very difficult to see the sled motor operate because it is mounted under the CD chassis. Often, a voltage and continuity measurement across the motor terminals will help you to identify an open feed or slide motor.

In some models, the slide motor has a motor pulley that drives a worm pulley to slowly move the pickup assembly with a motor drive belt. Check the belt for breaks or worn areas. Inspect the belt for oil spots if the movement is erratic. Replace the drive belt if it shows signs of slipping. The slide assembly will not move at all if the drive belt is broken.

Mechanical problems with the VCR or camcorder

Although the VCR and camcorder are two different electronic products, the cassette operation is quite the same in Beta, VHS, and 8-mm formats. Like the cassette recorder, the camcorder and VCR have supply and take-up reels or disks, rotating heads or cylinders, a loading motor, a capstan motor, a capstan flywheel, guide rollers, idlers, drive wheels, pressure rollers, different gears, and drive belts (Fig. 19-19). The camcorder might have additional focus, zoom, and iris motors.

■ **19-19** *Many moving components in the VCR must be cleaned when cleaning the video head.*

Drive belts

Several different drive belts are found in VCRs. The loading motor might drive a gear assembly with a small drive belt. The drive motor rotates the capstan-flywheel with a large belt and the idler wheel assembly. Erratic or no capstan operation might be caused with a cracked or broken drive belt. Inspect the belts and drive wheels if the unit has been taken apart for head cleaning. Suspect the worm drive wheels if rewind and fast forward operations are not normal. Clean the drive belts and wheels with isopropyl alcohol and cloth (Fig. 19-20).

■ **19-20** *Clean the drive belts, idlers, and moving parts with isopropyl alcohol and a cloth.*

Loading motor

The loading motor in the camcorder or VCR might eject, load, and unload the video cassette, releasing brakes, and engaging the fast-forward/rewind idler gears and playback gear. The loading motor in the camcorder might open the loading door by pressing the eject button. After loading the cassette, the door might be manually or electrically closed.

The loading motor is usually mounted off to one side in the VCR and might drive the opening mechanism by a drive belt or worm gear (Fig. 19-21). If the cassette will not load, check for a worn or broken drive belt. Sometimes foreign material might clog up the pulley and belt assembly, and destroy the belt. Erratic loading might result with a worn or loose belt.

In other VCRs, the loading motor might drive the loading mechanism with a worm gear assembly. Check for a binding worm gear assembly and for foreign material dropped inside the disc opening (Fig. 19-22). Check for correct voltage and continuity at the loading motor terminals if it will not rotate.

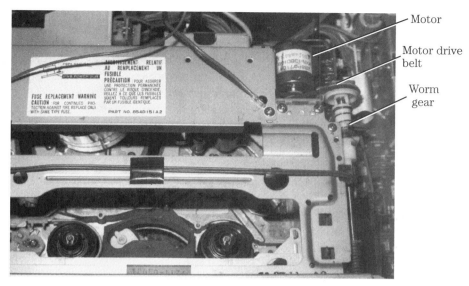

■ **19-21** *The loading motor in the VCR might have a drive belt, worm gear or both to complete the loading and unloading of a cassette.*

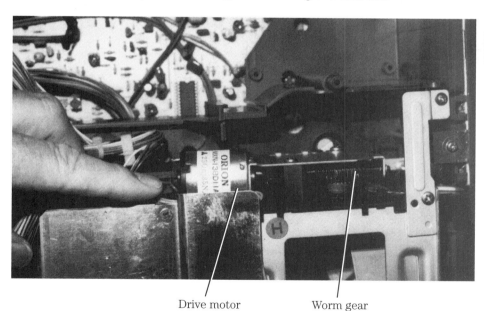

■ **19-22** *Check for foreign material in the worm gear or belt if the VCR will not load.*

Mechanical problems with the VCR or camcorder

Capstan motor

The capstan motor drives the flywheel or capstan with a pressure roller, and moves tape across the heads to the take-up reel. The capstan motor in the VCR might be controlled by speed and phase control circuits. The capstan motor in the camcorder might be belt driven to the various mechanical assemblies, which provides tape movement in play, record, rewind, fast forward, and search modes. These speeds are controlled by a servo, capstan speed, and phase-control systems.

Usually, the capstan motor drives the capstan/flywheel with a large rubber belt (Fig. 19-23). Oil or grease on the belt or too large of a belt might cause erratic movement and speed of the capstan. Visually inspect the belt, clean it, or replace it, if needed.

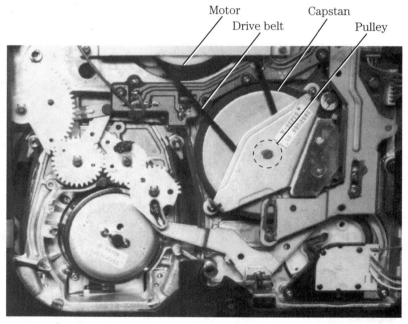

■ **19-23** *The capstan/flywheel is driven by a motor belt in the VCR mechanism.*

Cylinder or drum motor

The cylinder and drum motors are controlled by the servo circuits in the camcorder and VCR. The drum or cylinder consists of an upper and lower drum assembly. Operation of the cylinder or drum motor is quite complex in speed control, phase control, and

■ **19-24** *The video head is located in the top cylinder or in the drum of the VCR or camcorder.*

in the drum motor itself (Fig. 19-24). The tape head is part of the drum or cylinder that picks up the video and picture.

Focus motor

The focus motor operates the lens assembly automatically to bring the picture or view into correct focus. The focus motor is located on the lens assembly alongside of the zoom motor. The focus motor is driven with a dc voltage from a motor drive IC or from transistors. Check for voltage and continuity at the focus motor terminals if the motor does not rotate. Inspect the gear assembly for binding in slow operations. These motors stick out, so they can be easily damaged.

Zoom motor

The zoom motor brings the image or picture close up or far away from the lens in the camcorder. Both the focus and zoom motors are mounted on the lens board assembly (Fig. 19-25). The zoom motor drives the lens assembly with a worm gear. This motor might have a transistor or IC zoom motor circuit. Measure the dc voltage and the continuity at the motor terminals when the motor does not rotate.

■ **19-25** *The zoom and focus motor are located on the lens assembly of the camcorder.*

Iris motor

A meter system was used in early VHS camcorders to control the camera aperture. Today, the iris mechanism is controlled with an iris motor. The AIC (automatic iris circuit) controls the iris, according to the object brightness. The AIC circuit consists of an IC driver, AGC, and iris control circuit. The iris motor mounts directly on the lens and iris block assembly.

Supply and take-up reels or disks

Like the cassette player, the supply and take-up reels supply the tape that is pulled around the tape path and taken up by the take-up reel. The VHS reel or disk hubs are much larger and are inserted into the correct holes of the cassette. Both reels might be driven with friction mechanism with pulleys and idler wheels. Each reel disk might have a slip-type mechanism. Clean the drive areas of the reels with alcohol and a cleaning cloth.

The typical supply reel slip mechanism generates the torque necessary to eliminate tape slack and take-up the tape during reverse visual search and unloading (Fig. 19-26). The take-up reels or disk

■ **19-26** *The camcorder and VCR have supply and take-up reels that are somewhat like those of the cassette player.*

slip mechanisms work in the same way to eliminate tape slack and to take up the tape during the take up modes and during visual search. This allows the slip torque on the take-up and supply reels to be independently controlled in the camcorder.

Conclusion

Check Chapter 36 for additional mechanical and maintenance problems. Knowing how the mechanical components operate might help you solve both the mechanical and electronic problems.

Troubleshooting and repairing solid-state circuits

SOLID-STATE CIRCUITS ARE REPAIRED BY KNOWING WHAT the symptom is, isolating the circuit with block and circuit diagrams, and by pin-pointing the defective component with voltage, resistance, signal, and test instruments. Critical voltage and resistance tests on semiconductors can be used to isolate the defective component. Critical signal and waveform tests can help locate the suspected circuit. Knowing how to make these tests in the various electronic circuits provides critical test procedures for the intermediate electronic technician (Fig. 20-1)

■ **20-1** *The electronic technician is taking voltage and resistance measurements in a TV chassis.*

Critical voltage tests

Critical voltage tests on the solid-state component might indicate that the part is defective. Voltage measurements are usually taken from the component terminals with the red probe and the black probe clipped to the common ground or chassis. First, measure the collector terminal on the suspected transistor. Then take voltage measurements on the other two terminals. Likewise, first check the supply voltage on the IC. Compare these voltages with the schematic, if it is handy (Fig. 20-2).

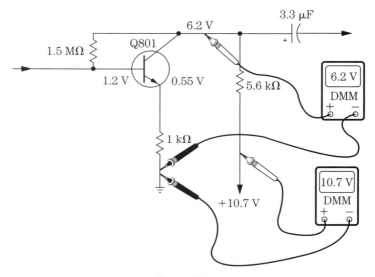

■ **20-2** Taking critical transistor collector and supply voltages in the AF transistor circuit.

You might assume that the transistor is leaky or has an improper supply voltage if the voltage measurement on collector terminal is low. Compare the base and emitter terminal voltage to the schematic. Very low voltages on all three terminals might indicate that the transistor is leaky. Do not overlook the supply voltage source. Measure the voltage at the collector load resistor that connects to the power supply. If the voltage is low at this point, suspect that the supply voltage is low. If the voltage is quite high compared to the voltage on collector terminal, suspect that a transistor is leaky (Fig. 20-3).

Remember, in some circuits, the signal or waveforms must be applied to the base of the transistor or the transistor might heat up

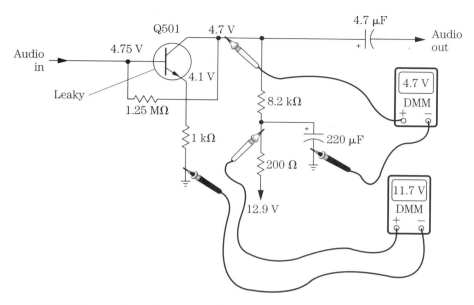

■ **20-3** *Suspect a leaky transistor if the collector voltage is lower than normal.*

and become leaky. If the horizontal drive signal or waveform is missing at the base of a driver transistor, both the driver and output transistors might be damaged in horizontal and vertical circuits. The same thing can happen in the directly coupled circuits of amplifier, VCR, CD, and cassette circuits. Determine if the supply voltage is normal before troubleshooting the rest of the circuits.

Power supply voltage

Usually voltage sources originate in the low-voltage power supply or from a battery source. The low-voltage source might start with a higher dc voltage, go through a dropping resistor or regulator for a lower dc voltage source (Fig. 20-4). Transistors, ICs, and zener diode regulators might be used to lower a regulated voltage. Simple dc or decoupling circuits might have a dropping resistor, filter capacitor, and zener diodes as regulator. These lower voltage sources are then fed to respective circuits. Transistor and zener diode regulators have a tendency to become leaky or go open.

Always check the low-voltage circuits if a lower-than-normal supply voltage is found in transistor or IC circuits. Check the $B+$ voltage across the large filter capacitors. Trace the voltage from the power supply to the voltage that is applied to the defective cir-

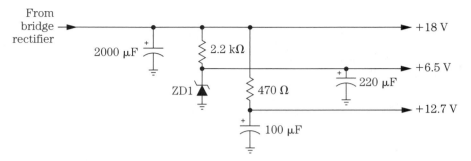

■ **20-4** *The different low voltage sources are produced with transistor, IC, and zener diode regulators.*

cuits. More than one stage might be affected by only one low-voltage source (Fig. 20-5). The different voltage sources are designed for different operating circuits.

Decoupling capacitors might become leaky and lower the voltages to other circuits. Often, burned resistors result from a leaky decoupling capacitor or another component in the overloaded circuit. A leaky part in another circuit might lower the voltage in the

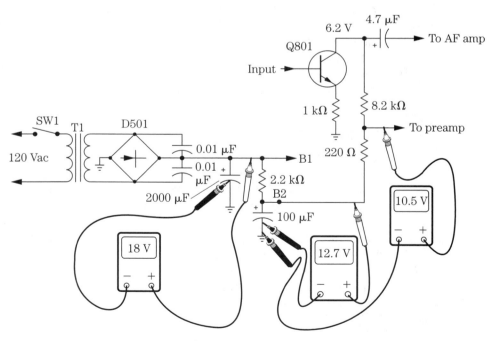

■ **20-5** *Check the voltage supply at the large electrolytic capacitor, voltage-dropping resistor, and where it runs to another circuit.*

circuit under test and cause the same symptom. Critical voltage tests on transistors and ICs might point to an improper voltage source found in decoupling power supply circuits.

Critical resistance measurements

Critical resistance measurements between transistor terminals and common ground might indicate that transistors, resistors, or diodes in the circuit are leaky or open (Fig. 20-6). Resistance measurements from IC and LSI terminals to common ground might indicate that an IC or a part tied to one of the terminals is leaky. Especially check the resistance of the voltage supply pin (V_{CC}) to ground if low voltage is found on the voltage source. Remove excess solder from the supply pin and pc wiring. If the voltage is normal and the resistance is below 1 kΩ at the supply pin terminal to ground, suspect that an IC or LSI is leaky.

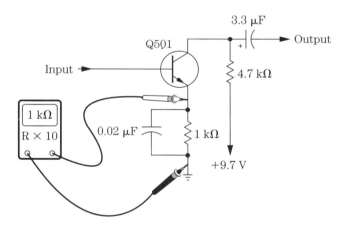

■ **20-6** *The critical resistance measurement from the emitter terminal and chassis ground might indicate a defective resistor.*

Critical resistance measurements on each IC or LSI terminal might indicate if the component is leaky. A lower-than-normal resistance might indicate that an IC is leaky. Follow the schematic diagram as each terminal is measured. If a very low resistance measurement is found, recheck the schematic for a possible low-resistance resistor, diode, or coil in that circuit (Fig. 20-7). Some manufacturers provide a resistance and voltage chart between terminals and common ground. Leaky capacitors and diodes have been found in the IC return circuits to common ground. When in doubt, remove the opposite end of the part from the circuit and take another

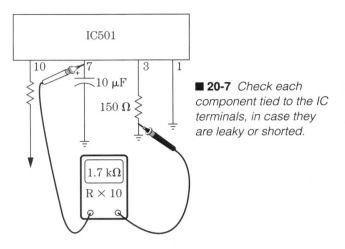

■ **20-7** Check each component tied to the IC terminals, in case they are leaky or shorted.

measurement. If the measurement is low in resistance with the component lead desoldered, the IC is leaky and should be replaced. Doublecheck the component resistance or continuity.

It's best to discharge the large filter capacitor that feeds the low-voltage circuits before taking critical resistance measurements. Voltage tests on a collector terminal might change as the voltage discharges through the meter and power supply circuits. The same might occur if resistance measurements are taken on the supply or higher voltage pins of an IC. When resistance measurements are taken with a VOM or FET meter, you can see the meter hand slowly lower as the voltage is discharged from these circuits (Fig. 20-8).

Transistor tests

Checking the transistor in or out of the circuit can indicate if the transistor is defective. First, check the transistor in the circuit and remove it for a more accurate test. If the transistor is intermittent, sometimes in-circuit and out-of-circuit tests will be fruitless. If the transistor tests open or leaky in the circuit, but tests normal the next time, suspect that a transistor is intermittent. Sometimes, just moving a transistor terminal might cause the transistor to act up. Replace the transistor if you are in doubt.

Always check the suspected transistor once again after removing it from the circuit. If the transistor is open or leaky, it will test the same out of the circuit. Test the new replacement before installing it. It is possible to receive a new defective part. Quick transistor tests can be made with the diode-junction tests of the DMM or with a regular transistor tester (Fig. 20-9).

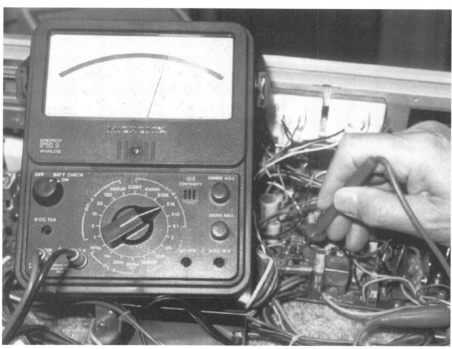

■ **20-8** *FET or VOM resistance measurements in the voltage circuits might decrease the hand slowly while electrolytic capacitors discharge in the circuit.*

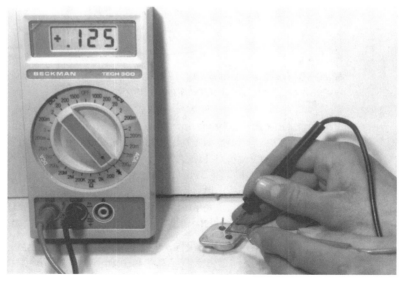

■ **20-9** *Quick transistor tests can be made with the junction-diode test of the DMM.*

Transistor bias voltage tests

Voltage measurements from base terminal to common ground and from emitter terminal to common ground should equal the bias voltage of an NPN or PNP transistor (0.6 V), if measured from base to emitter terminals. If an NPN transistor has 0.6 V between the base and emitter terminals, you might assume that the transistor is normal (Fig. 20-10). To verify, quickly take a voltage test to common ground of the base and emitter terminals. Now, subtract the difference between the two voltage measurements and it should be within a fraction of 0.6 V. If so, the transistor is normal. Of course, this bias voltage will depend on what circuit the transistor is used in. The germanium PNP transistor has a bias voltage of 0.3 V.

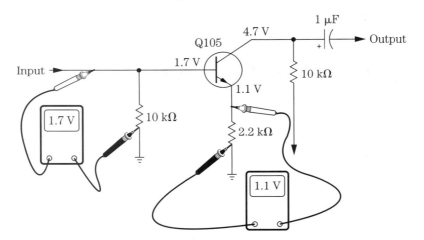

■ **20-10** *The transistor might be normal with a forward bias voltage of 0.6 V between the base and the emitter terminals.*

For instance, in a horizontal preamp oscillator circuit of the TV chassis, a PNP transistor has a high dc voltage applied to the emitter terminal (Fig. 20-11). The collector load resistor connects to common ground. The base voltage is 11.3 V, if it is measured to ground, while the emitter voltage to ground is 10.6 V. The difference between the base and emitter terminals is 0.7 V (0.1 V off of the silicon-biased resistor).

In a video emitter-follower stage, the collector terminal is at ground and the high dc voltage (5.5 V) is found at the emitter terminal. The base voltage to ground is 4.8 V (Fig. 20-12). Without looking at the schematic, you can see that the transistor is a PNP silicon tran-

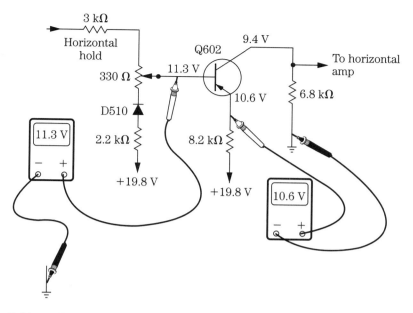

■ **20-11** *Q602 has a high dc voltage applied to the emitter terminal because it is a PNP with a collector voltage that is more negative than the voltages at the other two terminals.*

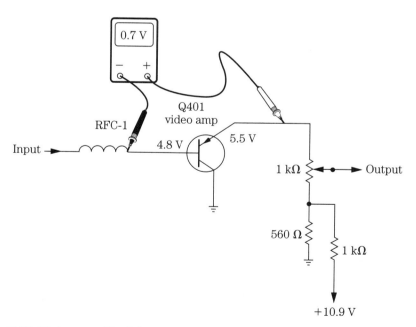

■ **20-12** *In an emitter-follower circuit, the video amp is a PNP with a 0.7-V forward voltage between the emitter and the base terminals.*

sistor because the collector voltage is negative compared to the base and emitter. Also, the difference in the bias voltage between base and emitter is 0.7 V, which indicates a silicon transistor.

IC voltage tests

Critical voltage tests on an IC might determine if IC is defective, if there is improper supply voltage, or if a leaky or open component is tied to one of the IC terminals. First, measure the supply voltage. Often, the highest voltage found on any terminal of IC or microprocessor is marked with a V_{CC} mark at the supply terminal. Notice if the voltage is quite low or is higher than normal. A higher supply voltage might indicate an open IC, but a very low voltage measurement might indicate that the IC is leaky.

In a horizontal/vertical deflection IC, the symptom might be no horizontal sweep or raster. A quick scope test on terminal 6 of IC501 has no horizontal drive waveform. Likewise, no vertical waveform is found at pin 13. Upon checking the TV schematic, the deflection circuit IC is powered from the low-voltage power supply source (135 V), not from a secondary winding of the flyback transformer. The voltage measured at supply pin 4 should be around 9.1 V (Fig. 20-13), but the voltage measured only 1.05 V.

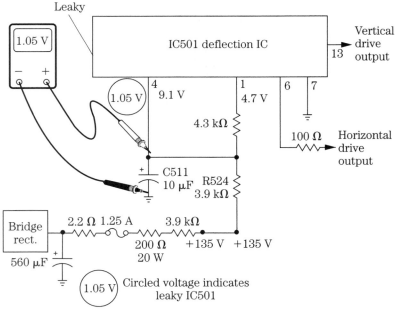

■ **20-13** *Leaky IC501 measured only 1.05 V at pin 4; it should be about 9.1 V in a deflection IC circuit.*

By removing the excess solder from around pin 4 from the pc wiring with solder wick, the voltage at C511 and R524 came up to 134 V. This indicated IC501 was leaky or that a component tied to it was leaky. A resistance measurement from pin 4 to chassis ground was only 154 Ω, which indicates that the IC is leaky. To make sure, critical voltage tests were made on each IC terminal and all were fairly close. No low-resistance measurements from any other pins to ground were found, except those that were wired directly to the common ground. Take critical voltage tests and compare those on the schematic of each IC terminal.

Troubleshooting various solid-state circuits

Learn how to troubleshoot the different solid-state circuits taken from the AM and FM radios, cassette players, audio amplifiers, CD players, camcorders, and TV circuits. The various circuits represent many different circuits that the electronic technician might encounter each week on the service bench. How to take the symptom, isolate the circuit, apply the correct test instruments, and locate the defective part is an every day occurrence.

Servicing the AM converter stage

In the early radio chassis, the AM converter circuits consisted of tubes, then transistors, and now ICs. The AM transistor or IC converter mixes two signals, such as those received and the local oscillator signal developed in the converter transistor, in a superhetrodyne receiver. The results of the two signals are the intermediate frequency (IF). The IF frequency in the table model radio is 455 kHz. The deluxe receiver might have a separate RF (radio frequency) amplifier, oscillator, and mixer stage. The converter circuit combines all three stages into one (Fig. 20-14).

The transistor converter circuit might consist of a tuning capacitor or permeability tuned system, antenna coil, transistor, and oscillator coil. The RF AM signal is picked up by the antenna coil and by the transformer coupled to the base of the converter transistor. The transistor amplifies the weak AM signal and mixes the oscillator frequency (emitter circuit) with the intermediate frequency found at the output of the collector terminal.

When the AM station is picked up at 1400 kHz, the oscillator frequency must tune at 945 kHz to pass the difference of 455 kHz to the IF transformer. The oscillator frequency is always lower than the signal received by the RF circuits in the AM radio. The oscilla-

■ **20-14** *The converter transistor is mounted near the variable capacitor and antenna rod that converts the incoming and oscillator signals down to 455 kHz.*

tor frequency can be heard in another AM receiver if it is placed close to the antenna and oscillator circuits.

Then, check the converter stage by placing another AM radio nearby. When the radio is tuned across the band of the other radio, you can hear a squeal or oscillating sound. A defective converter stage symptom might be that the stations are weak, only a local radio station can be heard, or that no stations can be received. Only a rushing noise might be heard in the speaker.

Check the AM converter transistor with voltage and transistor beta tests. Take a voltage measurement at the collector terminal (10.8 Vdc). If the voltage is higher than normal, the transistor might be open. When the collector voltage is quite low, suspect a leaky transistor (Fig. 20-15). Measure both the base and the emitter voltage

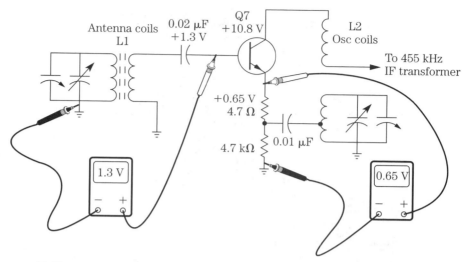

■ **20-15** *Take voltage measurements on all three transistor terminals and compare them to the schematic.*

to ground. The difference between the two voltages equals the bias forward voltage (1.3 to 0.65 V equals 0.65 V). Take a voltage measurement between the base and collector terminals (0.6 V). With these measurements, the transistor is normal. No voltage at the emitter terminal might indicate the transistor or emitter resistor (4.7 Ω) is open. If in doubt, test the transistor with a transistor tester or the diode-transistor junction test of the DMM.

A weak AM station can be caused by a broken antenna ferrite rod, coil wire, and open coupling capacitor (Fig. 20-16). Suspect that there are antenna coil problems if the test probe touches the base terminal and if the volume increases in the radio. A local radio station might come in loud if the probe touches the base terminal.

The AM IC RF/converter/IF IC might be a separate component or combined with the FM front-end circuits. The no-AM symptom might result from a defective IC or IF component. The AM RF signal is coupled through a 0.01-μF capacitor on terminal 1. The oscillator coil connects to pin 4, and the IF transformer is tied into terminal 7. The voltage supply pin is 10 (Fig. 20-17).

Go directly to the AM IC and take critical voltage measurements on terminals 10, 4, 3, and 2. If the supply voltage is very low, suspect that an IC is leaky or that the low-voltage source is improper. Check for a leaky IC if voltages are low on most terminals compared to the schematic. The AM IC is usually located close to the variable AM/FM capacitor and IF coils (Fig. 20-18).

Troubleshooting various solid-state circuits

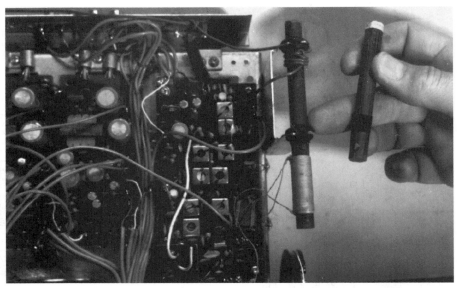

20-16 Weak AM reception might result from a cracked or broken antenna ferrite rod.

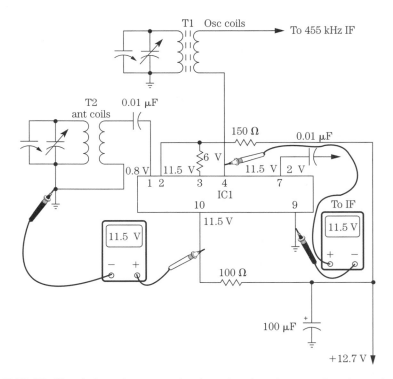

20-17 Check the voltage supply pin to the chassis ground to determine if IC1 is leaky or if there is an improper supply voltage.

Troubleshooting and repairing solid-state circuits

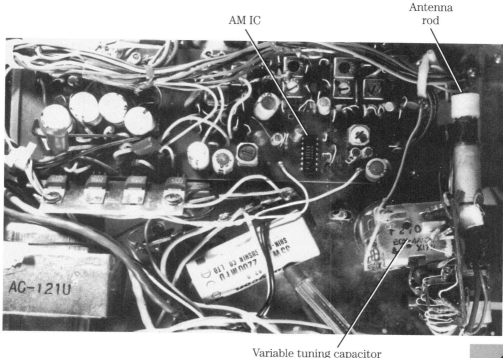

20-18 *The AM IC is usually located close to the variable capacitor and antenna rod.*

The FM radio RF circuits

The defective FM RF stage symptoms might result in weak-station reception or only the local FM station can be heard. Place the voltage test probe to the collector terminal of the FM RF transistor and if the volume increases, suspect that an RF stage is defective. The FM front-end coil might be connected to the outside FM antenna. Notice that the FM RF or oscillator coils might be self-supporting wound wire with only a few turns. Often, the separate transistor stages consists of an RF oscillator and mixer found in the front-end FM circuits.

Check the collector, base, and emitter voltages of the FM RF transistor (Fig. 20-19). Compare these voltages with the schematic. The open transistor will have a higher collector voltage than normal or an open emitter resistor. A very low collector voltage might indicate a leaky transistor. To make sure that the transistor is defective, make an in-circuit test. Replace the defective transistor with one of the same lead length and mount in the very same spot.

Troubleshooting various solid-state circuits

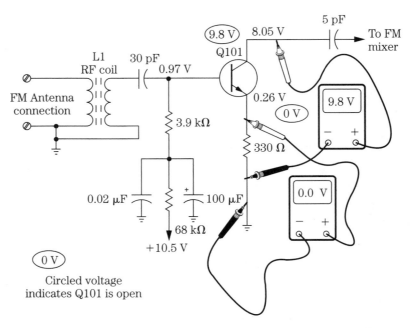

20-19 Check the collector and emitter terminals of the FM RF transistor to determine if the transistor is leaky or open, causing the weak FM reception.

Cassette head preamp circuits

The tape head preamp circuits might consist of one IC for both left and right stereo channels. The transistor preamp might contain one transistor or two directly coupled. In directly coupled circuits, one defective transistor might change the voltage on the other transistor. The tape heads might be switched into the base circuits of the first preamp transistor. Often, the base terminal is coupled to the head winding through a small electrolytic capacitor (Fig. 20-20). The collector of the first transistor couples directly to the second preamp transistor. The amplified audio signal is now capacitively coupled to the function switch or AF transistor.

Check the voltage on both preamp transistors. If the voltage is way off, test each transistor in the circuit. In case one of the transistors is leaky or open, voltage measurements on each transistor will change. Sometimes it is best to replace both transistors if one is found to be leaky or shorted.

In today's preamp circuits, one IC might serve both stereo sound channels. The left channel tape input signal is applied at pin 8 and amplified out of pin 6. Notice electrolytic coupling capacitors are

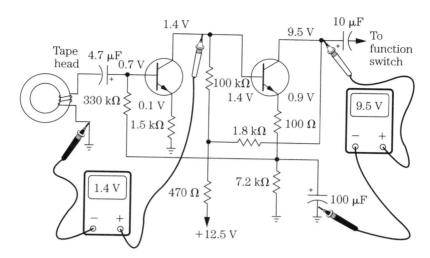

20-20 Check the voltage on both collector terminals of the two directly coupled preamp transistors to determine why the cassette output is weak.

found on both input and output circuits. The right channel input signal is found at pin 1 and the output is at pin 3 to the function switch or AF circuits (Fig. 20-21).

A weak or dead symptom of the preamp stages can be isolated if the radio and other stages are amplified out of the power audio circuits. Check by inserting a cassette tape. A quick audio signal test

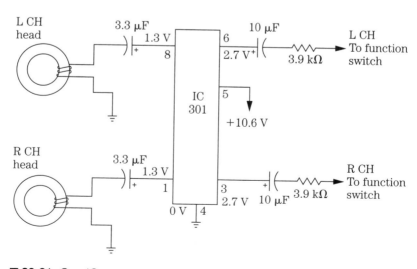

20-21 One IC serves as preamplifier in the tape head input circuits with pin 1 as the input and 3 as the output of the right channel.

Troubleshooting various solid-state circuits

at pins 8 and 1, or 6 and 3 might indicate that IC301 is defective. Take a quick voltage measurement on pin 5 to determine if the voltage source is normal. When the AF signal is heard at the input terminals and not at the output pins, suspect that the IC is defective. A weak channel might also be caused by a dirty, packed tape head.

Defective CD loading circuit

The loading motor circuit in the compact disc player is very simple with motor driver IC, mechanism control, and motor. Some loading motors are driven with one or two transistors. This loading voltage is sent from the mechanism control IC1 to pin 8 if the loading button is pressed on the front panel. IC403 provides a different loading motor polarity voltage if the signal is applied by IC1 (Fig. 20-22). To load, the loading tray is pushed out of the player and the disc is loaded. When the button is pushed again a different polarity of voltage is supplied to the small dc motor and the motor pulls in the tray assembly for loading on the disc platform.

Check the voltage across the motor terminals if the loading button is pressed. If nothing happens and there is no voltage, suspect IC403. Measure the voltage on pins 2 and 3 of the motor driver IC. Check the supply voltage at terminals 1 and 3 of the motor driver IC and the chassis ground. If the supply voltage is low or missing

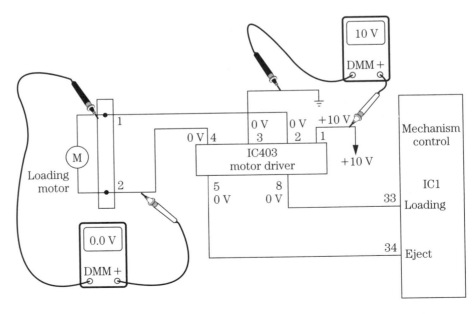

■ **20-22** *Check the supply voltage on pin 1 of motor driver IC403 and at the motor terminals if the loading motor will not rotate.*

check for a defective regulator transistor. When the supply voltage is normal and the signal is applied at pin 8, suspect that the motor driver, IC403, is defective. Check the motor continuity with an ohmmeter if the voltage is present on motor terminals and if there is no motor rotation.

Camcorder capstan motor circuits

The camcorder motor might be belt- or gear-driven to the various mechanical assemblies, providing tape movement in play, record, rewind, fast forward, and search modes. Because these different modes operate at different speeds, the dc voltage must be controlled by a servo, capstan speed, and phase control system. The system or servo IC might control the capstan motor through a capstan motor drive IC (Fig. 20-23).

When the capstan motor does not rotate, check the voltage at the motor terminals. Check the continuity of the motor if the voltage is found at the motor terminals. Suspect an IC or transistor capstan driver that supplies voltage to the capstan motor. Measure the supply voltage pin at driver IC902. Suspect that the capstan control (IC602) is defective if no control voltage is fed to the cap-

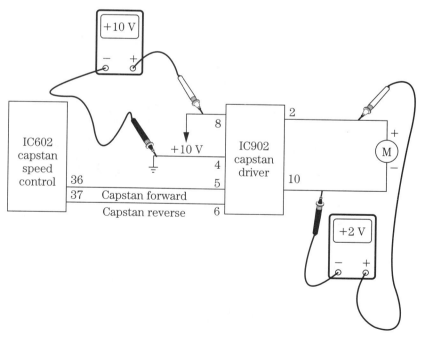

■ 20-23 *The speed control (IC602) controls the forward and reverse rotation of the capstan motor in the camcorder.*

stan driver IC. Replace all ICs with components that have the exact part numbers.

Red color missing at the output circuits

In the TV set, if one color is missing from the picture, try to determine what color is missing. When red is missing, the raster and picture might be a bluish-green. Usually, if one color is missing, the chroma IC signal is normal, but the signal from the IC to the CRT might be defective. The red output transistor, red color gun inside the picture tube, or circuits might result in no red in the picture.

Test the picture tube with the CRT tester to make sure that the red gun is not defective. Go directly to the red output transistor if the picture tube is normal. You might find that all three color output transistors on the CRT PC board are defective (Fig. 20-24). Locate the red transistor (Q703) on the schematic and PC board. Often, these output transistors are enclosed in a T-220 case.

■ **20-24** *The color output transistor circuits are mounted on the CRT PC board in this RCA color TV.*

Measure the voltage on the collector terminal of the red color transistor (136 V). The collector terminal is the metal end with a hole in it. These voltages are rather high to provide power to drive the color cathodes in the gun assembly of the CRT. Often, this voltage is fed from one of the flyback windings. Sometimes if the color output transistors become leaky or shorted the silicon diode in the +190-V supply might be damaged. If the voltage at the collector is low, suspect that a transistor is leaky or that a power source is improper (Fig. 20-25).

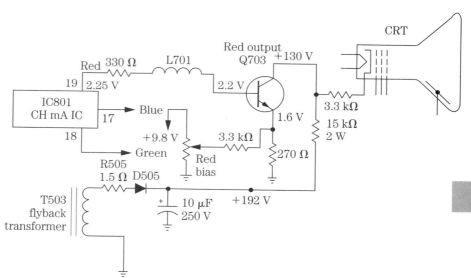

■ **20-25** *Check the collector voltage of the red output transistor if no red can be seen in the picture.*

Test Q703 in the circuit for leakage. Remove the transistor and test it out of the circuit. Before replacing a leaky transistor, check the isolation resistor R505 (1.5 Ω) and D505 for burned or open conditions. Make sure that the power source is functioning before replacing the transistor. Test the new replacement in a transistor tester before installing it.

Only a horizontal white line

If only a horizontal white line can be seen on the picture tube, there is no vertical sweep. The vertical sweep has collapsed down into a white line. Go directly to the vertical output circuits. The early vertical output circuits consisted of two power output transistors. Today, the vertical output circuits consist of one IC. Locate the vertical IC, which is mounted on a heatsink.

Scope the input terminal pin 4 of IC501 and output of vertical driver deflection IC (Fig. 20-26). If the drive waveform is present at pin 4, suspect a defective output IC501 or an open vertical yoke winding. Scope the output terminal pin 2 of IC501. If a large vertical pulse is found here, the yoke winding, coupling capacitor, or the ground return resistor might be open. Remember that the vertical output waveforms do not want to stand still and sometimes are difficult to obtain good waveforms.

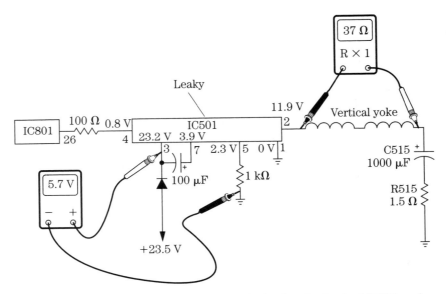

■ 20-26 *Scope the input and output terminals of vertical output IC501 and take critical voltage and resistance measurements to locate the defective component.*

Measure each terminal of IC for correct voltage and compare with those on the schematic. Make sure that the voltage supply source is found at pin 3. Low voltage here might indicate that an IC is leaky or that the dc supply source is defective. Take a resistance measurement on each terminal to determine if the IC is leaky or open. Remove one end of D503 and take a resistance measurement from pin 3 to ground. A low-resistance measurement under 1 kΩ indicates that an IC is leaky.

If the output pulse is found at pin 2 of IC501, check the vertical yoke windings (37 Ω). Next, check the return resistor R515 for open conditions. Shutdown the chassis and shunt C515 with another 1000-µF electrolytic capacitor. Often, if C515 dries up, you will still have insufficient vertical sweep. No vertical sweep occurs if C515 goes open. If C515 is open and if it is shunted, the vertical sweep returns. Replace C515.

Conclusion

How to take the symptoms, isolate the defective section, take critical voltage and resistance measurements, might quickly locate the defective component. Add scope waveforms and transistor tests and the electronic technician can locate the most difficult service problems. Here are only a few solid-state circuits that you must solve to become a successful intermediate electronic technician.

21

Correct part replacement

ALWAYS REPLACE THE DEFECTIVE PART WITH A COMPONENT that has the exact part number, when it is available. Exhaust all operations to obtain the correct replacement. The original replacement will not only function properly, but it will fit into the required space. Sometimes universal components are physically too large for the PC board (Fig. 21-1).

■ **21-1** *Replace the defective part with an exact or a universal replacement.*

Components with the exact part number should be used for critical specialized circuits. In some TV chassis, VCRs, and camcorders, critical circuits require an exact part replacement. But when the correct part can't be found, universal replacements must be used to put that electronic product back into service.

Original components are best

Choose components with the correct part number when replacing motors, video heads and drum assemblies, tape reels, audio heads, spindles, idlers, and mechanical levers in the VCR chassis. When exact part numbers can't be found through manufacturers' sources for the VCR or camcorder, check mail-order firms for exact replacements in idlers, rollers, clutch, cam gears, post assemblies, mode switches, gear assemblies, winders, and guide posts. Original pinch rollers, sensing lamps, pulleys, power transformers, belts, controls, relays, and RF modulators can be purchased from large mail-order firms or local electronics stores.

Critical parts found in the CD player are actuators, gear train assemblies, optical assemblies, drive motors, and mechanical assemblies. Large LSIs should be obtained from the manufacturer (Fig. 21-2). Check the manufacturer's service depots for critical components.

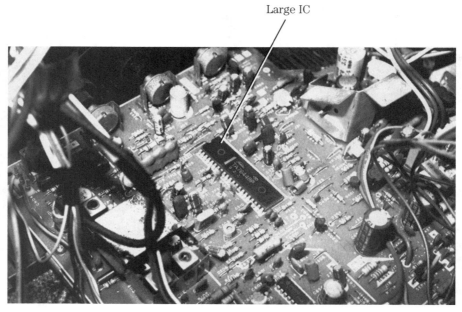

■ **21-2** *Large LSIs and microprocessors should be replaced with regular part number, but some ICs can be replaced with universal types.*

Most components found in the TV chassis can be replaced with original or universal parts. Some LSIs or critical microprocessors should be obtained from the manufacturer. Special flyback transformers, focus and screen controls, tuners, and transformers should be original part numbers. Check mail order firms and local electronics stores for original foreign TV components. Original Japanese semiconductors from NEC, Sanken, Fugitsu, Sharp, Toshiba, OKI, Sanyo, Sony, and Rohm can be obtained from electronics mail order firms.

Universal replacements

Semiconductors, such as transistors, ICs, LSIs, and SMDs, can now be replaced with universal parts for most electronic products and equipment. Some tape and VCR heads can be replaced with universal parts. Motor drive belts and tape idlers can be replaced with universal parts. Of course, resistors, bypass and electrolytic capacitors, choke coils, diodes, variable controls, and some SMD parts can be subbed with universal replacements. Each year, more universal SMD parts are becoming available. When the original component is not available, install the universal replacement (Fig. 21-3).

Varactor type
TV tuner

■ **21-3** *Large components, such as tuners, must be replaced with an exact part or sent back for repair. Small parts can be replaced with universal components.*

Transistor replacement

Just about every type of transistor has universal replacements. Of course, you might run into a special circuit, where the original transistor must be used to make the chassis perform (Fig. 21-4). In the TV, VCR, CD, and radio chassis, look for the transistor replacement in the parts list of Howard Sams Photofacts or check the original part number within the semiconductor replacement manual. Obtain the correct part number from the manufacturer's schematic.

■ 21-4 *The tuner and tuner-control modules in this RCA TV chassis must be replaced with an exact part or it must be sent in for repair.*

Most transistors found in auto and home radio chassis can be replaced with universal parts without any problems. Special ICs or microprocessors found in the present-day receivers might not be replaced with universal components. Often, the new AM/FM/MPX receivers will not have universal replacements on the market to replace the latest semiconductor (Fig. 21-5). In this case, try to

■ **21-5** *Small components within the AM/FM/MPX radio can be replaced with universal parts, except for the dual variable capacitor.*

obtain a component with the original part number. Brand new component numbers might have to be replaced with originals.

Transistors found in the TV chassis can be subbed with universal replacements without any service problems. Most semiconductors found in video, sound, AGC, IF, horizontal, and vertical circuits can be replaced with universal parts. Look up the part number found on the horizontal output transistor or in the schematic and compare to universal replacements in the semiconductor manual. Make sure that the figure and part number, and horizontal output transistor has a damper diode inside or outside in the horizontal circuits. Make sure that the replacement will fit in the circuit.

Typical horizontal output universal replacements are found in Table 21-1 and horizontal output universal parts are found in Table 21-2. Universal horizontal output transistors can be cross-referenced from RCA (SK numbers) to Phillips (ECG) and NTE replacements.

■ Table 21-1 Horizontal universal output and original transistors.

Transistor number	Voltage	Amps	Watts	Universal replacement
2SC1413A	1500 V	5 A	50 W	SK3115 ECG82
2SD870	1500 V	5 A	50 W	SK9119 ECG89
2SD871	1500 V	6 A	50 W	SK9119 ECG689
2SC1172B	1500 V	7 A	50 W	SK3710 ECG238
2SD822	1500 V	7 A	50 W	SK3115 ECG682
BUY69A	1000 V	10 A	100 W	SK3467 ECG283
2SD1453	1500 V	3.5 A	50 W	SK9422
2SD1455	1500 V	5 A	50 W	SK9422
2SD1497	1500 V	6 A	50 W	SK9476

SK=RCA   ECG=Sylvania

■ Table 21-2 Horizontal output transistors with damper diodes.

Transistor number	Voltage	Amps	Watts	Universal replacement
2SD900	1500 V	5 A	50 W	SK9119
SD1554	1500 V	3.5 A	40 W	SK9422
2SD1650	1500 V	3.5 A	40 W	SK9422
2SD1555	1500 V	5 A	50 W	SK9422
2SD1398	1500 V	6 A	50 W	SK9422
2SD1556	1500 V	6 A	50 W	SK9422
2SD1651	1500 V	5 A	60 W	SK9422
2SD1652	1500 V	6 A	60 W	SK9422
2SD1441	1500 V	4 A	70 W	SK9422
2SD1426	1500 V	3.5 A	80 W	SK9422
2SD1427	1500 V	5 A	80 W	SK9422
2SD1428	1500 V	6 A	80 W	SK9422

Correct part replacement

When replacing defective transistors with universal replacements, make sure that the working voltage, circuit application, and mounting applications are correct. Usually, with the correct part number, most universal replacement transistors can be found in the replacement semiconductor manual. If in doubt, try again to locate the original replacement (Fig. 21-6).

■ **21-6** *The horizontal output transistor located on heatsink can be replaced with universal transistors.*

IC replacement

Although all ICs and LSIs cannot be replaced with universal replacements, most can be replaced with universal replacements in the radio, cassette and tape players, and audio amplifiers. IC regulators can be replaced with universal replacements. Common ICs found in the TV chassis can be replaced with universal parts. Simply check the IC part number stamped on top of the IC. Always replace SMD LSIs and ICs with components that have original part numbers.

Diode replacement

Most silicon and germanium diodes can be replaced with universal types. Check the correct working voltage source of a zener diode replacement. If in doubt of the zener diode wattage, replace it with a 1-W or a 5-W diode. The larger the size of diode, the higher the wattage. Defective diodes in the low-voltage power supply can be replaced with 2.5-A diodes at 1 kV. Check the size of the silicon diode. Detection diodes are replaced with 1N34A and 1N60 types (Fig. 21-7).

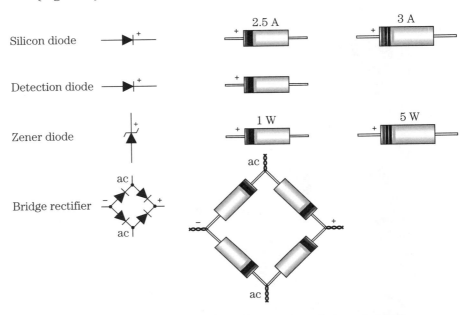

■ **21-7** *A list of the different diodes with correct symbols and polarity.*

The bridge rectifier might appear in full-wave or half-wave units. The 4-A bridge rectifier will replace most full-wave bridge circuits in the radio, tape recorders, and cassette and CD players. If a certain bridge circuit is not available, make the bridge circuit out of individual silicon diodes (Fig. 21-8). Make sure that the diodes have the correct peak inverse voltage and amperage. The 2.5-A 1000 piv silicon diode will replace most high-voltage, high-current diodes in power supplies.

SMD replacements

Always try to obtain the correct SMD replacement from the manufacturer outlets. Although there are universal resistors, bypass capacitors, and diodes at many electronics stores and mail-order firms that can be used as replacements. Critical LSIs and ICs should be replaced with original SMD replacements. Replace chip transistors and special digital transistors with original part num-

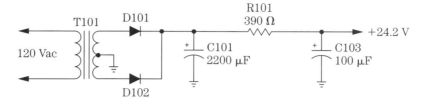

■ **21-8** *Silicon diodes in the full-wave rectifier circuits can be replaced with 2.5-A 1000-V diodes.*

■ **21-9** *Small SMD components are used underneath of the PC board in this 1994 RCA portable. Notice the round resistors.*

bers. Universal chip resistors might be of ½-W size and bypass capacitors up to 50 V (Fig. 21-9).

Resistor replacements

Universal resistor replacements come in fixed, variable, carbon, wire wound, power, and power standoff types. Fixed resistors can be wired in series to obtain the correct total resistance or paralleled for greater wattage. For instance, if you need a ½-W 1-kΩ resistor and do not have the correct resistance, connect an 820- and 180-Ω resistor in series. When you need a 1-kΩ 10-W wire wound resistor in a power circuit and have two 2-kΩ 5-W resistors on hand, wire the resistors in a parallel circuit (Fig. 21-10). Make sure that the defective resistor replacement has the same resistance and wattage as the original.

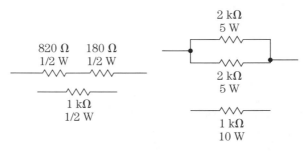

■ **21-10** *Resistors can be placed in series for additional resistance or in parallel for less resistance and greater wattage.*

Of course, you can replace a defective resistor with a higher wattage resistor, if there is room to mount one. For instance, you need a 10-kΩ ¼-W resistor for replacement. Can you replace the ¼-W resistor with a ½-W of the same resistance? Of course, a 10-kΩ ½-W resistor will replace the ¼-W resistor (Fig. 21-11). When replacing 5-, 10-, or 15-W resistors, mount them at least ½ inch above the PC board. Push the terminals through the board hole, form a small circle with long-nose pliers, and make a good soldered connection on PC wiring. Universal power standoff resistors can be replaced with regular power resistors for 5-, 7-, and 10-W sizes.

In critical RF applications, FM circuits use 1 or 5 percent tolerance resistors for replacement. When two resistors are wired in series in place of one resistor, make the soldered junction at the top so that each end of the resistor will fit in the correct holes (Fig. 21-12). Cut the resistor terminals for very short leads in critical RF, FM, and high-frequency circuits.

Variable resistor replacements

Variable controls can be replaced with universal types with a few minor changes. Select the correct resistance and taper in the universal replacement. Linear controls are used in most circuits, but audio taper controls are found in sound circuits. If the universal control shaft is too long, cut it to length with the hacksaw. Univer-

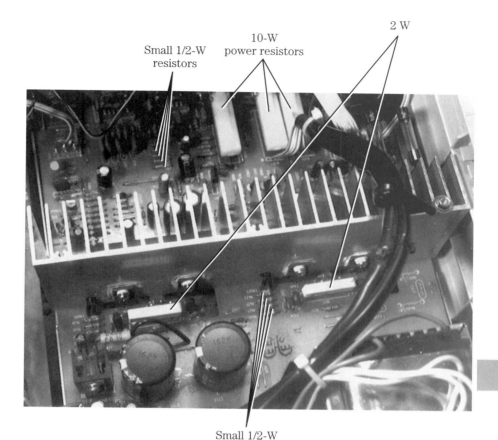

■ **21-11** *The different 10-W, 2-W, and 1/2-W resistors on a power amplifier PC board.*

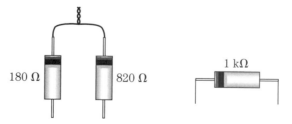

■ **21-12** *When wiring resistors in series to acquire the correct resistance, place the soldered junction at the top of the mounting.*

Variable resistor replacements

sal controls come in knurled, half round, and full round shafts for replacement. Round or grind off the rough corners. Make sure that the correct length is the same as the original from the base of the control (Fig. 21-13).

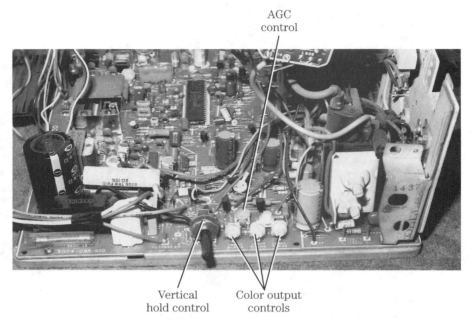

■ **21-13** *The different resistance controls, screen, focus, red, green, and blue color output bias controls, AGC, and vertical hold on the rear chassis of a color portable TV.*

Replace the combination on/off switch and control with universal parts. Most universal variable resistors have provisions to mount a switch on the rear of the control. Some universal controls have switches mounted on the rear of the control. Make sure that the resistance, taper, and wattage are the same as the original. PC-mount and variable trimmer potentiometers can be found at most electronics stores. Don't tie up the electronic product for a control replacement; replace it with a universal part.

Fuse replacement

Fuses are available in many different sizes and shapes. Some fuse types are regular, fast-acting, slow-blow, ceramic, pigtail, cartridge, pico/micro, chemical, and thermal fuses. Fuse amperage might start at $\frac{1}{16}$ of an amp up to 30 A with 32-, 125-, and 250-V rating, protecting electronic products. Plug-in and pigtail fuses are found in the TV chassis. Ceramic fuses are used in microwave ovens (Fig. 21-14). Thermal fuses are used in heat guns, hair dryers, transformers, and small household appliances. Subminiature pico/micro axial fuses are used where space is limited.

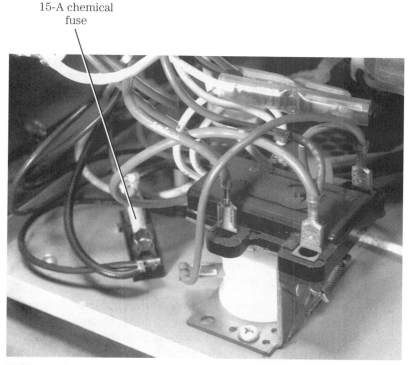

■ **21-14** *The 15-A chemical fuse in the microwave oven can be clipped into a fuse holder.*

Replace with the same type of fuse found in the electronic circuits. Of course, the pigtail fuse can be clipped into the same clips as a standard plug-in fuse. Make sure that the amperage is the same with correct working voltage. Replace the slow-blow fuse with one that can handle heavy current and fast turn-on voltages. Fast-acting fuses should be replaced in critical circuits for adequate protection. Do not wrap tin foil around any fuse when one is not available. Likewise, do not replace a higher amperage fuse with a lower amp fuse. You might quickly damage components in those circuits with improper fuse protection (Fig. 21-15).

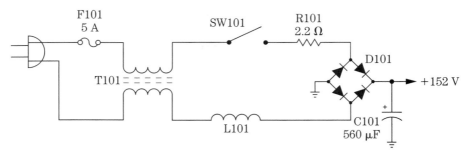

■ **21-15** *A line fuse of 4 to 5 A is connected to one side of the power line in a TV low-voltage power supply.*

Capacitor replacement

Capacitors used in electronic equipment have leads that are usually radially or axially mounted. The radial capacitor stands vertically, but the the axial capacitor lays down horizontally. Bypass, filter, memory backup, high-frequency/high-temperature, nonpolarized, tantalum, polyester film, mylar, ceramic, NPO, silver mica, disc, safety, surface-mounted (SMD), and metallized film capacitors are available. Electrolytic capacitors are found in filter, decoupling, and coupling circuits (Fig. 21-16). Memory capacitors are used in VCRs, clocks, and computer circuits, where power is removed from the circuits.

Ceramic disc, polyester film, and metallized film capacitors are found in bypass and coupling circuits. Mylar and nonpolarized capacitors are located in custom and speaker crossover systems. NPO and silver mica capacitors are used in stable high-frequency circuits. High-voltage safety capacitors are found in the horizontal output transistor circuits to prevent or to hold down the high-voltage circuits.

Capacitors can be paralleled to increase the capacity in a circuit, but capacitors are wired in series to decrease the capacity (Fig. 21-17). For instance, if you need to replace an electrolytic filter capacitor in the TV power supply of 680 UF - 250 V and have a dual 450 UF and 300 UF at 300 V, replace it with this capacitor. Observe the working voltage of electrolytic capacitors. The working voltage and capacity must be the same or higher. Never replace a 47-UF 50-V capacitor with a capacitor that is rated at 47 UF at 16 V. The capacitor might heat up, short out, and blow fuses. You can replace the 47-UF 50-V capacitor with a 100 UF at 50 V without any problems.

Correct part replacement

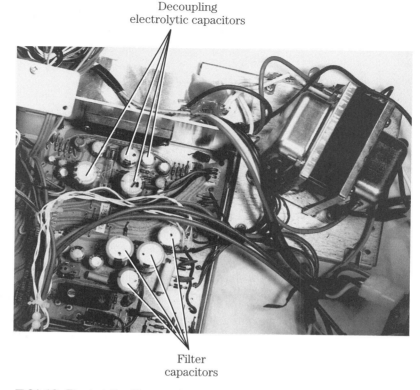

■ **21-16** *Electrolytic filter and decoupling capacitors are used in this AM/FM/MPX receiver.*

■ **21-17** *Connect the capacitor in parallel to add capacitance, and in series to decrease capacitance.*

Small bypass and coupling capacitors can be subbed in audio circuits if the voltage is the same or higher. Stock higher voltage capacitors with the same capacity for stability, breakdown, and no call backs. You can sub a 0.001-UF 1000-V capacitor with a 0.002-UF capacitor, or a 0.0022-UF 1000-V capacitor with no problems (Fig. 21-18). But do not change the capacity of NPO and silver mica capacitors in critical RF or high-frequency tuning circuits. Likewise you do not sub a regular capacitor in a nonpolarized

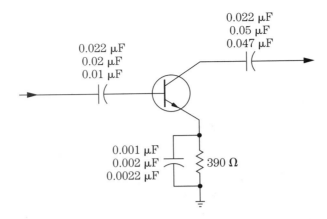

21-18 *The various coupling and bypass capacitors can be replaced with different values if parts with the exact capacitance are not handy. Make sure that working voltage is the same or higher.*

crossover speaker network. High-voltage safety capacitors should have the same capacity and at least 1.6 kV or higher voltage.

Axial electrolytic capacitors can be easily subbed for radial mounting capacitors. Bend over one terminal and place the correct polarity terminal close to the mounting hole. Larger radial or axial mounted capacitors can be subbed if there is enough mounting space. You can mount larger radial capacitors above the PC board with longer terminals and spaghetti used as spacers and insulators (Fig. 21-19). Bend larger radial capacitors over the top of other small components and hold them in place with a dab of glue or with rubber cement. Large electrolytic capacitors can be mounted on metal brackets above the chassis with extended hookup wire leads to the correct power supply circuits.

Belt replacement

Belts of every description are found in VCR, camcorder, CD, phono, and cassette players (Fig. 21-20). Besides motor drive belts, pulley and idler belts are found in most mechanical drives of electronic products. You can replace idler rubber tires on clutch, pulley, and take-up assemblies instead of replacing the whole assembly. Simply match the tire that you want to replace from the drawings that are found in most electronic catalogs. Try to replace with originals, if possible. Besides the manufacturers' exact part numbers, they might be secured from mail-order firms.

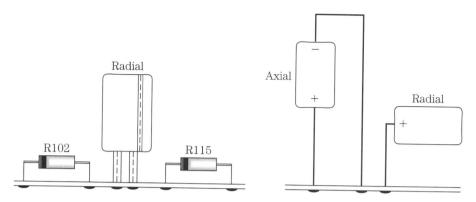

■ **21-19** *Larger radial and axial capacitors can be replaced on the PC board, if the space is limited.*

■ **21-20** *Here, three different drive belts are used in a J.C. Penney VCR.*

If the original belt is not available, identify the width and circumference as well as the part number. Lay the bad belt on the belt and idler drawings and order the exact universal replacement. To obtain belts for motor driven turntables, lay the old belt flat along a flat ruler and measure the inside diameter for the belt length (Fig. 21-21). You can pick up belt gauges for a quick and accurate way to measure the IC, thickness, and width of belts used in elec-

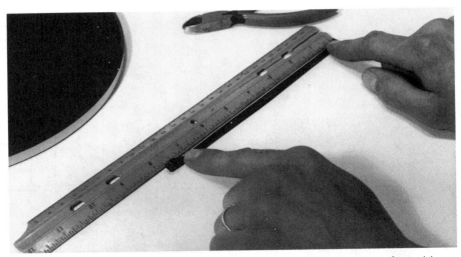

■ **21-21** *Lay the belt out flat and measure the inside diameter of an old belt to obtain the correct size of the belt.*

tronic equipment. VCR drives, and flat and square belt kits contain 90% of all belts used in electronic products.

Belt slippage can be caused by oil on the belt surface, glazing effects, and shiny areas on old belts, idler wheels, and pulleys. Besides alcohol, a rubber drive cleaner will clean and revitalize rubber belt areas. Liquid rosin might help to prevent belt and idler slippage.

Battery replacement

Alkaline and carbon batteries are replaced with AAA cell, AA penlite, C, and D cells at 1.5-V rating. Larger voltage cells of 6-V and 9-V cells are used to power battery radios and lanterns. Lithium batteries are small button-coin batteries for memory backups, hearing aids, cameras, and small radio applications. Nickel-cadmium batteries last longer, can be charged up over 1000 times, and are available in 1.2- and 9-V sizes. Although the nickel-cadmium batteries cost more they last for years, if they are charged properly. Nicad rechargeable communication batteries are used in hand-held transceivers and telephones.

A battery charger might have the capability to charge 4- or 8-V, D, C, AA, AAA, 6- and 9-V square nicad batteries (Fig. 21-22). You can purchase these chargers from $10.00 up to $25.00. If you have lots of projects or test equipment that operates from batteries, install nickel-cadmium batteries, and charge them up when needed with the battery charger.

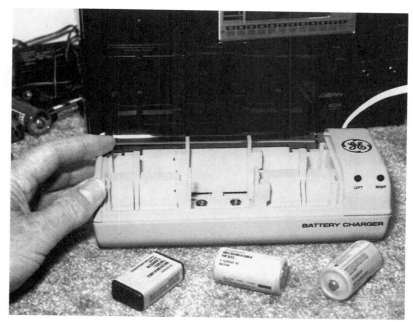

■ **21-22** *Replace all carbon and alkaline batteries in battery-operated test instruments that can be charged with a nicad battery charger.*

Mercury cells of 2.7, 5.4, 5.6, and 7 V consist of mercury oxide and are found in cameras and electronic toys. Sealed lead-acid cells provide reliable, maintenance-free, chargeable lead-acid batteries. Lead-acid cells are found in the field of security, computer, audio, video, telecommunications, communications, and cellular telephones.

The camcorder battery can be charged over and over again. Camcorders might have 6-, 9-, and 12-V chargeable batteries (Fig. 21-23). Let the battery run down, when the battery warning light comes on, or when the camcorder fails to function. Remove the battery and clip a 100-W light bulb across the battery terminals for complete discharge. First solder clip-on wires on the base of the light bulb. Then, recharge the camcorder battery in the camcorder battery charger. Most camcorder operators have an extra battery to replace the discharged battery, so as not to interrupt camcorder picture taking.

Switch replacement

Replace special switches (function etc,) with ones that have the exact part number. Toggle and slide switches can be replaced with

■ **21-23** *6-, 9-, and 12-V nicad batteries can be charged with the battery charger that came with the camcorder.*

universal types because they can be picked up at Radio Shack or most any electronics store. Rotary switches with several poles and positions should be replaced with original types (Fig. 21-24). High-current (20 to 25 A) toggle switches and auto speaker selector switches can be replaced with universal replacements. Handy selector switches for TV video games can be universal switches. Replace on/off switches that fit on the rear of volume and tone controls with universal components.

Speaker replacement

Large high-wattage speakers should be replaced with ones that have original part numbers. Most speakers can be replaced with the correct size of universal speaker, provided that the impedance, physical size, wattage, and magnet weight are the same. Speaker impedances might vary from 3.2, 4, 6, 8, 16, and 32 Ωs. The speaker size might start at 1 inch up to 18 inches. The speaker wattage might vary from 0.1 up to 800 W of power (Fig. 21-25).

The woofers range from 5 inches up to 18 inches in diameter. The woofer speaker might have a 4-, 6-, 8-, or 32-Ω impedance with the majority woofers rated at 8 Ω. A midrange speaker might be a horn, diffraction horn, or a midrange driver speaker. The midrange speaker might have a 8- or 16-Ω impedance. The tweeter type speaker might be round, square, horn type, piezo element in round and square types, with a 6- or 8-Ω impedance. When replacing damaged speakers with universal replacements, make sure that

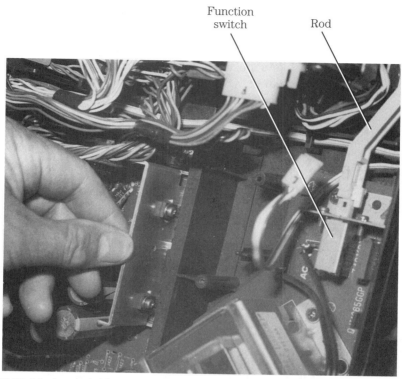

■ **21-24** *A special function switch operated by a plastic arm must be replaced with original part number in an AM/FM/MPX receiver.*

■ **21-25** *Here, a large 12-inch speaker is used as a woofer and 2-inch speaker is used as a tweeter in a speaker cabinet.*

the voice coil, physical size, wattage, and speaker magnet weight are the same as the original.

Auto speakers are available in many sizes and shapes. Some car in-dash speakers must be replaced with originals because of the shape of the speaker. Oval auto speakers are available in 4" × 6", 5" × 7", 4" × 10", and 6" × 9" units. The older auto speaker might have an impedance of 3.2 Ω, but the latest speakers have 4-, 8-, 10-, and 16-Ω voice coils.

Parts that can't be substituted

Large components (such as tuners, mechanical assemblies, digital control systems, and cabinets) must be replaced with original parts (Fig. 21-26). TV tuners can be sent in for repair at factory or tuner repair depots. Some large complete PC boards might be ex-

■ **21-26** *Tuners and tuner-control modules, mechanical-control systems, and cabinets must be replaced with exact parts in the TV chassis.*

changed or serviced by the manufacturer or by repair depots. Camcorder and VCR mechanical assemblies should be replaced by assemblies that have the original part numbers. CD player optical units must be replaced by the exact part number. Damaged cassette player assemblies might not even be available. Obtain special power transformers from the manufacturer. Some mechanical assemblies can be repaired by ordering the damaged original pieces.

Incorrect part numbers

Important part numbers are listed on the schematic parts list or on the component. ICs, transistors, and power transformers might have correct part numbers stamped on the body area. Obtain the correct schematic to find the correct part number. Howard Sams Photofacts has the correct part numbers listed with universal replacements. Check the manufacturers' service schematics for correct part numbers. You can check the different Photofacts for auto radios, VCRs, AM/FM receivers, TV sets, clock radios, and cassette players. If the original part number is not available, by trying many sources, place the electronic product back into operation with universal replacements. Do not tie the electronic product up for months or years; try a universal replacement.

22
Troubleshooting and repairing radio circuits

AM/FM RECEIVERS START WITH THE POCKET RADIO, CLOCK, earphone radios, deluxe table model receiver with cassette player, compact AM/FM/MPX units, shortwave, and ham radio receivers (Fig. 22-1). Small AM/FM radios appear in stuffed animals, on bikes, and even in peculiar places, such as in old-time Atwater Kent radio replicas. The AM/FM/MPX receiver might be included in boomboxes, CD players, compact stereo systems, and in high-wattage receiver designs.

■ 22-1 *Radio receiver circuits are used in tape decks, CD players, shortwave receivers, and amateur transceivers.*

Servicing the radio (large or small) is quite simple; only a few circuits are different compared to the TV chassis. Most service symptoms are related to the audio and low-voltage power supplies. Weak radio reception symptoms might relate to leaky RF or audio stages. If no stations can be tuned in with excessive noise, the mixer or oscillator circuit might be defective. Noise in the audio might be caused by pickup noise or defective components within the IF and audio circuits. Excessive distortion in the speaker might result from a defective speaker or from leaky audio stages. Usually, hum in the sound is caused by defective filter capacitors in the low-voltage power supply.

The portable radio

The small portable radio might contain only a few transistors or IC components (Fig. 22-2). Usually, a converter stage provides RF, oscillator, and mixes signals to convert the received signal down to 455-kHz IF frequency. One or two stages of transistors or one IC might amplify the 455-kHz IF frequency. The AM signal is detected by a detection diode and switched to the audio circuits. The FM 10.7-MHz IF signal is amplified and inserted into a multiplex circuit (MPX), detected, and applied with two FM stereo channels to the function switch (Fig. 22-3). Some portable radios do not have FM stereo reception.

■ **22-2** *The small portable radio has only a few transistors or ICs.*

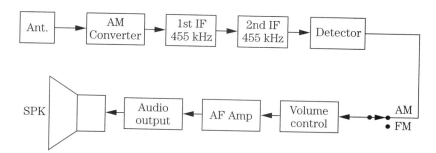

■ **22-3** *The FM/AM function switch selects either AM or FM to the audio circuits.*

The function slide switch switches to the AM or FM output circuits. One or two AF audio circuits might occur before the audio signal is controlled by the volume control. The AF driver transistor amplifies the weak audio signal to the power output stages. This output signal is capacitor- or transformer-coupled to the speaker and earphones. The earphone radio might have two small speakers used in the headset radio (Fig. 22-4).

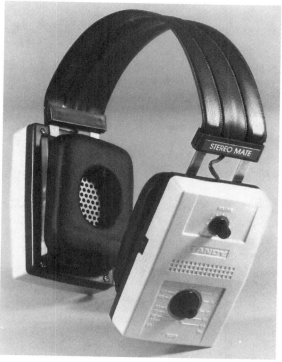

■ **22-4** *The portable stereo has a set of headphones.*

AM circuits (amplitude modulation)

The local AM radio broadcast station is picked up by a ferrite rod antenna that contains the antenna coil in the RF section of the radio. In early radios, tubes, and then transistors were king. Today, these small radios might have a few transistors, but most are controlled with ICs. The antenna coil or RF stage might connect directly to an RF, oscillator, IF, and detector IC.

AM converter

The AM converter stage in the small radio might still use a transistor or FET in a simple converter circuit (Fig. 22-5). The antenna coil is in the base circuit with the oscillator fed to a common AM/FM transistor or IC IF stage. The $B+$ voltage is switched into the collector or drain terminal with the AM/FM switch. The FET transistor has a high-impedance input compared to the regular transistor, with gate, drain, and source terminals. The antenna coil is tuned with a variable capacitor.

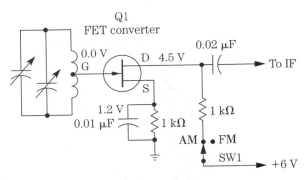

■ **22-5** *The AM converter stage might consist of an RF or FET transistor.*

The AM transistor converter might contain the RF, oscillator, and mixer circuits in one transistor. The antenna coil ferrite rod picks up the AM stations and is tuned with a variable capacitor. The oscillator circuit is found in the emitter circuit with the feedback coil and 455-kHz IF in the collector circuit. The 455-kHz signal might be tied to the AM/FM IF transistor or IC (Fig. 22-6). A variable capacitor tunes the AM and oscillator circuits. The deluxe AM receiver might have an additional RF tuned stage ahead of the converter transistor.

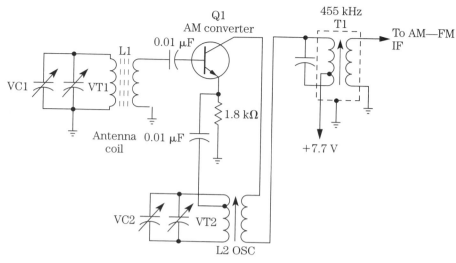

■ **22-6** *The AM converter transistor serves as RF amp, oscillator, and mixer circuits.*

FM RF and converter circuits

The FM (frequency modulation) RF transistor or FET might have a separate RF stage or combined with mixer and oscillator in the FM RF/OSC/mixer/IF IC. The FM signal is coupled to the FM converter circuit with C5. Notice that the FM signal on both RF and converter transistors enters the emitter terminals and the output at the collector. The FM oscillator circuits are connected directly to the FM 10.7-MHz IF stage (Fig. 22-7). The base terminals are tied to the *B*+ voltage source.

The FM RF amp/FM converter circuits might be included in one IC. The FM signal is picked up by a dipole antenna and fed through an input filter network and to pin 1 of IC1. The oscillator coil is taken from pin 7 through C1 (15 pF) and to the oscillator coil and variable capacitor. The RF coil is found at pins 3 and 4 of IC1. The latest AM/FM circuits might be tuned with varactor or varicap diodes. The 10.7-MHz IF signal at pin 6 connects directly to the FM IF transformer (Fig. 22-8).

Intermediate-frequency (IF) circuits

The AM IF circuits might feed directly into the second FM IF amp or into the IC FM circuits. The first AM IF stage might be separate or it might be combined with the first or second FM IF stage. Usu-

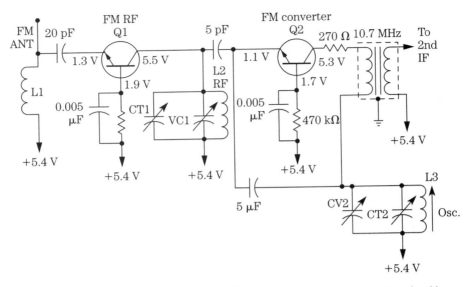

■ **22-7** The FM RF transistor amplifies the weak FM signal, mixes it with the FM oscillator, and feeds it to the 10.7-MHz IF transformer.

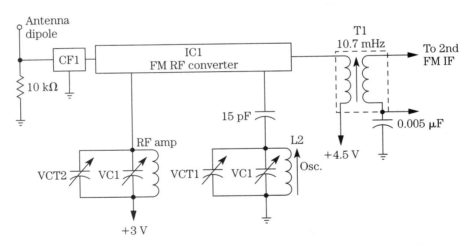

■ **22-8** The FM RF oscillator and mixer IC provides a 10.7-MHz resultant frequency to the first FM IF transformer.

ally, when both AM and FM circuits are dead or weak, suspect that an IF transistor or IC is defective. The AM and FM IF circuits are common to one another with the AM 455-kHz IF transformers in series with the FM IF (Fig. 22-9). The second FM IF transistor amplifies both AM and FM IF signals, with FM at 10.7-MHz and AM IF at 455-kHz frequency.

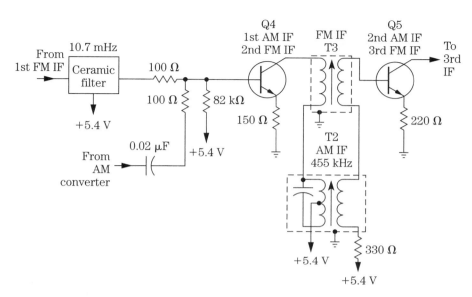

22-9 *The combination AM and FM IF circuits are amplified by Q4 and Q5.*

In FM IF circuits, one large FM/AM IF amp/detector IC might serve as both AM and FM IF circuits. The FM IF signal is coupled into pin 12 of IC101. The AM signal is fed into pins 13, 9, and 15, which serve as the 455-kHz IF stages. The FM tunable IF (10.7 MHz) is found at pin 6 (Fig. 22-10). The AM signal is detected inside IC101 and output is found at pin 2. The FM output is fed to the FM/MPX

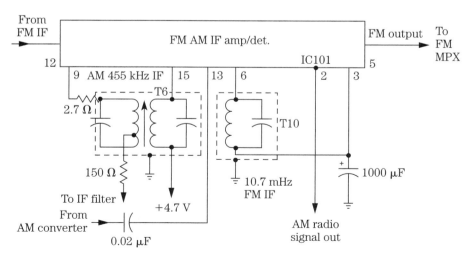

22-10 *The AM and FM IF circuits might be included in one large IC.*

Intermediate-frequency (IF) circuits

IC102. When the AM/FM broadcast stations are noisy, weak, or dead, suspect that the common FM/AM IF IC is defective.

FM MPX circuits

The frequency modulation (FM) multiplex (MPX) circuit is a special circuit found in FM receivers for stereo reception from a broadcast station transmitting an FM multiplex signal. Both channels (left and right) are broadcast on a single carrier wave. The FM band is at a higher frequency than the AM band—running from 88- to 108-MHz frequency.

The FM IF signal is fed into pin 2 of IC2. The FM stereo indicator light circuit is tied to pin 6 with pin 7 as the 19-kHz adjustment control (RV1). The left and right FM stereo sound output is found at pins 4 and 5. Electrolytic capacitors couple the FM audio signal to the respective FM stereo function switch terminals (Fig. 22-11). The multiplex IC brings in both FM stereo signals in one carrier wave and separates the FM signal to the left and right channels of stereo audio. The function switch applies the left and right audio signals to the left and right AF and audio output circuits.

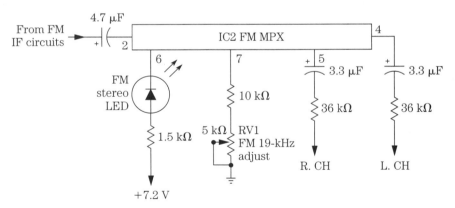

■ **22-11** *The FM/MPX circuits are included in IC2 with stereo FM output at terminals 4 and 5.*

The clock radio

The clock radio might have both AM and FM monaural radio reception. The clock table model radio might have features, such as wakeup settings, snooze, and sleep controls (Fig. 22-12). Some radios wake you up to an alarm, radio, or your favorite cassette music. For working couples, the clock radio might have two different wakeup times.

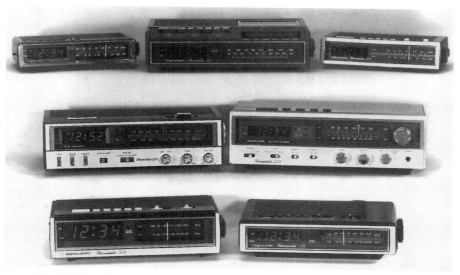

■ **22-12** *The clock table model radio has many different circuits and features.*

The AM/FM circuits in the clock or table model radio are similar to the small portable radios (Fig. 22-13). Transistorized circuits might be found in all stages or ICs with transistors in other radios. Usually, one AM converter serves as RF, oscillator, and mixer circuits. The AM converter signal is fed to the first AM IF stage coupled to the second FM IF transistor. In some clock radios, you might find only two AM IF circuits; in the deluxe clock radio, three 455-kHz IF stages are found. The simple IN60 detection diode rectifies the IF signal to audio.

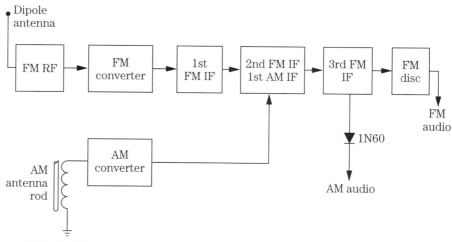

■ **22-13** *The block diagram of the AM/FM clock radio circuits.*

The FM clock radio circuits might contain separate FM RF and FM converter transistors. The FM IF circuits might contain three 10.7-MHz IF stages. The FM IF signal is applied to a discriminator stage that detects the FM audio signal (Fig. 22-14). The discriminator coil (T5) is adjusted for a clear and noise-free FM radio signal. The 1-kΩ resistor limits the FM audio signal fed to the AM/FM.

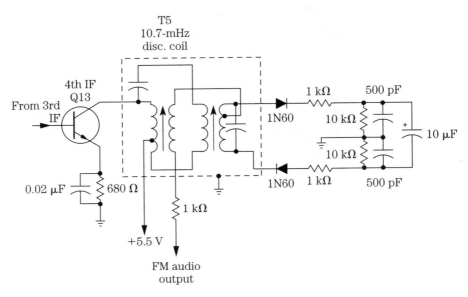

■ **22-14** *The typical transistorized FM discriminator circuits in the mono clock receivers.*

The early clock radio audio circuits might contain three or four audio stages with a speaker. Today, the clock radio circuits might contain one large audio IC. The audio IC amp input signal is controlled by the volume control (VR3) and capacity coupled to input pin 3 of the output IC. The voltage supply pin of IC5 is at pin 6. Audio output pin 5 is coupled to the earphone or speaker with C19. These speakers might have 8- or 16-Ω impedance (Fig. 22-15).

Shortwave radios

The shortwave radio has come a long way since Grandpa's shortwave radio project with tubes in 1934. Today, the shortwave radio might have digital tuning on all bands of the AM/FM/SW/LW frequencies. Besides digital tuning, some shortware radios have a dual-time LCD display and 45-station memory, a beat-frequency oscillator (BFO) to hear side bands (SSB) voice, and CW (code). Scan tuning and CB lis-

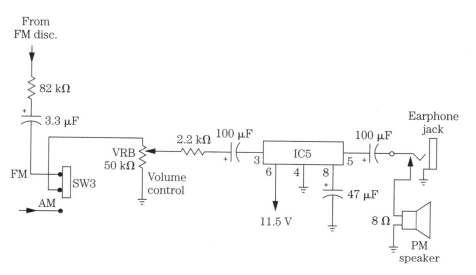

22-15 *IC5 amplifies the AM or FM clock radio signal. It is capacitively coupled to the speaker.*

tening might be the added feature in larger SW receivers (Fig. 22-16).

The SW receiver is built like any superhet with added features and switched coils to cover the AM/FM and SW bands. The 49-meter band covers 5.85 to 6.20 MHz, 41 meters (from 7.05 to 7.45 MHz), 31 meters (from 9.45 to 9.90 MHz), 25 meters (from 11.60 to 12.00 MHz), 21 meters (from 13.55 to 13.85 MHz), the 19 meters (from 15.10 to 15.60 MHz), 16 meters (from 17.45 to 18.00 MHz), 13 meters (from 21.45 to 21.95 MHz), and the 11 meters (from 25.70 to 26.15 MHz).

22-16 *The superhet shortwave receivers included many different bands with switched coils.*

The beat-frequency oscillator (BFO) circuit provides an oscillating frequency that is tuned to the receiver to receive sideband (SSB). Ham voice and continuous wave (CW) transmissions can also be received. Sidebands might appear on the upper and lower side of the main frequency of the ham operator's voice transmitters. Without the BFO circuits, the operator's voice might sound like a cartoon duck. When the BFO signal is "beat" against the IF frequency, the voice becomes distinct and clear.

Variable capacitor tuning

The small portable radio and early deluxe receivers have variable capacitors that can tune in the various stations of the AM and FM bands. The RF stage's variable capacitor tuned the incoming signal while the oscillator or converter circuits tuned the oscillator frequency with fewer metal plates (less capacity). For instance, if the AM broadcast station was tuned to 1400 kHz, the oscillator frequency would be tuned to 945 kHz, with resulting frequency of 455 kHz for the IF circuits. The RF variable capacitor has more capacity or meshed metal plates than the oscillator capacitor (Fig. 22-17).

■ **22-17** *The variable capacitor in the AM/FM radio might have an RF and oscillator section in the AM and FM section. The FM section has fewer plates, compared to the AM variable capacitor section.*

Likewise, the capacity of a variable capacitor is greater when all rotor plates are meshed and tuned to a lower radio station on the dial. If the plates are almost out of the station plates, the capacitor has less capacity, stations are tuned in higher on the AM broadcast band. Most AM radios tune from 530 to 1700 kHz. The deluxe receiver might have an added or separate RF stage for greater sensitivity and pickup with a 3-ganged variable tuning capacitor.

Permeability tuning

Along came permeability tuning where the inductance of the RF, oscillator, and mixer coils were changed with the magnetic core of each coil. The variation of the AM/FM frequency of an LC circuit can be changed by shifting or moving the magnetic core within the inductor. Besides front end radio tuning, permeability tuning is accomplished by changing or adjusting the IF coils, oscillators, filters, wave traps, and RF coils with screwdrivers or insulated tuning tools. This type of tuning is found in home and auto radios (Fig. 22-18).

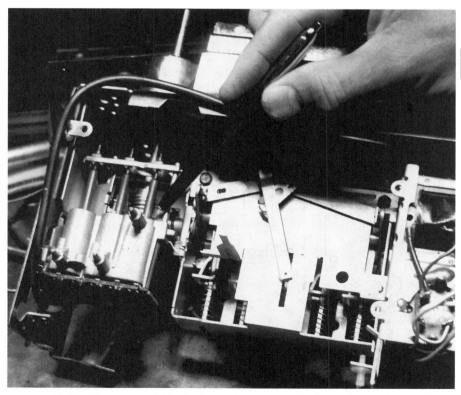

■ **22-18** *The permeability tuning coils are used in the auto receivers with pushbutton and manual tuning.*

In car radios, the RF, oscillator, and mixer coils are enclosed in a shielded assembly. Each coil has a magnetic iron core that enters into the coil area to tune the radio stations. The three magnetic cores are connected to a movable assembly that is turned by the tuning shaft of the radio. Each coil can be individually adjusted by the brass screw thread at each end of the permeability coil. When the magnetic cores are fully into the coils, the car radio is tuned to a broadcast station in the lower portion of the AM band.

Varactor or varicap tuning

Varactor tuning is found in many AM and FM receivers and TV sets. Varactor tuning is a method of tuning the RF, oscillator, and mixer coils with a varactor diode. The varactor diode acts as a variable capacitor. When a variable voltage is applied to the varactor diode, the diode will change in capacity (Fig. 22-19). The varactor diode is rated in picofarads (pF), just like the variable capacitor. The varactor diode has a limited range of capacity, compared to the variable capacitor.

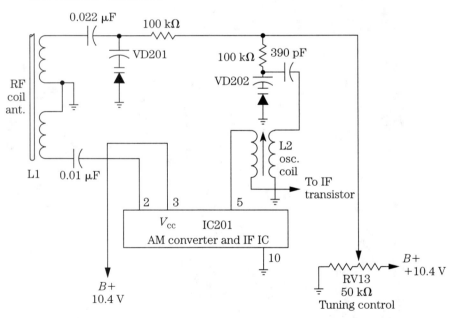

■ **22-19** *Varactor or varicap diode tuning is used in radio front-end circuits.*

In Fig. 22-19, varactor diode VD201 tunes RF coil L1 while VD202 tunes oscillator coil L2 in the AM receiver. Notice that 0.022-μF capacitor isolates the dc voltage from the ground in the RF varactor circuit, and the 390-pF capacitor does the same with VD202. These capacitors also couple the varactor diodes to the coil-tuning circuits. RV10 provides band tuning by placing a different voltage to the varactor diode, and changing the capacity in the RF and oscillator circuits. The radio tuning knob is fastened to RV10. Several varactor tuners are used in TV VHF and UHF tuners.

Digital tuning

The digital tuning system provides full electronic control of a varicap- or varactor-tuned AM/FM receiver. A block diagram of a typical digitally tuned receiver is shown in Fig. 22-20. The phase-locked loop (PLL) digital tuning system consists of two ICs, a controller and a PLL in a single chip, and a two-module prescaler. The controller chip (IC401) provides phase-locked loop capability with on-chip frequency division, a reference oscillator, whose frequency is controlled by an external crystal (4.5 MHz), and phase comparator circuitry.

The controller directly accepts the AM local oscillator signal and an FM signal from the two-module prescaler IC402, and the output controls signals for closer loop operation with these oscillators.

Low-pass filters TR407 and TR408 provide the dc voltage to the varicap or varactors in the AM/FM tuners. The controller IC also provides the signal to drive the display panel. The frequency of the tuned station is displayed on a multiplexed display (Fig. 22-21).

The typical AM station operation example with the PLL circuitry occurs when receiving a 1000-kHz signal, the VCO generates a 1450-kHz signal (1000+ 450 kHz IF). The AM VCO output signal (1450 kHz), which passes through buffer amp TR204, is applied to pin 28 of PLL IC401. This frequency is divided by N (=145). The resulting output will be 10 kHz. The reference oscillator 4.5 MHz is divided by 450, resulting in another 10-kHz frequency.

These two 10-kHz signals are fed to the phase detector. An error voltage is found on pin 1 of controller IC401 and passes through the low-pass filter (LPF), where the error voltage is integrated, and harmonics and noise are filtered out. The resulting dc voltage from the low-pass filter (LPF) is applied to varicap or varactor diode D210 (part of the VCO), whose capacity varies with the applied dc voltage. Now when the frequency of the VCO is corrected with the proper circuit design and precise adjustments, the VCO frequency is accurate and precise.

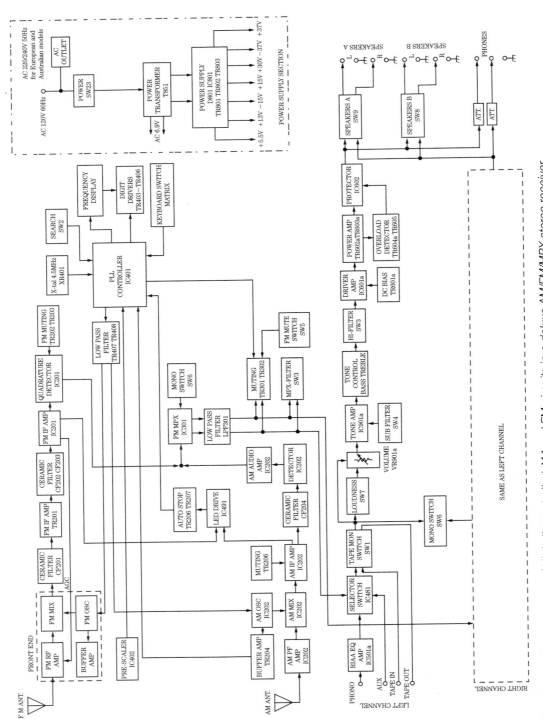

22-20 A block diagram of digitally controlled AM and FM circuits in a deluxe AM/FM/MPX stereo receiver.

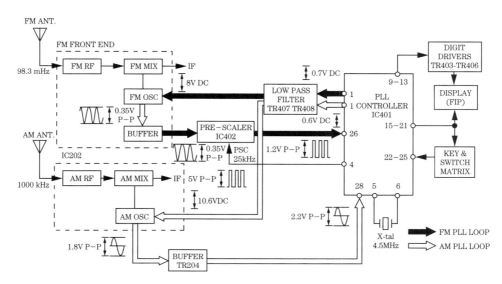

22-21 A phase-locked loop (PLL) block diagram on how the digital control supplies the tunable voltage to the AM and FM front-end circuits.

When the system is locked, the phase detector senses no phase differences and generates a frequency that is as accurate and stable as the reference crystal oscillator.

A typical example could occur when receiving a station at 98.1 MHz and the VCO generates 108.8 MHz (98.1+ 10.7IF MHz). The signal passes through the buffer amp and prescales (IC402) and is applied to pin 26 of PLL IC controller (IC401). This frequency is divided by N_1 and N_2 ($N_1 = 0$, $N_2 = 136$) together with pre-scaler IC402 and PLL, and controller IC40; the resulting output will be 25 kHz. The reference oscillator (4.5 MHz) is divided by 180 (again by IC401), which results in another 25-kHz frequency.

Now these two 25-kHz signals are fed to the phase detector. The error voltage is generated by the phase detector, which is proportional to the phase difference between these two 25-kHz signals. This error voltage is found at pin 1 of IC401. The voltage passes through the low-pass filter (LPF), where error voltage is integrated, filtering out noise and harmonics. The resulting dc voltage is applied to the varicap or varactor diodes (VCO), whose capacity varies with a change in applied dc voltage. Now, the output frequency of VCO is corrected and the VCO frequency is locked in. No phase differences are between the two 25-kHz signals and the VCO generates a frequency stable as the reference crystal oscillator. You will find that digital tuning does not separate broadcast stations like a variable capacitor.

Typical auto stop circuit

A typical automatic stop circuit in the deluxe AM/FM table model or auto radio consists of IC20, IC202, IC301, IC401, TR202, TR203, TR206, or TR207. When receiving an FM station, the IF signal from CF203 is applied to pin 1 of IC201 (Fig. 22-22). The input signal level: FM MUTE level = 15 PV. Pin 12 of IC201 will become low (from 3 to 0 V) and TR203 will be turned off. Under this condition, the high level (from 0 to 5 V) is provided to pin 7 of IC401. This causes the auto tuning mode in the microprocessor to stop.

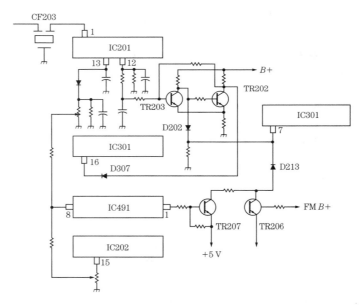

■ **22-22** *A typical auto-stop circuit in the deluxe table or auto radio receivers.*

Likewise, when receiving any AM station, the meter signal from pin 15 of IC202 is applied to pin 8 of IC491. Pin 1 of IC491 becomes the low level and TR207 will be turned on. Under this operation, a high level is provided at pin 7 of IC401, stopping the auto tuning from the microprocessor.

Monophonic audio circuits

Most clock radios have one source of audio, although you might find one or two models with stereo sound (Fig. 22-23). Although the clock radio chassis was transistorized a few decades ago, now one IC might contain the sound circuits. The AM/FM signal is applied to the function switch and to the volume control. C116 couples the audio to input terminal 3 of IC5. Pin 6 is the voltage supply pin while pin 5 contains the amplified sound coupled with C120 to the speaker. When audio stereo circuits are in the clock radio, both sound channels are identical (Fig. 22-24).

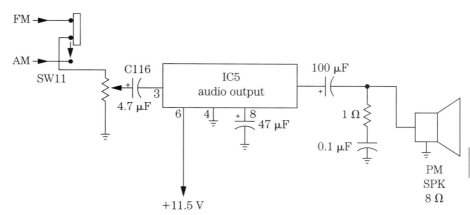

■ **22-23** *The simple stereo circuits might include one IC in each channel of the left and right audio channels.*

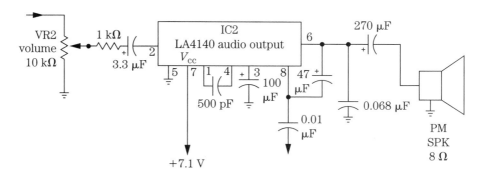

■ **22-24** *The sound circuits in a clock radio with one source of audio and single speaker.*

Dial cord problems

In clock radios that have variable capacitors tuning, the dial cord that rotates the tuning capacitor might break or begin to slip (Fig. 22-25). The broken dial cord should be replaced with regular dial cord or cloth fishing line. Notice how the broken dial cord lays before removing it. Usually, the dial pointer operates between two pulleys, with a knob and dial drum in between. Dial cords might pop off when a plastic bearing breaks or comes loose from the dial assembly (Fig. 22-26). The slipping dial cord can be repaired by removing the spring end, tieing one or two knots in the cord and rehooking the dial spring.

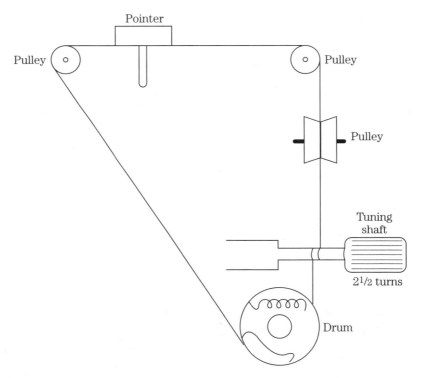

■ **22-25** *A simple dial cord stringing arrangement to rotate the variable capacitor and the dial pointer.*

AM/FM radio symptoms

The following symptoms are typical service problems that are related to the front end of any radio receiver.

■ **22-26** *The pencil points out a possible broken plastic shaft that might break; it will not move the dial pointer or variable capacitor.*

Symptom No AM, but the FM is normal. The FM or cassette player is normal.

Isolate Go directly to the AM converter circuit because the IF stages are common to the normal FM circuits.

Repair Place another AM radio close to the defective one and rotate the dial through the band and notice if a squeal is heard. No sound—go directly to the AM converter. Take the critical voltage measurement. Test the IC that you suspect in the circuit.

Symptom No FM, but the AM is normal

Isolate When no station can be heard on the FM band with a rushing noise, suspect the converter or oscillator transistor or IC. If the AM is normal, the IF stages are good.

Repair Check the voltage on the FM converter transistor or IC. Test the FM converter transistor in the circuit. Replace a leaky or open FM transistor with a component that has the same lead length and spot replacement.

Symptom Very weak AM reception

Isolate If only a local AM station can be tuned in, suspect a broken antenna ferrite rod or wire off antenna coil. Suspect the RF AM transistor in deluxe models.

■ **22-27** *Check the ferrite rod inside of the antenna coil or check the antenna coil wires for weak AM reception.*

Repair Check the ferrite rod in pieces and take continuity tests on the antenna coil terminals (Fig. 22-27).

Symptom Very weak FM

Isolate Check the RF or FET FM transistor for open or leaky conditions.

Repair Take voltage measurements on the RF or FET transistor. Test the transistor for open or leaky conditions.

Symptom Intermittent AM or FM reception

Isolate Isolate the trouble to the IF or stages that are in common with both AM and FM circuits.

Repair Suspect a broken IF coil or a poorly IF soldered terminal. Check for poor board connections. Check for intermittent leaky or open IF transistors. Monitor the forward bias between the base and emitter terminals for a change in voltage.

Symptom Erratic FM noises

Isolate Motor boating, whistling, and microphone noises can be caused by a defective transistor or high-end capacitor.

Repair Check all ground connections. Erratic FM noise with hum might result from dried-up electrolytic decoupling capacitors in the power supply circuits to the FM circuits.

Symptom No tuning in a digital radio

Isolate Determine if the varactor voltage is found at each varactor tuning diode.

Repair Measure the voltage on each varactor diode. If low or no voltage, check the low-pass filter transistors for the diode voltage. Test each transistor. Suspect the PLL controller if no voltage or signal is supplied to the varactor diodes. Check the voltage source on the PLL controller IC. Replace a leaky or suspected controller IC.

Symptom Loud tuning hum

Isolate Tunable hum on local stations might result from an improper voltage source or from poor connections on the tuning capacitor.

Repair Check the voltage supplied to the AM/FM front end circuits. If it is low, check the power source regulator transistor or zener diodes. Measure the resistance to ground between the rotor and common ground of the variable capacitor. Clean and resolder all ground connections around the tuner circuits.

Symptom Poor FM reception

Isolate Very weak, noisy, and garbled FM reception can be caused by the multiplex stereo circuits. Isolate poor radio reception to the FM stages.

Repair Does the stereo light come off and on? Check for leaky transistors or ICs in the MPX circuits. Check variable resistor controls in MPX circuits for noisy or poor contacts. Do not adjust any IF or MPX circuits without aligning equipment and using the correct procedures (Fig. 22-28).

Typical AM/FM alignment

Plug the radio to be repaired or adjusted into the isolation transformer. Allow the receiver to warm up 15 or 20 minutes. Connect the low side of the generator and indicator to ground unless it is specified differently by the manufacturer. Set the volume of the receiver at maximum. Keep generator signal output low enough to

■ **22-28** *Do not make any IF, trimmer, or coil adjustments without using proper alignment equipment.*

obtain a suitable response. Connect an ac multimeter across the speaker voice coil for an indicator.

Form several turns of hookup wire in a loop and connect an AM generator for AM and FM alignment (Fig. 22-29). Select for all coils and trimmer capacitors, G.C. Electronics alignment tools, 9089, 8276, and 5000. Follow the manufacturer's alignment procedure or follow the typical alignment in Tables 22-1 and 22-2. Check Fig. 22-30 and Table 22-3 for typical FM RF, IF, and MPX alignment procedures. Follow the manufacturer's alignment for specialized front-end circuits.

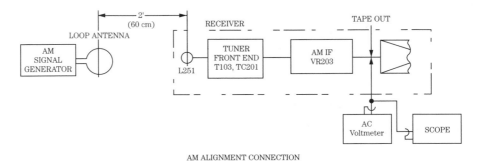

■ **22-29** *Connect the AM generator to a loop of hookup wire and place it near an AM antenna coil.*

Troubleshooting and repairing radio circuits

Table 22-1 A typical AM alignment with the selector in the AM position.

Generator frequency	Radio dial setting	Indicator	Adjust	Procedures
455 kHz to 400 Hz modulated	Tuning gang wide open	ac meter across voice coil	T7, T6, T5 IF Trans.	For maximum
600 kHz	600 kHz	same	L6 osc. coil	same
1640 kHz	1640 kHz	same	CT4 osc. trimmer	same
1400 kHz	1400 kHz	same	CT4 RF trimmer	same

FM IF alignment with AM generator: High side of generator through 0.001-µF capacitor to TP3 or emitter of FM converter transistor.

Generator frequency	Radio dial setting	Indicator	Adjust	Procedures
10.7 MHz	Where there is no interference	ac meter across voice coil	T4, T3, T2, and T1 of FM transformers	Adjust for Max.

Table 22-2 A typical FM RF alignment chart.

FM RF alignment: Selector in the FM position.
Place a 125-Ω resistor in each antenna lead from generator and connect it to the antenna terminals of the FM receiver (Fig. 22-30).

Generator frequency	Radio dial setting	Indicator	Adjust	Procedures
88 MHz modulated	88 MHz	ac meter across the voice coil	L4 and L2 L4 Osc. L2 FM RF	Adjust for maximum
108 MHz modulated	108 MHz	same	CT1 and CT2 CT1-FM RF CT2-FM osc	same

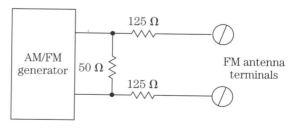

■ 22-30 *A typical FM dummy antenna alignment connections with generator and FM alignment.*

**Table 22-3 Typical FM RF, IF and MPX alignment procedures.
Follow the manufacturer's alignment if you are in doubt.**

Press FM switch.
Tune for 98.1 MHz on band.
Signal generator output level: 1000 µV   Deviation: 75 kHz, at 100% modulation of composite signal.
 (Fig. 22-31)
Connect signal generator to FM antenna terminal through FM dummy antena (300 Ω).

Step	19 kHz (Pilot signal) modulation level	Signal generator freq. set to	Output indicator connected to	Adjust Refer figure 7	Adjust for	Remarks
1	PILOT OFF	Carrier only	Frequency counter connect to TP (#17 pin) of PCB 0102 and ground	VR301	19 kHz	
2	8%	Composite 1 kHz R channel	ac voltmeter to TAPE OUT jack of R channel			Adjust input for audio output of about 0.7 V
3	8%	Composite 1 kHz L channel	ac voltmeter to TAPE OUT jack of R channel	VR302	Minimum	ac voltmeter reading should be at least 32 dB below reading in step 2.
4	8%	Composite 1 kHz R channel	ac voltmeter to TAPE OUT jack of L channel	VR302	Minimum	Same as step 3.
5	8%	Same as step 4 modulation level = 0	Same as step 4	VR303	Minimum	ac voltmeter reading should be at least 40 dB below reading in step 2.

If you did not obtain −32 dB readings in steps 3 and 4 (compared with step 2), readjust VR302 until you obtain −32 dB reading for both steps 3 and 4. (nominal 45 dB)

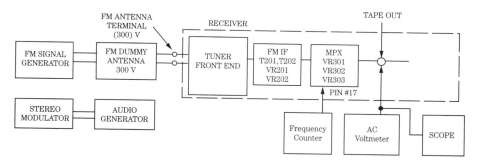

FM RF, IF and MPX ALIGNMENT CONNECTION

■ **22-31** *How to correct the scope, frequency counter, ac voltmeter, and generators for complete FM alignment.*

Typical AM/FM alignment

23
How to test and replace defective motors

PHONOGRAPH AND MICROWAVE OVEN MOTORS OPERATE directly from the 120-Vac power line. Motors found in the cassette, CD, camcorder, and VCR players operate from a low-voltage dc source. The combination cassette player can operate from batteries or an ac power supply. Motors in the camcorder might operate from a 6-, 9-, or 12-V large nicad rechargeable battery (Fig. 23-1). The CD and VCR machines operate from a dc source. Of course, the portable phono and CD player are powered with batteries or from a power pack.

■ **23-1** *The camcorder might have an auto-focus, iris, and zoom motor in the camera lens area.*

The motor is a machine that converts electrical energy into mechanical energy. These small dc motors have a permanent magnetic field with a rotating armature. The dc motor rotation can be reversed by changing the dc voltage polarity that is applied to it. The dc motor has a permanent magnetic field, armature, and brushes, which make contact with the commutator (Fig. 23-2).

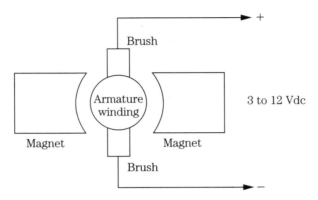

■ **23-2** *The dc motor used in many consumer electronic products consists of a permanent magnet, armature, and brushes.*

Within the cassette player, the dc motor rotates the tape, idlers, and capstan. A CD player might have four or five different motors. The auto cassette player and CD player operate from a 13.8-V car battery. The capstan motor in the VCR might rotate the tape, idlers, and reel assemblies.

Different motor problems

The defective motor might have an open armature, field coil, arcing brushes, flat armature, intermittent rotation, or squeaky and noisy bearings. The open field and armature windings might prevent the motor from rotating. Worn or poor brush contacts might make the motor excessively arcover or if a brush is hung up, it might not rotate. The motor with a so-called flat armature will not start, but it will rotate after it has been started. The intermittent motor might operate for long periods of time, stop, and not start up. Sometimes, tapping the motor end bell will start up the motor. Squeaky and noisy rotation might be caused by improper lubrication (Fig. 23-3).

Usually, a frozen motor has gummed bearings. The motor might be left on overnight or for long periods of time, become hot and have the bearings freeze when the motor cools. Extremely dry bearings

■ **23-3** *The small dc motor disassembled showing the various rotating components.*

might cause the armature and bearings to freeze. Sometimes a good clean-up and lubrication will cure the frozen armature problem.

Cordless battery-operated tools and soldering irons are powered with nicad batteries. These nicad batteries are also found in cordless drills, screwdrivers, saws, and battery-operated soldering irons. When the tool will not operate, charge up the nicad batteries. Most power tools on the electronic service bench have their own charging power device that plugs into the ac power line (Fig. 23-4). A low-voltage or improper dc voltage source might make the dc motor run slow or stop the rotation.

Voltage tests

The dc motor that you suspect to be defective might be located with improper or no voltage applied to the motor terminals. If the dc motor will not operate, measure the voltage across the motor terminals. A low voltage might make the motor run slow and drag. The motor is dead if no voltage is at the motor terminals. The dead motor with no operating voltage might result from a defective regulator dc source, power supply, or batteries. Make sure that the dc voltage output source is low before applying the VOM or DMM. Check ac motors with the DMM set on the 200-Vac scale.

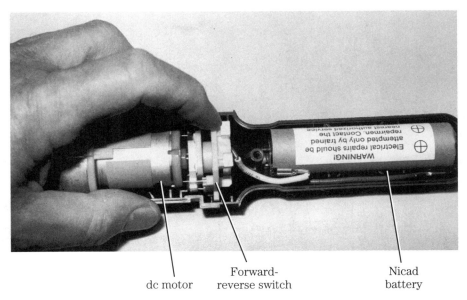

■ **23-4** *The dc rechargeable power tools have a small dc motor that rotates drills and power screwdrivers.*

dc motor | Forward-reverse switch | Nicad battery

Continuity tests

Check the motor terminals with the DMM set at the 200-Ω range. The continuity or resistance of these small dc motors is quite low. The motor resistance might be under 50 Ω. Usually, the continuity is less than 15 Ω. The small dc motor might have a resistance less than 1 Ω. If the DMM shows no measurement, the motor windings or armature is open (Fig. 23-5). Remove one terminal lead for correct continuity and resistance measurements.

If the motor terminals show that the windings are open, give the motor pulley a spin and see if a measurement is present. This action might indicate that the brushes are bad. In the small dc motor, the brushes are very small and might be used up in several years. Excessive arcing of the brushes when the motor is rotating indicates that the brush is hung up, that oil is on the armature, that the brushes are worn, or that the brushes are not seating properly. A dirty commutator might cause excessive arcing.

Intermittent motors

Suspect that a motor is defective or that a connection is loose if the motor will run, stop, and not start again. The next time that the motor is turned on, it seems to be normal. Sometimes you can

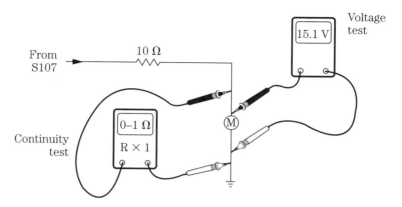

■ **23-5** *How to test the small dc motor for open continuity and for the voltage applied to the motor terminals.*

stop the motor by hand and the motor will not start. Make sure all motor terminals are soldered correctly. Monitor the voltage at the motor terminals. If the voltage is normal and the motor stops, replace the motor. Poorly soldered connections to the brush holder and armature might produce an intermittent operation. Make sure that the brushes are seated properly.

Overheated motors

Motors that begin to heat up after a few minutes of operation should be checked for a shorted field or armature windings. Check the motor terminal to the metal cover or ground of the motor with the 20-kΩ range. If any measurement is found, the field or armature winding is grounded. Overheated motors might be caused with shorted windings of field coils or armature.

Check the motor for dry bearings. Improper lubrication might cause the bearings to overheat. Often, a drop of oil on both front and rear bearings might cure the heavy friction damage (Fig. 23-6). Large motors have a bronze sleeve and metal plastic bearings. Plastic bearings are used in cordless power tools and small dc motors are used in cassette decks, CD players, and camcorders. Gummed up or dry bearings might cause a slow speed or prevent the small motor from operating.

Erratic motor operation

Erratic or uneven speed of small motors might result from poor brush contact, gummed up, or dry bearings. Wash out the bearings

■ **23-6** *Most motors used on the service bench have their own charging power supplies that plug into the power line and charge nicad batteries.*

with cleaning fluid. Lubricate the bearings with light sewing machine oil. Notice if the motor speed is restored. Remove the motor drive belt to see if the motor runs at a constant speed. Slow and uneven speed might result from a dry capstan flywheel bearing or idler wheel. Replace the motor if the speed is erratic.

Squeaky motor bearings

Dry or worn bearings might cause a squeaking noise as the motor rotates. Often, a squeaky or vibrating noise indicates that a bearing is dry or worn. Place a few drops of light oil on motor shaft and bearing as the motor rotates. It's best to remove all belts from the motor area, so as not to get oil on the belt area. Apply lubrication at both end bearings. Often the dry bearing at the drive pulley squeaks because pressure is applied by gears or drive belt. If lubrication will not help the squeaking bearing, replace the motor.

Flat armature

If the motor will not start with voltage applied, suspect that an armature bearing is flat or worn. Remove the drive belt or gears. Now see if this motor will start without a load. If not, give the motor pulley a spin. Suspect a flat armature when the motor will take off and run with external spin or pressure. A poor contact on the commutator and brush might produce the same problem. Replace the small motor if it is defective. Many of these small motors cannot be repaired because very small brushes and armatures are found. Once they are taken apart, the motor might be difficult to put back together again (Fig. 23-7).

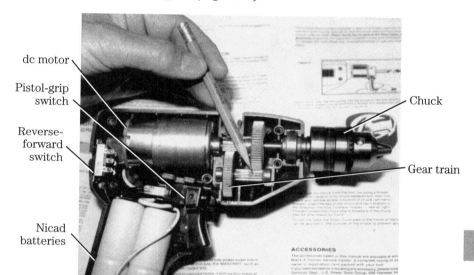

■ **23-7** *The rechargeable power drill and screwdriver with nicad batteries, dc motor, and gear train.*

The tap test

If the symptom is that the tape begins to speed up and then slow down, suspect a defective motor. Let the motor run and then tap lightly on the end bell. Give it a tap or two with the handle of a screwdriver. Smack the motor at the end bell, which is opposite from the motor pulley. If the motor speeds up, replace it. Excessive speed might also be caused by a defective motor or by the motor belt riding on the flange of the motor pulley. Monitor the voltage that is applied to the motor terminals. A higher voltage than normal might make the dc motor run faster.

How to locate and replace the cassette motor

The portable cassette player motor might operate from batteries or an external power pack, and the table or compact cassette player motor is powered from a low-voltage power supply (Fig. 23-8). The cassette motor voltage might be voltage regulated or it might come directly from a half-wave or full-wave bridge rectifier circuit. In compact combination models, the cassette motor might operate from a center tap of a power transformer in a half-wave silicon diode rectifier circuit (Fig. 23-9).

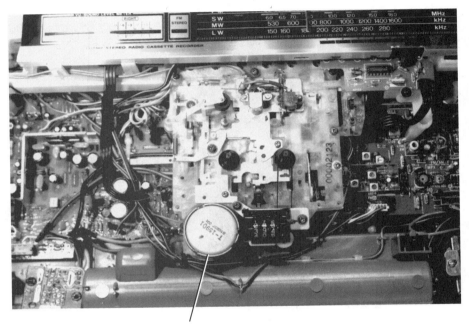

Cassette dc motor

■ 23-8 *The cassette motor is usually mounted on one end of the cassette mechanism assembly.*

In the bridge rectifier motor circuits, the dc voltage is supplied through cassette switch S701, through a small coil inductance to the motor terminal. One side of the cassette motor terminal is grounded. A bypass electrolytic capacitor (47 μF) shunts the motor terminals (Fig. 23-10). The voltage applied to these cassette motors from the low-voltage power supply might be from 14 to 18 Vdc.

The portable cassette player can operate from 3 to 9 V from batteries or with an external power pack. A leaf switch is used to turn on the motor and amplifier. Usually, these small motors operate at the total voltage of all the batteries combined. If the portable cassette has three C cells, the motor operates at 4.5 V (Fig. 23-11).

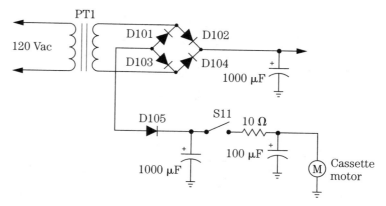

■ **23-9** *The portable cassette player might operate from batteries, but the table model is powered from the low-voltage power supply.*

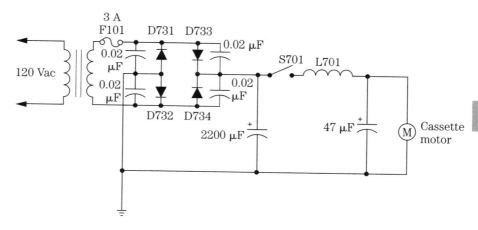

■ **23-10** *Here, the cassette motor operates directly from the full-wave bridge rectifier circuits of the low-voltage power supply.*

The boom-box or table model cassette player might operate from batteries or the ac power supply. If the ac plug is inserted at the rear of the boom-box cassette player, the batteries are switched out of the circuit. Likewise, when the cord is pulled out of the rear of the player, the dc power supply provides voltage to the tape motor circuits. The cassette motor voltage is activated with a pushbutton switch (S705) at the front of the cassette player (Fig. 23-12).

The dual cassette player boom-box has either separate motors or a single motor with two belts that drive each cassette player. If both players are driven by one cassette motor, an idler pulley is in-

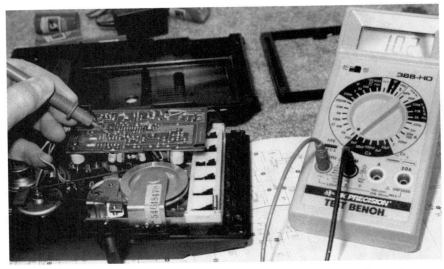

■ **23-11** *Check the total battery voltage to determine what the voltage should be across the portable cassette dc motor terminals.*

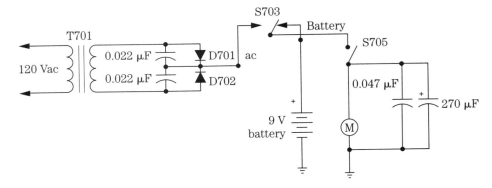

■ **23-12** *When the ac cord is removed from the boom-box or portable cassette player, batteries are automatically switched into the cassette motor circuit with S703 and start-up with S705.*

serted when both players are used at once. Usually, the motor operating voltage equals the total voltage from the battery source.

If the cassette motor will not rotate, check the voltage across the motor terminals with the cassette switch on. If the voltage is missing, suspect a defective or dirty on/off switch. Look for open low-ohm resistors or coils in the motor circuit. Suspect that the motor is defective if you find the correct voltage is across the motor terminals. Take a continuity check to make sure that the motor is open. Replace these motors and try not to repair them. Although some cassette motors can be replaced with universal types, try to replace it with a component that has the original part number.

Test and replace auto cassette motor

The auto stereo cassette motor might operate from the car battery, dc line, or from a transistor voltage regulator. The dc voltage across the motor terminals should equal the regular car battery voltage (13.8 to 14.4 V). Check for an open low-ohm resistor or choke coil in series with one motor lead with no voltage on the motor terminals. Take a low-ohm continuity check across the motor terminals with one lead removed from the circuit for a possible open motor winding.

Checking the phono motor

Most phono motors operate from the ac power line. You might run into some that operate from a dc power source. The old phono motor is switched on with the reject manual button and 120 Vac is applied to the motor terminals. After several years of operation or if the player is accidentally left on overnight, the motor might freeze and not rotate. Place a screwdriver blade next to the motor laminations and the blade will vibrate if voltage is applied to the motor winding (Fig. 23-13).

Remove the complete motor assemblies and dismantle them. These motors can be cleaned, lubricated, and placed back into operation. Remove both end bearings. Wash the bearings out with cleaning fluid to remove old grease and to tighten the bearings. Lubricate the bearings with light machine oil and reassemble the motor. Give the armature a spin to see if it is free and rotates. These small motors can be cleaned and repaired because they might no longer be available.

The dc motor can operate from batteries or from a low-voltage dc power supply. In stereo turntables, you might find that the phono motor has a voltage regulation circuit with variable speed adjustments. The step-down power transformers provide an ac voltage to a bridge rectifier circuit. That voltage is fed to a transistor voltage regulator (Fig. 23-14). The motor has five terminal wires with three speed-control resistors within the 33- and 45-rpm circuits. The rpm speed can be set with a stroboscope disc or pattern. Adjust VR2 until the stroboscope pattern is stationary. Likewise, adjust the 33-rpm rotation with VR1.

■ **23-13** *The old 33 1/3, 45, and 78 rpm phono turntable operate from 120 V.*

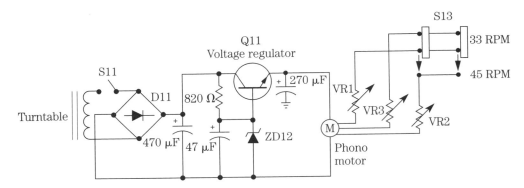

■ **23-14** *The dc motor used in the latest phono turntable operates from a dc source and applies a variable voltage to the motor for 33 and 45 rpm selection.*

How to test and replace CD motors

The portable CD player might have only a disc or sled motor, but the table-top CD player might have four or more motors. The auto CD player might have three or four CD motors, and the auto CD turntable might have four or more motors, a tray or loading turntable, elevation, a carriage or sled, and spindle or disc motors.

How to test and replace defective motors

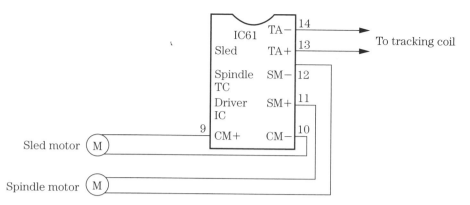

23-15 *The spindle or disc and carriage or sled motor might operate from the driver IC in the auto CD changer.*

The spindle and carriage motor might be driven from the same IC driver that drives the tracking coil (Fig. 23-15). The tray and elevation motor might be driven by a power IC driver.

The tray or loading motor pushes the tray out and when the tray button is pushed again, the tray motor pulls in the disc to load. The loading motor might have an IC or transistor driver circuit. The loading motor driver IC is controlled by the mechanism contact IC (Fig. 23-16). Usually, these small CD motors operate with under 5.5 V. Notice that the motor terminals are above ground and when the voltage is reversed, the loading motor will change direction.

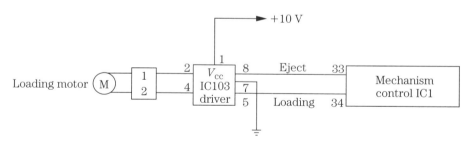

23-16 *The loading motor is controlled from the mechanism IC1, to control the load-and-eject signal of driver IC103 and the dc voltage applied to the loading motor in the CD player.*

Check the voltage across the motor terminals when the loading button is pushed. Suspect that the driver IC or transistors are defective if no voltage is on the loading motor terminals. Check the continuity of motor terminals if voltage is found at the motor terminals, but there is no rotation. Replace the defective motor with one that has the original part number. If no voltage is at the motor terminals, suspect that a driver IC is leaky or that no supply voltage is at pin 1 (V_{CC}).

The spindle or disc motor directly rotates the disc at a varying speed and it might be driven by several transistors or ICs. The motor drive transistors are controlled by the servo control IC. Both the disc and sled motors are controlled from the servo control IC (Fig. 23-17). Both motors will rotate in either direction. The voltage applied to the disc and sled motors might be under 5.5 Vdc.

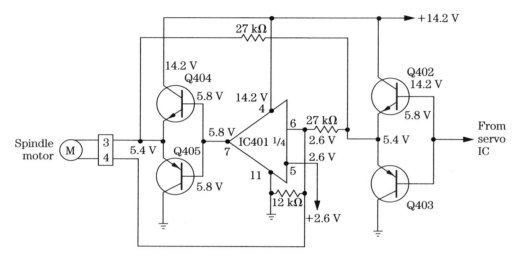

23-17 *The spindle or disc motor might operate from the transistor or IC drivers (or both) in some CD players.*

Check the voltage applied to the motor terminals and also the driver transistor and IC. Notice that the driver IC (IC401) might contain the driver in both the sled and disc (or spindle) motor. Look for a leaky transistor if an improper drive voltage is found at either motor. Check each transistor in the circuit for leakage or open conditions with the transistor tester. Make sure that the supply voltage is measured at pin 4 (14.2 V). Most motors found in the CD player are either controlled by a driver IC or by transistors.

The sled, slide, carriage (or feed) motor moves the optical pickup assembly across the disc from the inside to the outside rim of the CD. Usually, the sled motor is gear driven to a rotary gear that moves the laser beam down two sliding bars (Fig. 23-18). The slide motor might have a fast forward and rewind mode operation. The feed motor might be driven with transistors or ICs, like the disc

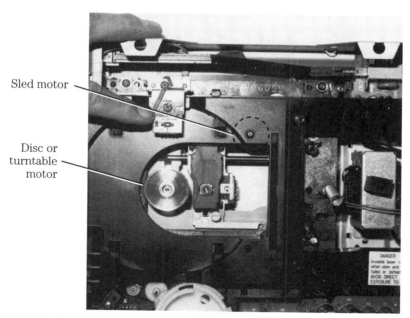

■ **23-18** *The sled or carriage motor might be located under the chassis with a worm gear that moves the optical assembly down a pair of sliding round bars.*

motor. Test the sled motor like any other motor found in the CD chassis.

Microwave oven motor tests

You might find up to seven different motors in the high-powered microwave ovens. In larger ovens, you might find a fan, turntable, convection, flapper, stirrer, timer, and vari-motor (Fig. 23-19). The fan motor keeps the magnetron tube cool and circulates air within the oven. The cooking time might be controlled by a timer motor. A stirrer motor spreads the RF energy out over the food in the oven cavity. The turntable motor provides even cooking by rotating the food in the oven. A convection motor circulates the heat provided by convection heating elements, while the damper or flapper motor opens the exhaust door.

Motors used in microwave ovens operate from power-line voltages. Check for a broken belt, gummed up bearing, or a stuck blade if the stirrer fan is not operating. The stirrer motor might have a higher resistance than other motors in the oven (300 Ω to 3 kΩ). Check for 120-Vac across the motor terminals and take a continuity test if the stirrer motor will not rotate (Fig. 23-20). Discharge

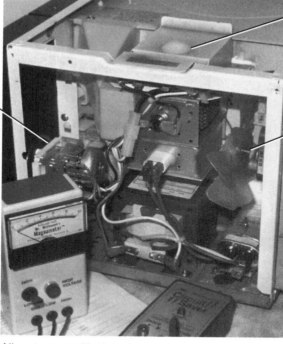

■ **23-19** *All motors used in the microwave oven operate from the 120 Vac.*

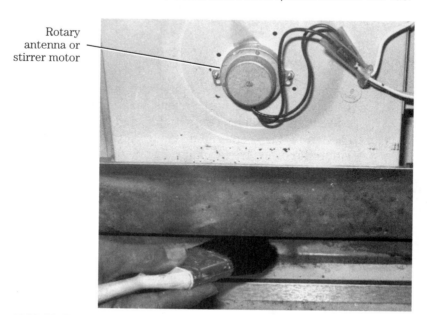

■ **23-20** *The rotary antenna or stirrer motor might be located at the top of the oven to move the radiation signal from the antenna of magnetron to the oven area.*

How to test and replace defective motors

the HV capacitor before checking any motor or component in the microwave oven.

If the oven shuts down or intermittently cooks, suspect that a fan blower is defective. The fan motor circulates the air on the magnetron and when the tube overheats, the thermal unit will shut the cooking process down until the thermal part cools down. Then it might start up again. Check the motor voltage and continuity if the fan will not rotate. Wash out motor and bearings if the fan blade is sluggish, then lubricate the bearings with light motor oil.

The turntable motor is geared downward to rotate a glass tray in the oven, so as to rotate food for even cooking. The new turntable motor might start up in either direction. If the turntable will not rotate, check the voltage across motor terminals and take a resistance measurement. The turntable motor resistance should be from 30 to 55 Ω across the motor terminals. If it is defective, replace it with a component that has the original part number.

Checking camcorder motors

The most important motors found in the camcorder are the loading, capstan, drum, auto-focus, iris, and zoom motors. The iris, auto-focus, and zoom motors are found on the camcorder lens assembly. The loading motor might eject, load, and unload the video cassette, releasing the brakes, and engaging the fast-forward/rewind and playback gear. The capstan motor might be belt or gear driven to the various mechanical assemblies, providing tape movement in play, record, rewind, fast-forward, and search modes (Fig. 23-21). The cylinder or drum motor rotates the large tape head and cylinder. The auto-focus motor keeps the lens assembly in focus at all times. The zoom motor brings the image close or far away from the lens.

The motors found in the camcorders are quite small and operate at low dc voltages. The early motor circuits in the camcorder were driven with transistor drive circuits. Today, most motors are controlled by a single IC or a combination of motors driven from one large IC.

Check the voltage on the motor terminals when the camcorder is operating. Most of these motors do not operate until a button or switch is pushed, like the loading and zoom motors. Measure the motor continuity to see if a winding is open. If no voltage is applied to the motor terminals, suspect that the driver IC is defective. Check for correct voltage on the driver IC at the supply terminal and also on the motor terminals (Fig. 23-22). Replace all camcorder

23-21 *The capstan motor in the camcorder might operate pulleys, idlers, and provide play, record, rewind, fast forward, and search modes.*

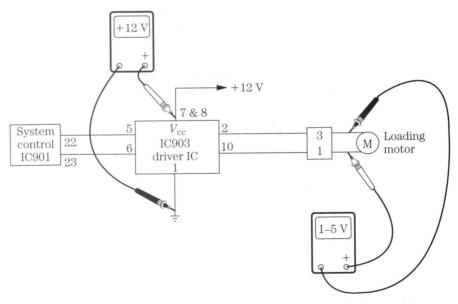

23-22 *Check the voltage across the motor terminals and the supply terminal of the driver IC if the loading motor will not rotate.*

How to test and replace defective motors

parts with those that have the correct part number. Do not try to substitute parts unless the part is made by the same manufacturer.

Motor test voltage supply

Most cassette motors operate from 12 to 18 Vdc, pulling less than 1 A of current. So, by building a dc fixed-power source, you can check out most motors. If the suspected motor is intermittent or operates fast or slow, it can be tested outside the motor circuit with a dc voltage source. Just check it on a 12- and 15-V source (Fig. 23-23). The voltage output of each source is regulated with 1-A IC regulators. If the defective motor pulls excessive current, the IC regulator will shut down the dc source. All parts for the motor tester can be purchased at Radio Shack or local electronic stores.

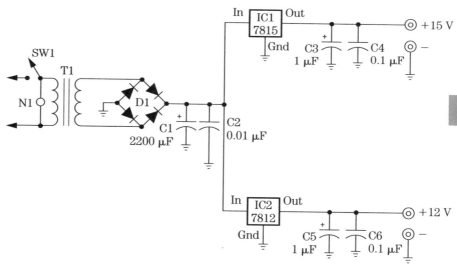

■ 23-23 *A motor test power supply of 12 and 15 V with 1-A output can check most cassette motors for intermittents, change of speed, and slow operation.*

The parts list for the motor test supply:

SW1	SPST (single-pole, single-throw) switch 275-1565
N1	120-Vac neon bulb indicator or equivalent
T1	18-V 2-A secondary power transformer 273-1575 or equivalent
D1	2-A bridge rectifier
C1	2200-µF 50-V electrolytic capacitor
C2, C4, C6	0.1-µF 100-V ceramic capacitor

C3, C5	1-µF 50-V electrolytic capacitor
IC1	7815 voltage regulator (15 V)
IC2	7812 voltage regulator (12 V)
Misc.	Two red and two black banana jacks, line cord, cabinet, nuts, bolts, solder, etc.

Low-voltage motor-test supply

Small motors that operate under 5 V can be checked on this voltage supply test. You can find these motors in battery-operated CD players and camcorder machines. Always remove the ungrounded lead from the motor to be tested. This dc power supply operates from three D cells. Choose a 100- to 150-Ω wirewound 5-W (or more) control to vary the dc source (Fig. 23-24). These wirewound variable resistors are often available at electronic surplus stores. Start with the control at 0 V and increase the voltage. Always keep the switch turned off when it is not in use because the variable control can run down the batteries.

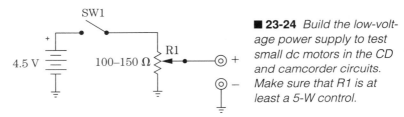

■ **23-24** *Build the low-voltage power supply to test small dc motors in the CD and camcorder circuits. Make sure that R1 is at least a 5-W control.*

Although motors do not cause as much trouble in VCRs, CD players, camcorders, and cassette players as other mechanical problems, the loading and cassette motor does create service problems. Knowing how each motor performs and how it is tied with either driver transistors or ICs can help you solve the next defective motor problem. The electronics technician should be aware that when mechanical and electronic components are placed in one unit, mechanical (not electronic) breakdown will cause more problems.

24
Servicing the cassette and phonograph circuits

TODAY, CASSETTE PLAYERS ARE AVAILABLE AS PORTABLE, walk-along, table model, compact, boom-box, and car units. The cassette player might have a mono or stereo amplifier. A portable cassette might have only playback, although most have playback and recording features. The dual system might have both, with one player used for playback and the other used for both playback and recording. Because the cassette player has mechanical and electronic operations, servicing problems are always present.

Circuit breakdown

The cassette player can be broken down into five sections: the tape head, preamp, audio output, motor, and power supply circuits (Fig. 24-1). The tape head picks up the music from the tape, transfers it into electronic energy and amplifies it with the preamp

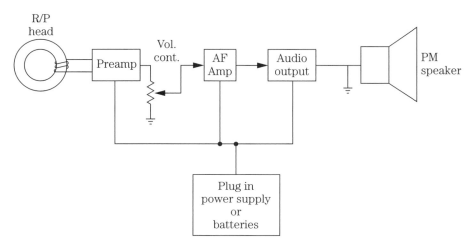

■ **24-1** *A block diagram of a simple portable cassette player.*

circuits. The AF and audio output circuits amplify the music for the headphones or to several speakers. The boom-box cassette player might have several stereo speakers, but in compact or combination units, the music is fed into high-powered amplifiers.

The mechanical section consists of a reversible dc motor, turntables or reels, idlers, pulleys, a capstan drive, a flywheel, pinch rollers, and switching systems (Fig. 24-2). The take-up reel takes the slack of tape out of the pinch-roller/capstan assembly and operates in fast-forward and rewind modes. The supply reel contains the bulk of the tape and it also operates in the fast-forward and rewind modes. Several idlers and pulleys are used in the different tape movements. The capstan-flywheel is belt or gear driven by the drive motor and the capstan drive shaft moves the tape past the tape head. A rubber pinch roller is pressed against the capstan drive assembly to keep the tape moving and it is not used when fast forwarding or rewinding.

Breakdown of cassette circuits

The monaural cassette player might have a transistor or IC preamp circuit with record and playback switching. The AF amp might consist of one transistor or a large IC serving both AF and output circuits. You might find one large audio IC in the small tag-along portables. Only one speaker is used in mono cassette players.

The stereo cassette player might have one IC that combines both preamp stages. The play/record switch is placed between the tape head and the preamp IC. Both channels are switched and amplified at the preamp IC. A preamp stage must be used to amplify the weak audio signal that is picked up by the tape head (Fig. 24-3). Usually, low voltages are found in the preamp terminals to prevent noise from entering the record and playback circuits. Both preamp circuits are identical. A play/record switch might be found at the output of the preamp that includes a Dolby sound circuit.

The tape head

The tape head is made up of a coil winding, magnetic metal core, and a smooth area with a gap that the tape moves against in recording and playback functions. The stereo tape head consists of both left- and right-channel windings (Fig. 24-4). The same winding is used for record and playback features. This winding is switched into the playback and record circuits. In the record circuit, a bias oscillator is switched in to erase the previous recording

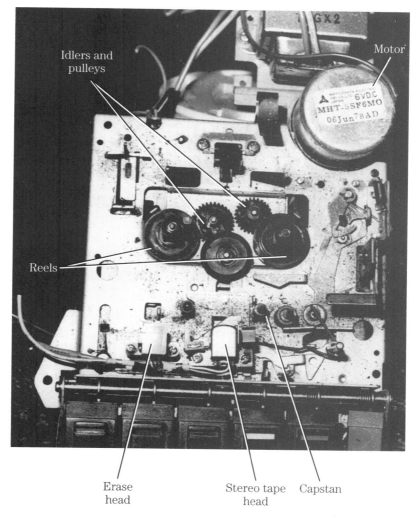

■ **24-2** *The top view of a cassette player, showing the motor, tape heads, system, reels, and idler pulleys.*

and to provide linear recording. The mono tape head has only one winding and the stereo has two windings in one tape head component (Fig. 24-5). You can recognize the stereo tape head because it has four connecting terminals.

Low-priced cassette players have a fixed mounted tape head. Other tape heads have an azimuth adjustment screw on one side of the tape head. When making azimuth alignment, the record and playback tape head gap is adjusted in parallel with the moving magnetic tape.

The tape head

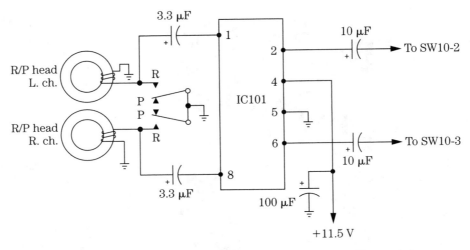

■ **24-3** *A schematic diagram of a preamp stage to amplify weak tape head audio.*

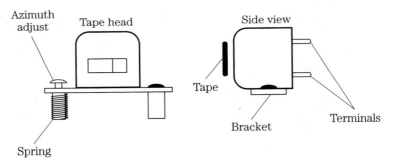

■ **24-4** *The tape head with stereo windings has four terminal connections.*

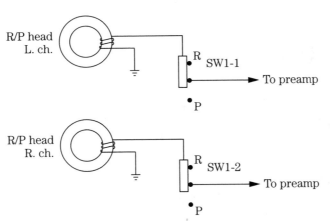

■ **24-5** *Both left and right stereo channel windings are used in one head component.*

The dirty tape head might cause dead, distorted, or weak recordings. Clean the head gap area with alcohol and a cleaning stick. The open tape head winding will not pick up the recorded music. Intermittent playback might result from a poor terminal connection of connecting cable or broken internal winding. A tinny or high-frequency sound might result from a worn tape head. The average tape head resistance is from 200 to 850 Ω.

Bias oscillator

The bias oscillator consists of a transistor, transformer, feedback, and switching circuits. The bias oscillator is used to erase prerecorded music and bias the tape heads for linear recording. In expensive recorders, the bias oscillator circuits might include two different transistors in the oscillator circuits. The bias oscillator is excited by switching the operating voltage to the oscillator circuits (Fig. 24-6). The bias waveform can be monitored at the stereo tape heads or at the erase head.

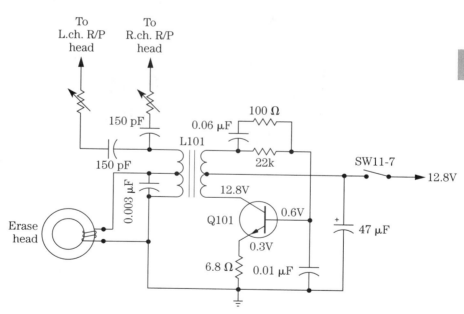

■ **24-6** *The bias oscillator is switched into the recording circuits to provide excited voltage to the erase and stereo tape heads.*

The bias oscillator provides oscillations in the secondary side of the transformer when $B+$ voltage is applied to the circuit. The primary winding picks up the bias oscillator signal and is applied to the erase and stereo tape heads. If the audio recording is poor, the tape head might be dirty or the bias oscillator waveform is not found at the tape and erase heads.

Most bias oscillator problems are caused by a leaky or open transistor, dirty record/play switch, and poor transformer connections. Clean the erase and tape heads with alcohol and a cleaning stick. Check the supply voltage source and collector voltage of the bias oscillator. Check Q1 with in-circuit transistor tests. Measure the continuity of each transformer or coil winding and resolder transformer PC board connections. Replace the coil or transformer if it has an open winding. The oscillator bias waveform can be checked on the erase head (Fig. 24-7).

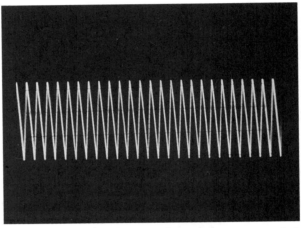

■ **24-7** Check the bias oscillator waveform at the erase and stereo tape heads.

Erase head

The erase head function is to remove any previous recording before the tape moves across the record/play tape head (Fig. 24-8). Jumbled or excessively distorted music can be caused with packed oxide dust on the erase head. If two or more recordings are found on the tape, suspect a defective erase head. The erase head is smaller than the record/play head. This erase head might be stationary or moved out of the way when playing on some cassette players.

Most erase heads are excited with a dc voltage or with an oscillator bias circuit. The dc voltage is switched in record mode to the erase head (Fig. 24-9). This dc voltage is switched from the dc

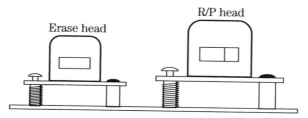

■ **24-8** *The erase head is mounted ahead of the R/P tape head to erase any previous recording.*

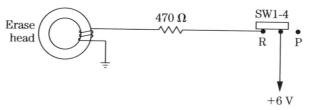

■ **24-9** *The erase head can be excited with a bias oscillator. In lower priced models, it can be excited with a dc voltage.*

power supply or batteries. The average erase head resistance is from 200 to 1000 Ω.

Erratic switching

If the cassette player fails to play or record with moving tape, suspect improper seating of the function switch. The function switch might be a rotary or sliding tape switch with many different contacts. Suspect that the switch is dirty if a warbling or screeching sound is heard when the function switch is moved or rotated. Clean all switching contacts inside of the entire switch assembly (Fig. 24-10). Place the nozzle of a cleaning spray can inside by each contact and spray all contacts. Work the switch back and forth to clean the contacts.

Preamp circuits

In early or low-priced cassette players, the preamp stage might consist of one or two transistors. Today, one IC might serve both preamp stereo circuits. One section of the play/record switch is located between the tape head winding and preamp circuit. In playback mode, the bottom of the tape head is grounded and a small electrolytic capacitor couples the tape head signal to the preamp

■ **24-10** *Clean the long function switch by spraying cleaning solution into the switch areas.*

IC input terminal with the top connection on the tape head (Fig. 24-11). If in the record mode, the top winding of the tape head is grounded and the bottom winding is switched to the record amp or output circuits of the amplifiers. SW1-1 switches both stereo tape heads into the record or playback circuits.

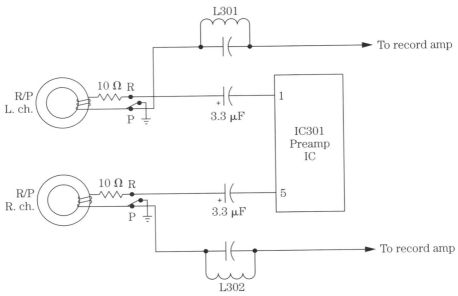

■ **24-11** *The preamp IC might include both R/P head inputs in record and play mode.*

Servicing the cassette and phonograph circuits

In playback mode, the picked up audio signal from its tape head is coupled by a 3.3-µF electrolytic capacitor to pin 1 of preamp IC1. The right stereo channel is also coupled by another 3.3-µF capacitor to pin 8 of IC1. After the audio signal is amplified, the left channel signal is found at pin 3 and the right channel output is at pin 6 (Fig. 24-12). The supply voltage source (V_{CC}) is applied to pin 4 (11.5 V) and pin 5 is grounded at preamp IC1.

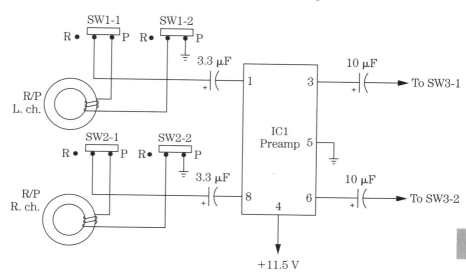

■ 24-12 *The amplified preamp audio is used at pin 3 of the left and at pin 6 of the right channel.*

AF amp circuits

The early audio frequency (AF) circuits were made with transistors, but now the AF and audio output circuits are found for both channels in one IC (Fig. 24-13). The AF amplifier and driver amplifier stages provided good audio to drive the push-pull output transistors. The two output transistors are coupled to a speaker through an output transformer.

The record/play head is switched into the first audio preamp stage in playback mode. The tape head is capacity coupled to the base of Q1. The base of Q2 is directly coupled to the collector terminal of preamp circuits. The volume control is located between the preamp and AF amp stages. The AF amp is capacity coupled to driver transistor Q4. Usually, the collector terminal of the driver transistor is connected to the primary winding of the interstage trans-

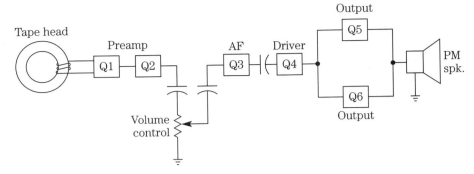

■ **24-13** *A block diagram of early transistor cassette player circuits.*

former. The interstage secondary winding is coupled to both base terminals of the output transistors.

IC output circuits

The AF and audio output circuits are often found in one large IC component. You might find a separate IC as a preamp, Dolby amp circuit, and power output amp in a stereo deluxe amplifier. In some cassette players, a separate IC is used in the record mode. The Dolby circuits are found between preamp and power output circuits (Fig. 24-14). In larger amplifiers, the audio output circuits for radio, phono, and cassette player might have high-powered ICs or transistors.

The dual stereo output IC might have the AF from the left channel enter C125 and be coupled to pin 5 of output IC303. The right-channel audio input is coupled through C225 to pin 8. The signal is amplified at the left channel and is coupled from pin 2 by C133 (1000 µF) electrolytic capacitor to speaker. The right output signal from pin 11 is coupled to the right channel speaker with C233 (Fig. 24-15).

Stereo amp circuits

One advantage in servicing stereo circuits is that the other channel can be used as a gain signal reference in each stage. If the left channel is weak and the right channel is normal, use the right channel to measure the gain of each transistor or IC. Check the signal at both volume controls and compare them. If the left channel is a lot weaker, check the preamp circuits. If both channels are fairly normal at both volume controls, suspect that the AF transistor or IC output stage is weak (Fig. 24-16).

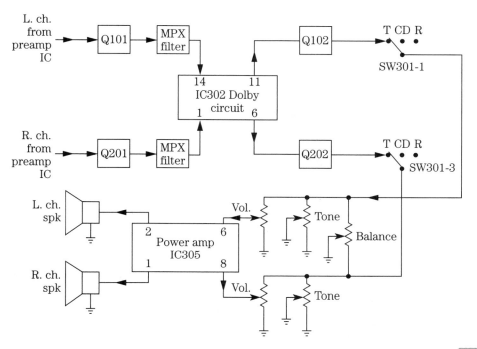

24-14 *The Dolby circuit might be used between the preamp and the power output IC.*

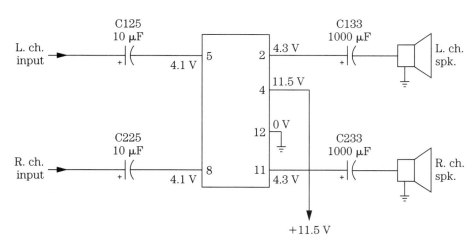

24-15 *C133 and C233 couple the power output sound from IC303 to the left- and right-channel speakers.*

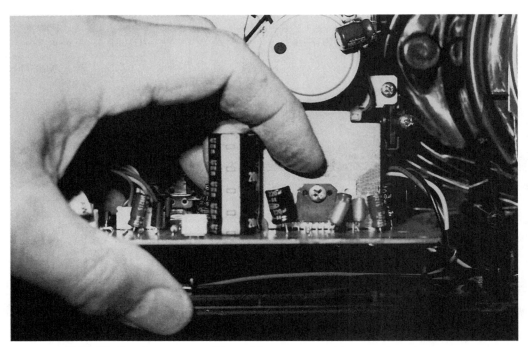

■ **24-16** *One large power IC provides power to drive speakers in a cassette/CD player.*

Transistors do not become weak; however, the circuit and components that are tied to the transistor or IC can create a weak audio channel. A change in the base and emitter bias resistors can create weak audio. Leaky or open components that are tied to the IC amp can produce weak sound. Doublecheck each electrolytic coupling capacitor that is between stages. The signal should be the same on both sides of the capacitor. A low power-supply voltage can cause weak sound.

Just about any component within the stereo circuit can cause a dead channel—especially test transistors in and out of the circuit. Poor switch contacts can kill audio. Suspect that the RP tape head is open if the stereo channel will not record or play. Often, a dead channel with excessive distortion can occur in the audio output circuits. Feel the power-output IC and notice if it is warm or red hot. The shorted or leaky IC will run hot. Both stereo channels might be dead if a common audio output IC is defective.

Before replacing a leaky or open output transistor or IC, check the bias resistors and capacitors. The shorted IC or transistor can cause bias or load resistors to run hot and smoke. If you do not have an exact schematic of the stereo cassette player, compare the defective channel resistors and capacitors to those in the good stereo channel. By taking a resistance test on each IC terminal to common ground might turn up a leaky capacitor or a change in resistance of a bias resistor. Check the operating voltages on each IC after you replace all defective parts and compare them to the normal channel (Fig. 24-17).

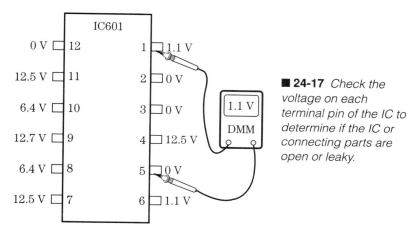

24-17 Check the voltage on each terminal pin of the IC to determine if the IC or connecting parts are open or leaky.

Recording circuits

The recording circuits in early cassette players contained transistors in the preamp and audio output stages. The same preamp circuits found in most cassette players are located in the recording circuits. Two or three transistors might be found in the preamp circuits. The recording circuits consist of a condenser or electret microphone, preamp, record amplifiers, and switching at the input and output circuits (Fig. 24-18). The bias oscillator is switched into the recording circuits.

A condenser microphone picks up the audio and is switched into the preamp circuits. The amplifier signal is applied to several transistors or one IC recording amp. The recorded signal is capacitively coupled through several resistor and capacitor networks and applied to the RP head and common ground. At the same time, the RP head is excited with bias voltage from a one- or two-transistor bias oscillator. The erase head is switched into the recording circuit to erase any previous recording. The erase head might be stationary or moved out of the tape path during playback in recent boom-box models.

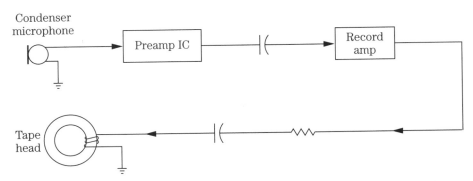

■ **24-18** *A block diagram of the recording circuits with a condenser microphone.*

Before attempting to record, clean the RP heads. Make sure that the RP switches are sprayed with cleaning fluid. If the deck does not record, check the cassette for a missing tab at the back. You might not be able to push down the record button with the tab missing. If the recorder operates normally when playing, suspect that the microphone or separate recording amp and switch is defective if there is no or poor recording (Fig. 24-19). Most tape cassette players use the same circuits for playing and recording.

■ **24-19** *A close-up view of the condenser microphone in the cassette player.*

Check all transistors in the recording circuits with a beta tester or with the diode-junction test of the DMM. Check each coupling capacitor with the capacitor tester. If a capacitor checker is not

Servicing the cassette and phonograph circuits

handy, shunt the same capacitance with the same working voltage across each small electrolytic capacitor. Poor switching contacts can prevent recording in either input or output recording circuits. If recording is jumbled and distorted, suspect dirty tape heads and improper recording bias at the tape heads (Fig. 24-20).

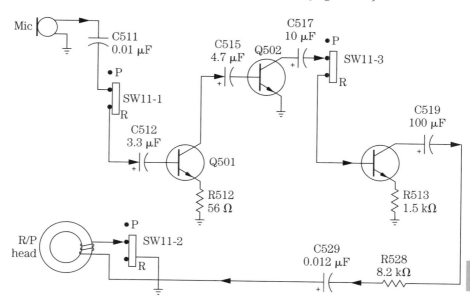

24-20 Follow the black arrows through the recording route from the microphone to the R/P tape head.

Remember, the microphone audio can be signaltraced through the preamp and record stages up to the RP tape head with an external amplifier. Place a radio near the microphone and record the voice or music while signaltracing the audio through the recording circuits. Insert a good recorder cassette and signaltrace the record stages with the external amp.

If excessively jumbled music or voice is heard, suspect a defective erase head or circuits. Notice if more than one voice or music is recorded on the cassette. The erase head might be excited with a dc voltage or from the bias oscillator circuit. Take a critical waveform on the ungrounded tape head terminal to determine if the bias oscillator circuit is working. Check the defective erase head for dirty oxide in the gap area, open head winding, and poor terminal connections. Make sure the erase head is pressed against the tape. A loose or missing mounting screw can prevent the head from touching the tape.

IC recording circuits

You might find IC circuits in both stereo preamp and output circuits. The left and right microphones are switched into the preamp IC circuits, with the left and right recording tape heads switched into the recording circuits. The preamp IC might serve as recording amp in low-priced and boom-box players (Fig. 24-21). Usually, the erase head and bias oscillator are switched into the circuit when recording.

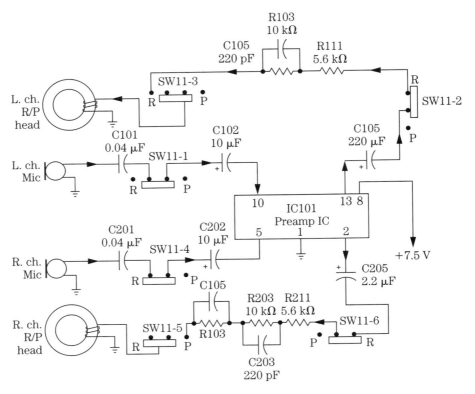

■ **24-21** *The preamp IC might serve as a playback preamp and a recording amp with stereo microphones and tape heads.*

The two stereo microphones are fed through an external mic jack and are switched to preamp IC101. The right-channel microphone audio is switched into pin 5 of the preamp. The left-channel audio signal is switched into pin 10 of the preamp IC. IC101 serves in both record and playback modes.

The left output signal from the preamp (pin 13) is coupled to SW11-2 with a 1-μF (C105) capacitor. The left channel output signal is fed through an isolation resistor to SW11-3. The left-channel RP head receives the recorded audio and then placed on the tape. The right output signal from the preamp (pin 2) is coupled to SW11-6 with C205. The right-channel output is fed through isolation resistor R211 to SW11-5. SW11-5 switches in the right RP head to record on the tape.

In larger recording circuits, both channels might be switched into a Dolby IC from the preamp circuits. The output of the Dolby IC might be capacitively coupled to an IC record amp. The output of the record IC amp is switched to the RP record head circuits (Fig. 24-22). Each Dolby circuit is in a separate IC, with separate recording amps in some cassette players. The Dolby circuits improve audio reproduction quality of a magnetic recording. Low-level sound is increased during recording and low-level audio is decreased when in the playback mode.

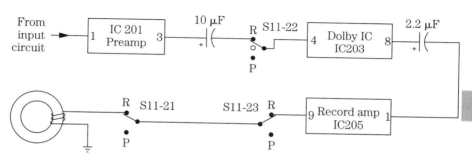

■ **24-22** *The Dolby IC might be switched into the recording circuit after the preamp IC.*

Troubleshooting IC record circuits

Determine if the playback circuits are normal before attempting to repair recording circuits. Most cassette players use the same circuits, except through a recording amp IC. If the playback circuits are normal, clean the record and erase heads. Clean any switches that are related to the recording circuits. A defective oscillator bias circuit might have distortion and weak recording. Check the bias waveform at the RP and erase head in record mode (Fig. 24-23).

If one channel is distorted and weak and the other normal, suspect that an IC recording amp is defective. Measure the supply voltage (V_{CC}) from the low-voltage power supply. If the voltage is low, the recording might be weak because of a leaky IC amp. Remove the supply voltage pin from the PC wiring by unsoldering and picking

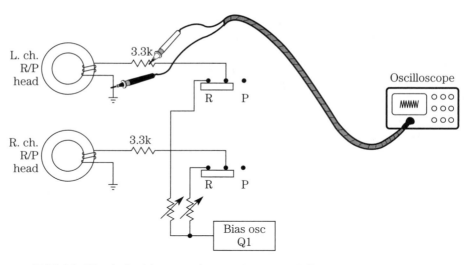

■ **24-23** *Check the bias waveform at the stereo R/P and erase head with the scope.*

up excess solder with solder-wick. If the voltage increases to normal, the recording IC amp is leaky, so replace it.

Check all voltages on each terminal of the IC and compare them to the schematic. If you have no schematic, compare the voltages to the normal channel of the IC. If the voltages are way off, check the components that are tied to the IC's pins for a possible leaky capacitor or for a change in value of a resistor. If the recording signal is coming into the IC, but is not found at the output terminal of that channel, replace the IC. Remember, the recording audio signal can be checked with the internal audio amplifier.

Check Table 24-1 for the electronic troubleshooting chart.

Tape speed adjustment

Insert a 3-kHz test tape (Teac MTT-111) or equivalent, to make speed adjustments. Connect a 35-Ω 5-W wire-wound resistor across a headphone jack and connect a frequency counter across the 35-Ω resistor load. Adjust the motor speed adjustment for 3 kHz on the frequency counter load test instrument (Fig. 24-24). If speed is above 3 kHz, the tape is running fast and when it is below, cassette player speed is slow.

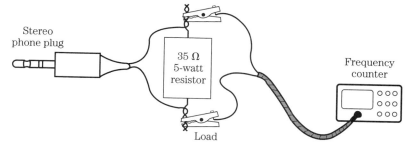

■ **24-24** *Connect a 35-Ω 5-W resistor to a stereo earphone jack to check the speed with a 3-kHz cassette and a frequency counter.*

■ **Table 24-1 A cassette player troubleshooting chart.**

Symptom	Troubles	Repair
Defective playback	No sound, $B+$ supply	Open primary winding of power transformer. Defective leaf on/off switch. Shorted $B+$ supply. Defective diodes in the power supply.
	Voltages on all transistors and ICs okay	Touch volume control for hum pickup. Turn the volume wide open and hear a loud rushing noise. Open tape head. Dirty R/P head. No hum at the volume control—defective amp. Bad speaker leads, defective output transformer, open coupling capacitors. Defective IC or transistors. Improper $B+$ voltage.
	No output, loud rushing noise.	Open R/P head. Broken wires on tape head.
	Defective tone	Shorted tone control. Bad leads on the control. Defective or dirty R/P head. Check the electrolytic capacitors.
	Weak volume	Check the voltages on the transistors and ICs. Defective coupling capacitors. Burned bias resistors. Defective output coupling capacitors.
	Low volume output	Dirty R/P head. Defective output IC. Defective output transistors. Check and compare the voltages to the normal channel.
Defective recording	No recording bias	Check the voltage on the bias oscillator. Defective R/P head. Defective leaf switch. Defective oscillator coil or transformer. Defective R/P switch. Dirty R/P switch.
	Bias okay; no recording	The level meter is okay, but there is no record. Dirty R/P switch. Defective R/P head. Open R/P head. Defective input jacks. Defective amp.
	Poor erasing	Clean the erase head. Check the waveform on the erase head. Open or shorted osc. coil. Defective bias transistor. Open or shorted erase head. Defective capacitors in the tank circuit of the bias oscillator. Defective oscillator circuit. Dirty switch to erase head. Improper supply voltage to the oscillator circuits.

Tape speed adjustment

Head azimuth adjustment

To align the tape head, insert a 10-kHz test tape (Teac MTT-14) or equivalent. Connect two 35-Ω resistors to a dummy stereo headphone plug, one for each stereo channel. Clip the ac meter, FET, or VTVM to one set of 35-Ω load resistors. A scope can also be attached to the other audio channel 35-Ω resistor. Adjust the azimuth screw along the RP head for maximum on the meter and scope. Check to see if both stereo channels are at maximum (Fig. 24-25).

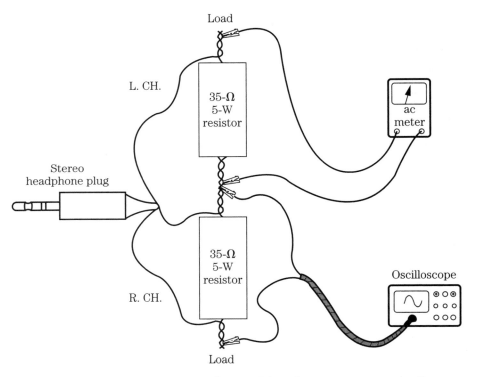

■ **24-25** *Connect the ac meter (low scale) and scope across respective 35-Ω 5-W resistors for head azimuth adjustment.*

Phonograph electronic circuits

The crystal cartridge mounted within the pick-up arm assembly picks up the sound from the record. Later on, a magnetic cartridge appeared with better fidelity and with less pick-up noise. The stereo cartridge contained two small crystal wafers that were attached to the needle or stylus picking up recordings to the left and right sides of the groove in the record (Fig. 24-26). The crystal cartridge has a higher voltage output compared to the magnetic

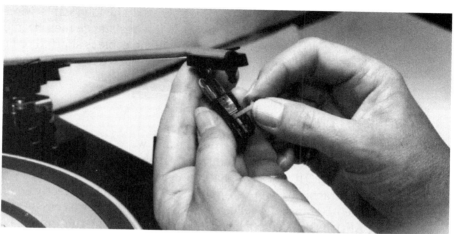

■ **24-26** *Installing a new stylus (needle) in a stereo cartridge at the tone arm of the phonograph.*

cartridge. The magnetic cartridge must have a preamplifier stage ahead of the AF amplifier while the crystal cartridge was directly connected to the AF circuits.

Check the magnetic cartridge or coils continuity on the RX10 range of ohmmeter while no measurement should be found on terminals of the crystal cartridge. This is because of small coils inside the magnetic cartridge (Fig. 24-27). Muffled sound might be caused by dust and particles picked up from the record and found piled on the stylus. Lightly press down on the cartridge with the record rotating. Select an old record for this test. Notice if the music cuts up and down. If the music is intermittent, replace the defective cartridge.

Broken or loose connections to the cartridge or pick-up arm might cause intermittent conditions. Inspect the phono cable and the

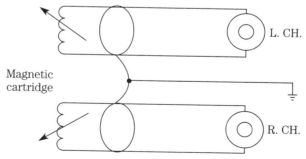

■ **24-27** *A stereo magnetic cartridge hookup with connecting leads to the stereo input jacks.*

plugs at both ends. Check for shorted shielded cable wires with the low-ohm scale of the DMM. A low-resistance measurement indicates that the insulation of the phono cable is shorted or broken. Remove the connection to the magnetic cartridge when measuring the resistance of a poor phono cable harness.

Transistor phono circuits

Usually, the crystal cartridge leads were connected directly to a volume control. The volume control audio is coupled to the base of AF amp Q1, with a capacitor and an isolation resistor. The two output transistors in push-pull operation might be directly coupled to the AF transistor collector terminal. Notice in this circuit the NPN AF transistor drives two PNP transistors in the output circuits (Fig. 24-28). The small speaker is coupled to the emitter circuit with a 220-µF electrolytic capacitor. The tone control is found in the collector and base circuit of Q1.

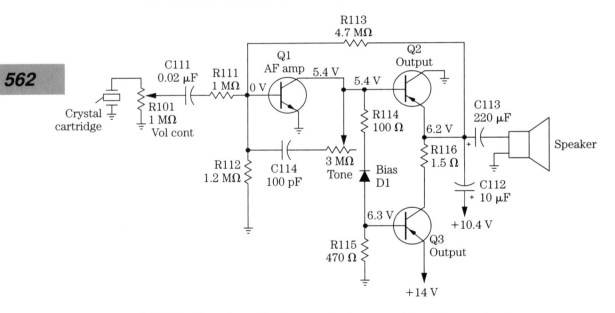

■ **24-28** *The schematic diagram of a low-priced transistor crystal phonograph player.*

The stereo IC circuit might consist of one large IC in the small record player. The dual crystal cartridge feeds directly into a dual 500-kΩ volume control. The audio volume is connected to pins 1 and 5 of IC105. Two identical electrolytic coupling capacitors couple the speakers to the output pins 7 and 8 of IC105 (Fig. 24-29).

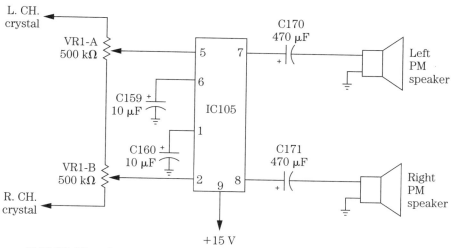

24-29 The phono IC circuit used in some low-priced players with capacitor coupling to the speakers.

The supply voltage (V_{CC}) on pin 9 is provided by a low-voltage transformer half-wave rectifier.

Distortion within the crystal phono IC circuits might result from a defective crystal cartridge, IC, coupling capacitors, and improper voltage supply. A lower-than-normal supply voltage might indicate that an IC is leaky. Open or dried up electrolytic coupling capacitors might result in weak and distorted audio. Replace the leaky output IC if you find normal audio at the input terminals and distortion at either or both output terminals. Weak and distorted audio can be signaltraced with the external audio amp. In higher-powered circuits, the phono input is switched into the AF amp circuit or amplifier.

Phono motor circuits

The low-priced phono player might have another winding wound on the same metal laminations as the phono motor (Fig. 24-30). The secondary winding acts the same as the secondary winding of a small step-down power transformer. D1 provides half-wave rectification with C2 as the filter capacitor. The phono motor circuit consists of the ac switch and motor, which operates from the 120-Vac power line. The motor resistance might measure between 35 and 350 Ω, depending on how it is wired and designed for a special circuit.

In some phono players, the speed of the turntable might be regulated with a motor that has five or six terminal wires. The dc mo-

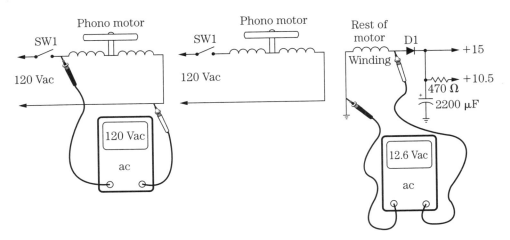

■ **24-30** *Another winding might be used around the motor laminations for a low-voltage power-supply circuit.*

tor is operated from a voltage transistor-zener diode regulation circuit with variable-speed resistors switched to 33⅓ and 45 rpm. The speed might be adjusted according to a strobe dot pattern found on the rim of the turntable (Fig. 24-31). By showing a neon light against the strobe pattern, the variable control is adjusted until the dots seem to stand still when set at a certain speed.

■ **24-31** *The strobe speed indicator might be used on the edge of a large phono turntable.*

Within the direct-drive automatic turntable, the dc servo motor might be controlled by a back electromotive-force frequency generator. Speed can be constantly monitored, and if there is a slightest change of speed, corrective force is immediately applied. The output frequency of the back electromotive force that is generated in the drive coil winding of the motor determines the speed of the turntable. By using a trapezoidal wave generator circuit, a pulse generating circuit, and a sampling IC, the back electromotive force generator output frequency is converted to a voltage, which maintains the rotation speed of the turntable (Fig. 24-32).

The operational control circuit functions as a control output voltage control keeping the rotation speed of the turntable constant. The op-amp active filter circuit provides ideal filter operation.

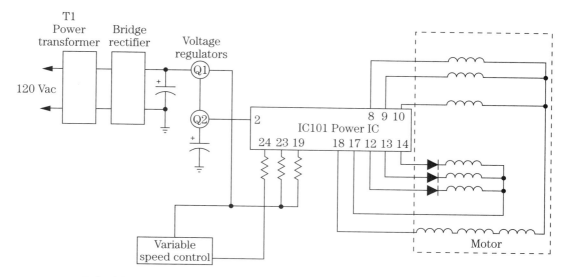

■ **24-32** *The direct fully automatic turntable might have a high-powered IC driving the dc motor with several windings.*

Voltage regulation is applied to the power drive IC from a bridge rectifier circuit. The power IC provides a starting torque for prompt starts. By means of the signal from the position signal coil, the starting circuit power IC operates, which results in smooth operation.

Conclusion

Although record player repairs are declining, every holiday seems to bring in some units for service. Servicing the cassette player can be fun and quite rewarding in solving the many speed and solid-state amplifier problems. The intermediate electronics technician should be able to service most cassette and record players that appear on the service bench.

25
Basic TV screen symptoms

MANY TV TROUBLESHOOTING SYMPTOMS CAN BE FOUND BY looking at the TV screen and listening to the sound. If there is no raster and no sound, there might be problems within the low-voltage power supply or horizontal output circuits (Fig. 25-1). The dark raster or screen with good sound might be caused with no high voltage or defective picture tube circuits. The white raster with normal sound might result from a video, IF, or tuner problem; and on it goes.

Block diagram

The front-end circuits in today's color TV chassis might consist of a VHF/UHF tuner, tuner control, saw filter, and IF stages. The video, AGC, sync, and luminance circuits might be contained in one large IC (Fig. 25-2). Often chroma and luminance circuits are found together, driving the color output transistors that connect to the picture tube circuits. The picture tube or CRT provides a display of the TV program channel that is tuned in. The AGC provides automatic gain control and the sync circuits contain pulses that lock the signals together.

Before any TV circuits can perform, the low-voltage power supply circuits must provide a working voltage source to each section. The dc power supply might have several different voltage sources with several regulated circuits. Today, most dc sources are derived from the secondary windings of the flyback or horizontal output transformer. In other words, the low voltage stage and horizontal circuits must operate before other circuits are supplied with the correct operating voltage.

About 85% of TV problems occur in the horizontal and vertical circuits (Fig. 25-3). Usually, the horizontal oscillator or countdown circuits are supplied a voltage from the low-voltage power supply,

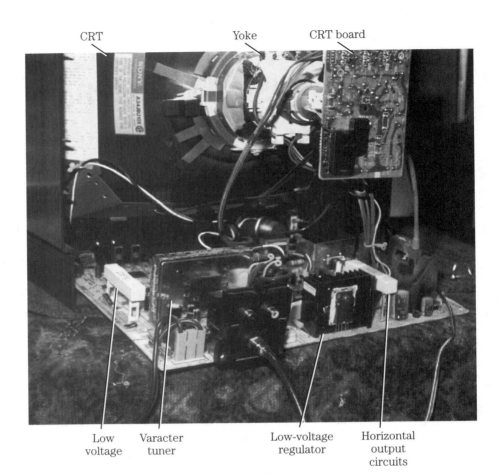

■ **25-1** *The different circuits and components that cause the most problems in the TV chassis.*

and in some chassis, the horizontal circuits will continue to operate with voltage derived from their own flyback circuits. Some TV chassis still supply a low-voltage source directly to the horizontal/vertical deflection IC from the low-voltage power supply. This type of horizontal circuit supplied by a low-voltage source is much easier to service than the derived voltage source. Just measure the supply voltage at the supply pin of horizontal deflection IC from low-voltage power supply.

The vertical countdown or deflection IC provides a horizontal and vertical sweep pulse or waveform to power the horizontal and vertical circuits. The deflection IC might also contain a separate horizontal and vertical amplifier that applies the signal directly to the

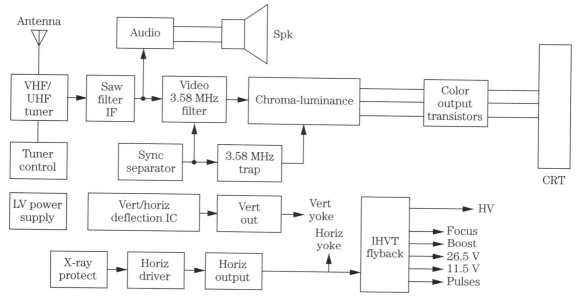

■ **25-2** *A block diagram of a color TV chassis.*

■ **25-3** *Most problems are caused in the horizontal and vertical circuits.*

vertical and horizontal circuits in some chassis. In the early TV chassis, the horizontal and vertical circuits contained their own oscillator section. Today, the countdown circuits might be controlled with a fixed crystal.

Besides providing sweep for the picture tube, the horizontal circuits help develop the high voltage (25 to 32 kV or higher) to the anode terminal of the CRT (Fig. 25-4). The horizontal pulse is found at the output pin of the deflection or countdown IC, is amplified, and is driven with a horizontal driver circuit. Although the horizontal oscillator circuits have varied throughout the years of television, the horizontal driver and output circuits are basically the same. The horizontal driver transistor is transformer coupled to the horizontal output transistor, which provides sweep and high voltage through a yoke assembly and a flyback transformer.

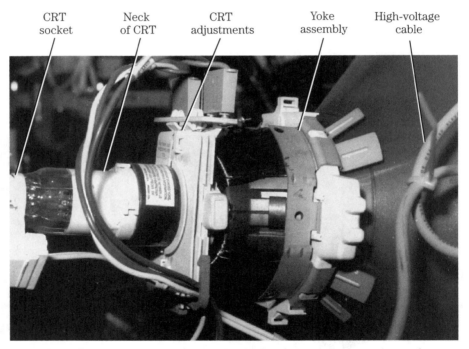

■ **25-4** *The horizontal circuits develop high voltage and other voltages for the picture tube circuits.*

The flyback provides high voltage for the CRT and several different voltage sources are taken from additional windings on the secondary circuits. Besides high-voltage, focus and screen voltages are developed in the higher voltage circuits. These low-voltage sources might provide working voltages for the picture tube, horizontal, vertical, video, luminance, front-end, and sound circuits. Often, a silicon diode rectifies the derived ac voltage, capacity filtered, and applied to the various circuits. Remember that a failure in the horizontal circuits might prevent other circuits from operating (Fig. 25-5).

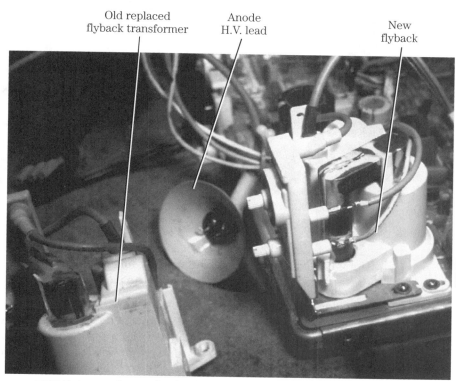

■ **25-5** *Low-voltage circuits from the secondary winding of the flyback might prevent other circuits from operating with defective horizontal circuits.*

A vertical amp or drive pulse from the vertical deflection IC provides vertical sweep through the vertical deflection yoke winding. Besides the countdown or vertical deflection IC, you might find a vertical amp or driver and two vertical output transistors in the vertical circuits. Some TV chassis might now have one IC chip that provides vertical sweep to the yoke assembly (Fig. 25-6).

The picture tube circuits receive focus, boost, and high voltage from the flyback transformer, video from the video circuits, color from the color output transistors, and variable controls that focus the picture and control the brightness. The CRT lights up with HV, control, and screen grid voltages applied to the tube elements. A separate fila-

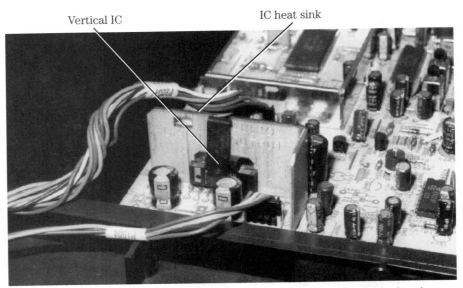

■ **25-6** *Suspect vertical output transistors (or IC) if there are vertical problems on the raster.*

ment or heater voltage for the CRT is taken from the flyback or horizontal output transformer. The raster contains the sweep, brightness, contrast, and color at the front of the picture tube.

The color circuits are controlled by a 3.58-MHz crystal connected to the color circuits inside of the luminance/chroma IC. The color output from the luminance/chroma IC is applied to blue, green, and red color output transistors, and is amplified and connected to each color cathode element of the CRT gun assembly (Fig. 25-7). A decade ago, transistors were used separately to acquire the various color circuits that now might be provided by one IC. Although the different color TV manufacturers might have their own circuits in the many different circuits of the TV chassis, they all provide a good TV picture.

The sound circuits are taken and separated from the video circuits after the IF circuits. Sound traps are provided to prevent audio from entering the video circuits. The early sound circuits consisted of a sound IF, detector, and amplifier, with audio output transistors feeding a speaker. Today, the sound circuits might be contained in one large IC. The 4.5-MHz sound signal is detected within the IC and is amplified (Fig. 25-8). The audio signal for stereo sound might be sent to stereo decoder circuits for processing. A dual speaker might be connected to the left and right audio output channels.

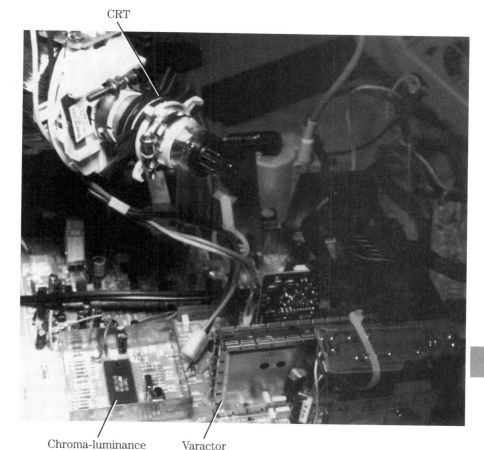

■ **25-7** *The color circuits might be in one IC, along with the luminance, horizontal, and vertical deflection circuits.*

The 14 most-common symptoms found in TVs

Dead TV

The most common symptoms found in the TV chassis relate to one or more circuits and might be heard through the speaker or seen on the TV screen or raster. The dead symptom indicates that there is no picture, no sound, and no low voltage. This type of problem might result from a blown fuse or a defective low-voltage power supply. Even the picture tube filament is black. In the dead chassis, it is very easy to locate the defective component, compared to the intermittent symptom.

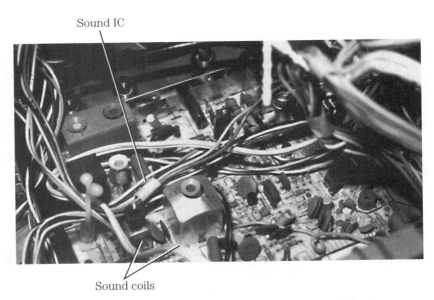

■ **25-8** *Today, one IC might contain most of the audio circuits in the portable TV.*

Black raster: no picture/no sound

This type of symptom might point directly at the low-voltage power supply or horizontal circuits. Often, with improper horizontal circuits, no sound is heard in the speakers. No raster or no light on the screen indicate no high voltage. The picture tube filaments will remain dark if the horizontal output and flyback circuits are defective. Check the fuse, low voltage, horizontal, and flyback circuits (Fig. 25-9).

■ **25-9** *A dark raster with no sound might result from a defective low-voltage power supply and horizontal circuits.*

White raster: rush in sound

The white raster indicates that the low-voltage, horizontal, and picture tube circuits are normal. A rush in the sound, when the volume control is turned up indicates that the sound circuits are normal. Of course, the picture is missing on the white raster. The raster might remain black if problems occur in the video circuits. Check the tuner, tuner control, saw filter, and IF circuits for a white raster.

Snowy picture: fair sound

Because the sound and picture are tuned in and amplified together in the tuner, saw filter, and IF section, a snowy picture and some sound might indicate a defective tuner or AGC circuit. In the early TV chassis, a leaky or open IF transistor might produce a snowy picture. Go directly to the tuner circuits and take critical voltage measurements (Fig. 25-10). Do not overlook the possibility that the antenna is poor or broken or that the antenna lead has cracked or come loose and is causing the snowy picture. Check the antenna and cable system with another portable TV set connected to them. The trouble lies in the front-end section of the TV chassis.

■ **25-10** *Check the antenna and tuner for a snowy picture.*

Intermittent picture: good sound

The intermittent picture might occur in the tuner, tuner control, saw filter, IF, and video circuits. If the antenna was defective, the sound might be weak and garbled. You know that the vertical, horizontal, HV, and picture tube are normal when the picture is good. Flashing and a negative picture might indicate trouble within the AGC circuits. Often, the IF and video circuits produce the intermittent picture. Look for defective AGC, sync, IF, and video ICs. Suspect that an intermittent picture is caused by a poor connection at delay line or open delay coil.

Excessively bright raster

A very bright raster with a picture in the background or no picture might be caused by defective video, luminance, or picture tube circuits (Fig. 25-11). Suspect that the picture tube is defective if the screen is excessively bright and the chassis shuts down. An improper or a missing boost voltage to the picture tube circuits might cause excessive brightness. Poor boost voltage might result from a leaky diode and open isolation resistor. Suspect a brightness reference or limiter transistor if the brightness is excessive. Do not overlook the possibility that a luminance IC might be leaky. Excessive brightness with retrace lines might result from a defective component in the picture tube circuits or CRT. Defective service switch terminals might prevent you from turning down the brightness.

■ 25-11 *Check the video, luminance, CRT circuits, and picture tube if the raster is excessively bright.*

Bright horizontal white line

Usually, the bright white line is in the center of the raster, indicating no vertical sweep. Go directly to the vertical output circuits (Fig. 25-12). Sometimes an insufficient vertical sweep is caused by a shorted output transistor. Suspect a defective vertical output circuit or IC. Scope the vertical pulse waveform at output terminal of the deflection or countdown IC. If they are normal, proceed through the vertical circuits with the scope. An open electrolytic coupling ca-

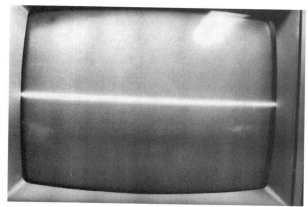

■ **25-12** *Check the vertical circuits if they have a horizontal white line or insufficient vertical sweep.*

pacitor to the yoke assembly, open yoke winding, or isolation resistor to ground might leave a white line on the TV screen.

Rolling pictures

Insufficient sync to the vertical circuits or defective sync, AGC, and IF IC might produce rolling pictures. When the picture cannot be stopped and locked in with the vertical hold control, check the vertical sync IC and low-voltage power supply. You might not find a vertical hold control in a newer TV chassis. Vertical rolling and crawling can be caused by a defective filter capacitor in the low-voltage power supply circuits. Scope the sync circuits and check the low-voltage power source feeding the vertical and sync circuits.

Insufficient height

Often, insufficient vertical sweep and linearity problems are located within the vertical output circuits. An improper drive voltage can cause insufficient vertical sweep. Poor vertical linearity at the top and bottom of the raster can be caused by a leaky top and bottom transistor. A leaky vertical IC or components tied to the IC can produce insufficient vertical sweep. An improper supply voltage to the vertical circuits can cause insufficient vertical height. Check for leaky regulator zener diode or transistors in the low-voltage circuits feeding the vertical circuits (Fig. 25-13). Readjust the vertical height and linearity controls in the chassis.

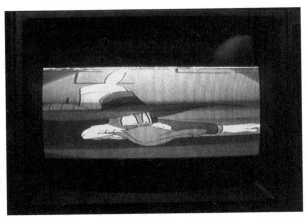

■ **25-13** *Check the vertical output circuits, transistors, and ICs for insufficient vertical height.*

Narrow picture

The sides might be pulled in or black might appear on both sides of a narrow picture or raster. The picture or raster has insufficient horizontal sweep. Check the high-voltage regulator circuits for poor width in the latest TV chassis. Defective high-voltage transistors, SCRs, and zener diodes in the regulator circuits can produce insufficient width. Poorly soldered connections to pincushion, regulator, and driver transformers might result in poor width. An improper low voltage applied to the horizontal circuits can result in insufficient width. Check for an improper high voltage at the anode terminal of the picture tube for poor width.

Poor focus

Poor focus results when the fine scanning lines appear blurred or will not focus in the sharpness of scanning lines. An improper focus might result from a low high voltage at the CRT. Check the anode terminal of CRT for correct voltage. Poor focus can be caused by an insufficient focus voltage (Fig. 25-14). Intermittent and poor focus might result from a poor focus pin at the picture tube socket. A defective focus control might produce erratic focus problems. A weak or defective picture tube might not focus properly. A defective IHVT flyback transformer can also cause poor focus problems. Check the focus voltage at the focus pin on CRT socket and see if the focus control will vary the focus voltage from 3.5 to 6.5 kV. The focus voltage is higher in the large picture tubes. Check the correct focus voltage on the schematic.

■ **25-14** *Poor focus in the picture or raster might result from defective focus circuits, low voltages, or picture tube problems.*

No color

Rotate the color control wide open. Almost any defective component within the color circuits can cause the color to fail. A defective 3.58-MHz crystal, a leaky transistor or chroma IC, defective bypass and electrolytic capacitors and resistors tied to the transistor or IC terminals (Fig. 25-15) might cause no color in the picture. An improper or low voltage from the power supply might also produce weak or no color. Check the voltages and scope each color pin if there is no color. If only one color is missing, check the

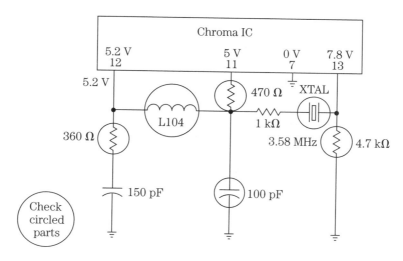

■ **25-15** *Check the chroma/luminance IC and the corresponding circuits for weak color or no color.*

color output transistors for leakage or an improper low voltage. Check the waveforms on the chroma IC and (especially) check the waveform from the flyback circuits.

No sound

The different service problems related to sound can be heard in the TV speaker. Most sound problems are easy to locate and repair. Often the defective component can be located with a few voltage and resistance measurements. Clip another speaker across the old one to determine if the speaker is open (Fig. 25-16). Signaltrace the audio with the external amp until you can no longer hear the sound. Critical voltage measurements on the transistor or IC terminals might help you to locate the defective component. If the horizontal output circuits are defective, sound might not be heard in some chassis.

■ 25-16 *No sound might result from an open voice coil in a speaker.*

Diagonal horizontal dark bars

Improper adjustment of the horizontal hold control might cause dark stripes or bars in the raster. The horizontal hold control might not be found in the latest TV chassis. You might hear a loud squealing from the flyback when this occurs. The horizontal oscillator frequency might drift off frequency and appear with slanted lines or horizontal bars in the raster (Fig. 25-17). Sometimes the oscillator frequency will drift off after several minutes of operation. Check the horizontal oscillator transistor or IC for leaky conditions. Replace the resonating and coupling capacitors in the emitter-tapped oscillator coil circuit in the early horizontal oscillator circuits.

■ **25-17** *Dark slanted bars on the screen might be caused by components in the horizontal oscillator circuits, by a defective IC, or by an improper horizontal sync.*

Horizontal slanted bars might appear in some chassis if the high voltage exceeds the shutdown circuits and disables the horizontal oscillator circuits. Horizontal pulling and drifting might result from poor filtering in the low-voltage source, which feeds the horizontal or countdown deflection circuits. Pulling at the top of the picture might be caused by a small electrolytic bypass capacitor in the horizontal circuits. If the horizontal lines cannot be straightened, suspect poor filtering in the low-voltage source or suspect that a high-voltage shutdown circuit is defective. Determine if the lines are caused by an HV shutdown by removing a transistor or diode terminal that connects to the horizontal circuits from the flyback and shutdown circuits.

Other TV symptoms

Although it is impossible to cover all of the various screen, sound, and chassis symptoms in this chapter, here are a few more symptoms that are listed alphabetically:

AGC problems

A white raster can be caused by a defective AGC tuner or video circuit. AGC problems might include a very dark and unstable picture. Check the AGC circuits if the picture is waving like a flag with a buzz in the sound. Excessive snow in the picture can be caused by defective AGC, tuner, or IF stages. Low contrast and erratic pulling of picture can be caused by improper AGC voltages. An intermittent picture and sometimes negative picture might be caused by bad connections within AGC circuits or module. Besides the AGC circuits, these same symptoms can be caused by defects in the tuner, IF, or video stages.

Bow in the picture

If the picture shows a bow, suspect that the pincushion circuit is defective. The pincushion circuits connect to the horizontal yoke and flyback circuits. In large 27- and 31-inch picture tubes, a separate circuit with transistors might be needed to keep the scanning lines horizontal, without bowing, at the corners of the picture tube. Check for leaky coils and pincushion transformers with improper vertical and horizontal linearity adjustments (Fig. 25-18). Look for an open pincushion output transistor for vertical bowing. Check both vertical and horizontal linearity with a color-dot bar generator with a crosshatch pattern. Resolder all pincushion transformer connections.

■ **25-18** *Suspect the horizontal pincushion circuits if the edges of the raster are bowing.*

Bunching lines

Often bunching of the scanning lines can be seen at the top of the raster. These lines might appear at once or appear after the chassis warms up. Sometimes an adjustment of the vertical height control might help. Check the bias resistors of the vertical output transistors. Unsolder one end and check it for correct resistance. Shunt the large electrolytic coupling capacitor between emitter terminals and yoke for bunched lines and insufficient height. Replace both vertical output transistors or the IC if all other tests fail to remove the bunched lines. Although these lines lie in the horizontal plane, trouble exists in the vertical circuits (Fig. 25-19).

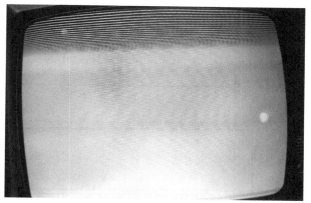

■ **25-19** *Bunching lines at the top of the screen might be caused by a defective transistor, IC, or yoke electrolytic coupling capacitor.*

Chassis shutdown

Just about any component in the horizontal circuits can cause chassis shutdown. The horizontal, low-voltage power supply, and vertical circuits cause the most chassis shutdown symptoms. The horizontal output transistor, high-voltage regulator, flyback, and oscillator or deflection IC are the most likely components. Chassis shutdown can occur at once or several minutes later. Check for arcing sounds in the flyback with the chassis shut down. The chassis might shut down with poor terminal connections on the horizontal driver transformer and loose transistor sockets.

Use a variable line voltage transformer to control the shutdown problem. Approach the shutdown problem as an intermittent or high-voltage shutdown. Take scope waveforms with critical voltage measurements. Isolate the high-voltage shutdown circuits temporarily to a defective chassis component by removing one lead of a transistor or diode from the high-voltage shutdown circuits. Do not overlook a defective CRT if the set turns on with excessive brightness and then shuts down the chassis at once.

Firing lines in the raster

Any component in the chassis arcing over might produce lines in the raster. A firing flyback transformer might cause lines in the picture—even if you cannot hear the actual arcing (Fig. 25-20). Poorly soldered connections on the flyback terminals can cause firing lines in the raster. Check the high-voltage cable to the CRT for breaks or a loose screw in the anode socket. If the HV lead is not snapped into the anode socket, firing lines can be heard and seen in the picture. Make sure that the picture tube shield or wires are grounded to the metal chassis and aquadag. Check to see if the high-voltage cable is clipped into the anode socket. Discharge the picture tube before handling the HV cable.

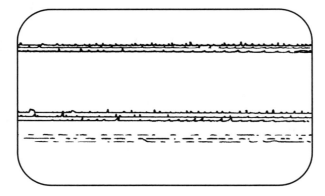

■ **25-20** *Check the HV circuits, flyback, focus control, picture tube, and pickup noise if there are firing lines in the picture.*

Remove the antenna lead to determine if the noise is picked up or if it occurs in the TV chassis. Excessive arcing inside the focus control can cause firing lines in the raster. Rotate the focus control and see if the lines disappear. Replace the focus assembly if it is defective. Sometimes the focus and screen controls must be replaced because they are mounted in one assembly. Do not overlook firing in the screen grid gaps off of picture tube elements. Arcing of tripler units to the chassis produce lines in the picture. Check for open filter capacitors in the $B+$ voltage regulation circuits, which might look like auto ignition pickup noise lines.

High-voltage arcover

High-voltage arcover can occur at the CRT, flyback, and tripler unit. Often, you can see and hear the arcover. Replace arcing flyback and tripler units (Fig. 25-21). Arcover between high-voltage capacitors molded inside the IHVT (integrated high-voltage transformer) can be heard if you listen closely. Sometimes the HV arcover at the picture tube anode terminal shows firing lines down to the banded area of the CRT. Arcover can occur at the spark gap assemblies, which indicates excessive voltage, dust in spark gaps, or that a defective component is nearby.

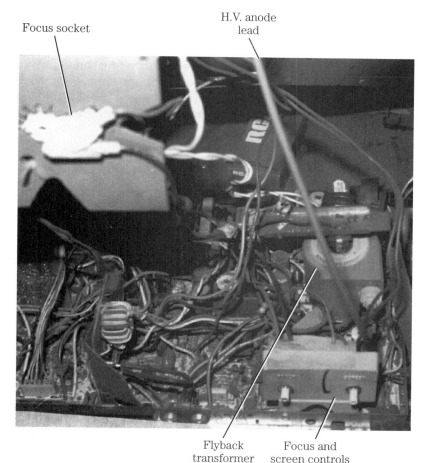

25-21 *These parts might cause high voltage arcover in the TV chassis.*

Usually arcover inside the picture tube gun assembly indicates that a filament or heater is open, or that a glass assembly is broken. You can hear the arcover occurring inside the yoke assembly with broken glass gun assembly. Check the $B+$ and high-voltage after all defective parts are replaced. Recheck the HV regulator or the low-voltage circuits, which produce excessive voltage at the picture tube. An open hold-down or safety capacitor on the collector terminal circuits of horizontal output transistor can cause the high-voltage to increase and arcover.

Horizontal foldover

Check the horizontal output circuits for horizontal foldover. Open bypass capacitors in the high-voltage transistor or flyback circuits can cause foldover. SCR horizontal circuits can produce horizontal foldover. Open terminals of reactors and regulation transformers can produce foldover. Check for poor connections and shorted turns in pincushion transformers for horizontal foldover. A split picture can be caused by an open coupling capacitor in the horizontal circuits of older TV receivers.

Hum bars in the picture

Hum in the audio with 120- or 60-Hz dark bars in the raster might indicate a defective filter capacitor or poor $B+$ regulator (Fig. 25-22). Readjust the HV or $B+$ control in the low-voltage power supply. Shunt a filter capacitor across the suspected one to isolate the defective capacitor. Most ac power-line chassis have only one large filter capacitor after the rectifiers. Shut the chassis down, discharge the old capacitor, clip a new capacitor in place, and observe the correct polarity. Try to raise or lower the voltage with the AVR or $B+$ control. Improper adjustment of the $B+$ or HV control might cause dark bars to appear in the picture; if it is turned up too high, it might cause HV shutdown.

Two dark lines that might appear at the top and bottom of the screen and rolls upward might result from leaky start-up diodes in the start-up transformer circuits. These bars might look like 120-Hz bars, but they aren't as wide. Check each diode in the start-up circuits for leaky conditions. Remove one end for correct leakage test. These diodes can be replaced with 2.5-A silicon diodes.

HV shutdown

The chassis might shut down with a high power-line voltage, a defective component in the TV chassis, or by the actions of the HV shutdown circuits. If the high-voltage exceeds the design of

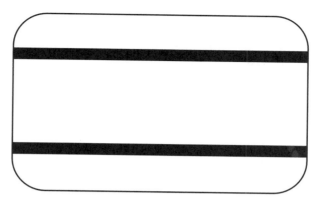

■ **25-22** *Hum bars in the picture and speaker might result from defective low-voltage filter capacitors.*

the shutdown circuits, the shutdown HV circuits go into action and shut down the TV chassis. Use a universal power line transformer to help solve the shutdown problem (Fig. 25-23). Slowly raise the ac voltage and notice at what voltage the chassis shuts down. Now bring the line voltage up once again, and keep the voltage setting just under the shutdown voltage to determine what section is causing the chassis to shut down. Measure the high-voltage at the anode of the CRT at this voltage setting to determine if voltage is too high compared to the ac line voltage.

■ **25-23** *Use the universal power-line transformer to service a shutting-down TV.*

Try to determine if the chassis is shut down by another component or if it is actually a high-voltage shutdown. Remove a diode or transistor terminal from the high-voltage shutdown circuits to see if HV is shutting down the chassis. Again raise the line voltage transformer voltage and see if the chassis will operate with the correct high voltage. Be very careful. If the HV shoots upward before you reach the power-line voltage (120 Vac), shut down the chassis at once. Monitor the high-voltage at the CRT with the HV probe while making these tests. Check the components in the high-voltage, horizontal, and low-voltage power-supply circuits for excessive high voltage.

Remember, an improper adjustment of the low voltage can cause high-voltage shutdown. An open safety or hold-down capacitor in the collector circuit of the horizontal output transistor might cause HV shutdown. Suspect a high-voltage SCR regulator circuit. Look for a defect in the high-voltage shutdown circuits if the high voltage is normal and the chassis shuts down.

Keeps blowing fuses

Suspect leaky or overloaded components in the low-voltage power supply or horizontal circuits if the fuse keeps blowing or opening. Check for leaky diodes and filter capacitors in the low-voltage circuits (Fig. 25-24). A leaky horizontal output transistor, damper diode, or flyback might blow the line fuse. Suspect that a low-voltage transistor or IC regulator is leaky in the low-voltage power-supply circuits.

If both the line and $B+$ fuses to the horizontal circuits are open, suspect a leaky horizontal output transistor or damper diode. If only the main fuse blows after replacement, suspect that a component in the power supply is overloaded or leaky. A quick resistance test across the horizontal collector terminal (body) and chassis might indicate that a transistor or damper diode is leaky. If the resistance is below 100 Ω, check the output transistor or damper diode. Remove output transistor from the circuit and perform leakage tests.

Check the low-voltage circuit with a resistance test across the main filter capacitor. Suspect the filter capacitor or overloaded components in circuits tied to the low-voltage sources. Quickly measure the resistance across each silicon diode in the power supply for leakage. Check for one or two leaky or shorted silicon diodes if the line fuse blows at once and the fuse appears dark.

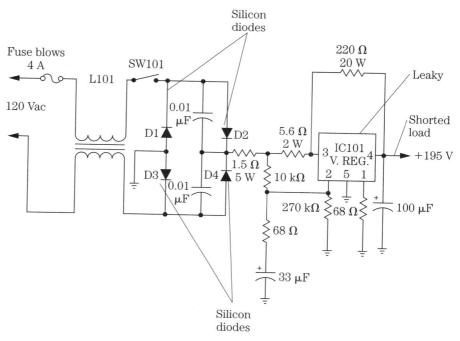

25-24 Check the components in the low-voltage power supply if the fuse keeps opening.

No red in picture

Check the color output transistors or picture tube if only one color is missing in the picture. The color output circuits can be checked for correct color waveforms or voltage measurements. Usually the color output transistors are found on the CRT socket board. If there is an open or leaky red output transistor or no heater voltage to the red gun assembly, red will be missing from the picture. Check the CRT with a picture tube tester to determine if the red gun is weak or open.

Weak color

Determine if the black-and-white picture is normal with correct brightness and contrast, by turning down the color control. Low or improper voltages from the power source or a leaky color transistor or chroma IC can cause weak color. Check for burned or open resistors and bypass capacitors in the bandpass circuits. Suspect that diodes are leaky or open in the AGC color circuits. Open coils and poor terminal connections of components on the PC board might cause weak color.

Weak sound

Weak sound can be caused by open or leaky transistors, ICs, and electrolytic coupling capacitors in the sound circuits. Weak and garbled sound might be eliminated by touching up the sound or discriminator coil. A speaker with a frozen cone will produce distorted and weak sound. Check the electrolytic coupling capacitor between the speaker and the output transistor for weak and intermittent sound. Take critical voltage and resistance measurements on the output transistors and ICs.

Vertical foldover

Check for leaky output vertical transistors, ICs, and a change in bias resistors. Most vertical foldover problems occur in the vertical output circuits (Fig. 25-25). Check for improper negative and positive voltage sources to the vertical circuits. Vertical foldover might occur at the top or bottom of the raster. Often, vertical foldover occurs with insufficient vertical sweep. Replace both vertical output transistors for vertical foldover. Suspect a leaky vertical output IC with foldover. Check the vertical pincushion circuits for burned resistors and transformer windings. Resolder the terminal connections of the pincushion transformer.

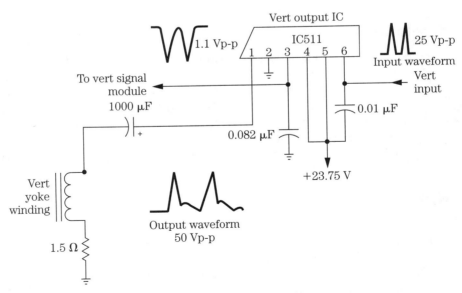

■ **25-25** *Check the vertical output and pincushion circuits for vertical fold over in the TV raster.*

Conclusion

Although many TV symptoms are found in this chapter, many others occur in the color and B&W TV chassis. Check the picture tube and speakers for others that might indicate a breakdown in the TV chassis. The intermediate electronics technician who can take these symptoms, check the block or schematic diagram, isolate, and then locate the defective component with test instruments, is on the right road to becoming an advanced or professional electronics technician.

26

Simple TV repairs

THE SIMPLE AND COMMON PROBLEMS FOUND IN THE TV chassis can be checked by the intermediate electronics technician. Replacing fuses in the low-voltage circuits; checking silicon diodes; and testing the capacitors, switches, and voltage regulators (all located in the low-voltage circuits) can be made with voltage and resistance measurements of the DMM or VOM. Simple continuity tests with the ohmmeter can locate a defective switch, diode, regulator, or electrolytic capacitor (Fig. 26-1).

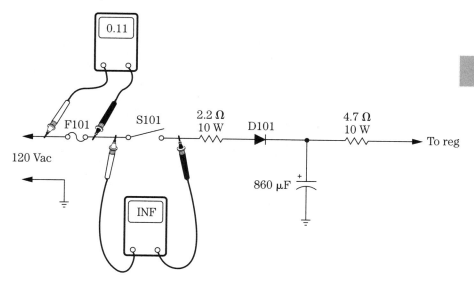

■ **26-1** *Check the fuse and on/off switch with the RX1 range of the VOM or DMM.*

Horizontal output transistor tests can be made with an in-circuit transistor test or with the diode test of the DMM. A quick resistance test from the metal body of the horizontal output transistor to chassis ground might indicate a leaky output transistor or damper diode. Simply remove the horizontal output transistor. Test the transistor out of the circuit.

Scope the horizontal and vertical deflection IC for correct horizontal and vertical drive waveforms. In some circuits, the driver or amplified horizontal circuits are contained in one large IC. Remember that the horizontal driver transistor might run warm and have no waveform, which is normal without a horizontal waveform found at the base of the driver transistor. Check the horizontal circuits from the deflection IC, driver, and waveform on the base of the horizontal output transistor.

No vertical sweep might result from a defective drive signal or from leaky and open vertical output transistors (Fig. 26-2). Check the vertical waveform at the output terminal of the deflection IC to the base of the vertical output transistors. Test the transistors in and out of the circuit. Check the vertical drive waveform at the input and output terminals of the vertical output IC up to the yoke winding. Vertical output waveforms are very unstable on the scope, compared to the horizontal waveforms.

■ **26-2** *A bright horizontal white line indicates no vertical sweep.*

Check the audio signals with scope or external audio amplifier. Substitute the suspected speaker with another for tests. Take audio transistor tests with the beta transistor tester or diode-transistor tests with the DMM on the transistors still in the circuit. Signaltrace the audio circuits from stage to stage for weak and distorted audio, and if no signal is on the external amp.

High-voltage and picture tube tests should be made with a high-voltage probe and CRT checker. Remember to connect the ground cable of the HV probe to the TV chassis ground to prevent a shock of 25 to 35 kV. Do not handle the high-voltage probe without first grounding it. The picture tube tester might indicate a shorted gun, weak emission, no heater or high resistance filament, and intermittent element tests. These simple TV tests might help you to repair a dead TV chassis.

Check the fuse

If there is no sound, no raster, and no picture, check the line fuse. You might find one or two fuse-protected circuits in the color TV chassis. Usually, the low-voltage power supply is protected with the line fuse, while the horizontal circuits contain a lower amperage fuse. The line fuse might be a 4- or 5-A and the horizontal circuits protected with a 500-mA to 1½-A fuse. You might not find a horizontal fuse in some TV chassis. Always replace these fuses with the correct amperage and voltage.

If only the horizontal circuit fuse is blown open, suspect overload conditions in the horizontal output transistor circuits. Sometimes the open fuse can be replaced and will operate with no further problems. But if the fuse blows at once, suspect a leaky output transistor or damper diode. If only the line fuse opens, check for problems in the low-voltage power supply.

Locate the fuse close to the line input power cord or at the rear of the chassis (Fig. 26-3). The horizontal circuit fuse might be lo-

Line fuse

■ **26-3** *Locate the line fuse at the rear of the chassis near the low-voltage power supply.*

cated in the middle of the chassis or next to the horizontal output circuits. Remove the power cord and check the fuse on the RX1 range of VOM or DMM. Make sure that the power cord is disconnected. The fuse might have long leads and be soldered into the PC board or clipped into a fuse holder. If the fuse is removable, remove and check it out of circuit. The normal fuse has a continuity of low resistance while the open fuse has no measurement. If the fuse opens after replacement, check the low-voltage power supply and horizontal output circuits for leaky components.

If the $B+$ fuse is normal and the line fuse blows or opens, suspect leaky silicon diodes, electrolytic capacitor, and voltage regulator. Instead of a $B+$ fuse, some chassis might have a low-ohm (10 to 15 Ω) isolation resistor. The resistor is connected between the low-voltage power-supply source and primary winding of the flyback or horizontal output transistor. Sometimes this resistor will open with a shorted horizontal output transistor (Fig. 26-4).

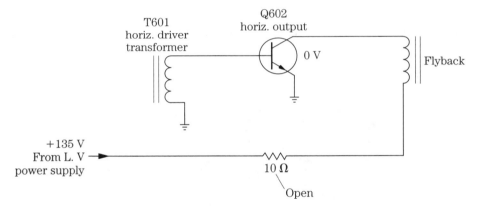

■ **26-4** *If no voltage is at the collector terminal of the horizontal output transistor, check for an open fuse or resistor.*

Check those silicon diodes

You might find a half-wave rectifier with one silicon diode, full wave with two diodes, and a bridge rectifier circuit with four diodes in the low-voltage power supply. Suspect leaky diodes or an electrolytic capacitor when the line fuse blows at once. If a low-voltage power transformer is found with a bridge or full-wave rectifier circuit, it's best to remove one lead of each diode for a correct leakage test. If one silicon diode or a bridge rectifier and transformer is found, check the continuity across each diode. You

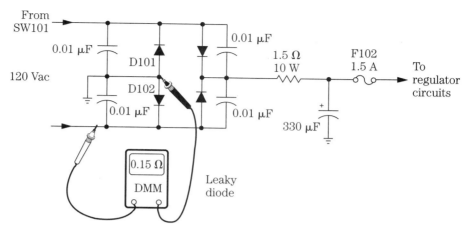

■ **26-5** *Check each diode with the diode test of the DMM.*

might find one or two diodes leaky in the low-voltage circuits. Replace each defective diode with a 2.5-A type (Fig. 26-5).

Silicon diodes in the low-voltage circuits might be from 1 to 3 A. Silicon diodes under 3 A, most technicians replace with the 2.5-A 1000-V diode. Doublecheck the schematic if larger diodes are found. The normal silicon diode will have a low resistance in one direction and infinity with reversed tests leads. The open diode has no measurement in any direction. The leaky diode might have a low-resistance measurement under 1000 Ω in both directions. The shorted diode might have a low resistance under 10 Ω in both directions.

On/off switch tests

The on/off switch might be an SPST or DPST of either a push-on or rotary switch. Some ac switches are turned on by electronic switches and standby power supplies. The ac switch might be included right after the line fuse in the TV power-line circuit. Sometimes a relay turns the chassis off and on in other chassis. Check the ac switch with RX1 continuity measurement across the switch terminals with the power cord removed from the outlet. Pull the ac power cord before making resistance or continuity measurements in any electronic chassis. Replace any switches that have signs of resistance or erratic measurement. Inspect the relay contacts and connections on the PC board. An open switch will have no measurement. Often, the ac switch is on the rear of a volume control and both must be replaced if the off/on switch is defective.

Capacitor tests

Electrolytic capacitors in the low-voltage power supply can be tested with a capacitor tester or with resistance measurements. Quickly discharge the main filter capacitor with a couple of screwdrivers and then take a resistance measurement across the capacitor terminals. If the resistance is below 500 Ω, suspect a leaky capacitor or overloaded power supply. Remove one lead from the capacitor and test again. Often, shorted or leaky filter capacitors will measure under 100 Ω (Fig. 26-6).

Main filter capacitor

■ **26-6** *Measure the resistance across main filter capacitor to determine if the capacitor is leaky.*

The VOM, FET, and VTVM are best to take resistance measurements across the capacitor. As the capacitor charges up, the meter hand will move upward and then discharge it very quickly. With the digital ohmmeter (DMM), the numbers will slowly charge up and slowly recede when the capacitor discharges. Sometimes the rotation of numbers are difficult to watch. The regular ohmmeter with meter hand is easy to watch. A small capacitor or one with lower capacity will not charge up very far and fall back very rapidly. It's best to check small bypass or disc capacitors in the capacity meter.

Disconnect the positive terminal of the suspected capacitor and check capacity with the capacity meter. The small capacity meter found with the volt-ohmmeter or DMM will only measure capacity up to 2 or 5 µF. Choose a capacity meter or tester that will at least test 1000- or 2000-µF electrolytic capacitors. Often, new electrolytic capacitors will test over the original capacity. Replace the electrolytic capacitor if it is well below the rated capacity. Of course, the leaky or shorted capacitor should be replaced without any further tests.

Voltage regulator tests

Today, voltage regulators might look like a power output transistor (TO-220), like the horizontal power output transistor with four terminals, and like a flat IC. The transistor type has three prong terminals and the metal body (IC-106) might have three terminals (Fig. 26-7). Other IC regulators look like power-output TO-220 transistors. The flat heatsink regulator might have four or more flat leads and look somewhat like a power output IC.

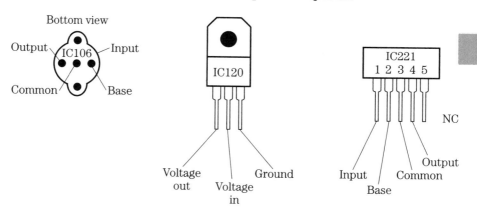

■ **26-7** *The voltage regulator might look like a transistor or IC.*

Voltage regulators are usually damaged by lightning or a power-line surge. If low or no output voltage is measured at the collector terminal of the horizontal output transistor, suspect a defective IC voltage regulator. Check the input and output voltage terminals (Fig. 26-8). The input voltage might be from +150 to +165 Vdc and the regulated output voltage might range from +115 to +135 Vdc. Check for an open or a change in resistance of resistors tied to the regulator terminals, after removing the defective regulator from the chassis. Voltage regulators can be replaced with universal types if original parts cannot be located.

■ **26-8** *The voltage regulator might have four or more terminals and be mounted on a heatsink.*

Leaky horizontal output transistor

If the power supply voltage is much lower than normal and there is no raster, check for a leaky horizontal output transistor. A quick voltage measurement at the collector terminal with the VTVM might indicate low or no voltage. A scope test at the collector terminal is best because many voltmeters will not measure the voltage at the collector terminal (metal body). Recheck the output transistor with a DMM test from collector (body) to common ground with the diode-transistor test (Fig. 26-9). Remove the ac cord from the receptacle. A low-resistance measurement in one direction indicates that the transistor is normal. If the measurement is under 10 Ω, suspect that a transistor or damper diode is leaky.

Check the schematic to see if the damper diode is inside the body of the output transistor. Only a resistance or diode test in one direction should be indicated if the damper diode is outside of the transistor. Remove the transistor and take another test on the collector mounting screw for the damper diode leakage tests. Test the output transistor while it is out of the circuit. A low resistance (under 10 Ω) with reversed test leads indicates that an output transistor is leaky or shorted. The metal case of the horizontal output transistor is the collector terminal, except in a few Sanyo models made for Sears stores.

■ **26-9** *Check the horizontal output transistor for leakage from the collector to the common ground.*

Heatsink   Horiz. output transistor

■ **26-10** *Older horizontal output transistors were mounted with two screws; newer horizontal output transistors are mounted with one screw.*

Most output transistors are mounted with two metal screws (Fig. 26-10). Except you might find one screw at the top with the new output transistors with a T0-220 mounting. Remove the transistor and take a voltage measurement on the collector screw terminal. Simply push the positive test lead into the collector screw hole for this measurement. If the voltage is much higher than the operating voltage, you can assume the low-voltage power supply is normal. Test the horizontal transistor out of the circuit for leakage or open tests. It is difficult to accurately check the output transistor for open conditions with the driver transformer winding in the base circuit.

Check the new output replacement before mounting it. You might receive a defective new component. Notice if a piece of insulating material is between the heatsink and the transistor. Apply silicone grease on both sides of the mica insulator before mounting the transistor. Doublecheck the resistance measurement between metal chassis and transistor for low-resistance measurements.

Damper diode tests

The damper diode prevents ringing in the power supply of a TV receiver and also protects the horizontal output transistor from overheating and damage. If the damper diode is left out of the circuit, the horizontal output transistor can be damaged in seconds. The damper diode might be located inside of the output transistor in some of the latest TV chassis. Always replace the output transistor with an exact replacement universal transistor, to include the damper diode (Fig. 26-11).

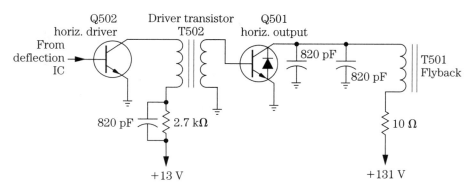

■ **26-11** *Check a defective horizontal output transistor with the damper diode built inside of the transistor case.*

The damper diode resistance measurement can be made to common ground with the DMM leads at the body of the output transistor. If a low resistance is noted in both directions, either the transistor or damper diode is leaky. Remove the metal screws from the transistor and take another resistance measurement or take a diode test between the collector screw and the chassis ground. A low measurement (0.556 Ω) in one direction indicates that the damper diode is normal. If low resistance (under 10 Ω) is measured in both directions (reversing test leads), the diode is shorted or leaky. Replace it with a universal replacement if the original part is not handy.

Driver transistor tests

Quietly take a voltage measurement on the collector (metal end) terminal of the driver transistor to determine if the voltage is low, missing, or high. An open driver transistor will have a high collector reading. The leaky driver transistor might have low voltage at the collector terminal or the transistor might have no drive waveform or pulse from the horizontal deflection IC. Next, scope the deflection IC horizontal waveform at pin 21 of IC501 (Fig. 26-12). If there is no drive signal here, the deflection IC might be defective or no voltage is being supplied to IC501 from the derived secondary voltage of the flyback. All of the horizontal circuits must perform before low voltage is applied to the voltage pin of the deflection IC.

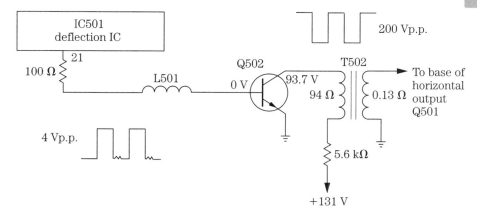

■ **26-12** *Scope the horizontal waveform at pin 21 of IC501 to the base of driver transistor Q501.*

If no drive waveform is available at the base terminal of the horizontal driver (Q501), the transistor will run quite warm with low collector voltage. The voltage-dropping resistor that provides voltage to the driver transformer primary winding also operates quite warm. This is normal without any drive pulse. The secondary winding of T501 connects directly to the base of the horizontal output transistor. Check the driver transistor in the circuit for leakage or open conditions. If it is left on too long with no drive pulse, T501 and Q501 can be damaged. Scope the horizontal deflection pin 21, the base terminal of Q501, and the horizontal driver waveform.

Intermittent TV start-up can be caused by poor terminal connections of driver transformer (T501). Resolder these connections and take resistance measurements on both windings. The primary winding might be between 45 and 100 Ω and the secondary winding is under 1 Ω. If the primary resistance is quite low compared to the schematic, the leaky driver transistor might have damaged the winding. Check both the primary and the voltage-dropping resistor for correct resistance after removing a leaky driver transistor. Scope the horizontal circuits after repairing them to make sure that all waveforms compare with those on the schematic.

Scope deflection IC

If no horizontal or vertical waveforms are found at the deflection IC, suspect the IC. Both horizontal and vertical drive waveforms are found at the countdown or deflection IC. Check the schematic for correct pin numbers of both horizontal and vertical output waveforms. These drive waveforms might be included in one IC with the video, chroma, sync, and holddown circuits (Fig. 26-13).

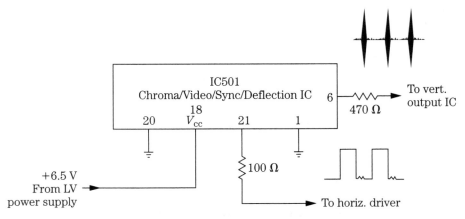

■ **26-13** *Scope the deflection IC horizontal and vertical output waveform to determine if the IC is normal.*

Go directly to the supply voltage source pin terminal and check for normal voltage. If low or no voltage is measured at pin 18, suspect a defective IC or an improper voltage source. Determine if the supply voltage is fed from a secondary voltage from the flyback circuits or from a low-voltage power supply. If the voltage source comes from the secondary circuits of the flyback winding, all horizontal circuits must function before a supply voltage source is developed. Check the low-voltage source and IC if the supply pin receives voltage from the low-voltage power supply.

Vertical output tests

If the vertical sweep is insufficient or if there is a horizontal white line, scope the vertical waveforms from the deflection output pin to the vertical output transistors or IC. You might find two vertical output transistors in older TVs and one power output IC in the modern TV. Although vertical waveforms are not too stable, you can scope the input and output vertical circuits to find the defective component. Scope the input and output terminals of the vertical output IC (Fig. 26-14). If there is an input waveform and no output, take critical voltage and resistance measurements on the IC terminals.

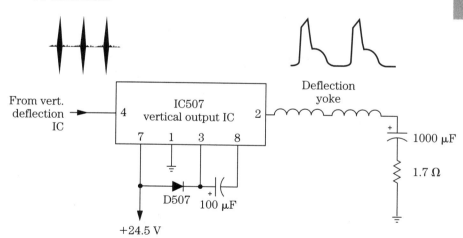

■ **26-14** Scope the input terminal (4) and output terminal (2) to determine if IC507 is defective.

Check the voltage on both vertical output transistors if a normal drive pulse is found at the output of pin 10 of IC310. Often, one of the output transistors might be open or leaky; sometimes both transistors must be replaced. If the vertical output is intermittent, replace both output transistors. Test the transistors in and out of circuit with the beta and DMM tests. While the transistors are out of the circuit, check the bias and load resistors for correct resistance (Fig. 26-15). Do not overlook the 3-Ω return resistor in the leg of the deflection yoke assembly.

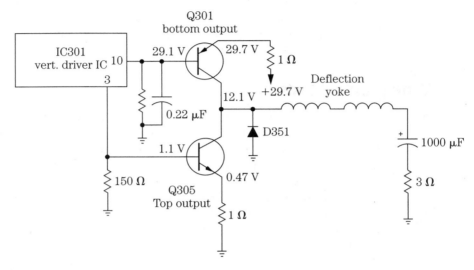

■ **26-15** *Check the bias diodes and resistors while the vertical transistors are out of the circuit.*

Replacing vertical output transistors

You might find the vertical output transistors on separate heatsinks anchored to the PC chassis. Collector voltage measurements can be taken from the metal heatsink if the transistor is not insulated from it. Check the collector voltage on the metal screw of the vertical transistor (T0-220). Sometimes these transistors are insulated from the heatsink with a piece of mica. If not, measure the voltage from the heatsink, if it is in the middle of the PC board. Unsolder all three terminals on the PC board and remove the mounting screw. Remove the suspected transistor from the heatsink (Fig. 26-16).

Sometimes the transistor will test normal outside of the circuit and test leaky or open in the circuit. If heat is applied to the transistor terminals, the transistor might restore itself. Replace the transistor. If one vertical output transistor is leaky or open, replace both output transistors. Before replacing them, smear silicone grease on the back side of the transistor. If a piece of mica insulation is between the transistor and the heatsink, apply silicone grease to both sides of the insulator. Check the new replacement

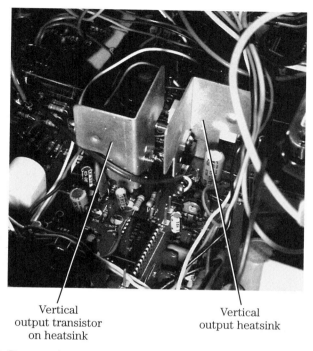

Vertical output transistor on heatsink

Vertical output heatsink

■ **26-16** *Remove the vertical output transistor from the heatsink by removing the top metal screw and unsoldering the three terminal leads.*

before installing it and check it again with the transistor tester after replacing it. Make sure that the insulated transistor is not grounded to the heatsink.

Yoke tests

Take critical low-resistance measurements on the vertical and horizontal windings if the circuits are normal feeding the deflection yoke. The deflection yoke might short between vertical and horizontal windings with firing inside and a curl of smoke might be seen from the deflection yoke assembly. Sometimes the yoke winding will short against another layer of coil wire and produce insufficient sweep. The yoke winding might open up with a break near the soldered yoke cable wires (Fig. 26-17).

Simply remove the red wire from the yoke assembly with a narrow width raster or low voltage at the collector terminal of the horizontal output transistor. A shorted or leaky yoke can load down the output circuits and not generate a high voltage. The high voltage might come up from the flyback with lower HV than normal.

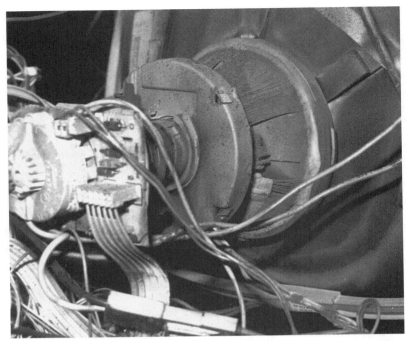

■ **26-17** *The defective yoke might collect dust and spilled liquid, then arcover between the windings.*

Just unsolder the red lead and see if voltage increases. This voltage will still be lower than normal with the yoke out of the circuit.

Take critical resistance measurements on both vertical and horizontal windings. The average resistance measurements of the horizontal winding might be around 2 to 6 Ω, and the vertical winding resistance is from 10 to 20 Ω. These measurements should be checked against the TV schematic for exact resistance. Resistance measurements will indicate continuity, but will not indicate shorted turns in each winding. Check the resistance between the vertical and horizontal windings. Remove the red and common ground wire from the circuit and take a 100-kΩ measurement. Replace the yoke if a lower reading is found.

Sub the speaker

Check the speaker if it has intermittent audio, mushy sound, distortion, or no sound. Simply clip another speaker across the suspected speaker. Remove one lead if the sound is distorted and is tinny. Any size of speaker can be used for this test (Fig. 26-18). Remove the suspected speaker and check for a dropped cone. No-

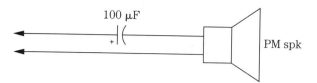

■ **26-18** *Check the suspected speaker by clipping another across it.*

tice if the speaker has a dual cone. Replace it with a speaker that has the same size, power rating, and correct speaker impedance. Do not replace a 3.2-Ω speaker with one that has an 8-Ω impedance because the volume will be less than normal. Conversely, do not replace a 32-Ω speaker with an 8-Ω speaker or you could damage it.

Transistor output sound circuits

In today's lower-priced TVs, transistors might still be used in the sound circuits. Usually, an audio preamp, IF amp, and detector IC provide audio to the first audio driver transistor. The sound is coupled in series with a small resistor and electrolytic capacitor to the base of the audio driver transistor (Fig. 26-19). The output from the collector terminal is directly coupled to the base of the audio output transistor or diodes might be inserted in the base circuits. Q103 and Q105 work in push-pull operation to drive the speaker. The audio output is taken between two identical emitter resistors

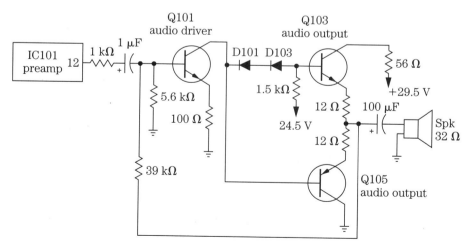

■ **26-19** *Take critical voltage measurements and test the transistors in-circuit with a beta transistor tester.*

with an electrolytic capacitor. The speaker voice coil might have an impedance from 8 to 32 Ω.

Look for leaky or open transistors in the audio circuits if the sound is weak and distorted. Weak audio might result from dried-up or defective electrolytic capacitors. Check the 100- to 1000-µF speaker coupling capacitor and transistors for intermittent audio. Leaky diodes, transistors, and bias resistors might cause weak and distorted sound. Check the audio stages with an external amp to locate the defective stage.

IC sound output circuits

The audio output might contain one IC as an SIF, AFT, detector, and audio output. In the sound circuits, a single IC might be used as the sound output. With no sound, distorted and weak sound, take a quick test on the voltage supply pin (9) of the IC. Suspect a leaky IC or improper voltage source if the voltage is very low (Fig. 26-20). Disconnect pin 9 from the PC board by removing the excess solder around the supply pin. Make sure that the pin is loose from the PC wiring. Notice if the voltage increases at the PC wiring terminal. Suspect that an IC is defective if the voltage returns to normal or is a few volts higher (18.5 V).

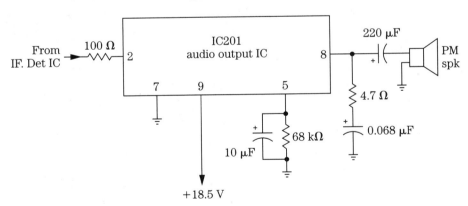

■ **26-20** *If pin 8 of IC201 is quite low, suspect a leaky IC or an improper voltage source.*

Before removing the IC, take critical resistance measurements on each transistor terminal pin and at the common ground to locate a leaky resistor or capacitor. Sometimes, a poorly soldered pin connection might cause weak, distorted, and intermittent audio. Check each measurement against the schematic. Make sure that all components tied to the IC are normal before removing and replacing the sound output IC. Transistors and ICs in the sound circuits can be replaced with universal parts.

Signaltracing audio circuits

The audio signals can be signaltraced with the external audio amp for dead, weak, and distorted signals. Start at the detector stage and signaltrace the audio up to the speaker. If audio is present at the volume control, start here and trace the audio into the input circuits of the transistor or IC. Signaltrace the input audio signal at the audio output IC and output terminal of the IC.

In transistor circuits, signaltrace the audio signal from the base to the collector and to the next base terminal. Look for weak audio after an electrolytic coupling capacitor. If the signal is normal going into the sound IC and weak or no audio is present at the output terminal, suspect the IC or the voltage source. Check the supply voltage source on the V_{CC} terminal—usually the highest voltage to the sound IC.

Tuner tests

Suspect that the tuner is defective if the picture is snowy or missing, the raster is white, and there is no sound. Check the voltages on the tuner assembly and compare with those on the schematic. The early turret tuner might have +10 to 15 V with a negative voltage from the AGC circuits. A new varactor tuner will have a supply voltage (V_{CC}) and also a tunable voltage (Fig. 26-21). The tunable

■ **26-21** *The varactor tuner consists of small components on top of the board and SMD components on wiring side.*

voltage will change as each station is selected with the tuner control. If not, check the tuner control IC.

The defective tuner might provide a normal picture and then appear intermittent. The picture might become snowy and finally disappear. The next time, the tuner might operate perfectly. Monitor the voltages to the tuner assembly or module. If the voltages are constant if the picture disappears, suspect that a tuner is defective. Remove the module or soldered bottom terminals from the PC board wiring. These tuners can be sent in for repair, interchanged, or you can purchase a new one. Complete tuner replacement is rather expensive, however. The old turret tuner might just need a good cleanup (Fig. 26-22).

■ **26-22** *Older turret tuners can be cleaned by removing the bottom metal cover and spraying the contacts with tuner cleaner spray.*

Picture tube tests

The defective picture tube might have a black screen, blotched color areas, poor focus, negative picture, and internal arcover. Turn up the contrast control and notice if several areas in the picture have a shiny or colored area. If the picture tube takes a long time for the picture to clear up with enough brightness, check the tube for weak emission. Sometimes tapping the end of the CRT socket might show lines and a negative picture, indicating that the gun elements are intermittent. Check the tube filament and see if it lights (Fig. 26-23).

■ **26-23** *Check the HV voltage on the picture tube with a CRT tester.*

Take a high-voltage test at the anode terminal with the HV meter. Be careful not to receive a shock. Make sure the ground wire of the HV meter is grounded to the metal chassis. Test the CRT on a regular picture tube tester. Low emission in any one of the gun assemblies might cause the picture to have poor color. If one gun is very weak or shorted, the picture tube should be replaced. Before the picture tube is replaced, check the price of replacement and provide a written estimate for the customer.

Replacing the picture tube

27

THE DEFECTIVE PICTURE TUBE MIGHT BE GASSY AND leave blotched areas within the color picture. A weak CRT takes a long time for the brightness to come up on the screen. If the CRT is very weak, it might remain dark. The intermittent picture tube might have flashing lines, tearing, or a negative picture. The intermittent picture might result from particles scaling off of the cathode element and floating between the grid elements. If the picture tube socket is tapped, the picture might have lines that come and go. A defective picture tube socket might produce an intermittent picture (Fig. 27-1).

■ **27-1** *Make all necessary tests on picture tube before attempting to replace it.*

If one color is weak or missing, the color gun assembly might be defective. If one color takes a long time to appear, the color emission is weak and needs to be balanced with a rejuvenation process or the picture tube needs to be replaced. A blob or section of color found on the screen might result from improper purities. The screen might be magnetized if strong speaker columns are nearby or if you shut off a carpet sweeper in front of the TV set.

Poor tube socket connections

The corroded picture tube socket might cause an intermittent, out-of-focus picture, or no picture. Notice if the heaters or filaments are lit inside of the gun assembly. Sometimes moving the picture tube socket might let the heaters come on with a picture. Replace the defective socket because poor contacts might be burned, resulting in poor heater contacts (Fig. 27-2). In the early TV chassis, the large picture tube socket elements were larger and could be repaired by soldering the pins to the CRT socket. Replace today's sockets with parts that have the original numbers and clean the tube prongs for greater contact.

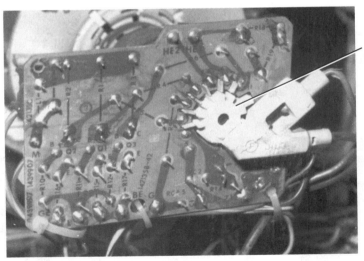

■ **27-2** *Intermittent picture might be caused by a defective or corroded tube socket.*

Poor picture tube focus might be caused by a poor connection of the focus pin and socket, a broken socket pin, or a corroded socket pin and socket. In the early chassis, the focus pin might corrode so badly that the socket pin would crumble, resulting in no focus pin contact. Remove the plastic cover or PC board chassis to see if the focus pin is corroded. Check the socket focus pin with the large black or white cable coming from the focus-control assembly on the chassis. Replace the socket and clean the focus pin. The focus-screen adjustments might be mounted at the rear of the CRT or behind the flyback (Fig. 27-3).

■ **27-3** *The focus and screen adjustments are used on the end of an RCA CRT socket.*

If the picture tube brightness is very bright with a chassis shutdown, suspect a defective picture tube or arcing in the spark gaps. The spark gaps are found off of the different grid pins in the schematic. Most grid spark gaps are found inside the CRT socket and if excessive dust has collected, the gaps will fire over and shut down the chassis. Blow out all dust from the CRT socket. Remove the picture tube socket cover if it is found on the rear of the picture tube. Suspect a broken glass picture tube assembly if you hear a loud arcing noise inside of the picture tube. Often, the glass neck of the CRT will crack and break inside of the yoke assembly and the high voltage will arcover. Of course, the picture tube must be replaced.

Excessive brightness that cannot be controlled or turned down might be caused by a defective picture tube or improper screen and grid voltages. Check the screen grid terminal for correct boost voltage. If the voltage is low or missing, the brightness control has no affect on the bright raster. A defective picture tube can be detected by visually inspecting it, testing the tube, and measuring the voltage.

High-voltage problems

Besides visual symptoms, low or no high voltage applied to the picture tube will cause a no raster symptom. Often, higher voltage can be detected at the CRT by holding your forearm near the TV screen. The hair on your arm will stand up. Another method is to hold a small piece of paper next to the screen and watch the paper pull toward the screen. High voltage can be detected if you hear the deflection yoke expand (when the receiver is turned on) and collapse (when TV is turned off). High voltage should be measured with the HV probe at the anode terminal of the picture tube (Fig. 27-4).

■ **27-4** *Check the HV with a high-voltage probe connected to the chassis and the anode terminal.*

For successful picture tube and corresponding circuit tests, the following instruments are required:

- ☐ *CRT tester* The picture tube tester checks the performance of each individual gun for proper emission. The CRT tester is a must-have instrument for the complete service shop or establishment. For a few more dollars, purchase a CRT tester with a rejuvenation process so that each gun can be charged, balanced, with normal gun emission (Fig. 27-5).
- ☐ *VOM, DMM, FETVM, and VTVM*

■ **27-5** *The color picture can be repaired by charging or cleaning each cathode terminal.*

☐ *A separate high-voltage probe that will measure up to 40-kV dc voltage with a 25-kV factory-calibrated accuracy*

☐ *VTVM with high-voltage probe that contains multiplier instrument ranges by 1000* Choose a probe that can be used with a high impedance voltmeter such as a VTVM or FETVM.

CRT components

The picture tube components consist of a glass envelope, electron gun assembly, focusing system, deflection circuits, and the phosphor screen. The black-and-white and color picture tube are available in many different sizes and shapes. Besides high voltage applied to the anode terminal and electron gun assembly, a deflection yoke provides vertical and horizontal sweep of the raster. Static magnets and dynamic convergence circuits with three color gun assemblies provide color pictures on the screen. Each color, red, green, and blue, has a separate color gun assembly with cathode and heater terminals.

The latest color picture tubes might have a beam bender instead of dynamic coil convergence circuits (Fig. 27-6). You might find separate trim magnets on the corners to provide normal purity on the 27-inch and larger picture tubes. On some picture tubes, the deflection yoke assembly might be welded or glued to the bell of the tube and must be replaced with the same brand of picture tube with yoke assembly. These are flat front, glass screens with a greater viewing area.

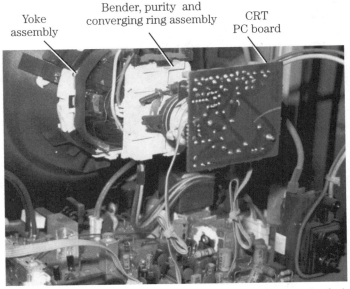

■ **27-6** *In the CRT tube bender, purity rings are used on neck of picture tube.*

Larger glass picture tubes

Picture tubes are getting bigger each year. A few years ago, the 24-inch tube with a 25-inch diagonal design around the corners was the largest size. Today, 26-, 27-, 31-, and 35-inch TV screens are available. The 27- and 31-inch screens are quite popular and are used in table models and consoles. The 27-, 31-, and 35-inch tubes have a flat viewing surface with square corners. Most picture tubes are measured diagonally from corner to corner.

Although the larger screens might have the same scanning lines as the 9- or 10-inch screen, the 27- and 31-inch tubes still have an excellent picture. Very little distortion is found in the larger tubes, with special pincushion linearity circuits. The larger the screen, the heavier the TV. It takes two people to lift, remove, and replace the larger picture tubes. Also, the cost of replacing the 31- or 35-inch CRTs might exceed the original cost of the whole TV set.

Picture tube circuits

The color picture tube has a heater or filament for each gun assembly, cathode, grid, screen grid, and anode elements. The outside circuits connected to these elements must perform to make the picture tube function. In today's chassis, the heater voltage is developed with one or two turns of wire that are loosely coupled around the core of the flyback or horizontal output transformer. A small voltage-dropping resistor is used in some heater circuits (Fig. 27-7). All three color cathodes are tied directly through a resistance or choke to the color output transistors. The cathode and color output voltage is supplied from the boost circuits of the flyback winding (+170 to 190 V).

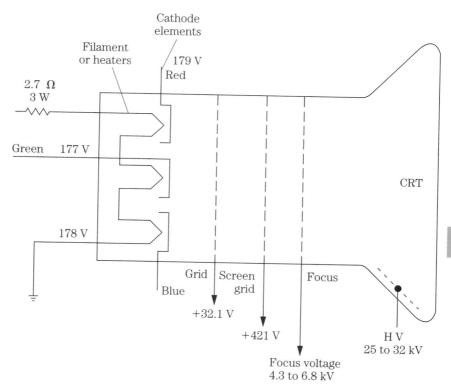

27-7 *The heater voltage comes from a winding on the flyback with low-resistance isolation and a voltage-dropping resistor.*

Usually, the fixed grid voltage is tapped from the boost or low-voltage winding of the flyback with a very low voltage (32 V in a 27-inch tube). The screen grid voltage is quite high (350 to 450 V) to pull the electrons toward the face of the picture tube and the voltage is supplied from a screen grid control located with the focus control. The focus element voltage connects to the focus control divider network and supplies a variable voltage (4.3 to 6.8 kV) in large picture tubes. The focus voltage in smaller CRTs might vary between 3.5 and 5.5 kV. The anode voltage is developed in the HV and flyback circuits and is tied to the anode button. Nominal high-voltage for the RCA CTC140 chassis (with maximum beam current) is 27.5 kV on a 27-inch picture tube. Check the manufacturer's service literature for the correct anode voltage.

Horizontal and vertical sweep is applied to the picture tube through the deflection yoke. Any sweep problems in the picture at the top and bottom might occur in the vertical section, while improper horizontal problems are caused by improper width or no horizontal sweep. The deflection yoke cables are connected respectively to the vertical and horizontal output circuits (Fig. 27-8).

■ **27-8** *The deflection yoke cables might plug into the horizontal and vertical output circuits.*

Safety factors

If not careful, you can get shocked very easily while working on the color chassis. Keep your hands away from the HV rubber anode connections (Fig. 27-9). After being shocked once or twice, the electronics technician knows what areas to touch and to avoid. Be careful when taking high-voltage measurements on the CRT, so as not to pull off the rubber socket while inserting the HV probe tip. Make sure the ground wire of the high-voltage probe is grounded to the chassis or the ground wire of the CRT. Check to see if the picture tube ground wire lies against the black aquadag coating on the outside of the picture tube (Fig. 27-10). If it doesn't, firing lines will result and white interference lines will show across the TV screen.

Wear safety glasses when handling or moving a picture tube. A slight bump or something falling against the neck of the picture

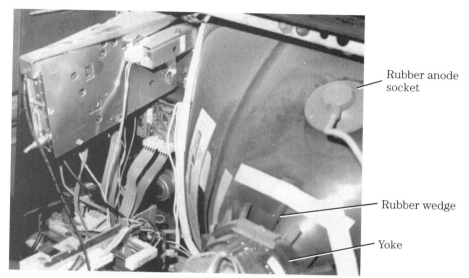

27-9 *Do not touch the TV anode rubber socket while the TV is operating.*

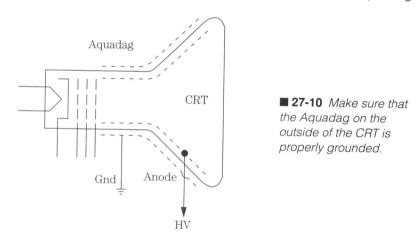

27-10 *Make sure that the Aquadag on the outside of the CRT is properly grounded.*

tube might make it implode, throwing pieces of glass everywhere. The picture tube neck can accidentally be broken off if you are removing the back cover on the TV and let the cover fall down on the neck of the tube. This just about happens to everyone working on TV sets. But if you do it once, you will probably never do it again. Because of these dangers protect your eyes at all times.

Discharge the high voltage at the picture tube when replacing or working around the high-voltage circuits. Remember, the picture tube aquadag on both sides of the glass tube acts as an HV capacitor and can hold a charge for months. Stay alert when discharging or working around high-voltage circuits to prevent shocks or causing someone else serious injuries.

In the old black-and-white tube chassis, the picture tube was discharged from the anode button connection to the chassis ground. Do not attempt to discharge the picture tube in this manner. You can damage transistors and ICs in the solid-state chassis. Always discharge the HV at the anode socket to the black aquadag on the outside of the picture tube. Use a long-bladed, well-insulated screwdriver to get under the rubber socket connection, and ground a similar screwdriver blade to the outside ground of the CRT (Fig. 27-11). Touch the metal areas of the screwdrivers together to discharge the picture tube. Before removing the picture tube from the chassis, ground the anode button once again.

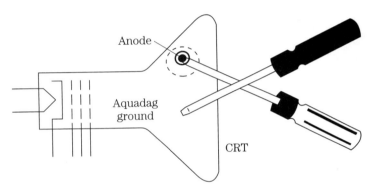

■ **27-11** *Use two screwdrivers to discharge the picture tube at the anode and Aquadag ground area.*

Testing the CRT

The defective picture tube can be located with a CRT tube tester and critical voltage measurements. Check for correct high voltage at the anode terminal. Visually inspect the neck of the tube for light in all three gun filaments or heaters. Check the picture tube in a CRT tester. Test each color gun assembly for balanced emission. If one of the gun assembly emissions tests quite low, a CRT rejuvenator tester might charge up or restore the poor emission problem.

The picture tube can be checked in the home, in the carton, or on the service bench with the CRT tester (Fig. 27-12). Sometimes intermittent conditions might not be located with the picture tube tester. Tapping the end of the tube might turn up a shorted gun assembly and you might notice flashing in the short or leaky indicator light on the tester. Visual inspection, and high- and low-voltage

■ **27-12** *Check the CRT with a picture tube tester. Use the rejuvenator to balance each color gun assembly.*

measurements on the tube elements can help you to discover that a picture tube is intermittent. For difficult or suspected intermittent picture tube problems, connect the chassis to the bench tube for observation.

CRT repairs

Operation of the weak or intermittent picture tube can be extended by applying a tube brightener or charging the tube with a picture tube charging test instrument. The tube brightener plugs on the end of the picture tube with the CRT socket plugged into the end of the brightener assembly (Fig. 27-13). The brightener raises the heater voltage, which increases emission and brightness. Although increasing the tube brightener is not a permanent repair, the life of the tube might be extended for several months. Most technicians do not believe in temporary repairs and will not fool around with brighteners.

The best method to repair the defective CRT is by charging each gun assembly with a picture tube rejuvenation tester. Charging the picture tube might prolong the life of a CRT for several years. The test instrument renews each gun assembly by removing bom-

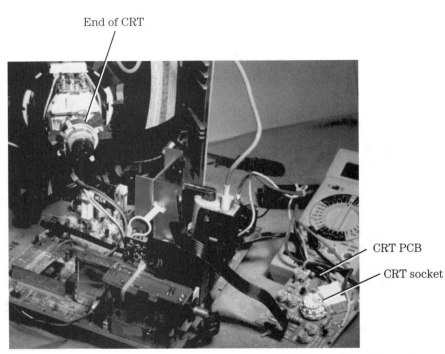

■ **27-13** *The tube brightener plugs onto the CRT base. Then it can be plugged into CRT socket.*

barded ions from the cathode area of the heater gun assembly. These ions form on the cathode element, which eliminates emission from the surface of the cathode element.

The CRT restorer test instrument makes emission tests, leakage tracking, focus continuity, short removal, gun cleaning balancing, and cathode rejuvenation. This test instrument has several picture tube sockets to fit the old and new tubes, such as CR1 (90% in-line RCA), CR3 (90% RCA button base, large Trinitron), CR4 (90% GE in-line button base), CR5 (small Trinitron), CR6 (110% B&W RCA button base), and CR20 (Zenith Tri-potential). Other adapters can be added as new sizes and bases are required in the TV field (Fig. 27-14). You can see the charging or firing as ions are removed or stripped from the cathode of each gun assembly to restore the picture tube.

Removing the picture tube

Make sure that all tests point toward a defective picture tube before attempting to remove the tube. Remember, all picture tubes use an internal vacuum, so handle the tube with care. Do not re-

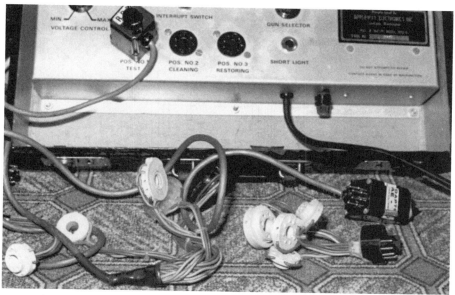

■ **27-14** *The picture tube tester and rejuvenator contain several extra sockets to test most CRTs.*

move or install the picture tube without wearing protective goggles to protect your eyes from possible implosion and flying glass. Replace the CRT with the exact replacement. Do not handle the picture tube by the neck. Grab the picture tube with both hands on its heavy glass front. Never place the flat area of the picture tube against your stomach while moving the tube from set to service bench. Remember, some in-line picture tubes have the deflection yoke that is permanently attached. Remove the TV chassis before attempting to remove the picture tube.

After determining that the picture tube is defective and must be replaced, check all components found on the neck of the CRT. Place a ruler alongside of the neck of the tube and check the position of the purity ring and the beam magnets. Measure the distance between the magnets from the end of the tube for future reference to install the new tube (Fig. 27-15). Although this distance is approximate, the magnets might have to be moved slightly for the different adjustments. Notice what parts must be removed before removing the CRT. Remove the degaussing coil and holders.

Lay the front of the TV set down on a blanket or padded cover to protect the cabinet and picture tube. Remove the screws or bolts in each corner to free the bracket or shell that holds the degaussing coil. Sometimes the degaussing coil is held in place at the top

Removing the picture tube

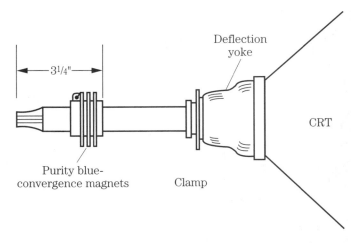

■ **27-15** *Measure the purity and convergence ring assembly from the base of CRT to install the new components.*

and bottom with plastic ties or metal screws. Remove the metal shield from around the picture tube.

Remove the screws or bolts on each corner of the picture tube (Fig. 27-16). It is best that two people lift the CRT, especially the

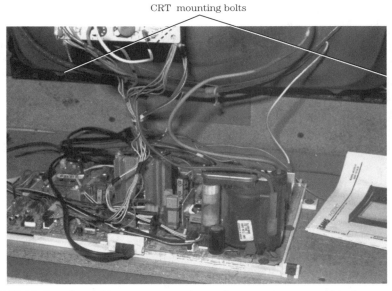

■ **27-16** *Lay the cabinet on a rug, piece of carpet, or on a padded cover and remove the bolts in each corner of the CRT.*

27- and 31-inch tubes. Place the defective tube on a padded bench before removing components from the picture tube. Mark down where the different components are mounted. Always wear safety glasses or goggles when removing and installing the new tubes. Simply reverse the procedure in picture tube replacement.

Picture tube replacement

Remove the metal band and CRT brackets at each corner of the picture tube. Place these brackets in the same position on the new tube. Do not tighten brackets too tight as the tube might be shifted after the tube is placed in the opening. Make sure that the CRT is flush with the front CRT mask area and use a light from behind to see if it is flush on all sides. Now tighten up the corner bolts on mounting corner brackets.

Replace the degaussing coils and the metal shield if one is mounted over the picture tube (Fig. 27-17). Make sure that the ground spring or straps bond against the CRT bell area to properly

■ **27-17** *The degaussing coils are used around the bell area of the picture tube.*

ground the aquadag of the picture tube to ground and chassis. Replace all components on the right places of the neck of the tube. Slip the chassis into the cabinet and connect the HV cable to the anode button. Make sure that the ground strap is in position, grounding the picture tube to the metal or common ground of the TV chassis. Install the CRT socket, plug-in yoke assembly, and remaining wires to speaker, degaussing coil, etc. Doublecheck all cables before firing up the chassis. Do not fasten the chassis down until all picture tube adjustments are made.

Degaussing coil

Although the picture tube has a degaussing coil mounted around the bell of the tube, an outside degaussing coil works best to clean off any impurities or magnetized areas on the face of the picture tube. The hand-held degaussing coil consists of many turns of wire wound in a circle (Fig. 27-18). This degaussing coil plugs into the 120-Vac power line and will remove any impurities or colored areas on the face of the picture tube. The manual degaussing coil is much stronger than the one built inside of the TV set.

■ **27-18** *The manual degaussing coil consists of many turns of copper wire. It plugs into the power line to degauss the front of the picture tube.*

Suspect front screen magnetization if color areas appear in corners of the TV raster. Try to degauss the picture tube by shutting off the TV for several minutes and then turning it back on. Usually, poor purity is noticed on new tubes and the manual degaussing coil must be used.

Plug in the ac cord of the degaussing coil and bring it close to the face of the CRT. Go around and around the frame of the picture tube and across the center area. Especially cover the area that has been magnetized. After a few trips around the tube, slowly rotate the degaussing coil and back away from the TV screen. Keep rotating the coil in a circle as you back away. If you are 8 to 10 feet from the TV set, turn the coil perpendicular to the TV screen and unplug the ac cord of the degaussing coil.

Check for patches of color or color areas. Often, the corner areas are where the unwanted color begins to appear. Sometimes the yoke assembly will loosen up and cause poor purity or improper convergence. After a new picture tube has been installed, automatically degauss the front screen area.

Poor purity

The different color areas in the TV raster (if the shadow mask has been magnetized) are called *poor purity*. Proper operation of the color tube depends on each color gun assembly striking its own color dot at the front of the phosphor screen. Improper yoke setting, convergence, and a magnetized screen can produce poor purity. Sometimes you might never eliminate a small patch of impurities or have complete convergence in some tubes. This might be caused by a misaligned gun assembly, improper convergence assembly, or a loose shadow mask inside of the picture tube.

Correct degaussing and yoke settings might take care of the most difficult purity problems. First, degauss the entire front area and sides of the picture tube. Now check for impurities in the raster. Turn or pull the service switch to the service position to collapse the screen down to a fine line, to aid in convergence, and to eliminate the incoming signal. Turn down all screen and bias controls. Turn up the red screen control for a sharp red line. Check the screen by pushing in the service switch.

For a red screen, move the yoke back and forth until the whole screen is red. Try to keep the yoke as level as possible throughout these adjustments. A touch-up of the purity ring might be needed to help in getting a clear red screen or raster (Fig. 27-19). Now

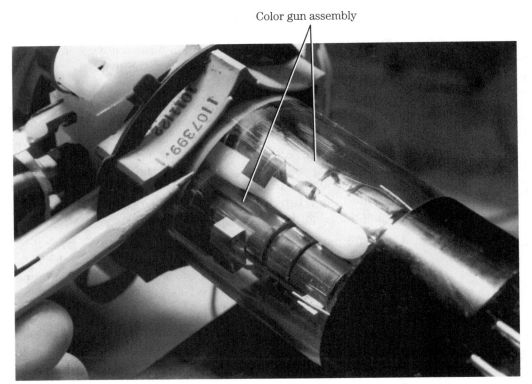

■ **27-19** *The early purity ring fits over a targeted area in the picture tube. Notice the color gun assemblies.*

check each color on the raster by turning up the screen controls. Doublecheck the level of the yoke assembly. If the yoke is attached to the new picture tube, no yoke adjustments are needed. If the yoke assembly pops loose from the bell of the tube, complete purity and convergence adjustments must be made.

Typical purity adjustments

Let the TV warm up for 20 to 25 minutes before making purity adjustments. Set the red screen control fully clockwise. Rotate green and blue color controls fully counterclockwise. Move the deflection yoke against the bell of the picture tube. Center the vertical band purity magnet to center the red on the screen (Fig. 27-20). Move the yoke back and forth to produce a uniform red raster. Now, individually turn up the blue and green controls to check the purity of each blue and green screen. Snug up the deflection yoke tightening bolt to lock the yoke into position. Check to see if the

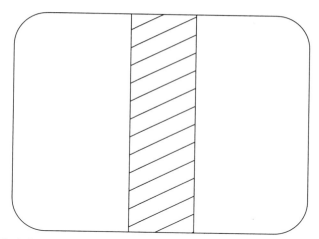

■ **27-20** *Adjust the yoke and purity magnet to form red purity in the center of the picture.*

yoke is level. Replace the rubber wedges behind the yoke and hold it in place with sticky tape.

Large screen purity adjustments

After 20 to 25 minutes of warm-up, disconnect the ac plug and loosen up the yoke screw so that the yoke assembly can be moved. These adjustments do not apply with bonded yoke picture tube assemblies. Remove the rubber wedges from behind the yoke and push the yoke assembly fully forward. Face the TV set with the screen turned toward the north because of the magnetic north pole. Degauss the front of the TV screen. Obtain a red raster by turning up the red screen control. Adjust the beam bender to obtain correct purity (Fig. 27-21). Tighten the yoke clamp. In some TV purity adjustments, outside magnets might be used to complete the purity adjustments. Leave the trim magnets in the corners for larger TV screens. Follow the manufacturer's complete purity and B&W adjustments.

Typical convergence adjustments

Usually, a typical convergence adjustment is needed after purity adjustments are made—especially while the TV chassis is warm. If not, allow the TV to warm up 20 to 25 minutes. Connect a dot-bar generator to the antenna terminals. Tune in a crosshatch pattern. Rotate and spread the tabs of the 4- and 6-pole magnets on the rear of the tube to converge the red and blue lines at the center of

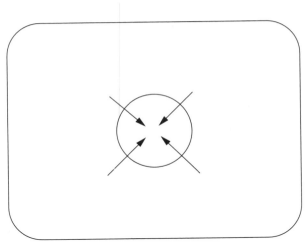

■ **27-21** *Rotate the red screen control and adjust the beam bender to obtain red purity in the center of the picture.*

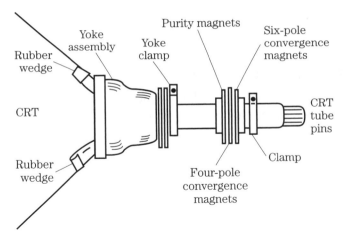

■ **27-22** *The mounted position of the yoke assembly, purity, and convergence magnets on the end of the picture tube.*

the TV screen (Fig. 27-22). Rotate and spread the 6-pole magnets to converge red/blue lines with the green lines at the center. Tilt the deflection yoke in both directions to converge the edges of the screen. Replace the rubber wedges.

Convergence problems

If convergence is completed, the three color beams must pass through the proper holes or shadow mask, striking the exact color dot. Sometimes the picture tube glass area is not perfectly flat, and the color beams must be corrected with static and dynamic convergence methods. Poor convergence results if colored lines around the subject can be seen at all times. Usually, improper convergence is noted in the corners, top, and bottom of the raster (Fig. 27-23).

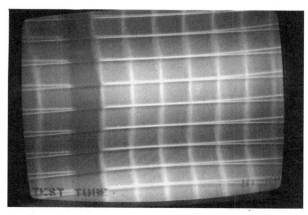

■ **27-23** *With the crosshatch, adjust the convergence so that lines will come together at the end and corners of the CRT.*

Always use the color dot-bar generator for convergence adjustments. Do not try to set the convergence magnets without the test instrument. Just connect the dot-bar generator to the antenna terminals, rotate the tuner to channel 3 or to the manufacturer's specifications. Turn the generator to the crosshatch symbol. Convergence adjustments should only be made after purity adjustments.

Notice which colors need to be converged together. Sometimes one or two colors might be out of line. Notice that the R-G control at the top will correct the red and green lines at the top of the raster. Shift the blue line with the blue lateral magnet. Rotate the R-G control at the bottom to converge the red and green lines at the bottom of the raster.

Typical slotted shadow mask convergence

First, degauss the front of the picture. Obtain a green raster by turning up the green bias control with the red and blue bias controls turned down. Loosen the yoke assembly and pull it backward. Rotate the purity rings to obtain a uniform green color at the center of the screen. Slide the yoke forward to obtain a uniform green raster. Check for a clean red or blue screen by alternating the red and blue bias controls.

Rotate the generator to a crosshatch pattern for static and dynamic convergence. Make sure that the black-and-white adjustments are good. Rotate the magnet to converge the vertical red to the green bars. Rotate the magnet to converge the horizontal red and blue to the green lines.

For dynamic convergence, tilt the yoke up and down, and right and left to converge the outside lines. Place the rubber yoke wedges for proper dynamic convergence (Fig. 27-24). Secure the rubber wedges with strong adhesive tape. Recheck the purity and touch it up, if needed.

■ **27-24** *Replace the rubber yoke wedge at 2, 7, 9, and 12 o'clock after the yoke adjustment.*

Typical black-and-white adjustments

Color temperature, or a normal B&W raster, must be achieved for a good color picture. Tune in a local TV station and properly adjust brightness and contrast controls. Pull out the service switch (if one is on the chassis), turn the color bias controls counterclockwise, and turn the drive controls fully clockwise. Some manufacturers might require a jumper wire or voltage injection to obtain B&W set-up. Rotate all screen controls counterclockwise.

Adjust the red screen control to produce a thin horizontal line. Adjust the screen and color bias controls to produce a white horizontal line. You might find one screen control fully clockwise. Return the service switch or ground jumper to the normal position. Readjust color drive controls to obtain a warm B&W picture. The color bias controls adjust low light or dark areas, and the color drive controls adjust the high or light areas of the raster. The intermediate electronics technician should follow the manufacturer's service literature for B&W set-up, purity, and convergence adjustments.

28
Servicing remote-control devices

JUST ABOUT EVERY TV SET SOLD TODAY IS CONTROLLED with a remote-control transmitter. Servicing the remote-control circuits is just as easy as repairing the sound circuits of a TV receiver. With additional cable and satellite stations, the remote-control transmitter and receiver have taken over in most homes. You can operate the TV with the remote from your easy chair (Fig. 28-1). Like any electronic product, the remote transmitter receiver circuits break down and need repair because of rough treatment and electronic failure.

■ **28-1** *The early remotes were based on supersonic, RF, and infrared transmission systems.*

The remote-control circuits consist of a remote transmitter and a receiver located inside of the TV set to operate its various functions. In the early remotes, the mechanical transmitter operated the on/off, volume, and channels up and down with a tone sound. Next, the RF remote picked up transmitted signals operating the various functions with several dc motors. Later, the mechanical transmitter was replaced with a supersonic model using a transducer. Today, the infrared transmitter has many different operations, including scanning stations up and down, selecting individual channels, on/off, volume selection, and muting sound. The new remote might include operating a VCR, CD player, and a stereo receiver (Fig. 28-2).

■ **28-2** *The digital control remotes can control TVs, VCRs, and stereo equipment.*

Basic transmitters

One of the first remotes generated a supersonic wave with a mechanical hammer that would strike three different metal rods of different lengths. The frequency or tone of each rod in turn controlled the on/off switch, channel up, and channel down. This supersonic frequency was picked up by a remote transducer at the receiver, which controlled the various operations.

Another method to transmit or control a remote signal was with an electronic signal through a speaker (Fig. 28-3). Although the supersonic transmitter was a big improvement over the mechanical or tone transmitter, there were many drawbacks. In apartments or town houses, the transmitter might operate several different TV sets. Your TV set might be controlled by the next door neighbor. Signals from various electrical and RF-generating devices could trigger or control the RF remote receiver.

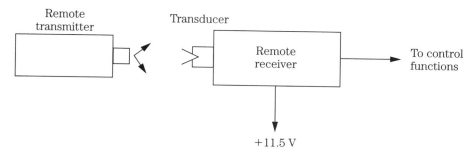

■ **28-3** *The supersonic remote signal is picked up by a transducer and is fed to a remote receiver to control various functions.*

Infrared remote

Today, the infrared remote solves all outside frequencies, supersonic and RF methods from interfering with other TVs. In fact, you can have a TV set in the kitchen, den, or bedroom that will not be affected by the infrared remote. The infrared remote must be pointed directly at the TV. The TV will not respond when someone walks between the remote transmitter and the TV. The infrared remote operates in the line of sight between transmitter and TV set (Fig. 28-4).

The infrared remote transmitter might have few or many buttons, depending on what it is made to control. The digital remote is easy to operate. In fact, these remotes might control the TV channels, sound, on-screen display, of setting the brightness, contrast, color, and tint controls. The RCA Dimensia digital remote operates the TV, VCR, receiver, phonograph, tape deck, and CD player from one remote transmitter.

Instead of a transducer element, the infrared remote uses an LED that transmits the infrared signal to the TV infrared receiver. Some infrared remotes might have two different LEDs. The early infrared remotes were all controlled with transistors. Today, the complete remote functions operate from one large IC. Some of these remotes are even crystal controlled and might operate from two AA cells (Fig. 28-5). Most remotes operate from 3, 4.5, and 9 V.

Infrared remote tests

A defective remote might provide intermittent, weak, or no operation of the TV set. The intermittent remote transmitter might operate intermittently with dirty battery contacts, weak batteries, and a defective circuit board. Weak batteries might let the trans-

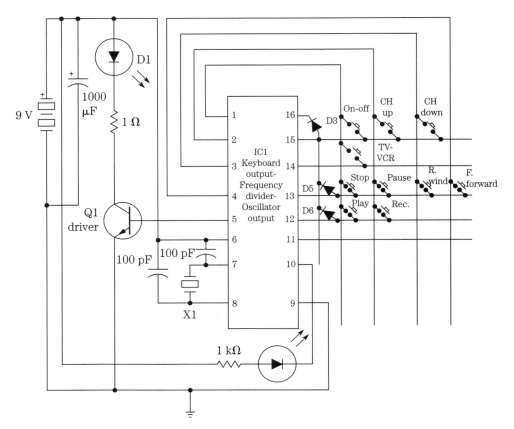

■ **28-4** *The infrared remote sends out an infrared line-of-sight signal to the infrared receiver in the TV.*

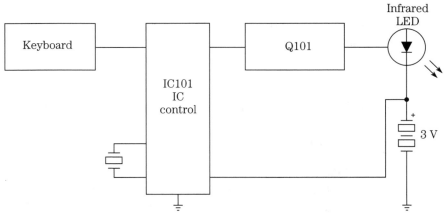

■ **28-5** *This infrared remote operates from two small AA batteries (3 V).*

Servicing remote-control devices

mitter operate when it is pointed directly at the TV set from a short distance. Sometimes weak batteries might operate only a few functions, but the other operations are dead.

Check the output signal from the infrared remote to determine if the remote or receiver inside of the TV, VCR, or CD player is defective. The output signal of the remote transmitter can be checked with any radio. Turn the radio on and place the remote close to it and press each individual button. You should hear a tone or gurgling sound in the radio speaker. Another method is to test the infrared remote with an infrared indicator card. The infrared indicator card will change color when the card is held against the front of the remote.

Battery tests

Check and replace any battery if it drops below 0.5 V. If one or two batteries do not test up, replace all batteries inside of the remote. Be sure and check each battery for the correct battery polarity when it is inserted. Clean off the battery contacts with a cloth, paper towel, or pant leg by rubbing ends against the cloth several times. It's best to check batteries under load or in a regular battery tester. When under load (push any function), the single battery will test lower than by testing each separately with the VOM or DMM (Fig. 28-6). Dirty battery terminals and one weak battery might cause intermittent remote operation.

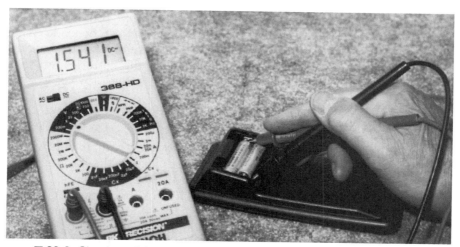

■ **28-6** Check each battery with a VOM or a DMM to determine if they are weak or dead.

Infrared remote tester

You can construct your own infrared tester with electronic parts from the local electronics store. Just place the remote a few inches away from the tester and press each button on the remote control. You should hear a chirp or tone when any button is pressed (Fig. 28-7). This remote will check all infrared remote transmitters for the TVs, VCRs, CD players, and receivers.

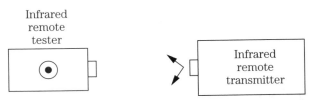

■ **28-7** *Test the remote transmitter on an infrared indicator card or on a remote tester.*

Move the remote in front of the tester until you can hear the loudest sound. Check the remote by slowly pulling the remote away from the tester to determine the distance of operation. With this tester, some of the stronger remotes can be heard up to 2 or 3 feet away. Check the remote for weak reception by moving the unit away from the tester.

Check each button for a tone in the remote tester. Suspect a dirty button if the remote is erratic or has an intermittent sound in the tester. If the remote does not work after being dropped, remove the battery cover and clean each battery connection. Bend the battery tabs out so that a greater contact can be made. When you might have to press hard on the button and it takes a long time for the remote to function, suspect that the batteries are weak. No sound indicates that the battery is defective or weak.

How the tester operates

The infrared tester consists of only a few components wired in series. The infrared phototransistor picks up the infrared signal and operates a flashing LED or piezo buzzer. You can use a regular low-voltage LED or a blinking LED. Check the component polarity for the infrared tester function. The 9-V battery positive terminal must be connected to the (+) positive terminal of the LED. The positive terminal of the piezo buzzer must be connected to the negative terminal of the LED (Fig. 28-8).

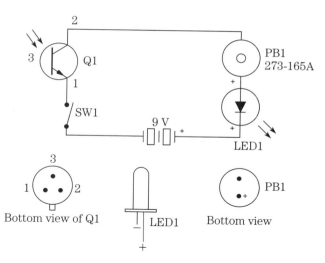

■ **28-8** *The schematic diagram of the infrared remote tester.*

The piezo buzzer is small in size and can be mounted on the top side of the plastic case. The buzzer operates on 3 to 30 Vdc and has an operating frequency of 2.8 kHz. Notice the positive sign on the buzzer case. The blinking LED combines a MOS IC driver and red LED within a plastic LED case. The blinking LED pulls approximately 20 mA at 3 Vdc.

Parts list

Q1	Infrared phototransistor detector 276-142
SW1	SPST toggle or slide switch 275-645
PB1	1 to 6.5-kHz piezo buzzer 273-065A
LED1	Blinking LED 276-06C or 276-020
Batt	9-V battery
Cabinet	3½"-×-2½" plastic box
Misc	Battery clip, harness, solder, PC board, or predrilled board, etc.

Laser remote tester

The LPM5673 Saniwa laser power meter might be used to test the laser beam of the CD player and remote-control transmitters. Most laser power meters can test the remote control with infrared emission. The three measuring ranges are 0.3, 1, and 3 mW, which can be used to check all infrared light sources. The meter displays include three different measuring ranges (Fig. 28-9) of measured

■ **28-9** *A laser infrared tester indicates the laser power of a portable CD player.*

light output with upper scale (0 to 1 mW), intermediate scale (0 to 3 mW), and the lower range (0 to 0.3 mW).

The laser power meter might be used to check out the infrared remote transmitters. Place the black pickup assembly against the end of the remote using the 1- or 3-mW ranges. You might want to measure the infrared emission of the remote a few inches away from the pickup. Comparison measurements can be made of several different new remotes and register the readings to determine what scale or reading to use in further remote measurements. Of course, a weak or defective remote might show a very low measurement or none at all. Besides checking the laser beam of a CD player, put the meter to work checking the low-powered remote transmitters (Fig. 28-10). Move the meter pickup around the remote LED to get the highest reading.

Universal remote control

You can pick up a universal remote control in just about any electronics store, including K-Mart and Wal-Mart. These remotes have many different operations to control the TV, VCR, CD player, au-

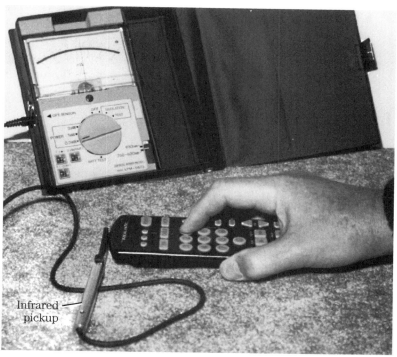

■ **28-10** *The infrared laser tester can also be used to test the remote controls.*

dio, and video components. The General Electric RRC500 does the work of three remotes, and the RRC600 does the work of four remotes. The RRC600 controls up to four infrared audio/video products, with over 200 key combinations, program sequencing, LCD display, and low battery indicator.

The Realistic 3-in-1 easy remote has large numbers and letters for easy operation. This remote will operate cable, TV, and VCR infrared devices. The remote operates from four AAA cells. The set-up codes for different TVs might be three numbers. For instance, the 3-in-1 remote code of the RCA TVs are 018, 019, 038, 047, and 135. Start with the first code numbers and proceed until you have located the correct digital code. Start and press TV-A-B-C-0-1-8. When the red LED blinks twice in the remote, the control is ready. Aim the 3-in-1 remote at the TV set and press the power key. If the TV does not turn on, set the code of the next set of numbers. Code 047 responded to the 27-inch RCA TV. Check each universal remote set-up with the manufacturer's instructions.

The universal remote operates from batteries as the regular remotes. Check the universal remote with a remote tester or laser infrared meter. Inspect the battery terminals for dirty or intermittent connections. Most manufacturers request that the universal remote be sent back to the dealer or factory for major repairs.

Remote-control receivers

The RF remote receiver might include a transducer or sensor with a remote preamp feeding into the main remote receiver. The channel-up operation is controlled by one frequency, channel down by another frequency, and on/off and volume are controlled by another frequency. The amplified RF signal is applied to a relay or motors controlling the tuner or various TV circuits. Although each manufacturer has a different operating remote system, they all do the same thing: make TV operation easier for the operator.

In some TV sets with tuner and control modules, the remote receiver is contained inside of the control system module. A simple remote receiver might contain only one IC amp, infrared LED, and very few connecting components. The LED or phototransistor picks up the infrared signal and is fed into the IC amplifier (Fig. 28-11). The output of the IC amp connects to the system control microcomputer within the control module.

The deluxe or large infrared remote receiver might have the infrared LED pickup device coupled to three preamp transistors to just amplify the remote pickup signal. Besides the pickup infrared signal, the manual TV controls are fed into a remote decoder IC. The LSI decoder might be crystal controlled. A separate volume, volume emitter-follower, on and off switch, and on/off inverter transistors are controlled from the IC decoder (Fig. 28-12). When the defective infrared receiver is enclosed inside the tuner control

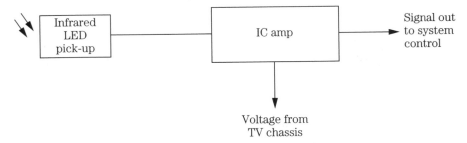

■ **28-11** *The infrared signal is picked up by an LED and is fed to an IC amp and the output is sent back to system-control circuits.*

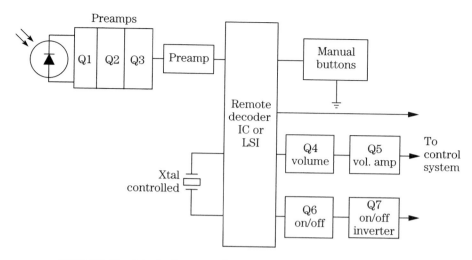

■ **28-12** *The block diagram of an infrared preamp IC decoder with crystal control.*

module, the whole module must be replaced. Send the module in for repair or order a new control module.

The IR detector might feed into an IR preamp and main amp before the infrared signal is applied to the main interface circuit LSI. The keyboard buttons and system control microcomputer provide signals to the interface LSI in some TV models. The keyboard code might consist of four bits because each bit can be high or low. The different bits identify the function of the individual keys at the keyboard matrix. When changing the state of bits, determine the function of each key. The microcomputer outputs the correct control signals to the interface LSI or IC during tuning to provide band switching and tuning voltages (Fig. 28-13).

Testing remote transmitters

Determine if the remote is weak or is not sending out an infrared signal with the laser or remote tester. Check the batteries. Insert a new set of batteries, if you are in doubt. Watch for the correct polarity. Take voltage measurements on all transistors and ICs. Check the transistors with the beta tester or with the diode-transistor junction diode test of the DMM. Test each diode, including the infrared LED with the regular DMM diode test. The keyboard assembly should be replaced instead of attempting to repair it. Send the remote transmitter in for repair or exchange at the TV repair depots or to the manufacturer, when simple repairs will not cure the malfunction (Fig. 28-14).

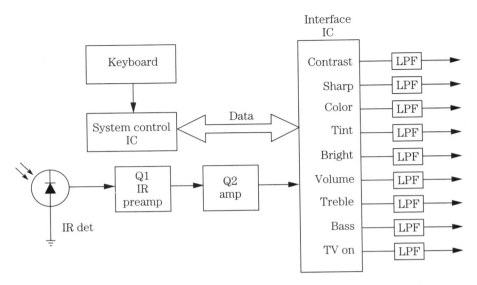

■ **28-13** *The keyboard and remote receiver circuits feed into an interface IC that controls the different TV functions.*

■ **28-14** *Check transistors, LEDs, and diodes, and take critical voltage tests inside of the defective remote.*

Testing infrared receivers

Make sure the infrared remote transmitter is operating before attempting to repair receiver circuits. Place the infrared remote close to the transducer, sensor, or LED sensor. If the remote receiver is inside a control module, replace the module. Check all transistors, LEDs, and diodes in the receiver circuits. Take critical voltage tests on transistors, IC or LSI. Determine if the supply voltage is weak or low. Sometimes low supply voltage might let the tuner operate at only the high end of the TV channels.

In transducer receiver circuits, the pickup element acts like a condenser microphone. Measure the dc voltage applied to the high end of the transducer. This negative or positive voltage might come from a separate diode rectifier in the low-voltage power supply (Fig. 28-15). Check the infrared sensor LED with the diode test of the DMM. A leaky LED will show a low resistance in both directions. Unsolder one lead of the LED and make another test.

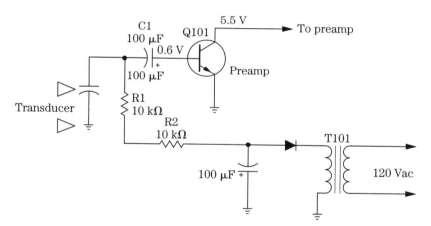

■ **28-15** *The early transducer is energized with a negative voltage source in the preamp circuits of the receiver.*

Servicing interface circuits

If the keyboard is functioning and there is no remote-control action, you know part of the interface circuits are performing. The trouble must lie in the remote, remote amp, and receiver. Determine if either the remote or keyboard is operating; then isolate the circuits. Check the remote transmitter on an infrared tester to make sure that it's functioning. If the remote and keyboard are operating, suspect the IR detector, IR preamp, and output amp (Fig. 28-16).

If the keyboard is not functioning, check the keyboard and system control IC. If only one key will not function, check for a dirty key or a broken cable wire. All keyboard switches can be checked with

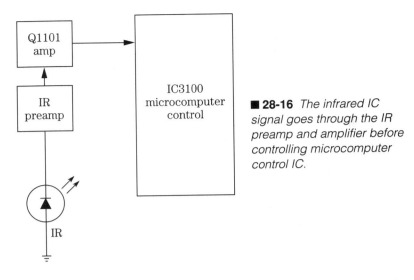

■ **28-16** *The infrared IC signal goes through the IR preamp and amplifier before controlling microcomputer control IC.*

the ohmmeter while holding down each key or button. Check the signal output of the system control IC if the keyboard seems normal. Do not overlook the possibility that there might be low or no voltage at the supply pin of the system-control IC. If both remote and keyboard systems do not operate, suspect a defective microcomputer IC or power supply.

Troubleshooting the keypad

The early keypad might consist of only a few operations, while in others the keypad will do all the major functions that the remote will operate. Usually, the keypad operates directly to a control module or control system IC. The IR receiver connects to the same system. The keypad consists of a number of momentary switches that plug (cable) into the system control microcomputer (Fig. 28-17).

If one of the keypad switches does not function, suspect that a switch is defective. If all functions of the keypad will not operate, check the cable socket and microcomputer IC or LSI. Here, all of the various switches (SW1 through SW5) are tied to ground. If one or two switches fail, take a resistance test between that switch and ground. For instance, if the on/off switch (SW1) sometimes turns on the TV and other times doesn't, check to see if the remote control will turn the TV on and off.

If the remote turns the TV set on and off without any problems and the keypad buttons do not, suspect poor switch contacts on SW1 or a bad solder cable connection. Place the low-resistance

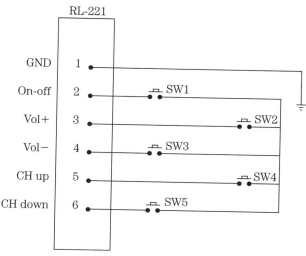

■ **28-17** *Test the defective keyboard switch with the low-resistance range of VOM or DMM.*

scale of the DMM on pin 2 and ground. Press SW1. If contacts are erratic or if there is no continuity, the contacts are dirty or corroded. Some keypad switches can be cleaned with cleaning contact spray. If the contacts are enclosed, replace the whole keypad. Make sure that the cable connections are good. Resolder the cable wire at pin 2 of RL-221. Sometimes these contacts are pressed together against the bare wire, without solder, and after some time, it makes a poor contact.

Standby circuits

The standby circuits provide voltage to the remote circuits when the TV set is turned off. The standby circuits are on all the time while the TV circuits are turned off. This keeps the TV chassis alive for remote operation. Standby circuits of some sort are used in the latest TV remote chassis. Always use an isolation transformer while servicing the standby TV circuits.

If the remote on button is pressed, the voltage supplied by the standby circuits activates the remote-control receiver circuits to start the TV chassis. Suspect a defective remote transmitter, remote amp and receiver, or improper or no standby voltage if the keypad on/off controls operate and the remote operation is dead. Go directly to the standby circuits if the remote transmitter is normal, but there is no remote operation (Fig. 28-18).

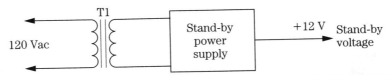

■ **28-18** *The standby power supply is on at all times so that the remote control circuits will function.*

In older TV sets, the standby power was turned on with a relay from a separate power supply. Today, the standby power might be switched on with a relay or switching transistor or IC circuits. The standby voltage might be supplied from a separate power transformer with a voltage regulator and filter network. Most standby power-source problems are caused by regulator transistors, zener diodes, diodes, switching transistors, and open transformer terminals.

The simplest standby voltage source consists of a separate power transformer, bridge rectifier circuit, filters, and transistor voltage regulators (Fig. 28-19). D1 through D4 rectify the stepdown ac voltage, which is filtered by a 1000-μF electrolytic capacitor. The dc voltage is fed to the collector terminal of Q1. ZD101 and Q101 determine the regulated 12-V source for the standby voltage. This 12-V source is on all of the time for remote-control operation.

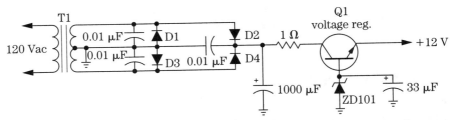

■ **28-19** *Check the standby power source for the standby output voltage (12 V).*

In the latest RCA standby circuits, the chopper transformer is on all of the time with a secondary winding providing a dc voltage to the standby circuits. If the output chopper transistor is turned on by the regulator IC, current flows through the primary winding, transferring energy to the secondary winding of the transformer. The ac secondary voltage is rectified by a silicon diode, which supplies 15 Vdc to the 5-V standby regulator IC (Fig. 28-20). The standby voltage is supplied to the on/off switch transistor and horizontal circuits.

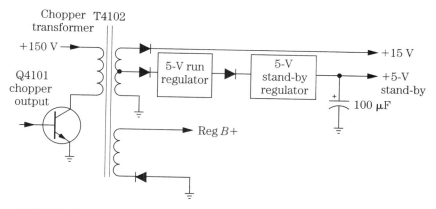

■ **28-20** *The block diagram of a chopper power supply. It is on at all times to supply the voltage to the 5-V standby power source in an RCA TV.*

Standby problems might result from dirty relay contacts or from an open relay winding. Clean the switch with a piece of cardboard and cleaning fluid. Place the thin cardboard between the switch contacts and move it back and forth. Then, spray on the cleaning fluid. Check the solenoid winding with an ohmmeter. Check for a standby voltage at the transistor or IC regulator. If there is no voltage, check the input voltage. Next check the ac secondary voltage of the transformer. Suspect an open primary winding of the transformer or poorly soldered board connections. Check each diode with the diode test of the DMM.

Conclusion

The intermediate electronics technician will find that more problems are found with the remote-control transmitter than with the remote turn-on and operational circuits. Isolate the remote transmitter and standby circuits to determine what circuit is malfunctioning. If the keypad circuits operate the TV set, check the remote, remote receiver, and standby circuits. Before checking the remote receiver circuits, make sure that the standby voltage source is available. If not, check the standby transistor or IC regulators.

29
How to locate and repair intermittent problems

INTERMITTENTS WITHIN THE COLOR TV, CD PLAYER, cassette deck, VCR, camcorder, and amplifier produce more headaches for the consumer electronics technician than any other type of symptom. It takes more time to service the intermittent chassis. Sometimes you cannot charge enough for all of the time spent locating the intermittent component. The intermittent chassis might turn up as one of the toughest-tough dogs in the consumer electronics field (Fig. 29-1).

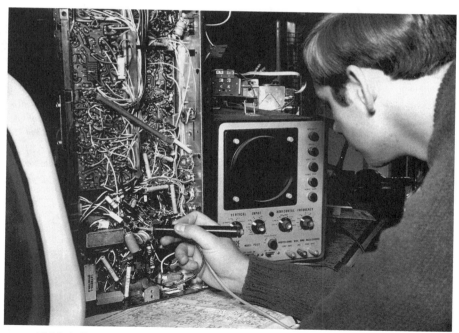

■ 29-1 *The electronic technician is taking a waveform measurement in the color TV chassis.*

A connection covered by a blob of solder and poorly crimped socket connections are difficult to locate. Improperly tinned resistors, capacitors, or component leads also add up to a lot of wasted time. A broken eyelet or cracked PC board wiring produces many intermittent joints that cannot be seen with the naked eye. Fine cracks in PC wiring produce a share of intermittent problems. Improperly soldered joints and connections on the PC wiring add up to extra time. Poorly soldered SMD terminals produce more difficult intermittent problems, mounted on the double-sided wiring (Fig. 29-2). Double-sided wiring doubles the many different intermittent problems.

■ 29-2 *SMD components mounted underneath of the chassis on the PC wiring might cause more intermittents.*

Hot and cold applications

The older experienced electronics technician might handle the intermittent chassis with kid gloves, so to speak, not to upset the intermittent problem. Let the chassis operate until it breaks down. Of course, just transporting the intermittent chassis from the home to shop might cause the set to not act up. The intermittent chassis should be left in the home if it only becomes intermittent about once a month. If the chassis acts up two or three times per day, you might be able to find the intermittent. It's very difficult to locate the intermittent part or connection if the chassis acts up once or twice a month. The intermittent will act up more often, just give it time; they always do.

Intermittent connections, poor terminal leads, transistors, ICs, and diodes might act up after heat is applied within the cabinet. Sometimes just removing the chassis from the cabinet might let the set cool down so that it never acts up again. Heat-related components might break down after the temperature keeps rising (Fig. 29-3). Cover the chassis with a blanket, comforter, or pad to contain heat inside of the cabinet or over the chassis to make it act up. It might help to spray a heat gun on a suspected semiconductor. The intermittent component is easier to locate if the chassis becomes intermittent without anyone touching it.

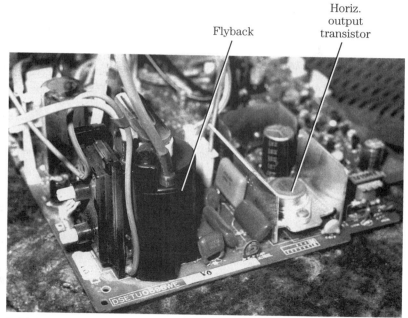

■ **29-3** *Heated-related components, such as horizontal output transistors, flybacks, and ICs might cause intermittent problems in the TV chassis.*

Applying heat with hot air spray and then coolant spray on the suspected parts, soldered connections, and cracked boards might cause the intermittent to return. Several coats of coolant might be needed with cold spray to cool the component that you suspect. A coolant spray might generate cold up to −97°F (−71°C), excellent for locating thermal intermittent components. Poor internal transistor junctions or connecting terminals might cause the intermit-

Hot and cold applications

tent transistor to act up with heat or coolant applied (Fig. 29-4). The intermittent IC might begin to act up after it operates for several hours, and then you spray the whole body with coolant. Solid-state components have a tendency to become intermittent with a change in temperature.

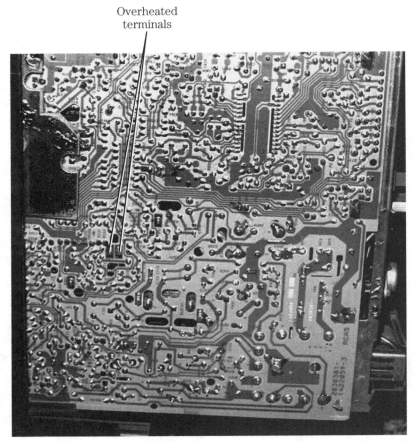

■ **29-4** *Overheated terminals might produce a dark area around the PC wire terminals on the large PC board.*

Raising and lowering the power-line voltage with a variable ac isolation transformer might cause the intermittent to act up. The lower or excessive voltage applied to the intermittent component might cause it to break down. Be careful not to destroy original parts with too much voltage applied to the chassis.

The intermittent chassis seem to come in bunches. In fact, you might see several months' worth in one week. Sometimes the intermittent might occur in the home and not the shop. A defective antenna system or an improper line voltage can cause the intermittent. Make sure that the electronic product operates correctly for 3 or 4 days, return it, and check for conditions that might cause the chassis to be intermittent. Intermittents can cost time and many hours of brain power. Also, you might find more than one intermittent in the same chassis.

Approach the intermittent with care (Fig. 29-5). You might disturb the intermittent and it might not act up again for months. Let the intermittent act up on its own after two or three full days of operation. Sometimes intermittents take a couple of hours or a full day before acting up. You might have to return the chassis or electronic product to the customer to operate it until it acts up. Have them call the next time that it does.

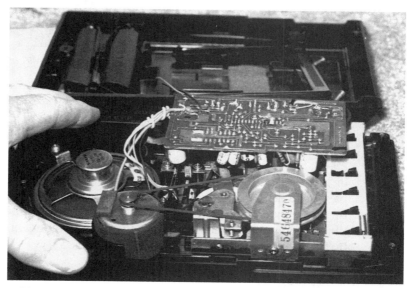

■ **29-5** *Handle the intermittent with care. Carefully remove the cabinet or pull out the chassis so that you can monitor the intermittent with a test instrument.*

Moving components

The chassis might become intermittent and then restore to normal if you prod the parts with an insulated tool (Fig. 29-6). Transformers and larger components can be moved with fingers to check for poor terminal connections. Small components, such as transistors, diodes, capacitors, and resistors, can be moved slightly with the insulated tool. Warping and bending the main chassis or top-mounted subchassis might indicate an intermittent soldered con-

■ **29-6** Move small components around on the chassis with an insulated tool or a pencil to make the intermittent act up.

nection or poor connections on surface-mounted components. Be careful if SMD components are involved.

Just nudging the small resistor might cause the set to act up (Fig. 29-7). Sometimes these resistors might be overheated and crack or have poorly tinned connections. Intermittent coupling and bypass capacitors might indicate a poor terminal connection or internal connection by prodding and moving the capacitor. Poorly tinned component leads might act up if the component is moved by the insulated tool.

Intermittent boards

Sometimes you can locate a loose or cracked board connection with coolant or by moving and prying on sections of the PC board. Be very careful if the SMDs are mounted on the opposite side. Try to locate the section of the board that is defective by prodding and pulling the board upward. Try to isolate the intermittent to one section of the PC board (Fig. 29-8).

Whatever the intermittent condition, the cracked board, poor socket connections, and PC wiring will always bother technicians. A double-sided PC board with fine cracks is especially susceptible to breaks and intermittent symptoms. A poorly soldered connection or tinned electronic part leads might leave oxides that will

■ **29-7** *Prodding and moving small resistors or diodes around might turn up a broken or cracked resistor and terminal.*

■ **29-8** *Moving the board up and down might turn up a poor terminal lead, damaged PC wiring, or disconnected ground lugs.*

■ **29-9** *Isolate the section of the PC wiring by pushing up and down on the large PC board.*

tion or tinned electronic part leads might leave oxides that will cause the break to malfunction several years later (Fig. 29-9).

PC board connections and broken wiring have always caused intermittent problems. Sometimes the chassis might be dropped in shipment or handled roughly, a break in a connection between the component and the PC wiring occurs, and acts up later. Small cracks around transformer lugs or connections might cause intermittent operation (Fig. 29-10). These you can see, but breaks around small capacitors, diodes, and resistors might not be visible and will not show up until they are moved. Check the PC board wiring for breaks around plastic or metal stand-offs. Sometimes the boards will warp and crack the fine PC wiring.

After isolating the intermittent to one section of the board, carefully inspect the etched wiring with a strong light behind the board and a magnifying glass. Check for fine breaks around feed-through eyelets or component terminals. Sometimes these breaks are so fine and small that you can't see them with a magnifying glass (Fig. 29-11). If your eyes are weak, try continuity tests.

Locate the breaks in the PC wiring by taking low-resistance continuity between sections of PC wiring. Measure the resistance from

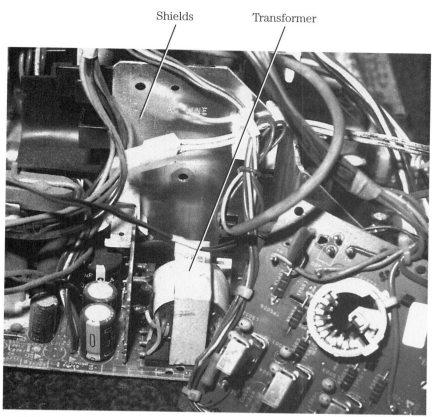

■ **29-10** Check for PC board cracks around heavy objects, such as transformers and large choke coils, shields, and pressure stand-offs in the TV or VCR chassis.

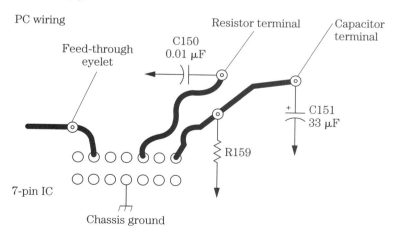

■ **29-11** Check for fine breaks with a magnifying glass and strong light in the PC wiring. Especially check around eyelets.

Intermittent boards

one terminal to the next to locate the break in wiring. Most breaks occur right around the terminal area. Very seldom does the break occur in the middle unless the entire board is cracked. Set the ohmmeter at RX1 and take PC wiring resistance measurements between the two different points (Fig. 29-12). Of course, this takes a lot of time, but you can locate the fine break in the PC wiring.

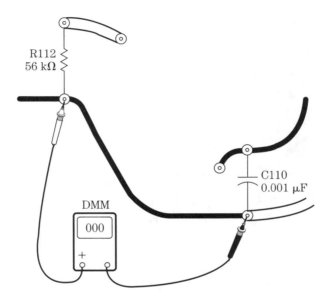

■ **29-12** *Measure the resistance between two PC terminal connections to locate a break in the wiring.*

PC wiring breaks can be found by taking voltage measurements between common ground and at the terminals on the same piece of wiring. If the voltage is not reaching a component terminal, check the voltage at the part terminal and source feeding it. The voltage should measure the same at both ends of the PC wiring, unless the foil is cracked or broken (Fig. 29-13). Then, verify the break after shutting down the chassis and taking a resistance measurement between the break in the wiring. No measurement indicates broken wiring, but an intermittent continuity measurement shows up as an intermittent break in the PC wiring.

Locate and isolate the section

Locate the section with given symptoms and compare the same section on the chassis (Fig. 29-14). Check each component in that sec-

■ **29-13** *Take critical voltage measurements between two terminal points in the PC wiring and check the voltage and resistance to locate broken wiring or a poorly soldered terminal.*

■ **29-14** *After locating the intermittent problem within a certain section of the PC board, take voltage and resistance measurements at each component.*

tion for intermittent operation. Probe the small parts and move the larger components to determine where the intermittent component is located. If you move the large components and the set becomes intermittent, resolder the terminal connections. Notice which way you move the transformer or choke component to make the set act up. You might find a break in the PC wiring close to the large component with poorly soldered terminal connections.

After isolating the intermittent to one section of the board, carefully check the etched wiring with a light and magnifying glass. Large cracked boards are easy to find, small cracks are difficult to locate. Check the area for large blobs of solder over several connections. Melt down the section with the soldering iron and remove excess solder with solder wick. Check the component terminal lead for poorly tinned terminals.

Touch up all soldered connections with the soldering iron and feed-through eyelets. It's best to feed a small bare piece of hookup wire through the eyelet and bend over the ends of the wire to the PC wiring (Fig. 29-15). Resolder the connecting wires on both sides of the board. Usually, feed-through eyelets only have solder tying both sides of the wiring together. This method ensures a good wiring connection.

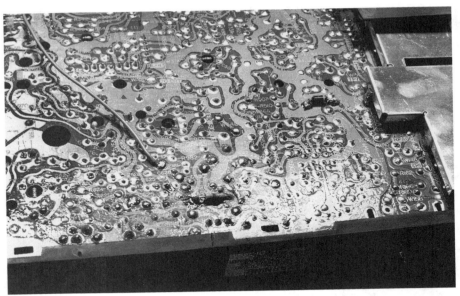

■ 29-15 *Connect the short broken wiring sections of the PC board with bare hookup wire. Then feed it through the PC wiring and solder in hookup wire.*

Sometimes the intermittent might take several hours, days, or weeks to act up. Monitor the intermittent section with voltage measurements and scope waveforms. Work nearby on the service bench and if the chassis acts up, notice the voltage and scope waveforms. If the voltage decreases or if there is no voltage with no waveform, the intermittent component is nearby. Take critical voltage measurements in that area to locate the intermittent part. Spray related devices with coolant and notice if the electronic product begins to operate.

Broken boards

A cracked board can be salvaged by connecting the broken wiring with regular hookup wire (Fig. 29-16). Do not just apply solder across the broken ends of wiring. Short pieces of bare hookup wire can be bridged and soldered across the broken wiring. Scrape off any wiring coating on each piece of wire to make a good soldered contact. Clean off the wiring with a heavy-duty defluxer spray. Make sure that the component lead is soldered back to the broken wiring.

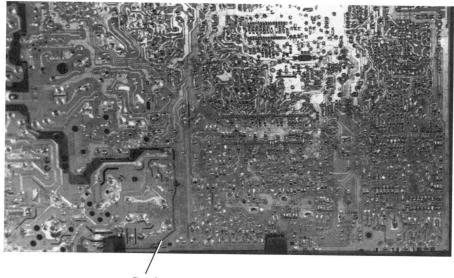

Crack in PC board

■ **29-16** *Check the PC board for large cracked areas indicating that the electronic product was dropped or was treated roughly.*

It's best to run longer pieces of hookup wiring from the part terminal to another terminal, across the broken area. Scrape off the wiring connections around the terminal connection. Form a loop in the hookup wire and solder around the terminal lead. Make sure that the connecting hookup wire is across the same broken piece of wiring. Doublecheck it after the connection is soldered. If the PC board is broken off, try to order a new board. Sometimes these boards are not available.

Shotgun terminal approach

If a certain board section acts up and you cannot locate the defective connection or component, soldering all terminal connections might solve the intermittent problem (Fig. 29-17). Be careful not to apply excessive solder that might run into another PC wire or component terminal. Go over each terminal with a hot soldering iron and add solder. Leave the iron over areas with large solder blobs or larger solder joints a little longer to ensure well-bonded connections.

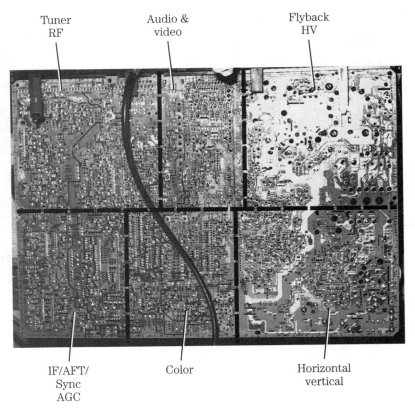

■ **29-17** *After locating the intermittent section, solder each terminal to uncover poorly tinned component leads or solder connections.*

Inspect the wiring with a magnifying glass after each terminal is soldered. Doublecheck each connection to make sure that a good soldered connection is made. Be very careful around transistor terminals to prevent excessive heat from destroying the transistor. Likewise, use a small low-wattage iron on IC terminals (Fig. 29-18). Move the board to verify if the intermittent problem has been corrected with the chassis operating. If the intermittent still exists, suspect that a component is intermittent.

■ **29-18** *Solder IC terminals with a low-voltage battery-powered iron.*

Intermittent components

Intermittent components break down with heat or excessive voltages. After isolating the section where the intermittent might occur, spray the transistor or IC with coolant. Apply several coats on the transistor before going to the next one. Several minutes of heat from a hair dryer might make the transistor or IC act up.

A quick voltage measurement between base and emitter terminal might indicate if the transistor is leaky or open (Fig. 29-19). You should have 0.6 V with an NPN silicon transistor and 0.3 V with the germanium transistor. Sometimes the intermittent transistor might test normal if the test probes are touched on the transistor terminals.

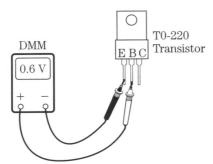

■ **29-19** *A quick voltage bias reading between the base and the emitter terminals of a transistor will indicate if the transistor is open or leaky.*

Check for poorly soldered connections around start-up or driver transformers that might cause intermittent operation. After soldering all connections and the transformer seems intermittent if it is moved, suspect poor internal connections. Remove the suspected transformer from the PC board. Remove the outer cover to uncover the terminal connections. These connections might not be soldered and after a while, they will become oxidized. You might find that the enameled insulation was not entirely removed from the transformer terminal wire and it makes a poor connection. Scrape off the transformer wire from the winding and tin the end. Wrap the lead around the flexible connecting wire and solder it.

Volume and tone controls might have worn or cracked areas that cause intermittent sound. Replace any worn controls. Replace any burned or dirty bias controls. High-voltage focus and screen controls might become intermittent with large carbon areas on a ceramic surface (Fig. 29-20). These control connections have a tendency to corrode with high voltage attached, and cause intermittent focus and brightness. Check for green corroded areas around the connections, inside of the control. These high-voltage controls have a tendency to open and cause intermittents.

Intermittent problems in radio circuits

Intermittent problems within the radio are caused by transistors, ICs, coupling capacitors, and IF transformers (Fig. 29-21). Try to isolate the intermittent component to one section of the chassis. Determine if the intermittent section is in the front end or in the audio section. Check the front end at the volume control with the external audio amplifier.

Spray the transistors with heat and coolant to determine which transistor is intermittent with the receiver operating. Apply several coats of coolant on the ICs to cease operation. Prod each tran-

■ **29-20** *The defective focus and screen controls might cause an intermittent raster with corroded internal wiring on the ceramic control area.*

■ **29-21** *Wiggling the small electrolytic capacitors and IF transformers might help you turn up the intermittent component or PC board connection.*

sistor with an insulated tool to locate a defective component or a poorly soldered terminal connection. Monitor the voltage to the intermittent section to determine if the voltage regulator or power source is intermittent.

Clean the AM/FM switch for poor and intermittent contacts (Fig. 29-22). Check for broken wires on the switch terminals. Likewise, clean the noisy volume control for intermittent audio. Replace the volume or tone controls if they are worn or broken.

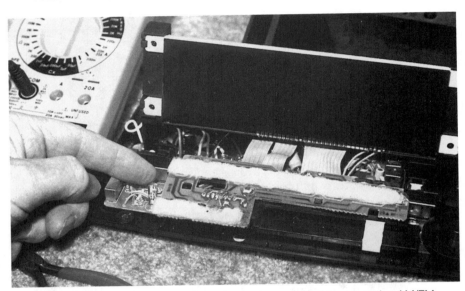

■ **29-22** *The front-panel PC board must be removed to get at the AM/FM switch in this table model radio.*

Broken board or poor terminal connections can cause intermittent reception in the radio circuits. Push or pull on the IF transformers for a poor terminal connection. Sometimes, a bare wire or terminal inside of the IF section might produce intermittent reception. Check for intermittent FM padder capacitors on the FM tuning capacitors. Simply touch the padder with an insulated tool.

Intermittent hum in the audio section might result from bad input cables and poor shielded ground. Improper shielded areas might cause intermittent hum in the radio. Move the small coupling capacitor for internal or terminal breaks. Defective power output transistors and ICs might run excessively warm and intermittent. Do not overlook the possibility that a voice coil inside of the speaker is intermittent—especially if greater volume is applied.

Intermittent amplifier problems

Most intermittent components in the audio amplifier are caused by power output transistors and ICs, coupling capacitors, and bias and volume controls. The audio sections can be monitored to determine what section or channel the intermittent sound is in. Use the external amp and voltage tests to locate the intermittent section. Prodding and prying the various audio components might make the music begin and quit. Poor board and component terminal connections might cause intermittent audio.

The noisy or intermittent volume control can be quieted with cleaning spray. Excessively worn or cracked controls should be replaced. Do not overlook a corroded bias or level control. These small wire-wound controls have a tendency to provide oxide contacts between the wiper blade and they might cause intermittent sound.

Check each coupling capacitor between the stages for intermittent conditions (Fig. 29-23). Burned components or a change in resistance in the power output transistors and ICs might produce intermittent sound. Determine which channel the intermittent audio is in and go directly to the output stages. Power transistors running quite warm might become intermittent. After the intermittent audio begins, spray the output transistor or IC with several coats of cold spray to cool it. If the intermittent audio goes away, replace the defective transistor or IC. Poor output speaker connections and cables might cause intermittent sound.

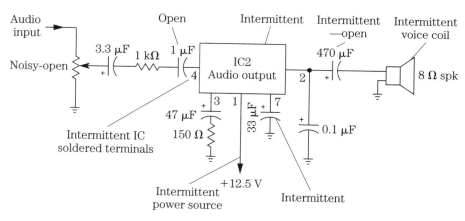

■ **29-23** *Components that might cause intermittent audio in the small audio amp circuits.*

Intermittent TV circuits

Intermittent problems within the TV circuits can occur at any section of the TV chassis. With screen symptoms, waveforms, and voltage measurements, try to isolate the intermittent problems to a given circuit or section. For instance, if the audio is intermittent, go to the sound circuits. For intermittent rolling or vertical height problems, check the vertical section. If the raster comes and goes, suspect that the intermittent is in the horizontal, horizontal output, high-voltage, or picture-tube circuits. Isolate the defective section with symptoms on the screen, speaker, power-supply circuits, and check the block diagram or schematic for the probable intermittent section.

Intermittent sound

Like any sound circuits, worn volume control, coupling capacitors, function switches, transistors, ICs, speakers, wires, and component board connections can produce intermittent sound (Fig. 29-24). First, determine if both picture and sound are intermittent. If the intermittent is only in the sound circuits, try to isolate the sound stages by checking for intermittent sound at the volume control. Signaltrace the audio circuits with the external amp. Notice if the sound is normal at the top of the volume control and not at the wiper (center) terminal. Sometimes volume controls will short inside and produce intermittent sound.

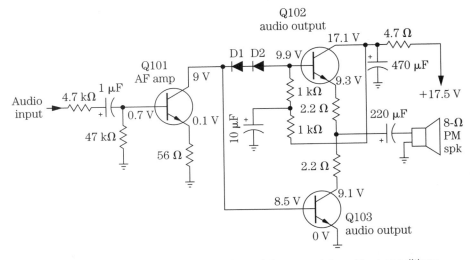

■ **29-24** *The various components that might cause intermittent conditions in the TV sound circuits.*

Now, go from base to the collector terminal of each AF amp, driver, and output transistors until you notice the audio acting up. Check the output IC for intermittent reception. Trace the signal in and monitor the audio at the output terminal. Believe it or not, transistors are still being used in the present-day TV chassis.

Check for poor board wiring or bad connections for intermittent sound. Prod and move the sound board area and notice at the points of pressure that cause the sound to act up. Apply pressure to the speaker cone area to see if the braided wiring to the speaker voice coil is broken. Do not overlook the possibility that a defective speaker coupling capacitor is producing distortion, or weak or intermittent sound. Doublecheck all electrolytic coupling capacitors for intermittent audio. Check the sound coils for poorly soldered leads and defective bypass capacitors inside of the shielded coil form.

Intermittent tuners

An old turret tuner with dirty contacts will produce intermittent reception until it is cleaned with cleaning fluid or spray. A VHF/UHF varactor tuner might appear intermittent or drift off channel. To determine if the tuner is defective, sub a tuner-subber in its place, attached to the IF cable. Doublecheck the antenna lead-in terminal and cables. Make sure that the varactor tuner has the correct supply voltage. Monitor the supply tuner source to see if the voltage cuts up and down.

Usually, varactor tuners cannot be repaired at most service tuner establishments. Remove the tuner and order a new one with the exact part number or send it into the manufacturer's repair depot or to tuner repair centers. The varactor tuner might contain common components on one side of the PC board and miniature SMD parts on the other (Fig. 29-25). Save time by sending the tuner in for repair.

Intermittent TV power supply

The low-voltage power supply with intermittent conditions is easy to service. Simply monitor the low-voltage source to a given circuit. From the symptoms, notice what stage or section acts up or goes dead. If the different voltages are taken directly from the power supply, monitor or check each voltage source. Bad component connections, poor wiring connections, and faulty transistor and IC voltage regulators cause the most problems.

■ **29-25** *The intermittent varactor tuner might house SMD components on one side and standard parts on the other side.*

You must determine if the horizontal circuits shut down if the voltage source is taken from the secondary windings of the flyback transformer. If all sections are intermittent or shut down, the defective component might occur in the horizontal section. If only the video section was intermittent with a normal raster and sound, the trouble lies in the video circuit or supply source from the secondary winding of the horizontal output transformer. Monitor that voltage source to the video circuits.

Sometimes, by tapping or prodding around on the chassis, the intermittent begins to operate and then quits. Check for a loose fuse holder. Suspect defective relays or relay contacts in remote-controlled relay circuits. Check the interlock cord and the plug. If the plug has been cut loose from the back cover, make sure that the plug is inserted properly. Always monitor the voltage source from either the low-voltage power supply or flyback voltage circuits.

Intermittent vertical problems

The intermittent vertical raster might intermittently collapse down to a white line or with insufficient vertical height. The TV set might have to run several hours before the intermittent vertical problem acts up, or it might collapse at once. The intermittent symptom can be caused by transistors or ICs breaking down, poor transistor terminal connections, and problems with many other components. Suspect poor board connections, loose components, poorly soldered eyelet or griplet connections if the raster collapses when the chassis is moved. Narrow the intermittent down to the vertical oscillator or output sections. Clip a scope lead to the vertical countdown IC or oscillator and notice if the waveform disappears if the raster collapses. If so, the trouble lies in the IC or voltage source feeding the IC. Monitoring the vertical circuits with the scope and voltmeter can help locate intermittent components or stages.

If the raster or picture goes to a horizontal white line and the scope waveform is normal at the oscillator or countdown IC, suspect that the output transistor or IC is defective. Take a quick voltage test on each vertical output transistor. If the raster flips to normal, suspect the transistor or IC of which voltage tests were being made. Spray the suspected transistor with coolant. If either transistor does not act up, apply heat to the area with a heat or hair dryer. If the picture returns, spray coolant on each transistor. Replace the defective transistor if the picture collapses. Sometimes it's best to replace both vertical output transistors.

Sometimes both of these vertical output transistors must be replaced if either one is intermittent. Do not overlook the separate vertical driver transistor for intermittent vertical sweep. Check the vertical oscillator or subchassis that sticks above the chassis for intermittent board and tie-in PC wiring contacts (Fig. 29-26).

■ **29-26** *Check the vertical module or subchassis for intermittent vertical collapse.*

Poor socket or cable connections on the vertical yoke assembly might result in no vertical sweep. Look for an open electrolytic or low-ohm resistor in the return yoke winding to chassis ground.

Intermittent raster

The intermittent raster might result from a defective component in the horizontal or high-voltage circuits. A defective picture tube can produce an intermittent picture or raster. Intermittent voltage source to the horizontal circuits might cause the intermittent raster. Poor board connections around the horizontal section and flyback transformer result in an intermittent raster. Sometimes the set will shut down and not come on with an intermittent component in the horizontal output circuits.

Monitor the horizontal oscillator or countdown waveform and waveform near the flyback with the scope to determine if the horizontal circuits are at fault. Monitor the high voltage with a high-voltage probe at the anode socket on the CRT. If the raster shuts down, notice if the high voltage is missing and waveform at flyback is not present (Fig. 29-27). When HV is found at the CRT, check the video circuits for intermittent video component.

If the horizontal circuits are intermittent, check the horizontal driver, driver transformer connections, horizontal output transistor, and HV SCR regulator for horizontal breakdown. Look for poor transistor plug-in socket connections on transistors and SCR. Besides voltage and waveform monitoring, the intermittent compo-

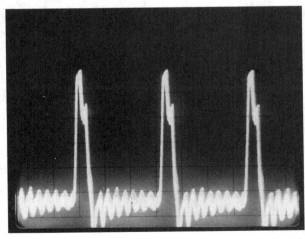

■ **29-27** *The horizontal waveform of the flyback can easily be checked by holding a scope probe next to the flyback transformer to determine if the horizontal section is intermittent.*

nent might be found by cooling and heating areas. Moving components around on the PC board might help you to locate poorly soldered terminal connections.

It is a great feeling of relief and satisfaction to repair an intermittent chassis. Step back and take a couple of deep breaths. The intermittent chassis requires a lot of precious servicing time for the electronics technician. Intermittent problems in CD players, VCRs, and camcorders are found in respective chapters.

Troubleshooting directly coupled circuits

DIRECT COUPLING IS A METHOD BY WHICH ONE CIRCUIT IS connected to another to transport the signal without the benefit of a coupling capacitor or transformer. You might find that the circuit handles both ac and dc signals. The directly coupled preamp or audio circuits are wired directly in the input and output circuits without a transformer. Directly coupled transistor circuits might be more difficult to service because a defective transistor might upset and change voltages on the other transistors and components.

The directly coupled circuits are found in the preamp and audio stages of cassettes and boom-box players. The AF amp or driver transistor might couple directly to the audio output circuits. You might find direct-coupling methods in the vertical and horizontal circuits of the TV chassis. The sound output stages in the TV or audio amp might have direct-coupling circuits. Direct coupling might also be found in the video circuits of the TV chassis.

Directly coupled circuits

Many types of circuits transfer signals besides those that are directly coupled. A bypass or electrolytic capacitor can be used to couple the two stages together. Capacitor coupling transfers ac energy between two different circuits. A capacitor is used to couple two audio stages together and prevent or block dc voltage from entering another circuit (Fig. 30-1). Electrolytic capacity coupling is used between the audio output IC and the speaker. Capacitor coupling is used in many stages of the audio, video, horizontal, vertical, and tape-head circuits.

Transformer coupling uses electromagnetic induction to transfer electrical energy from one circuit to another. Interstage and output transformers were used in early audio stages between the AF or driver transistors and from the output transistor to the speaker to

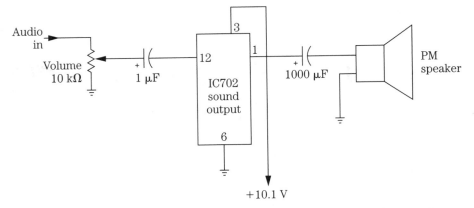

■ **30-1** *In many audio circuits, electrolytic capacitors are used to couple the sound between the stages and the speakers.*

transfer the audio signal (Fig. 30-2). Transformer coupling is used between the horizontal driver transistor and the horizontal output transistor in the horizontal circuits of the TV set. Inductive coupling is sometimes used in radio and TV IF transformers. Transformer coupling is much more expensive than directly coupled circuits. Link coupling is a low-impedance coil that couples two RF

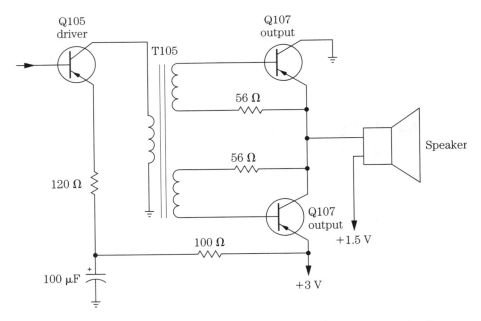

■ **30-2** *Transformer coupling might be used in some audio output circuits.*

circuits together. The impedance-coupled amplifier uses a coupling capacitor and single-wound transformer for output load coupling.

Directly coupled preamp circuits

The directly coupled preamp tape-head circuits consist of a tape head working directly into the base circuit of a transistor or IC and coupled directly to another preamp transistor. You might find a combination of capacitor and direct coupling in the preamp circuits (Fig. 30-3). Here, the tape head is capacitor coupled to the first preamp transistor (Q101) and the collector terminal is connected or wired directly to the base terminal of D102. Again, the output signal from the collector terminal couples the preamp signal to the volume control or to the audio AF amp. Notice that both transistors are NPN types. In other audio circuits, one preamp might be an NPN type and the other is a PNP transistor.

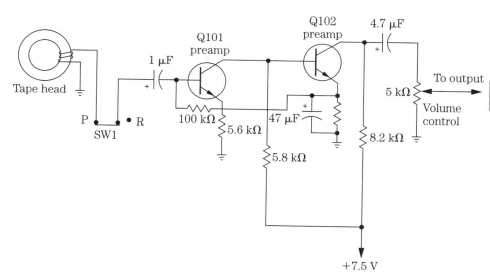

■ **30-3** *In the preamp tape head circuits, both capacitor and direct coupling connects the tape head and the transistors together.*

Q101 bias voltage is provided through the 100-kΩ resistor to the emitter terminal of Q102. Both the collector terminal of Q101 and the base of Q102 are at 1.3 V, obtained through the load resistor (5.8 kΩ). The collector voltage for Q102 is supplied through the 8.2-kΩ resistor. The weak signal from the tape head is amplified by the preamp audio circuits and is capacitor coupled to the volume control by a 4.7-μF electrolytic capacitor.

Some AF circuits are directly coupled, with two transistors coupled together with a small resistor in between (Fig. 30-4). Here, the AF amp collector terminal is tied directly to a resistor between the collector of Q203 and Q204. Resistance coupling is usually combined with resistance-capacitance, with the capacitor as a voltage-blocking capacitor. The input audio signal is taken from the volume control and is capacitor coupled (1 µF) to the base terminal of Q203. The audio output is taken from the emitter terminal of the emitter follower (Q204) of an NPN transistor. The low-impedance emitter circuit might feed to a headphone output or to a low-impedance transformer.

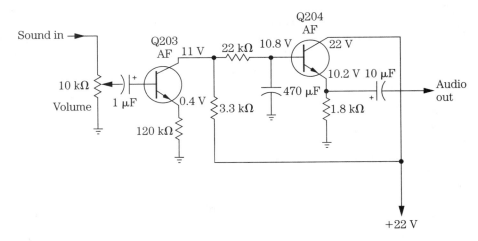

■ **30-4** *Direct coupling with a small resistor between the AF and the driver transistor provides amplification for the audio output circuits.*

NPN and PNP dc circuits

Direct coupling is found in audio-amp power-output circuits. Here, the audio is taken from the volume control and capacitor coupled to the preamp (Q403) with a 1-µF electrolytic capacitor (Fig. 30-5). The amplified signal from the collector terminal of Q403 is directly coupled to a power driver transistor (Q405). Notice that Q403 is a PNP and Q405 is an NPN transistor. The power audio from the driver transistor is directly coupled to the power output transistor (Q407 and Q409) in push-pull operation. Notice that in the output transistors, Q407 is an NPN and Q409 is a PNP power-output transistor. The high-powered sound is capacitor coupled by C419 to the speaker output terminals. You might find the same type of circuits in high-wattage amps, which have higher operating voltages.

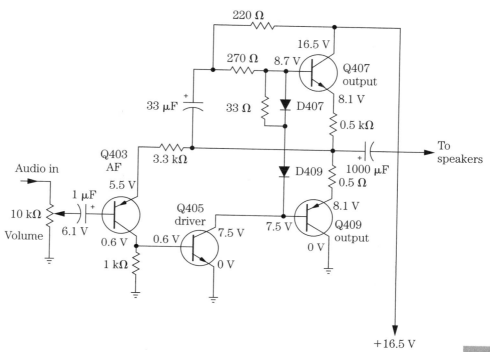

30-5 *Capacitor coupling is used in the AF amp audio stage and direct coupling with diodes is used in the output circuits.*

Transistor interaction

When either directly coupled transistor in the circuit goes open or becomes leaky, voltages will change on each transistor in directly coupled circuits. Often, emitter bias resistors will overheat, run hot, burn, and go open when either power-output transistor becomes leaky. If Q209 shorts, R221 and R223 are burned (Fig. 30-6). If Q209 opens up between the base and emitter terminals, the collector voltage will go higher (37 V) and no voltage will be at the emitter terminal. Very low voltages are located at the base terminal of Q209 and at the collector of the preceding driver transistor.

If a driver transistor (Q208) becomes leaky or shorted, Q212 might be damaged and the bias resistors (R220 and R224) might be burned. Voltages are high at the emitter and base terminals of the driver transistor (Q208 and Q212) (Fig. 30-7). In push-pull output circuits, it is best to replace both output transistors if either one is open or leaky. Sometimes, both transistors are damaged, with one a dead short and the other open. If the driver

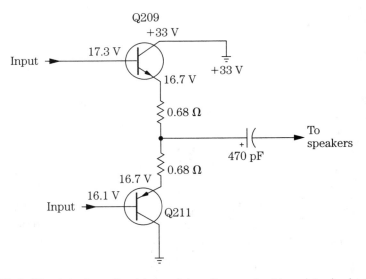

■ **30-6** Check both emitter bias resistors if one output transistor is shorted or leaky.

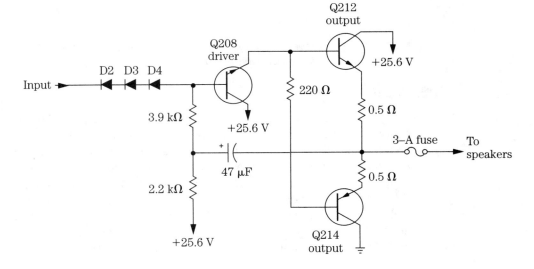

■ **30-7** Voltages might be high on Q208 and Q212 if the driver transistor (Q208) is leaky.

transistor appears to be leaky, replace both the driver and the corresponding output transistor.

In this directly coupled audio output circuit, notice that Q212 and Q208 are supplied with a high positive voltage and Q214 has a negative collector voltage (−26.2 V). Two different voltage sources are used in some high-powered amplifiers. NPN transistor Q212 has a positive collector voltage and PNP transistor Q214 has a negative applied voltage to the collector terminal.

In early negative power sources, a PNP transistor (Q115) might have a negative voltage (−15 V) with the NPN transistor in push-pull operation (Fig. 30-8). The collector terminal of Q117 is at ground potential with a negative bias at the base and emitter terminals. The audio signal was coupled to the speaker between the two emitter resistors. If Q113 and Q115 both became open, −15 V can be found on the collector of Q113, and also on the base and collector terminals of Q115. The base of Q117 might also be −15 V.

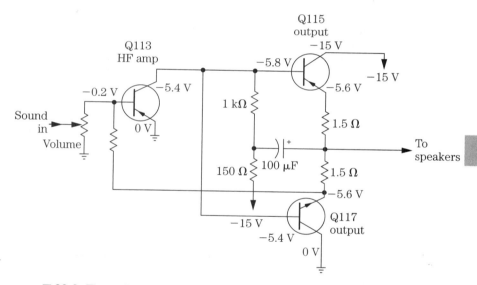

■ **30-8** *The collector terminal (Q117) of the output transistor is grounded or positive with a negative supply voltage at the collector of Q115.*

With directly coupled transistors, one transistor might go open or become leaky and change the voltage on the other transistors in the circuits. Likewise, if a bias resistor changes resistance or goes open, voltage will change on the transistor. If Q117 goes open, −15 V might be at the emitter terminal of Q115. Likewise, if the emitter resistor of Q115 goes open, a higher voltage will be at the emitter terminal of Q115.

Always replace both output transistors if one is leaky or open. Replace the driver and both output transistors if the driver transistor appears to be open or leaky. If the output transistors are out of the circuit, check both the emitter and the base bias resistors for the correct value. These resistors might overheat with a shorted or leaky output transistor, and open and change in resistance. Do not overlook the possibility that the base bias diodes are leaky in the early sound circuits to change voltages on directly coupled transistors.

Directly coupled IC circuits

In most audio circuits, capacitor coupling is found at the input and output circuits. You might see directly coupled circuits in the CD and VCR motor circuits. The loading and spindle motors are directly controlled with IC or transistorized directly coupled circuits. In fact, the mechanism or system-control IC is directly coupled to the IC or transistor drivers with the motor coupled to the driver IC (Fig. 30-9). The motor voltage is controlled directly by the driver IC or transistors.

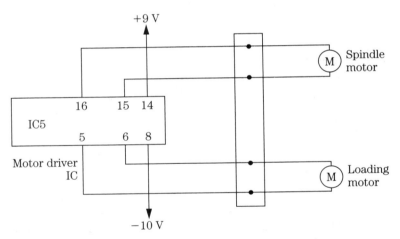

■ **30-9** *Directly coupled motor circuits run from the system-control IC to the motor driver (IC5) to the spindle and loading motors in most CD players.*

Directly coupled vertical and horizontal IC circuits are used in the TV chassis. The vertical countdown IC might directly couple to the output IC through a small resistor. Actually, the vertical output IC couples directly to the vertical winding in the deflection yoke, although the yoke return winding might pass through an electrolytic capacitor or through a small resistor to the chassis ground (Fig. 30-10).

In the horizontal circuits, the countdown deflection IC might couple the drive waveform through a small resistor, choke, or through

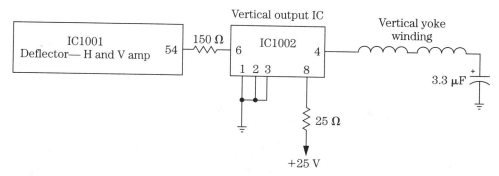

■ **30-10** *Directly coupled circuits are used in some TV IC vertical circuits, between the deflection IC and the vertical output IC.*

both to the base of the driver transistor. The choke coil prevents the unwanted signal within the deflection IC from being passed on to the horizontal driver transistor. Of course, the output of the horizontal driver transistor is transformer-coupled to the horizontal output transistor. Besides the deflection circuits, you will find directly coupled circuits in the audio and video circuits of the TV chassis.

Solid-state in-circuit tests

Transistors can be checked in the circuit after critical voltage tests indicate improper voltages. Sometimes, if the test clips of the transistor tester are clipped to the suspected transistor, the intermittent or open transistor might test normal. Often, leaky transistors can accurately be checked in the circuit. Remove the suspected transistor and test it out of the circuit on a beta transistor tester (Fig. 30-11). High leakage or no measurement (which indicates that the transistor is open) can be seen on the transistor tester.

It's best to remove the suspected transistor after finding a leaky or open transistor within the circuit. If the transistor showed signs of being open or if there is no reading and leaky conditions in the circuit, but it tested normal outside the chassis, spray the transistor with several coats of coolant. Notice if the transistor appears to be open or leaky. If not, try blowing hot air from the heat gun or hair dryer on the transistor and notice the change. Then, apply cold spray to the overheated transistor to show up an open transistor.

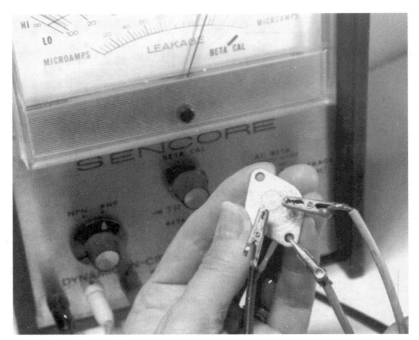

■ **30-11** *Remove suspected transistors and make out-of-circuit beta tests with the transistor tester.*

Critical voltage and resistance tests

Like any electronic circuit, critical voltage and resistance measurements can be used to locate a defective directly coupled transistor or IC. Remember, the voltage measurements of each transistor might be off several volts with an open or leaky transistor in directly coupled circuits. In fact, voltage measurements on all transistors in that directly driven circuit might not correspond with the schematic.

Resistance measurements within the directly coupled transistor or IC circuits might uncover an open, burned resistor, or a change in the resistance value of the defective resistor. High-value resistors (in the MΩ range) might change to a higher resistance. Unless the resistor is burned, cracked, or overheated, the resistance will always increase. Burned or overheated resistors might decrease in resistance. A leaky diode or open resistor in the input or output circuits of a directly coupled transistor audio circuit might change voltages on all the transistors within that circuit. Of course, open resistors and leaky bypass capacitors in the IC output circuits might change several voltage measurements on critical terminals.

Troubleshooting cassette preamp recording circuits

In most tape head transistor circuits, an electrolytic capacitor couples the weak tape head sound to the base of a preamp transistor. You might find the same type of capacitor coupling in the preamp IC circuits. In a few IC preamp circuits, the tape head winding is switched directly to the preamp IC through a low-ohm resistor (Fig. 30-12). Here, the tape head winding is switched by SW1 and SW2, through a 1-kΩ resistor to pin 5 of the preamp IC (IC101). Usually, the preamp IC output signal is capacitor coupled to the audio AF or power output circuits.

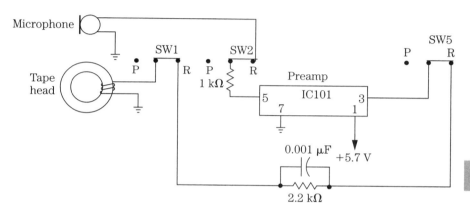

■ **30-12** *In recording circuits, directly coupled circuits might be used with resistors in series with the microphone and tape head.*

Within the cassette recording circuits, the same audio output signal from the preamp IC might be switched directly or through a resistor to the tape head winding. Here, the microphone is coupled directly through the same switched 1-kΩ resistor to the input terminal of IC101. The output preamp circuit is directly coupled to the tape head winding with several resistors in between.

If the cassette will not record or play, suspect a tape head, IC preamp, switches, or an improper voltage source. Check all voltages on the IC and compare them to the schematic. The tape head signal can be signaltraced with the external audio amp. If the cassette player will play and not record, check the different recording switches and clean them with cleaning spray. Scope the tape head for the bias oscillator waveform. No waveform indicates that the bias oscillator circuit is defective. Check the bias oscillator circuits.

Repairing directly coupled sound circuits

Low-priced radios and cassette players might still use three transistors in the audio output circuit. The audio circuit might consist of an audio amp or driver and two transistors in a push-pull operation in the output circuits. The AF amp (Q201) might be directly coupled to the output transistors with two diodes in series to the base of Q202. The input base terminal of Q201 is coupled to the input audio circuit with an electrolytic capacitor (Fig. 30-13). The small speaker is coupled to the emitter resistors with an electrolytic capacitor. C207 serves as a voltage-blocking and coupling capacitor in the audio circuits.

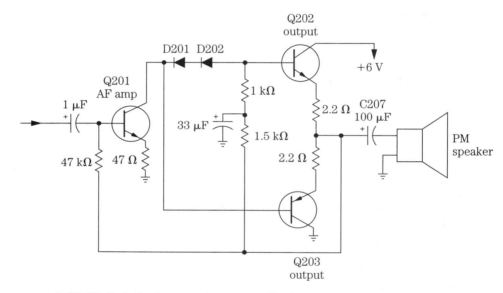

■ **30-13** *Both the input and output audio signal have capacitor coupling with direct coupling between Q201 and Q203.*

Suspect that the output transistor is leaky or open if the audio is weak and distorted. A quick voltage test at the collector terminal of Q202 will indicate if the power source is normal. If the voltage is higher at the collector terminal than normal, Q202 or Q203 might be open. Test both transistors in the circuits. If one is leaky or open, replace both output transistors. If a suspected transistor is intermittent, but the readings return to normal after being removed, change both output transistors. Check both emitter resistors (2.2 Ω) for an open or a change of resistance. If either D201 or D202 become leaky, low audio distortion can be heard in the speaker.

Troubleshooting directly coupled audio circuits in TV chassis

Transistor directly coupled audio circuits are still found in the lower priced TV portables and table models (Fig. 30-14). The first transistor audio amp is capacity coupled to an IF/SIF/Vert/Horz/AFT IC. Q201 amplifies the weak audio signal through directly coupled diodes to the base of Q202 and directly to the base of Q203. NPN and PNP output transistors are used in push-pull operation. The 32-Ω speaker is coupled to the output circuit with a 100-µF electrolytic capacitor (Fig. 30-15).

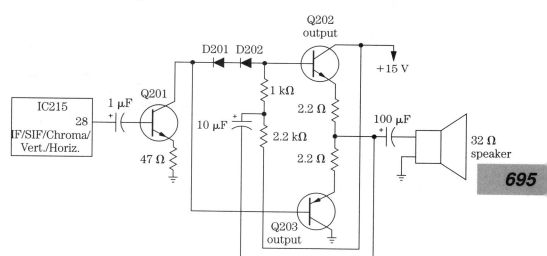

■ **30-14** *Look for directly coupled audio output circuits within the low-priced TV chassis.*

Low distortion might be found in the speaker if one of the diodes becomes leaky. It is best to remove one end of each diode for a correct diode test. If tested in the circuit, the leaky diode might have an incorrect measurement. If one of the output transistors appears intermittent, you might have to replace both output transistors and diodes. High distortion with weak sound might result with a leaky Q202, emitter bias resistor, and a leaky diode. Replace both output transistors if one is open or leaky. Doublecheck the low-ohm bias resistors (2.2 Ω) for correct resistance.

■ **30-15** *The three directly coupled transistors are mounted close together in the sound circuits of a TV chassis.*

Servicing directly coupled vertical circuits

Like the audio circuits, three transistors might be used in the vertical output circuits. A vertical driver is directly coupled to the two vertical output transistors. The yoke winding is coupled with an electrolytic capacitor (470 to 1000 µF) from the emitters of the output transistors. If the driver transistor has a vertical and horizontal deflection IC with vertical amp inside, the IC is capacitor coupled to the base of the vertical driver transistor.

If the drive signal or the waveform from the deflection IC, driver transistor, and either output transistor goes open or leaky, the raster will go to a white horizontal line. Insufficient vertical height might result from either a leaky output transistor or bias resistors. Vertical rolling is caused by poor vertical sync inside of the deflection IC or sync input transistors. Some push-pull output circuits contain a transistor that is marked *top* and the other one marked *bottom*. Of course, the top output transistor affects the top half of the raster and the bottom (marked *transistor*) results in improper sweep of the bottom part of the raster.

In later vertical circuits, the deflection vertical IC might contain the oscillator, vertical amp, and driver circuits. Only two vertical output transistors are used with this type of directly coupled circuit (Fig. 30-16). The output terminal 1 of IC502 feeds a vertical pulse to the vertical output (Q501), directly through a 100-Ω resistor. The output of Q501 (NPN) is directly coupled to an NPN output transistor with a fixed diode. Instead of using a PNP and an NPN transistor, both transistors are NPN types. C509 couples the vertical output pulse to the vertical yoke winding with a small resistor returned to the chassis ground.

Scope pin 1 of IC502 with a white horizontal line or with an improper vertical sweep. If the vertical pulse is abnormal, take critical voltage tests on IC502. If it has a normal input waveform, check the voltage on the collector of Q502 and Q501. Suspect a leaky Q502 and Q501 or an improper voltage source if there is low voltage at the collector terminal of Q502. For improper height, check Q501, Q502, D501, and D502, and the 47-Ω supply source resistor. A higher-than-normal voltage will be found on the collector of both Q502 and Q501 if Q501 goes open. Improper or no vertical sweep might be noted if the 470-μF yoke coupling capacitor goes open or dries up.

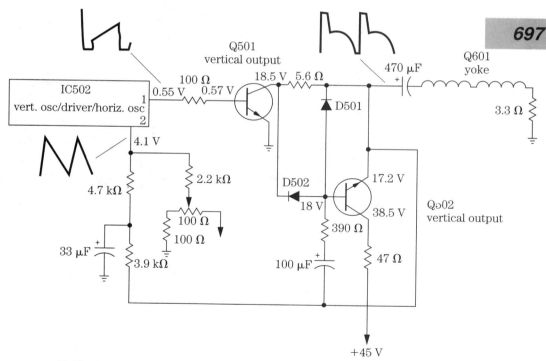

■ 30-16 *Only two vertical output transistors are used with the driver or pre-amp inside of the deflection IC.*

Troubleshooting directly coupled horizontal circuits

Horizontal driver circuits might be directly coupled to the horizontal deflection IC through a choke coil or a low-ohm resistor (Fig. 30-17). The horizontal drive pulse is amplified by the horizontal driver transistor and the transformer, which is coupled to the base of the horizontal output transistor. A square-wave pulse is found at terminal 55 of IC401, directly coupled to the base of the driver transistor, where it is amplified and inverted at the collector terminal.

■ **30-17** *The horizontal driver output transformer is located between the deflection IC and the flyback circuits.*

Scope pin 55 of IC401 to determine if the horizontal deflection circuit is normal. If there is no signal, suspect that IC401 is defective, improper voltage source that might come directly from the low-voltage power supply or flyback circuits. Scope the base and collector terminal of Q401 with a normal waveform at pin 55 (Fig. 30-18). With no output waveform take a collector voltage measurement on the collector terminal of Q401. A voltage measurement above 100 V indicates Q401 is open. Voltages below 25 V indicate Q401 is leaky. Check resistor R411 and T401 for correct resistance

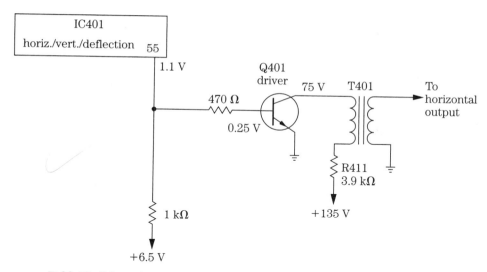

■ **30-18** *If there is no raster and HV, check the horizontal waveform at pin 55 of the horizontal deflection IC.*

if Q401 is shorted. Suspect improper lead transformer terminal connections on T401 if the chassis shuts down after turn-on.

Servicing dc video circuits

The base of the video amp might be directly coupled through a coil and resistor or both. You might find a 4.5-MHz sound filter to bypass sound from the video circuits in the directly coupled circuits. Often, the sound and video is taken from the IF/SIF/chroma/AFT IC1001, directly coupled through the resistor and coil or 4.5-MHz sound trap to the base of the first video amp (Q301). The video output signal is fed from the emitter terminal to the video IC circuits (Fig. 30-19).

Troubleshoot the video amp by taking a waveform at the output of IC1001 (pin 45), with no video picture. Scope the base terminal of the video amp. It's best to connect a color dot-bar generator to the antenna terminals to scope the video and chroma circuits. Then, scope the output video signal from the emitter terminal of the video amp. The signal will stop at the base terminal with an open video amp. If the video amp (Q301) becomes leaky, the picture might be negative or very dim. Take critical voltage and transistor measurements on a video amp to verify a defective Q301. Suspect a low-voltage source with no voltage at the collector terminal.

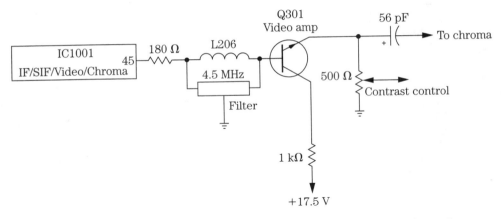

■ **30-19** *The video amp transistor might be directly coupled through a coil and low resistance from the IF IC (IC1001).*

Wholesale part replacement

Sometimes if the audio output or vertical circuits are intermittent with three transistors in the output, you might have to replace all three transistors. In directly coupled circuits, it is difficult to locate the intermittent component because most voltage measurements will change if one transistor opens or becomes leaky. Likewise, intermittent diodes or a change in resistance of bias resistors might foul up the voltage measurements. Determine if the output circuits are defective. Make sure that the input sound or vertical drive waveform is present.

You might find that one output transistor is leaky and the other is open. Sometimes the AF or vertical drive amp becomes leaky, changing voltage measurements on the output transistors. Leaky input diodes can cause distortion in the sound and a collapse of the vertical raster. It might be necessary to change both output transistors and emitter resistors with a shorted driver or output transistor. Do not overlook the possibility that there is a poor board connection. Directly coupled circuits might provide improper voltages in the transistor output circuits for the electronics technician, but the toughest repair problem can be licked with wholesale part replacement.

External power source

If the TV chassis goes to a vertical white line before the chassis shuts down, suspect problems within the vertical circuits. In TV sets where the power-supply voltage for the vertical circuits is scan derived, with chassis shutdown, troubleshooting the vertical circuits is very difficult. If the vertical circuits are fed from the secondary windings of the flyback circuits, and you suspect the vertical circuits are defective, service the vertical circuits with an external voltage power supply (Fig. 30-20).

■ **30-20** *Use a universal power supply with two variable output voltages to inject voltages into the vertical circuits to determine if the circuits are normal.*

Check the supply voltage source feeding the vertical circuits at the collector. By inserting the same supply voltage, you can service the vertical circuits with an external voltage source. Two different voltage sources might be needed to power the vertical circuits. External voltage sources are desired. You can use a battery or a couple in series to acquire the correct voltage. Check the schematic for the voltage source of the vertical countdown IC and vertical output circuits. Connect one voltage source to the IC and the other source to the transistor output circuits. Tack a piece of hookup wire on the PC board if the clips will not stay in place in the PC wiring source (Fig. 30-21).

Now scope the vertical circuits starting at the output terminal of the vertical deflection IC. Proceed to the vertical output circuits to scope the various waveforms. Check the output waveform at the coupling capacitor ahead of the yoke winding. If no waveform is present, suspect problems in the vertical circuits. With a normal vertical output waveform, suspect that a yoke winding or yoke re-

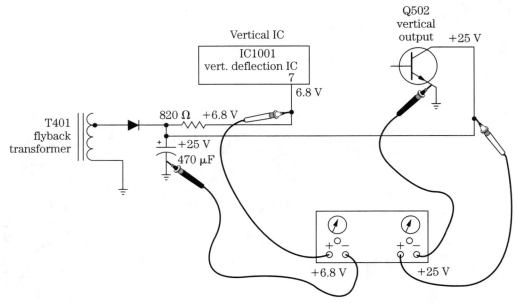

30-21 Here, the vertical deflection IC operates at 6.8 V and the vertical output transistor is at +25 V from the derived-scan power sources in the flyback secondary circuits.

turn resistor is open. External voltage sources can be used to repair the vertical, horizontal, sound, color, and audio circuits of the TV chassis, if it is taken from a scan-derived voltage source of flyback.

How to service consumer electronics without a schematic

SERVICING THE ELECTRONIC CONSUMER PRODUCT IS QUITE difficult with even the correct schematic. Trying to repair a TV chassis without the schematic is most difficult, but the electronics technician does it every week, year in and year out. It is impossible to secure a schematic for every TV, radio, VCR, and camcorder. Of course, the technician is expected to repair any electronic device that crosses the service bench (Fig. 31-1).

The electronics technician must isolate, locate, remove, and replace the defective component. Servicing the defective chassis is

■ **31-1** *The electronic technician is expected to service any TV that appears on the service bench.*

much easier if you have a block or wiring diagram to see how the parts are tied together and where they are located. What do you do if no schematic is available at the moment or if it is doubtful that the correct schematic will ever be located? The technician can take many avenues to repair a product in the shortest time possible. Time is the technician's worst enemy. Solving intermittent or tough-dog problems can bring a sigh of relief and many cherished moments of service satisfaction.

Visual inspection

Take a good look at the defective chassis for signs of overheating, burned or smoking components, cracked boards, arcing parts, excessive hum, and components that are too hot to touch. Do not be afraid to work in the chassis. Before digging in, take a second look at the parts.

Arcing between the parts on the PC boards might result from liquids accidentally spilled inside of the chassis. You can see the arc-over of a tripler unit or CRT spark gap with the naked eye. Firing or arcover inside the picture tube gun assembly might result from a broken neck of the CRT inside the yoke assembly. Smoke or an acrid smell in the chassis might indicate that there are overheated components.

If condensation forms on the carbon resistor, overheated components are nearby. The plastic blown from the center of a power IC might indicate that an IC is leaky or shorted. Several burned resistors might point to a leaky transistor in the audio output circuits. Dark brown areas around the component leads of the PC wiring might indicate that a resistor is overheated with a leaky component connected. Spend several minutes isolating the problem before jumping to conclusions.

Listen to the music

A mushy speaker sound might result from a dropped cone or a frozen voice coil (Fig. 31-2). Vibrating audio might be caused by a loose speaker cone or with several holes punched into the paper cone. Check the electrolytic filter capacitors if excessive hum can be heard in the speaker. Some hum with distorted music might be caused by leaky output transistors or ICs. Low hum might be created by dried-up decoupling electrolytic capacitors. Pickup hum might result from a poorly grounded microphone, an ungrounded base circuit, or a poor chassis ground.

■ **31-2** *A dropped or frozen speaker cone might cause distortion and mushy sound.*

By holding your head near the TV chassis, you can hear the high voltage come up and (if the chassis is shut off), the collapse of the deflection yoke. A tic noise in the flyback might indicate that there are horizontal problems. High-voltage arcover inside of the horizontal output transformer can be heard (Fig. 31-3). You might be able to hear the horizontal oscillator frequency with a defective horizontal circuit. Loose particles within the flyback might indicate that there is a defective flyback or a loose mounting. Diagnose the possible problems by sight, smell, and sound.

Part numbers

You might be able to locate a defective part by glancing at the chassis. The part number of the transistor or IC stamped right on the body area can tell you what circuit, correct polarity, and the correct terminal numbers. Look up the solid-state component in a universal replacement manual, which will give you the correct outline, voltage and current ratings, and what circuit the component will operate in.

For instance, a Goldstar CMT-2132 TV portable might have a dead chassis, no sound-no picture and no raster. From past experience, the symptom points out either a dead low-voltage power supply or

■ 31-3 *The flyback might cause arcing sounds inside of the transformer as the enclosed HV diodes arcover.*

horizontal output circuits, or both. Both fuses were checked and the 1-A fuse was replaced with open conditions. Because no schematic was handy, you must locate the low-voltage power supply and horizontal circuits. The main TV chassis shows five different ICs.

You can locate the fuse and power-supply circuits with a large filter capacitor (470 µF 200 V) with a five-legged IC nearby. You might assume the filter in the low-voltage circuits and IC regulator has been located (Fig. 31-4). A voltage check on the regulator indicates 157 V on terminal 3 and 125 V on terminal 4. This means the normal power voltage source (125 V) should feed the horizontal output transistor. TR402 was located at the rear right of the chassis with no voltage on the collector terminal.

A quick ohmmeter test from the collector (body) terminal of the horizontal output transistor to the 125-V circuit was open. No doubt a small isolation resistor was open on the primary winding of the flyback transformer. A continuity check from the collector to the flyback indicated that there was continuity between terminals 10 and 3. Finally, R408 (0.47 Ω) was located between the primary winding

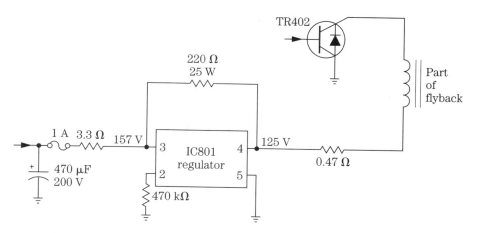

■ **31-4** *The large regulator IC supplies high dc voltage (115 to 135 V) to the flyback isolation resistor (0.47 Ω) that was open.*

of the flyback (3) and pin 4 of the IC regulator. Obviously, the high current of the output transistor knocked out the isolation resistor.

Upon testing the horizontal output transistor, the transistor showed leakage between the base and the emitter, and between the emitter and the collector. A low resistance between the emitter and the collector with reversed test leads indicated a leaky output transistor. The 2SD1555 transistor was looked up in the universal replacement manual and was replaced with an RCA SK9422 transistor (Fig. 31-5). The T-048 outline and application indicated that there was a damper diode inside of the transistor, from the emitter to the collector terminal.

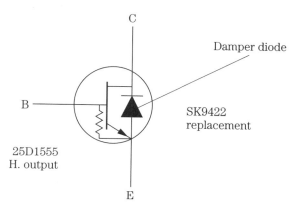

■ **31-5** *The 25D1555 horizontal output transistor was replaced with a universal SK9422 with a built-in damper diode.*

After replacing the open 1-A fuse, 0.47-Ω resistor, and output transistor, the chassis was tested once again. A high-voltage probe monitored the HV and a voltmeter connected to pin 3 of the flyback checked the low-voltage source. The scope monitored the base terminal of the horizontal output transistor (TR402). Slowly, the variable line transformer was raised to 60 Vac with no waveform on the base of the horizontal output transistor. TR402 was overheating and operation shut down. Perhaps the horizontal driver or countdown deflection IC was defective.

IC501 was located close to the horizontal circuits and flyback transformer. If IC501 (LA7626) was checked in the universal replacement manual, the replacement (SK9750) was checked out to be the horizontal drive, x-ray, horizontal countdown, and vertical deflection IC. Pin 17 was the horizontal drive output that fed to the horizontal drive circuits (Fig. 31-6). Of course, no voltage or waveform was found here. The supply voltage terminal from IC501 replacement outline indicated that terminal 16 had no voltage. No doubt, IC501 supply voltage source was fed from a scan-derived flyback circuit. A resistance measurement from pin 16 to the chassis was above 1 kΩ.

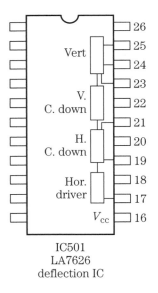

■ **31-6** *Here, the universal SK9750 IC is replacing the original LA7676 chip.*

The external supply voltage was applied to pin 16 and the chassis ground. Pull the ac cord out of the wall socket for this test. Slowly, the voltage source was raised to 12 V. Again, no scope waveform was found at pin 17. All horizontal and vertical pins were checked for leakage to the common ground and most were normal. Even pin 16 was removed from the foil pattern and it showed no sign of leakage to ground (Fig. 31-7). IC501 was believed to be open or defective and was replaced with an RCA SK9750 horizontal deflection IC. The chassis sprang to life with a drive waveform at pin 17.

The dead symptom for the Goldstar portable was repaired by locating the flyback, horizontal output transistor, isolation resistor, and deflection IC from the universal replacement manual. Replacing the unknown horizontal output transistor (2SD1555) and deflection IC (LA7626) with the semiconductor manual solved the dead symptom without a schematic diagram.

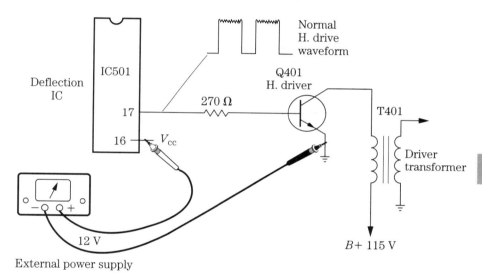

■ **31-7** *The external voltage power supply was used to check the condition of IC501 with no drive waveform at pin 17.*

The new chassis

You might find a fairly new Panasonic CTL-1031R portable required service with no schematic. By looking for the TV model number in Howard Sams Photofacts, you will find that the chassis is located in set 2749. If there are no Panasonic dealers in town, order a schematic from a Sams wholesale dealer or from the Panasonic service depot.

Two large ICs are located in the TV chassis. IC101 contains the IF/SIF/Video/Chroma/Vertical/and Horizontal deflection. Because the symptom was a completely dead chassis, you suspected the low-voltage power supply or horizontal circuits. After replacing a

3-A fuse, a normal 115-V source was found at pin 4 of the regulator and no voltage was found on the collector terminal of horizontal output transistor R806. A fused resistor was found open at the 115-V source. This resistor was replaced with a 2-Ω 10-W resistor.

Very little voltage was measured at the collector terminal of Q502 when the line voltage was slowly raised, indicating a leaky output or that there was no drive voltage. Q502 was tested in the circuit and appeared to be normal (Fig. 31-8).

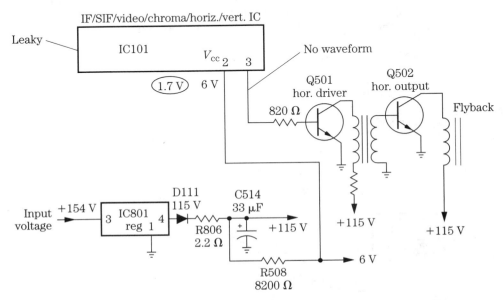

■ **31-8** *A leaky horizontal deflection IC (IC101) caused R508 to overheat in the low-voltage power supply.*

The supply source (pin 2) of the deflection IC was low at only 1.7 V with a variable 60-V line voltage. Undoubtedly, IC101 was leaky. R505 (8.2 kΩ 2 W) was running quite warm as it fed the 6-V source to IC101. A resistance measurement from pin 2 of IC101 to common ground showed 49 Ω leakage. Pin 2 was isolated from the PC wiring with solder wick and a soldering iron. Again, pin 2 was measured at about 50 Ω, which indicated that IC101 was leaky. Even the power supply voltage raised higher than a normal 6 V.

After checking deflection IC101 in the semiconductor parts list of Sams folder 2749, no universal replacement was found for the AN5156K, within the RCA SK series or the NTE universal replacement manuals. This only means that the part must be obtained from the manufacturer or parts depot. You will find that many parts found in a new chassis cannot be replaced with universal replacements.

Different face, same chassis

Several chassis might be manufactured by only one electronics firm. The TV circuits might be the same for 10 years or more (Fig. 31-9). You can use the same schematic to service many different TV sets. Like the camcorder, only a few manufacturers actually make these units, but several different names appear on the same product. The TV, camcorder, and VCR might be made by one manufacturer, with several different names of other electronic appliance firms.

■ **31-9** *Use the same RCA CTC157 chassis schematic to service the CTC156, CTC158, and CTC159 chassis.*

Today, the TV chassis is fairly basic and the only changes made recently are in the stereo sound, on-screen display, and small added features. Of course, the cabinet is changed each year. The horizontal and vertical circuits are practically the same in all TV chassis, with the exception of added ICs to the deflection circuits. The new TV chassis might have one IC as the vertical output, instead of transistors. Try to use a schematic that is similar to the chassis found on the service bench, instead of waiting for months for the exact diagram.

Service notes

Before attempting to service the chassis without the exact schematic, look up the case histories or service notes on other chassis made by the same manufacturer. You might find a case history that has the same symptoms in an early chassis that has the very same circuits. Look up the manufacturer's service notes if you subscribe to their literature and schematics. Most service shops that provide warranty repair for the manufacturer receive service notes at all times.

If a breakdown occurs in one chassis, look for the part to break down in other chassis. The component that breaks down the most in the TV chassis is the horizontal output transistor. You can make money on these repairs because they occur over and over. Remember to write down each unusual breakdown on the schematic or have a simple card file of service notes. A little time spent here might make a lot of money later on.

Schematic comparison

Some manufacturers have their own "pet" circuits and parts. But you can take a similar schematic and use it to service another TV chassis by the same manufacturer. You might find that the very same part number of a color or deflection IC that has been used for several years can still be replaced by the part that has the same number. Part numbers and letters found on ICs and transistors will help you to locate the correct circuit and component, and sometimes operating voltages. Look up the part number stamped on the body of the component and find it in the RCA, GE, Sylvania, and NTE semiconductor universal replacement manual.

Locating the various circuits

After determining the correct symptom, try to locate the possible defective part on the chassis. Without a block diagram or schematic with a parts layout, it might take a little longer (Fig. 31-10). To diagnose the trouble, start with the symptom and try to isolate the defective part. Make a note of the possible circuits that the defective component might be located in. Then, take critical voltages, scope readings, and resistance measurements.

You can find the different circuits on the TV chassis by checking where the large parts are mounted. For instance, the low-voltage circuits are located near the large filter capacitor and several sili-

■ **31-10** *The low-voltage power supply is usually near the electrolytic filter and decoupling capacitors in the separate car radio chassis.*

con diodes (Fig. 31-11). Check around the flyback for the horizontal output transistor mounted on a heatsink. The driver transistor is often located next to the driver transformer. The vertical output circuits might be located with transistors mounted on separate heatsinks or at the output IC on a heatsink. Small transistors might be marked right on the PC board.

The varactor tuner is almost alone, except for the shielded IF circuits nearby (Fig. 31-12). In fact, the antenna 75-Ω cable will connect directly to the tuner. If the tuner is mounted off of the chassis, follow the IF cable to the input stages. Sound circuits might be located around the audio output IC or the detector shielded coil. Trace the speaker wires back to the audio circuits, if need be.

The color or chroma IC is always located around the 3.58-MHz crystal. The chroma circuits might be included in the IF/SIF/video/chroma/luminance IC. Today, the chroma circuits might be located with the deflection horizontal and vertical IC. In the latest TV chassis, you might find only two large ICs mounted on the PC board chassis. Of course, one is the computer chip that controls the many operations of the TV chassis. Locate the IC with

Locating the various circuits

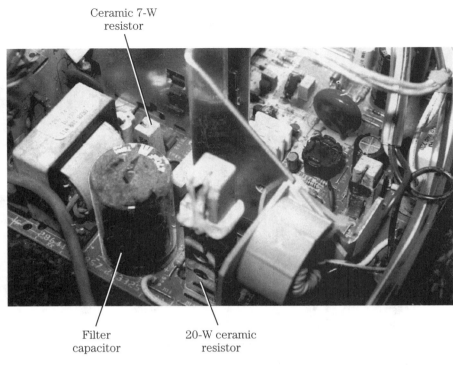

■ **31-11** *Check around the large filter capacitor and ceramic high-wattage resistors for the low-voltage circuits in a TV chassis.*

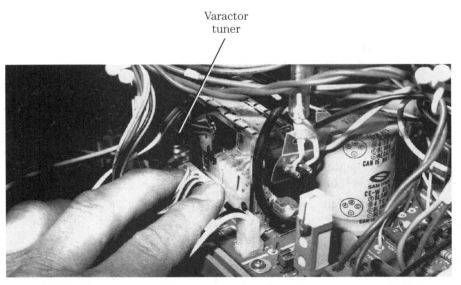

■ **31-12** *The antenna cables connect directly to the varactor tuner, which is soldered to the PC board chassis.*

the 3.58-MHz crystal nearby for chroma and luminance problems (Fig. 31-13).

The color output transistors might be located on the CRT socket PC board. Besides color transistors, the various spark gaps and boost voltages are found on the small board. Locate the screen grid terminal with cable wire from the screen grid control. Often, the screen grid and focus control are on one component. The picture tube voltages can be checked right on the CRT socket.

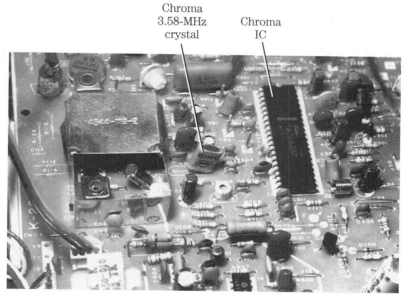

■ **31-13** *The chroma IC has the 3.58-MHz crystal nearby.*

After locating the suspected component, take critical voltage and resistance measurements (Fig. 31-14). Naturally, it is normal to check voltages on solid-state devices before checking any other components. Likewise, take critical scope waveforms in and out of the IC to determine if the part is defective. If the horizontal and vertical deflection circuits are performing, scope waveforms might be taken from the countdown or deflection IC to the horizontal driver, output, and flyback circuits. In vertical circuits, the drive waveform can be scoped from the countdown IC to the vertical output IC or transistors. Although the horizontal waveforms are very steady, most vertical waveforms jump around and sometimes are difficult to read.

■ **31-14** *Take critical voltage and resistance measurements after locating the suspected part.*

Ten service tests

Most electronics technicians start with the symptom and proceed from there. The cassette player symptoms might be that the player is eating or spilling out tapes. So, check the take-up reel for stoppage or slow movement, and clean the tape heads, capstan, and pinch roller. Distortion found in speakers of the right channel of a boom-box player guides the electronics technician to the right channel audio output stages. With weak sound in the audio amplifier, check for a defective component with the external amp or sine- and square-wave input signal (using the scope as an indicator) (Fig. 31-15).

In the TV chassis, a white horizontal line indicates problems within the vertical output circuits. No sound/no picture/no raster might result from a leaky horizontal output transistor or a defective component in the low-voltage power supply. A dark screen with normal sound and no raster might indicate a leaky video transistor and picture tube circuits. Improper width indicates problems within the horizontal or pincushion circuits, or the voltages might be too low. By looking at the TV screen and listening to the speaker, the electronics technician can tell what stage is malfunctioning.

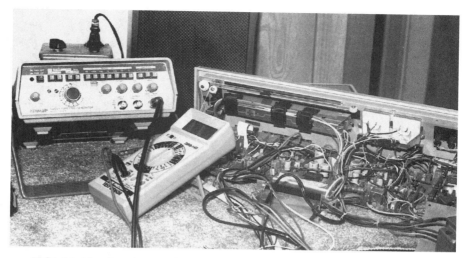

■ **31-15** *Use the sine- and square-wave generator to locate the weak component in the preamp and audio circuits.*

The electronics technician spends more time locating the defective component than replacing it (Fig. 31-16). Locating the defective part might take several hours, while replacement takes only minutes. Always test the new component before installing it. By checking it, you can save a lot of valuable service time later on. The following 10 steps should be followed when locating, repairing, and/or replacing the defective component.

1. What is the symptom?
2. Isolate the correct circuit.
3. Locate the suspected component.
4. Take critical voltage and resistance measurements.
5. Make necessary waveform tests.
6. Remove the defective part.
7. Make sure that the component is defective.
8. Check the new part before installation.
9. Install the new component.
10. Doublecheck and clean up.

Off the main chassis

Sometimes parts mounted off of the chassis or main PC board might be ignored when servicing the electronic chassis. You might find that the horizontal output transistor and low-voltage regula-

■ **31-16** *Taking various voltage, scope waveforms and resistance tests require more time to locate the defective component.*

tor are mounted on a separate PC chassis. PC boards that are separate or are mounted on the chassis might contain the defective part. Different types of transformers are found mounted off of the main chassis. The defective component might be hiding under a shielded area (Fig. 31-17).

The defective yoke assembly might cause the picture to collapse with an open vertical winding or a poor yoke socket and board connections. A poor speaker connection or cable, a blown voice coil terminal might produce intermittent sound. If red is missing from the color picture, do not overlook the possibility that the red output transistor is defective (Fig. 31-18). Flashing video might result from an intermittent or corroded picture tube socket, and on it goes. Always check for the defective part on the subchassis, on the CRT PC board, if it cannot be found on the main chassis.

Part replacement

Try to replace the defective component with parts that have the original part number. Critical parts listed with stars along the side should be replaced with originals in safety-type circuits. Not only

■ **31-17** *Look for the defective component on separate sub-boards or under shielded areas.*

will the part perform well, but it fits properly in the required space. If it is going to take weeks or months to receive the part, substitute a universal component.

Substituting universal components, such as transistors, ICs, electrolytic and bypass capacitors, resistors, and large-wattage resistors are fairly easy to locate (Fig. 31-19). Transformers, coils, motors, tape heads, and VCR and camcorder parts should be replaced with the original-type components. Of course, defective tuners can be sent in for repair.

Doublecheck the soldered connections after installing the new part. Use a magnifying glass for a close-up view of the connections. This is especially true when you are soldering in the IC. Sometimes one of the small terminals might double-up and remain under the mounted IC. Count all connections and doublecheck each terminal before soldering it. Make sure that the new IC or processor has all pin terminals before mounting it.

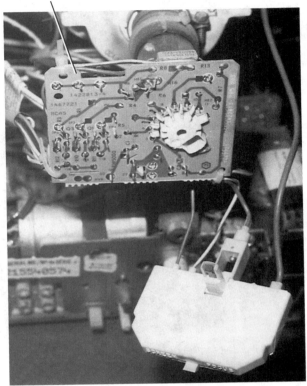

■ **31-18** *The color output transistors are located on the picture tube board in many TV chassis.*

After replacing an IC or transistor, take critical resistance measurements to the common ground from each terminal (Fig. 31-20). This will ensure that all bias resistors are intact. If the resistance reading is way off, doublecheck it for possible damage or if another component is defective. Make this a regular service routine when replacing output transistors or ICs in the audio circuits. Critical resistance measurements might save damaging a new replacement. Clean the PC wiring around the component terminals with heavy-duty rosin defluxer.

Chemical products

The various chemical products that the electronics technician uses every week are coolant or cold spray, defluxer, and dust remover. The dust-all spray is perfect for cleaning dust, lint, and small parti-

■ **31-19** *Transistors, ICs, capacitors, resistors, and diodes can be replaced successfully in the audio amp circuits.*

721

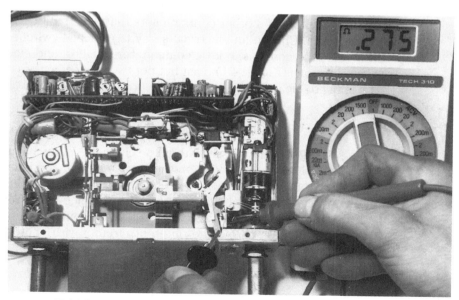

■ **31-20** *After replacing components in the electronic chassis, take critical resistance measurements to locate other leaky or shorted parts in same circuits.*

Chemical products

cles from hard-to-reach places in electronic circuits. The dust clean-off spray removes dust without scratching delicate surfaces. You might have to remove a layer of dust and lint to take a peek at the component that you suspect is defective. Choose a dust spray with a non-residue formula.

Cooling spray chills components so that you can isolate heat-related problems. Some coolants are colder than others. Coolant spray is an essential tool for locating intermittent capacitors, resistors, transistors, ICs, and LSIs.

Sometimes you can locate hairline cracks in PC boards. You can spray coolant on connections or components to protect them against damage while soldering. Spray on the coolant after touching up the solder connection so that you don't overheat the component.

Choose a heavy-duty defluxer spray to remove all activated and nonactivated rosin flux. Apply the defluxer before trying to mend or solder the PC wiring and connections. No scrubbing or brushing is necessary. Choose one that is harmless to metal and most plastics, and is fast drying.

Spray cleaners are also handy to have on the bench. Foam cleaners are useful for cleaning exteriors of electronic equipment and computers. These sprays remove fingerprints, lipstick, and ink stains. The head and disc cleaner spray removes deposits of oil, dirt, nicotine, dust, and other contaminants that have built up on the tape heads. Choose an electrical spray that cleans switches, volume and tone controls, relays, commutator-brushes, and electrical connections (Fig. 31-21). Last but not least, select a glass or plastic spray to clean the TV screen and portable cabinets.

Conclusion

Servicing the electronic chassis without a schematic is much easier after several years of working on the service bench. You know the various symptoms, where to locate the suspected component and how to take critical voltage, scope, and resistance tests. You soon learn how to locate the defective stages and component on the main chassis. Comparing the different circuits with another manufacturer's schematic becomes a habit.

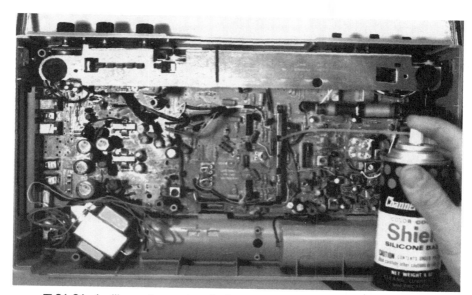

■ **31-21** *A silicone base chemical can be sprayed into sliding switch areas to clean the noisy or intermittent switches.*

Conclusion

Servicing electronic imports

TWENTY YEARS AGO, JAPANESE SOLID-STATE RADIOS, TV, tape, and record players were commonplace in the U.S. marketplace. In fact, major American manufacturers were importing or selecting Japanese firms to make such units to be marketed under famous U.S. brands. Many Japanese manufacturers were importing Japanese units through their own U.S.-based operations. Those different brands include Hitachi, Panasonic, Sony, Sharp, Sanyo, and Matsushita. Later on Toshiba, Samsung, Goldstar, and Tatung was added to the list.

Today, these same manufacturers have plants, U.S. distribution headquarters, and service depots in the U.S.A. These names are so well known that they might not be considered imports any more. Of course, many of the U.S. manufacturers still have some units and components made in the import marketplace and assembled for the consumer electronics field.

Now the import products and parts might be shipped in by Bohsei, Blaupunkt, ASTI, Funai, Fujitsu, Daewoo, Hakko, OKI, Rohm, Sanken, and Taiko, to name a few. The electronic imports might come from China, Thailand, and Malaysia. A Brooklyn AM/FM antique radio and the Fiction 5 TV (ACN3505) are made in China. The Emerson and Orion small-screen color TVs and radios might be manufactured in Thailand and Malaysia (Fig. 32-1).

The biggest drawback in servicing the recent imports continues to be securing replacement components. Japanese parts are now fairly easily obtained from the various service and parts depots that ship nationwide. Today's imported semiconductor products might present a problem for the electronics technician. If the electronics technician is to service these imported products successfully, to make a profit, quick and efficient repairs must be made. If replace-

■ **32-1** *A small-screen TV with a built-in VCR unit.*

ment parts are hard to come by, the profitable service repair is more difficult.

Troubleshooting foreign devices

Although the exact component might be difficult to obtain, in a reasonable length of time, the technician can service these products by subbing American and Japanese components. Some small clock radios might have a tiny schematic included with the operation manual. Of course, no part numbers are found on the diagram. The electronic imported product can be serviced without the exact schematic, semiconductors, and small components. Large parts, such as tuners, IF transformers, LSIs, and power transformers, might total out the electronic product.

Sometimes it might seem impossible to service the import without a schematic. The electronics technician must repair several every week. Imports, no matter where they are manufactured, use essentially the same circuits as domestic versions. Most circuits are very simple. Although you might find some interesting innovations, most circuits are very similar.

You can find a schematic of a radio, cassette player, boom-box, TV, camcorder, or VCR that is similar to the import on the service bench (Fig. 32-2). Of course, servicing the VCR or camcorder might take more time than usual if you don't have a schematic. The more complicated the unit, the more difficult it is to service. Select a comparable schematic and go to work. The less you know about a product, the more time is lost.

■ **32-2** *Sometimes the clock radio might have a small schematic attached to the operator's literature.*

Imported solid-state components

You might not find markings or part numbers on imported transistors or ICs. Even if you do, they might not be found in a cross-reference chart. Comparison parts were once difficult to obtain, but now there are many different American and Japanese components available for replacements. First, determine if the transistor is a PNP or an NPN. Then, figure out the operating voltage and what circuit the component functions in. With the change from germanium to silicon, the solid-state devices are lower in price, with universal replacements available that can provide a large range of frequency and power requirements.

The Japanese transistor manual and transistor substitution manual might help in selecting the correct import replacement. A Japanese transistor manual might list the many different types of transistors with operating voltage, current, and frequency data. Although the Japanese writings and headers might be difficult to understand, the voltage, current, and frequency data is very clear and applicable. Besides the operating data, transistor outlines and terminal elements are easily readable. Instead of transistor terminals marked EBC, numbers 1, 2, and 3 are found on the transistor outline. The numbers correspond with the emitter, base, and collector in the transistor chart. The Japanese transistor manual lists the various circuits that each transistor is found in.

A Japanese transistor substitution manual is quite valuable to locate certain transistors that might not be crossed-over in the domestic universal transistor replacement manuals. The original part number can be subbed with Sanyo, Toshiba, NEC, Hitachi, Fujitsu, Matsushita, Mitsubishi, and Rohm transistors. Take the transistor number and compare it to the transistor manufacturer that is readily available at wholesale resources. You might have many transistors in stock that were purchased for a certain chassis and can be used in the many different Japanese and import chassis with the transistor substitution manual as a guide. If the IC part number is found on the component, look up the number in an IC cross-reference book.

Semiconductors and parts are key items cross-referenced of the VCR and camcorders in the RCA/GE guide from Thomson Consumer Electronics. This guide is an excellent source for crossing brand names and others to the RCA/GE numbers. The ISCET VCR cross-reference information for 50 brand names contains both model and part numbers (Fig. 32-3). Most of these service guides and manuals can be obtained through the local electronics wholesaler or mail order from MCM Electronics, 650 Congress Park Drive, Centerville, OH 45459-4072, (800) 543-4330.

Correct part replacements

Sometimes it is very difficult to obtain the correct part number in the unknown chassis. Correct part numbers can be secured in most Japanese imports because various service depots and wholesale part distributors specialize in providing NEC, Fujitsu, Oki, Sanken, and Rohm original Japanese semiconductors.

■ **32-3** *VCR components might be located with RCA/GE guide and ISCET VCR cross-reference of information for 50 different brands.*

Video products and components found in VCRs and camcorders might be obtained with original part numbers of gears, clutches, rollers, cam gears, post assemblies, idlers, take-up assemblies, phototransistors, pinch rollers, loading gears, motors, and batteries and can be secured from mail-order sources. These electronic and mechanical parts are original part numbers from many imported and Japanese manufacturers. Exact replacement video heads can be obtained for many different VCR and camcorder models. If a component with the exact part number is not available, substitute a universal part.

Identifying components on the PC board

The various components found on the main TV chassis might be marked clearly as other domestic chassis or with very few marked parts. You can recognize most large components on the TV chassis and become mystified by others. You might find bulky heatsinks in the TV chassis. IF coils might be mounted under the chassis with long adjustment screws on the top side. Flyback transformers, and screen and focus control assemblies should be replaced with original parts.

Often, the TV chassis, radio receivers, and cassette player components are easily recognized on the PC board. Small electrolytic capacitors and resistors might have a symbol number to match the schematic. Very seldom are the part values stamped on the PC board. You should have no problems replacing resistors and capacitors because they are color-coded and capacitors have values stamped on the body. Defective tuners, ICs, sound coils, and variable capacitors are difficult to obtain. If you can't find a suitable replacement, you will have to scrap the electronic product.

Critical radio circuits and components

Servicing small portable or clock radios is not profitable unless you are repairing the radio for a good customer who has several TVs, VCRs, and camcorders in the household. Solid-state products costing $29.95 or more might require only a quick visual inspection, voltage and resistance tests to quickly uncover the defective component. If a schematic is not available, spending time prodding through the chassis might be unprofitable. Of course, servicing the large AM/FM/MPX receiver can be profitable for the electronics technician.

The AM/FM/MPX stereo circuits might be found in table-top receivers, auto radios, boom-box players, and high-priced stereo components. Often the imported radio receiver circuits are comparable to the American and Japanese receivers because many are manufactured in the Far East. Many domestic electronics manufacturers have AM/FM/MPX receivers made for and assembled with Japanese components. Besides the Japanese firms, radio receivers might be assembled in China, Singapore, Korea, Thailand, and Malaysia.

Like many of the TV and electronic circuits, transistors, and IC components produce most radio service problems. Analyze the solid-state chassis to determine in what section or stages the trouble would occur. For instance, if the sound is normal in the boom-box player with cassette or CD operation, but there is no FM operation, go directly to the FM circuits (Fig. 32-4). Likewise, if the audio is normal in the left channel and distorted and weak in the right channel, check the audio output transistors and ICs.

Pinpoint the trouble in one stage to a single component using waveform, voltage, and transistor testing, and signaltracing methods. Some technicians prefer to check the suspected transistors or ICs in the circuit. Transistors can easily be checked in-circuit with

FM circuits

■ **32-4** *Locate the FM circuits on the main PC board of a boom-box radio, CD, or cassette player.*

the beta tester or diode-junction diode test of the DMM. Sometimes the intermittent transistor might test good if the test prods are placed on the transistor terminals. Suspected ICs can be quickly checked with the voltage, resistance, and signal in and out tests.

Remove the suspected transistor from the circuit board and test it out of the circuit. Check the transistor for NPN or PNP polarity, the voltage on the collector terminal, and what circuit the defective transistor is operating in. Most radio receiver transistors can be replaced with American and Japanese replacements. Check for a similar schematic if one is not handy. Locate the correct circuit and transistor number on the schematic for correct replacement. If no transistor markings are found on the transistor or if it has a smeared part number, refer to the domestic semiconductor parts manual or to a Japanese transistor replacement and substitution guide.

For instance, in an FM/AM/MPX auto radio and cassette player, the audio is completely dead. The model number and make is nowhere to be found to secure a schematic diagram. Voltage and resistance tests on a large stereo IC that contains both stereo channels was found to be leaky. The part number on the body of the output IC was LA4480. This was the only part number to go on because the parts list for most electronic products is found with the schematic (Fig. 32-5).

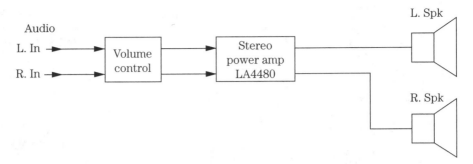

■ **32-5** *Check for part numbers on the body of the stereo power IC amp and check the various supply sources and mail-order firms.*

No replacement for this part number was found in the RCA SK series or in the NTE semiconductor replacement manuals. However, this same part number was found listed in a mail-order wholesale catalog for $3.49. By calling the 800 toll-free number of the electronics distributor, the component was received within one week. If import ICs and transistor part numbers cannot be found locally, try the mail-order route. ICs found in microcomputer and multiplex radio import circuits might be difficult to find in the universal replacement manuals. Try to secure the make and model number and obtain the part through the manufacturer's outlets, if possible.

TV import circuits

Instead of a complicated low-voltage power supply, the TV import chassis might have half-wave rectification with high line-voltage regulators. One large IC power regulator might provide 120 to 125 Vdc to the horizontal driver and output circuits (Fig. 32-6). At first, many of these high-voltage regulators were difficult to obtain, but now they are stocked by most electronic distributors. Components in the low-voltage power supply can be substituted with universal replacements or original parts.

Horizontal import circuits

You might find a few horizontal deflection or countdown circuits operating from a voltage source of the flyback circuits, except that most import chassis operate from the low-voltage power-supply sources. A voltage-dropping resistor lowers the 120- to 125-V source to the required operating voltage of the deflection IC (Fig. 32-7). These circuits are much easier to service for the electronics technician. You do not have to second-guess what stage is functioning.

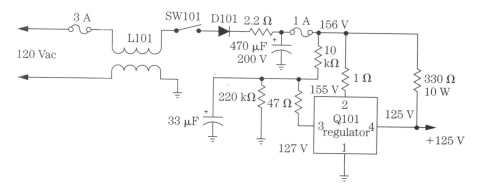

32-6 *The import low-voltage power supply might consist of one silicon diode, capacitor filter input, and large IC voltage regulator.*

32-7 *Locate the low-voltage power supply and horizontal circuits on the chassis for horizontal sweep problems.*

A quick voltage measurement at the supply pin of the deflection IC might determine if the power source or IC is defective with no horizontal sweep. Often, if the deflection or countdown IC becomes leaky, the voltage-dropping resistor (R402) will run quite warm and show signs of overheating. Simply remove supply pin (V_{CC}) from the PC board with a solder-sucking tool or mesh and take a measurement to the common ground for a low-resistance measurement (Fig. 32-8). Suspect a leaky IC with resistance less than 1 kΩ, with the terminal pin removed.

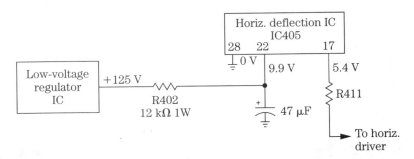

32-8 *Measure the supply voltage source pin of the horizontal deflection IC to determine if the IC or if the supply is defective.*

Scope the horizontal drive pin terminal (pin 17) for a square-wave pulse or waveform with the normal power supply voltage. If there is no waveform, it might indicate a defective IC or connection tied to the pin terminals. In some circuits, the horizontal amp might be included inside the deflection IC. Another waveform taken at the base terminal of the horizontal output transistor indicates if the sweep is applied to this point. Sometimes it's difficult to get at the base terminal because of the mounting of the output transistor, so take the waveform from the secondary side of the driver transformer.

The horizontal driver and output circuits are similar to the American or domestic chassis, except that a damper diode might be found inside of the case of the horizontal output transistor (Fig. 32-9). Of course, the defective output transistor should be replaced with an enclosed damper diode. Universal hot transistor replacements function very well in these horizontal circuits.

You might run into a horizontal output transistor (STD580) that has unusual terminal connections. The STD580 has the same appearance as any output transistor, except that the collector and emitter terminals are switched. In domestic and Japanese output transistors, the metal body of the transistor was always the collector terminal. Here, the body or metal case is the emitter terminal (Fig. 32-10). This transistor is not listed in any universal replacement manual and it can only be obtained through Sanyo outlets.

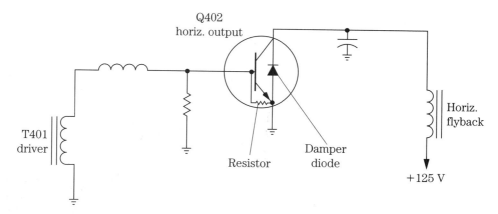

■ **32-9** *Damper diodes might be located inside the hot transistor case of the TV chassis.*

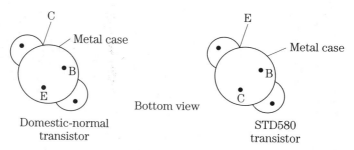

■ **32-10** *The unusual STD580 horizontal output transistor has the body (metal) as the emitter terminal instead of the collector.*

The STD580 output transistor might be found in a Sanyo 91C58UA TV or in some Sears TV chassis.

If testing the STD580 transistor with resistance measurements, you might accidentally discard a normal transistor. A small resistor is found inside from the base to emitter terminals. The resistance between the base and emitter might indicate a leaky transistor. Remember that the damper diode is found internally between the emitter and the collector terminals. A diode resistance measurement from the emitter to the collector will show a low resistance in one direction. If a low resistance is found in both directions between collector and emitter, suspect a leaky hot transistor. Do not discard the transistor with a low-ohm measurement between the emitter and the base terminals.

Vertical import circuits

Like the early domestic TV chassis, the vertical circuits consisted of a deflection vertical IC and two vertical output transistors. Today, the vertical output circuits consist of one large IC supplied with a dc voltage from the scan-derived flyback circuits or from the low-voltage power-supply source (Fig. 32-11). The vertical output drive is taken from pin 6 of a multipurpose IC201, including horizontal and vertical countdown or deflection drive circuits.

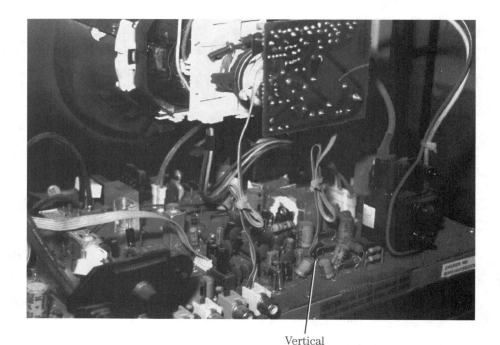

Vertical IC

■ **32-11** *The vertical output might be one large IC in the TV chassis.*

The vertical drive waveform might be found at input 6 of the vertical output IC301 amplifier. IC301 amplifies and drives the vertical sweep to a yoke deflection winding. Notice that the yoke winding is connected directly to pin 4 (output) of IC301. The yoke return winding proceeds through a 10-Ω resistor and 0.022-μF capacitor to the common ground. C338 provides a return sweep path and also blocks the dc voltage from the ground (Fig. 32-12).

Check the vertical circuits with a scope for waveform at pin 6 (IC201) and at the input pin 6 of IC301. Scope the output pin 4 for

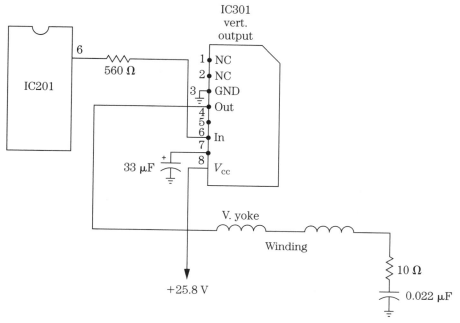

■ **32-12** *Take vertical input waveforms at pin 6 of the output IC (IC301) and output terminal of pin 4 to determine if the IC is leaky or open.*

a sawtooth pulse of around 50 Vp-p. Take a critical voltage test at supply pin 8 of IC301. The supply voltage for most vertical output ICs is around 25 V. Normal waveforms at pin 6 and a lower voltage at pin 8 might indicate that IC301 is leaky with abnormal output sweep waveform.

Domestic and import replacement components

Try to locate the defective component on the schematic for a correct part number. Determine if the product is listed in Sams Photofacts index for a correct schematic. If not, check the next electronic product with the same manufacturer. Also check the manufacturer's outlets for the required schematic. If no schematic is available, check the part number on the solid-state component. Look up the part number in American universal replacement manuals, such as RCA and NTE.

Then, check the Japanese transistor and substitution manuals for possible replacement. Look up the IC part number in Howard Sams cross-reference manual. Large transformers and flybacks must be ordered from the manufacturer outlets. Send in the defective tuner to tuner repair depots. Always try to locate the defective part through the manufacturer channels. Besides local electronics distributors, check the mail-order firms who might have exact Japanese and import components for TVs, VCRs, camcorders, and cassette players.

Distributors of imported electronic parts

Apollo Wholesale Electronics
1944 W. Shady Grove
Irving, TX 75060
1-817-877-5854

Audio Video Parts, Inc.
1071 South Labrea Ave.
Los Angeles, CA 90019
1-800-999-4555

B-B&W Electronics
4320-H Yale, NE
Albuquerque, NM 87107
1-800-367-2114

Consolidated Electronics, Inc.
705 Watervliet Ave.
Dayton, OH 45420-2599
1-513-252-5662

Dalbani Corp.
2723 Carrier Ave.
City of Commerce
Los Angeles, CA 90040
1-213-727-0054

Diversified Parts
2114 S.E. 9th Ave.
Portland, OR 97214
1-800-338-6342

East Coast Distributors
2 Marlborough Rd.
W. Hemstead, NY 11552
1-516-483-5742

Eiger Electronics
70A East Jefryn Blvd.
Deer Park, NY 11729
1-800-835-8316

Electronics Warehouse Corp.
1910 Coney Island Ave.
Brooklyn, NY 11230
1-718-375-2700

Fox International LTD, Inc.
23600 Aurora Rd.
Bedford Heights, OH 44146
1-800-321-6993

HBF Electronics, Inc.
6900 New State Rd.
Philadelphia, PA 19135
1-800-426-4230

ICM International
1-800-748-6232

Mil Electronics, Inc.
1500 Main St.
Waltham, MA 02154
1-800-343-4608

Mills Electronics, Inc.
2026 McDonald Ave.
Brooklyn, NY 11223
1-800-346-8994

Ness Electronics
441 Stinson Blvd., NE
Minneapolis, MN 55413
1-800-331-7617

Pacific Coast Parts
15024 Staff Ct.
Gardena, CA 90248
1-800-421-5080

Panson Electronics
I-80 & New Maple Ave.
Pine Brook, NJ 07058
1-800-255-5229

MCM Electronics
550 Congress Park Ave.
Centerville, OH 45459-4072
1-800-543-4330

Distributors of imported electronic parts

Union Electronic Distributor
18012 So. Cottage Grove
So. Holland, IL 60473
1-800-648-6657

Vance Baldwin, Inc.
2207 South Andrews Ave.
Fort Lauderdale, FL 33316
1-800-432-8542

Tough repair problems

33

A TOUGH REPAIR PROBLEM CAUSES A LOT OF WASTED TIME and money. A tough repair problem might take over three hours or more (even weeks) to finish the job. Some electronic tough repair problems require help from several people to locate the defective components. Those tough repair problems seem to come across the service bench in bunches (Fig. 33-1).

The tough repair problem might occur in any electronic chassis or product that is found in the consumer electronics field. The more complicated the product, the more time is lost. These tough repair problems seem to wait until the bench is full and the back room is stacked up before "pouncing" on the technician.

■ **33-1** *Tough problems might become more difficult when the TV chassis is hit by lightning.*

The tough repair problem might be a shutdown or intermittent chassis. Intermittent noise in the high-power audio amplifier might extend beyond allotted time to a tough repair problem situation. Multiple component breakdown in the low-voltage power supply results in extra time lost. Horizontal repair problems in the TV chassis might be difficult to repair. Trying to service an intermittent or shutdown in the CD player or VCR without a schematic haunts the electronics technician. Just bear down, think clearly, use all of the technical information available, and if needed, call for outside help.

Early to bed

Time is precious and rewarding for the electronics technician. Too much time servicing any electronic product might end up in lost revenue. You can only charge so much for a given repair or you might exceed the cost of the unit. Sometimes a repair might be a tough repair problem for one technician, but the next technician might locate it at once. In fact, some electronics might be taken to several service establishments before being completely repaired. By then, there might be some added problems that require service to fix errors someone else committed (Fig. 33-2).

■ **33-2** *Be careful when taking voltage measurements so not to short out the test prod across IC pins or transistor terminals.*

If more than two hours of service are spent and the problem still exists, set the chassis aside and tackle the next repair. Sometimes intermittent chassis take a lot of time and must be monitored with the scope and voltage tests. Try tackling that tough repair problem early in the morning when your mind is sharp and clear.

From a different angle

After spending several days working on the chassis off and on and not knocking out that tough repair problem, ask for help. Check with the electronics distributor for help. Call the manufacturer's servicing depot for added help. Usually they have service help available. Contact the factory source representative or national service manager. Just a simple telephone call and 15 minutes of your time might save several hours of frustration. Check with the local technician down the street who services that brand of electronic product. It pays to be friendly with your competitors. Remember, there is someone out there who can help you with this tough repair problem.

Multiple problems

You might have a chassis that is intermittent, goes dead, and takes weeks to act up again. Sometimes you repair one problem, then the next moment something else breaks down. It's possible to have distorted, intermittent, and noisy audio in a cassette player with two or more defective components. Locating each defective component takes more time to finish the service problem (Fig. 33-3).

The TV chassis might break down every two or three days, and the next time it's turned on it plays for weeks. The symptoms might include an intermittent, shutdown, and improper sound. You might wind up replacing an intermittent capacitor in the horizontal output circuits, a bad socket connection for the high-voltage SCR, horizontal output transistor replacement, and when the raster was completed, you thought that a bowing picture existed. Repairing the pincushion circuits with a pincushion output transistor cured the last service problems, you hoped. Multiple problems within the electronic chassis takes a lot of valuable service time if there are several different defective parts.

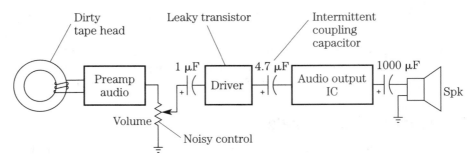

■ **33-3** *Locating the tough dog component is very difficult when several different components might be defective.*

Intermittents

The tough intermittent is the most difficult service problem found on the service bench. Sometimes the intermittent might not act up for several days or weeks. When it does, the chassis does not stay in the intermittent state long enough to locate the trouble. Return the TV or product to the customer until it does.

Intermittents can be located if you are armed with cans of coolant and a heat gun. These products are excellent for locating components that are intermittent because of heat-induced failures. Apply the heat and coolant right on the component (Fig. 33-4). Also, try raising and lowering the input ac voltage to the intermittent product.

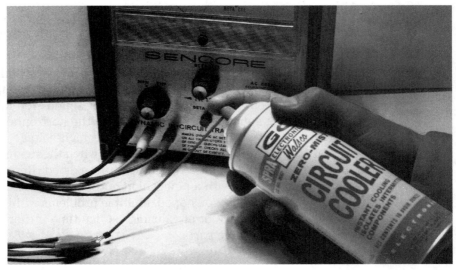

■ **33-4** *Apply several coats of coolant right on the suspected part with intermittent conditions.*

Try it again

After several different days of trying to locate the service problem, let the chassis operate for a few days. Start early in the morning. Try different servicing methods. Check all solid-state components while it is intermittent. The next day, check all PC board connections. Another morning, check the value of each resistor, capacitor, and diode. Don't give up; try another approach.

The adjustment of the various controls on the TV chassis or any electronic chassis is a simple test, but it can cause problems. With these adjustments you might uncover an erratic or defective control, but when the control ends at a different position, it can cause other problems or obscure the original one. If the symptom or defect triggers a certain amplitude, the adjustment might stop the symptom temporarily. So, don't make unnecessary adjustments (Fig. 33-5).

■ **33-5** *Do not touch any controls as it might upset the intermittent mode of the tough dog problem.*

If a part is unsoldered during the process of trying to locate the defective component, or a new part is installed, make sure that the part terminals are placed in the correct holes of the PC board. Be careful!

Tough repair problems are sometimes produced by lack of proper test equipment, test methods, electronic knowledge, and schematics. Other problems are more difficult to determine. They just happen. You cannot see a broken gear, like the auto mechanic, but must rely on test equipment to locate the tough repair problem (Fig. 33-6).

■ **33-6** *Besides critical test equipment, coolant and heat application might help to locate the intermittent dog.*

Defective new part

One of the routine lessons that you learn in electronic troubleshooting is to always test each new component before installing it (Fig. 33-7). You might have picked up the wrong one, placed it in the wrong box, or the resistance might be way off, compared to the circuit diagram. Likewise, make sure that the electrolytic or bypass capacitor has the same or higher capacitance with adequate working voltage. Some electronics technicians stock only 1000-V bypass capacitors and diodes, which cover most circuits,

■ **33-7** *Remove one lead or remove resistor completely from the circuit for accurate resistance measurements. Test the new component before installing.*

except for specialized high-voltage circuits. Check that capacitor on the capacity checker for the correct or higher capacity. Above all, check for correct terminal polarity.

Test each new transistor with a beta tester or with the diode-junction test of the DMM before installing it (Fig. 33-8). Make sure that it is not leaky or open. Recheck the substituted part number in the universal semiconductor manual. You might find that the wrong transistor or IC was sent for replacement. It is difficult to test ICs out of the circuit. Make sure that pin 1 is at the right corner and that all pins have been pushed through the holes in the PC board. Be very careful in replacing and installing the new component; you want it to cure the tough repair problem.

Noisy audio

A low frying noise in the audio might result from noisy transistors or ICs. Check the output stage if it contains distorted music and a loud hum. Suspect a defective audio output IC if there is a constant frying noise and weak sound. Check for low frying or hissing noise within the preamp or AF amp circuits (Fig. 33-9). Besides solid-state components, check the noisy motor if noise is heard only

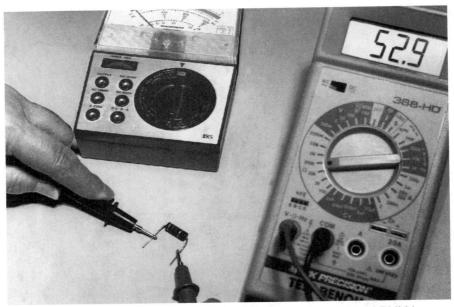

■ **33-8** *Test each transistor with beta or diode-junction tests of DMM before installation.*

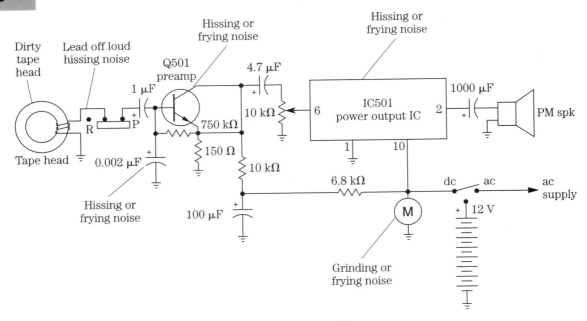

■ **33-9** *Check the following parts for low frying or hissing noise in the cassette player.*

Tough repair problems

when the cassette motor is rotating. Do not overlook the possibility that there is a poor wiper blade contact on variable bias resistors or ceramic input capacitors and there are frying noises. Single-ended audio output transistors might produce a popping sound.

Notice if both audio channels are noisy and distorted. Turn down the volume control and listen for a frying noise. If the noise is present with the volume control turned down, suspect the AF or driver amp, output transistors, or power ICs. Replace the noisy power IC with the volume turned down. You cannot locate a noisy transistor or IC by transistor, voltage, resistance, and continuity tests. Replace the component if you suspect that it is defective.

If the noise disappears after the volume has been turned down, suspect a noisy AF transistor or preamp IC (Fig. 33-10). This means that the noise is occurring in the front end of the amplifier. Start at the front end and work toward the volume control. Replace the preamp IC if both audio channels are noisy.

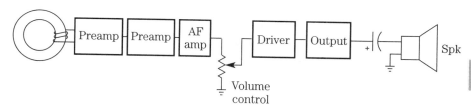

■ **33-10** *Determine if noise is in the front-end or output audio circuits by turning down the volume control.*

One method to locate a noisy transistor is to take and short the base and emitter terminals together. Do this only on preamp, AF and driver amp transistors. Do not short out the emitter and the base terminals in direct-coupled power output stages. Another method is to disconnect a coupling capacitor between the stages to see if the noise has quit or is still present. If the noise stops, the noisy component is still ahead.

Take a small 10-µF electrolytic capacitor and place a grounded alligator clip on the grounded (–) side of the capacitor and a test prod clipped to the positive (+) terminal (Fig. 33-11). Place the probe to each base terminal of the transistor or to the input terminal of the IC and notice when noise quits. As you proceed through the circuit, go to the next transistor and if noise returns, check the preceding noisy transistor. Discharge the capacitor each time that it is probed in the circuit.

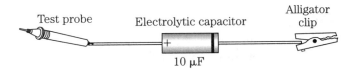

■ **33-11** *A small electrolytic capacitor with wire clips can be used to locate a noise stage at the base of each transistor.*

A low frying noise was heard in a Sony CFS-66 AM/FM/MPX radio and cassette player. Determine if the noise is in the radio or amplifier section. Rotate to FM radio and check for noise. The noise could not be heard with a local FM station tuned in. But when turned to tape function, the low noise was present. To determine if the noise is in the front end of the amplifier, turn the volume control way down. No frying noise could be heard.

The low frying or hissing noise must be from the tape head to the function switch. If the base and emitter terminals of Q201 were shorted, the noise was present, but it was very low. C203 was found between the base of Q201 and S303 (Fig. 33-12). S303 placed the play head winding into the circuit. The switches were cleaned, but there was still a hissing noise. If the 10-μF capacitor with test leads was shorted to chassis ground at the tape head, the

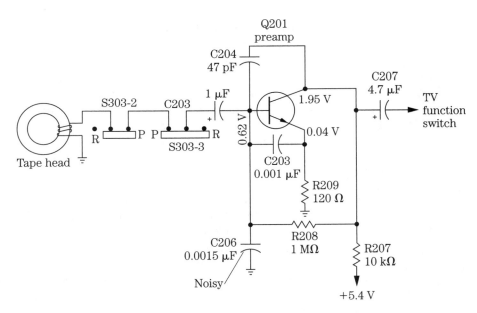

■ **33-12** *A noisy bypass capacitor caused a constant frying noise in the preamp stage of a Sony CFS-66 cassette player.*

Tough repair problems

noise was still present. But when the probe was placed on the positive side of C203 (1 µF), this noise disappeared.

Coupling capacitor C203 was replaced and the noise was present. The hissing noise only would appear after the set was on one or two hours. Voltage, continuity, and resistance measurements were taken on Q201 without any luck. The preamp transistor (2SC1362) was replaced with an SK3124A universal transistor. The noise was still there. Each time that the chassis was shut down, the frying noise took time to act up again.

Q201 terminals were resoldered to the PC board with no results. The next step was to disconnect the noisy component. Before removing any other components, I decided to see if the heating or coolant on any component would cause the hissing noise to quit. After the hissing noise started up again, coolant was sprayed on each component (Fig. 33-13). When the coolant spray hit the C206 (0.0015 µF) bypass capacitor the frying noise ceased. To prove that C206 was defective, heat was applied and the capacitor started hissing. Coolant was applied and the noise quit. Replacing the bypass ceramic capacitor C206 solved the low frying noise.

Usually, ceramic capacitors might cause a frying noise with fairly high voltage connected to one end. Here, the highest voltage was only 0.62 V. The defective capacitor showed no signs of leakage on

■ **33-13** *Try locating the frying noise in the speaker by spraying each ceramic capacitor with coolant.*

a 100X ohmmeter range. Besides solid-state components, ceramic capacitors in the front end of the audio stages might produce a low frying noise in the speaker. The third attempt was needed to solve the noisy problems.

Intermittent recording

Sometimes the right channel recording in a J.C. Penney model 1770 combination radio, cassette, and 8-track player and phonograph would play for hours and the next time you attempted to record, the recording became intermittent. The intermittent chassis provides plenty of trouble, but with cassette players, expect double trouble. You must record and play back after each test, and this adds up to extra time.

Before tackling the intermittent recording problem, the tape heads and the function switch contacts were cleaned up. You might find 24 to 36 switch contacts in some of these function switches (Fig. 33-14). Silicon cleaning spray can be placed right into the switch area with a plastic application tube. Insert the plastic tube at each slotted area and spray cleaning fluid into the opening. Work the switch back and forth to clean all switch contacts.

■ **33-14** *Clean up dirty or erratic contacts of the large switches by spraying cleaning fluid into the open areas of the switch.*

Again, another recording was made with only a hissing noise and low garbled music from the microphone or circuits. Because the cassette player would play back with no problems, the recording circuits must be defective. To prove that playback operation was normal, a commercial music cassette was played several times without any problems. Another new cassette was inserted for another recording. To record, music was picked up from an FM radio station at the microphone.

After checking the schematic diagram, the recording circuits consisted of IC302, Q302, and Q306, and the bias oscillator, Q307. IC302 was eliminated because the IC is used for both playback and recording (Fig. 33-15). The recording circuits were fairly simple, consisting of only two transistors, Q302 and Q304. When the player was recording, the forward bias voltage between base and emitter terminals of both transistors was a normal 5.7 V.

Trace the circuit out from the cassette right CH input jack through R304 and C302 to the base of recording amp Q302. The collector terminal of the automatic level control (ALC) transistor (Q304) is tied into the input of the recording amp to provide level control. C304 couples the collector output of Q302 to S7-6 and switches to the input of sound IC302 through C310 (4.7 µF). The rest of the circuit must be normal with correct playback.

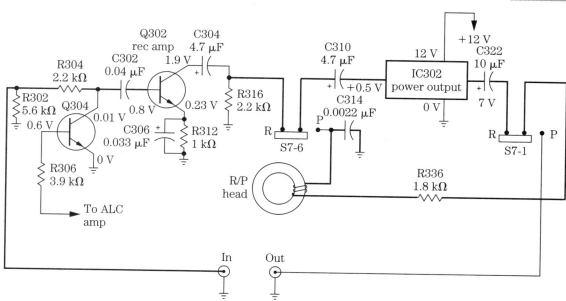

■ 33-15 *Suspect the recording circuits that are not used in playback if there are intermittent recording problems.*

Voltage measurements were normal on Q302 and Q304. Several resistance tests of transistors and capacitors were fairly normal. Because the recording circuits can be checked with an external audio amp, the input and output of Q302 audio was fairly normal. Often, you can see if the recording circuits are normal by watching the recording meters. Of course, this chassis only had a recording LED.

Remembering that poorly soldered contacts can cause recording and playback trouble, the bottom connections were checked. To make sure, all switch contacts were resoldered (Fig. 33-16). The results were the same.

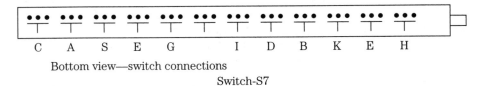

■ **33-16** *Resolder all large switchboard connections if there is intermittent operation.*

To make sure that the bias oscillator amp (Q307) was functioning, the scope probe was clipped to one side of the tape head. The oscillator bias waveform was normal through three different changings of cassettes. All of a sudden, the waveform would come and go. There was no doubt that the bias oscillator circuits were intermittent.

Q307 was tested with in-circuit beta and bias voltage tests, with normal measurements (Fig. 33-17). The dc voltage was monitored

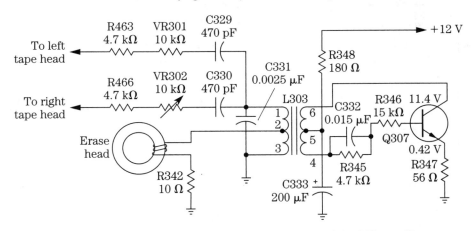

■ **33-17** *A poorly tinned transformer or coil terminal of the bias oscillator created intermittent recording in this cassette player.*

at the collector terminal. The recording was started up again and it played for hours. Finally, one morning, no voltage was measured at the collector terminal of Q307. C333 has +11.5 V at the positive terminal, indicating that voltage should be entering L303 to the collector of Q307. Then, the cassette player began operating once again. A poorly soldered connection on the bias oscillator transformer solved the recording problem.

Insufficient vertical height

The vertical height in a J.C. Penney model 2009A chassis was only four inches high. All of the vertical sweep or raster was below the center of the picture tube, indicating that the top half of the vertical deflection was missing. Insufficient vertical height is usually caused by vertical output transistors, bias resistors, an improper supply voltage, and poor vertical drive. Remember that in direct-coupled circuits, one transistor voltage can affect the other.

According to the schematic, Q21 (driver) and Q22 (top output transistors) produce the top half of the picture height (Fig. 33-18). Although they were suspected, when checked, both were good. Next, all vertical deflection transistors were tested in-circuit by the diode test function of a digital multimeter (DMM), showing that Q19 and Q20 were leaky. Both were removed and tested leaky. Original RCA transistors were installed and the receiver op-

■ **33-18** *Locate the vertical output transistors to make voltage and beta tests.*

erated normally for about two minutes before the vertical deflection stopped, leaving just a horizontal white line.

The retrace-switch transistor Q20 was extremely hot and very leaky, with a 43-Ω short. Q20 was replaced again and it operated for a few seconds. Then, there was a serious overload. Although power had been applied for only a few seconds, Q20 was ruined, and a new replacement was installed. Before firing up the chassis again, resistance measurements were taken. R120 (680 Ω and thermistor PTC1) were checked and were found to be normal (Fig. 33-19).

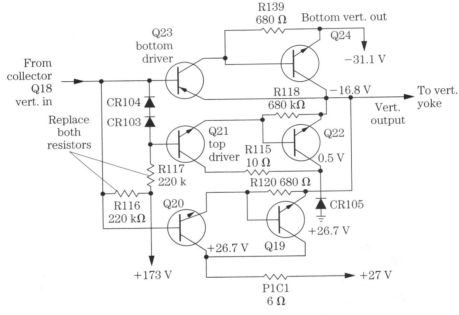

■ 33-19 *R116 and R117 caused insufficient vertical height in this J.C. Penny 2009A.*

No doubt, any obvious defect called for a review of troubleshooting methods. Some transistors fail under load, but check perfect with test instruments. Resistors should be disconnected from the circuit before the tests can be made accurately.

As a test, the top-driver transistor (Q21) and top-output transistor (Q22) were replaced. Q20 continued to operate very hot. Next, one lead of each vertical resistor was disconnected from the circuit and the volume was checked according to the schematic before the resistor was resoldered. Although R117 (220 kΩ) had checked very close in the circuit, it was found completely open, when one end was removed from the circuit. Installation of a new 220-kΩ (R117) restored full height with good linearity and stopped Q20 from overheating. Both R116 and R117 (220 kΩ) resistors were replaced.

TV chassis shutdown

Most TV chassis shutdown problems are caused with overloaded circuits in the secondary flyback and horizontal sweep circuits, and improper low-voltage sources. The primary low-voltage power supply must provide voltages from 115 to 135 V for the low-voltage regulator and sufficient $B+$ voltage for the start-up circuit to run the horizontal countdown, oscillator, and driver circuits for a brief second. Likewise, the low-voltage regulator circuit must now provide a $B+$ voltage to the horizontal output transistor through the primary winding of the horizontal output transformer (Fig. 33-20).

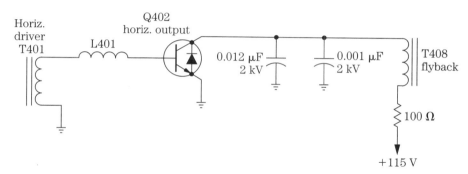

■ **33-20** *The horizontal output transistor receives supply voltage through the primary winding of the flyback transformer.*

The horizontal countdown IC circuits provide a pulse to the horizontal driver, then to the output transistor that now pulses the flyback, which will produce a magnetic field around the core of the transformer. The secondary windings of the flyback will produce voltages to the various scan-derived voltage sources. If one winding or circuit is connected to these voltage sources and produces a short or overload, the magnetic field will saturate the core of the flyback and act as a short, shutting down the chassis. The output of all scan-derived voltage sources is now zero. An open or shorted secondary scan circuit will affect the output voltage of all other secondary voltage and circuits (Fig. 33-21).

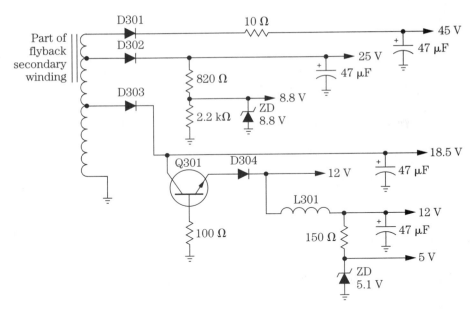

■ **33-21** Scan-derived voltages from the flyback or horizontal output transformer might provide voltages to several different circuits in the TV chassis.

Once that the primary winding of the flyback is pulsed, the secondary windings will provide a B+ voltage to run the horizontal countdown IC, oscillator, and driver circuits. Every shutdown circuit relies on one or more secondary windings, regardless of whether it is at the high- or low-voltage shutdown circuit.

The high-voltage shutdown circuit will shut down or kill either the horizontal countdown, oscillator, driver stage, or low-voltage regulator, if the output of the flyback secondary voltage increases above the HV limit. A high-voltage shutdown circuit connects to the base of either the driver or the oscillator stage via a zener, resistor, or a regular diode. The diode end is connected to the secondary winding in series with a resistor (Fig. 33-22). If the secondary voltage becomes very high, this voltage exceeds the rating of the zener diode, and shuts down the horizontal circuits.

The low-voltage shutdown circuit will kill either the horizontal countdown, oscillator, or driver transistor. Now, the chassis is completely shut down and must be started up again by turning on the on/off switch or remote. Any defective component within all circuits tied to the flyback will automatically shut down the horizontal circuits. All circuits must function or the TV chassis will shut down (Fig. 33-23). This type of horizontal circuit is more difficult to ser-

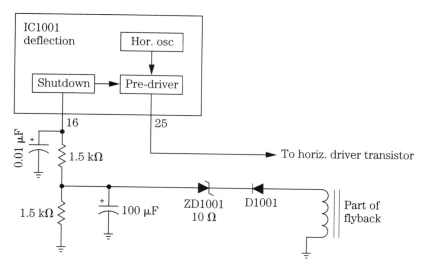

■ **33-22** *The high-voltage shutdown circuit is taken from a winding on the flyback and it is fed back to shut-down the horizontal oscillator or driver stage.*

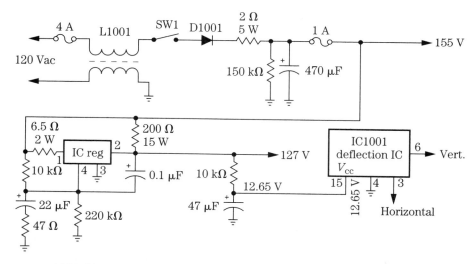

■ **33-23** *Horizontal circuits are much easier to service with the supply voltage taken directly from the low-voltage power supply instead of from the flyback scan-derived voltages.*

vice than the low-voltage power supply feeding a direct voltage to the horizontal countdown or oscillator circuits.

To determine if the horizontal oscillator or countdown circuit is operating, you must inject a dc voltage source to the horizontal supply voltage pin (V_{CC}). This dc voltage can be supplied from batteries or from a variable dc power supply. Then, take a scope waveform from the IC pin, which feeds the horizontal output driver transistor to determine if the oscillator and countdown circuits are normal. The ac power cord is removed for this test.

The chassis came up and shut down in a Wards model GBN-17632A TV. B+ voltage was found at the horizontal output transistor, directly from the low-voltage source. No supply voltage was found at the horizontal countdown IC700 (pin 9). Checking the schematic revealed the 25-V source was fed from the scan-derived secondary of the flyback (T402) transformer (Fig. 33-24). Sometimes the chassis would stay on a few minutes and then shut down. No screen symptoms were indicated when the raster collapsed.

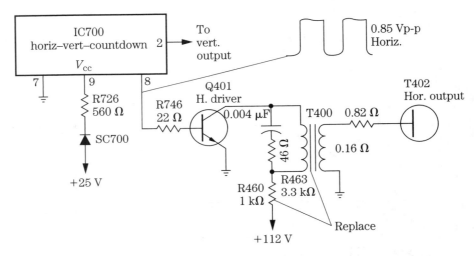

■ **33-24** *This Montgomery Wards TV had a shorted winding of the driver transformer and isolation resistor.*

To determine if the horizontal and vertical countdown IC was performing, a 9-V external voltage was applied to pin 9 of IC700. The horizontal scope waveform was fairly normal at pin 8. The IC700 horizontal output circuits were fairly normal. The drive waveform was traced up to the base terminal of the horizontal driver transistor (Q401).

Firing up the chassis once again indicated lower collector voltage at Q401. The voltage was checked at the junction of resistors R463 (3.3 kΩ) and R460 (1 kΩ) and was 97 V. Then, the chassis shut down. When in shutdown, the horizontal driver transistor and resistor R460 began to run warm. The voltage was under 33 V at the collector terminal of Q401. This was normal with no drive waveform. Because the driver transformer connections have caused shutdown problems in other chassis, T400 transformer connections were resoldered. In fact, when solder was sucked up from a poorly tinned primary transformer lead was located (Fig. 33-25). Removing the transformer and cleaning the leads did not cure the shutdown problem. Replacing T400's 27-Ω primary winding and replacing R460 solved the problem. T400 should have a resistance of around 46 Ω.

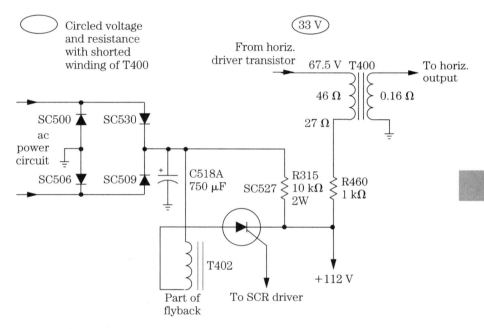

■ **33-25** *Usually, the driver transistor will pull heavy current if the horizontal waveform is not at the base terminal; sometimes the driver transistor, transformer, and the voltage-dropping resistor are destroyed.*

Conclusion

The tough repair problem might be difficult for one technician and easy for the next one. Intermittents with multiple defective parts are the most difficult. The technician who tries to learn new servicing techniques with correct test equipment can service most tough repair problems. Further your knowledge by attending manufacturer's service meetings to keep up to date with new products in the commercial electronics field.

How to repair microwave ovens

34

SERVICING MICROWAVE OVENS IS RIGHT DOWN THE ALLEY, so to speak, for the electronic television technician. Just about every major appliance manufacturer is building them. The microwave oven can be found in kitchens, motels, restaurants, and lunch rooms. Microwave ovens don't require the same amount of service as a TV, but they do break down and require service.

Microwave ovens are available in many power sizes, from 450 W up to 1300 W (Fig. 34-1). The small ovens are often used in campers and motel rooms, the 450- to 750-W ovens are found in the home, and 1000- to 1300-W ovens are used in restaurants. The oven might be controlled with a manual timer and automatic pushbutton control boards. The early ovens had doors that opened horizontally and some of the new doors might be opened downward like a regular kitchen oven.

■ 34-1 *The microwave ovens used in the home use anywhere from 450 to 750 watts of power.*

Power is applied to the magnetron tube through a sequence of buttons, interlock switches, oven relays, and triac assemblies. The magnetron tube radiates RF energy that is tunneled through a waveguide into the oven and cooks food placed in the oven cavity. Some ovens have a rotating stirrer fan that circulates the RF energy to cook all portions of the food. In other ovens, a turntable motor rotates the food on a glass tray, for even cooking.

Microwave energy

The microwave electromagnetic waves of energy are similar to light, radio, and heat waves. They travel in a straight line, can be generated, transmitted, reflected, and absorbed. The magnetron tube generates the microwaves. They are transmitted into the oven cavity and then absorbed within the food that is in the oven cavity; the food is cooked within seconds or minutes.

The magnetron tube has a cathode in the center of the magnetron and a filament or heater provides heat to emit electrons from the cathode element. The cathode is connected to the negative side of the power supply, which has a potential of 4000 V or more, with respect to the anode, which is connected to the positive side. The high voltage is produced by means of a high-voltage transformer winding, operating in a voltage-doubler circuit action of an HV capacitor and diode (Fig. 34-2).

■ **34-2** *The magnetron, HV transformer, HV capacitor, and diode, work in a voltage-doubler high-voltage circuit.*

Electrons are negative charges, which means they are attracted to the positive anode element and are repelled by the cathode. The electrons travel directly from the cathode to anode at a high rate of speed with a high positive voltage at the anode element. Two permanent magnets are found at the top and bottom of tube elements, resulting in the electrons traveling in a fast circular motion to the anode terminal. The anode element consists of many cavities. This circular motion by the electrons induces alternating current in the cavities of the anode.

When an electron is approaching one of the segments between the two cavities, it induces a positive charge in the segment. As the electrons go past and draw away, the positive charge is reduced while the electron is charging the next segment. The anode cavities resemble a form of a spoke-wheel pattern. The high-frequency energy, produced in the resonant circuit or cavities, is then taken out by the antenna and fed into the oven cavity through the waveguide.

The magnetron tube consists of heater, cathode, and anode elements. The heater or filament voltage (3.3 Vac) from another winding on the power transformer lights the heater element, which boils off the electrons from the cathode element (Fig 34-3). The elec-

■ **34-3** *The magnetron tube consists of heater, cathode, and anode elements with powerful magnets mounted on each end of the tube.*

trons are collected by the anode terminal and with the strong magnetic field of magnets and cavities pull the electrons at a rapid rate to the anode terminals, which is at ground potential.

The microwave wavelength is rather short compared to light. Three frequency bands are allotted for microwave operation by the Federal Communications Commission: 915 MHz, 2450 MHz, and 5500 MHz. New microwave ovens operate at 2450 MHz.

Oven components

The average microwave oven might consist of a blower motor, high-voltage capacitor, cook relay or triac, cook switch, buzzer, timer or control board, interlock switches, 15-A fuse, sensors, oven light, magnetron, HV diode, bleed resistors, stirrer motor or rotating turntable, thermal cutout, thermistors, transformer, and waveguide. The blower motor brings in fresh air and blows it across the magnetron to keep it cool during the cooking operation (Fig. 34-4).

■ **34-4** *The fan blower keeps the magnetron tube cool during the cooking process.*

HV diode and capacitor

The high-voltage capacitor and diode provide a voltage-doubler circuit, supplying voltages from a high-voltage winding of a power transformer, which doubles the raw ac voltage and applies a high negative dc voltage to the cathode terminal of the magnetron (Fig. 34-5). Both the HV capacitor and diode might break down and provide high-voltage leakage, which would blow the fuse. The secondary winding of the transformer, HV diode and capacitor, mag-

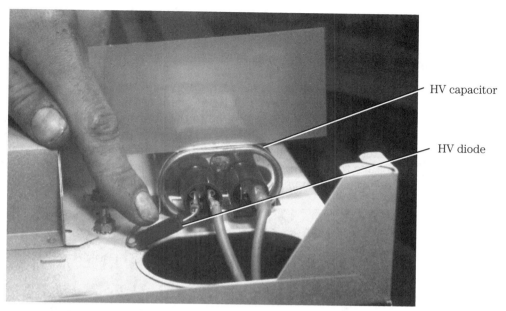

34-5 *The HV diode and capacitor with the secondary HV winding of the transformer doubler at the output voltage of the transformer.*

netron, and filament winding are found in the high-voltage circuits. Bleeder resistance can be found across the HV capacitor or diode to leak off high voltage when the oven is shut down.

The 15-A fuse protects the whole oven circuits and opens with a shorted monitor switch, leaky magnetron, and HV capacitor and diode (Fig. 34-6). The ceramic chemical fuse is a cartridge type

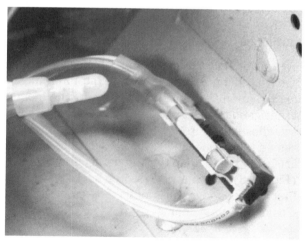

34-6 *The 15- or 20-A chemical fuse protects components in the oven circuits from a leaky or shorted component.*

that plugs into a fuse holder. Most oven fuses are 15- or 20-A chemical fuses.

Cook relay or triac

A cook relay or triac applies the 120 Vac to the primary winding of the power transformer (T1). Some ovens use a cook relay instead of a triac. Usually, triacs are found in the later microwave ovens. The cook relay voltage operates across the power line while the triac switching is controlled. A front control-panel circuit supplies gate voltage to make the triac elements to conduct applying 120 Vac to the primary winding of the power transformer (Fig. 34-7).

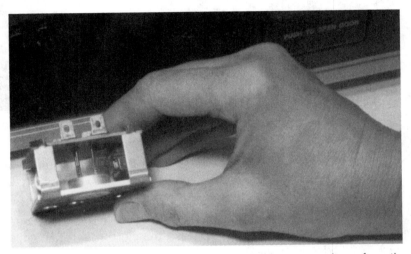

■ **34-7** *The triac assembly is controlled with a low gate voltage from the control panel board.*

Interlock switches

You might find three or more interlock switches mounted around the front door, to protect the operator and the circuits from receiving radiation from inside of the oven. The monitor or safety switch is designed to short out the 15-A fuse if one or more of the interlock switches hangs up or becomes shorted. Interlock switches are designed to provide electrical voltage to the oven when the oven door is closed and to open the electrical circuit when the door is opened (Fig. 34-8). One interlock switch backs up the other if one fails. The first and secondary interlock switches are activated by opening and closing the oven door.

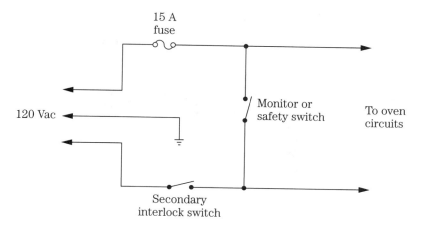

■ 34-8 *To protect the operator, interlock switches are mounted around the door area to shut down the oven if the other switches and interlocks fail to open.*

Switch or not to switch

The cook switch is pushed to start the oven process of cooking, which energizes the oven relay or triac assembly. Some ovens are controlled by a timer that can be set manually and will shut the oven off, ringing a bell. A piezo buzzer or speaker sounds when the cooking is finished. Besides these switches, the oven relay contacts and thermal cutout unit is in the low-voltage circuits. The thermal cutout might be found along the outside of the oven cavity or on the body of the magnetron (Fig. 34-9). Oven heat and gas sensors might be used in larger ovens.

The magnetron

The magnetron tube is the most important component in the oven and is the most expensive to replace. The magnetron filament might go open and shut down the flow of electrons. Higher dc voltages are found on the cathode element with open heaters, tube elements, or poor heater connections. If the magnetron shorts or becomes leaky, very little voltage is found at the cathode and a heavy load is placed on the high-voltage transformer. Sometimes the fuse will open and other times it won't, depending on the amount of leakage. No heat or no cooking is a common symptom of a defective magnetron tube.

Suspect that the magnetron is overheating and there is no fan blower operation if the oven cuts in and out. If the magnetron gets too hot, the thermal cutout mounted on the body of the magnetron

■ **34-9** *Here, the thermal unit is used outside of the oven cavity to shutdown the oven if the oven cavity overheats.*

will open the ac circuit to the power transformer. Suspect a defective magnetron thermal component if the oven operates intermittently (Fig. 34-10).

Stirrer or turntable motor

The stirrer motor is located at the top of the oven, operating a fan to stir or move the microwaves around the oven cavity. The stirrer fan might be operated from a separate motor or be driven by a belt from the fan blower assembly. Slow cooking occurs when the stirrer fan is not rotating. Check for a broken belt or a defective motor.

A turntable motor is located under the bottom area of the oven cavity, that rotates a glass tray with food. This provides even cooking in the oven cavity. The turntable motor operates in a reduction gear assembly and slowly rotates or turns the food. In the new ovens, the turntable motor might start up in either direction (Fig. 34-11). The gear-driven turntable motor might jam gears or contain a defective overheated motor.

HV transformer

The high-voltage transformer consists of a primary winding, high-voltage and filament secondary winding. You might find in the older ovens, a separate filament transformer besides the high-volt-

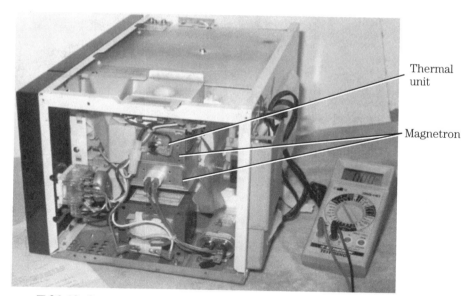

■ **34-10** *Suspect a defective thermal unit if the oven operates intermittently.*

■ **34-11** *The new turntable motor is mounted under the oven cavity and it might start up in any direction of rotation.*

age transformer. T1 provides heater and high voltage to the voltage-doubler circuit. The defective transformer might short out windings in the HV winding. You will find that the primary and filament windings are wound with large-diameter copper wire (Fig.

■ **34-12** *The defective high-voltage transformer might open or short coils inside of the HV winding.*

HV transformer

34-12). Poorly soldered or no soldered connections (where the primary leads attach to the coil winding) might go open. Poor heater leads on the magnetron heater or filament terminals might result in slow cooking.

Divide and conquer

The low-voltage circuits can be separated from the high-voltage circuits for easy troubleshooting. The low-voltage circuits consist of the fuse, interlock switches, timer, oven relay or triac, thermal cutout, and oven sensors (Fig. 34-13). An open fuse or interlock switch prevents the ac voltage from being applied to the primary of the power transformer. Likewise, a defective oven relay or triac might not close contacts or provide an ac voltage to the transformer. The open thermal cutout on the magnetron prevents ac voltage from reaching the primary winding.

Besides these components in the low-voltage circuits, there are the blower motor, oven light, turntable motor, cook light, and stirrer motor. Heating elements and a power relay are used in the low-voltage circuits of a convection oven. The convection oven might have a humidity sensor, heater fan motor, thermal heater cutout, and a latch switch. Of course, these components might become defective and cause the oven to shut down, but they do complete the low-voltage ac circuit. All components in the low-voltage circuits operate from the ac power-line voltage (Fig. 34-14).

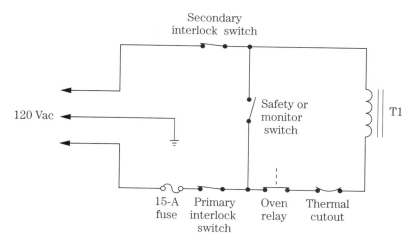

■ **34-13** *The low-voltage circuits consists of parts operating at 120 V to the primary winding of T1.*

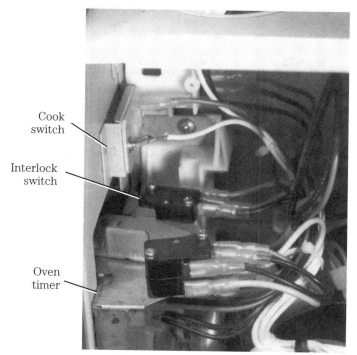

■ **34-14** *The oven start and cook switch operates in the low-voltage circuit with timer unit.*

High-voltage circuits

Components in the secondary winding of the power transformer might be called *high-voltage circuit parts* (Fig. 34-15). When 120 Vac from the low-voltage circuits is applied to the power transformer, the magnetic transfer of energy is applied to the filament and high-voltage secondary windings. Most magnetron heater voltages are around 3.3 Vac. The high ac voltage from the secondary winding forms a voltage doubler network with the high-voltage capacitor and rectifier.

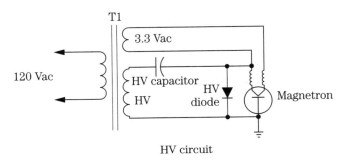

■ **34-15** *The high-voltage circuits include the HV winding of T1, the HV capacitor and diode, the filament winding, and the magnetron.*

Two separate voltages are applied to the magnetron. Low filament and high negative voltages are fed to one side of the heater terminals and a positive voltage is fed to the anode terminal. Actually, you are measuring the high negative voltage with the positive terminal at ground potential. Do not attempt to measure the filament or high voltage with the oven operating, without proper test instruments.

Discharge

Always discharge the high-voltage capacitor before attaching instruments or placing your hands and tools in the oven. This HV capacitor holds a charge that can knock you down and cause injuries to you and other nearby personnel (Fig. 34-16). Technicians who service ovens regularly should have a sign above the work bench to discharge that HV capacitor! You can get a terrible shock from the capacitor with the oven turned off.

Heavy amperage, large current, and high voltage is found in the microwave oven high-voltage circuits. Remember, high voltage is stored within the HV capacitor. Discharge the HV capacitor before and after each time the oven is fired up. Never attach test in-

■ **34-16** *Discharge the HV capacitor before attaching test instruments or placing tools in the oven circuits.*

struments while the oven is operating. The voltages found in the HV circuits are not as high as those found in the TV chassis, but the high current can cause serious injuries. Again, discharge the HV capacitor!

Required test instruments

Most of the equipment found on the electronics technician's service bench can be used in microwave repair. Besides a good Phillips screwdriver, some ovens require a Torx or LASTIX screwdriver for moving covers, power cords, and components. A handy light tester made with a 100-W bulb and pigtail socket can be used to check the low voltage at the primary winding of the power transformer. Clip it across open switches and cutout elements, and test the low-voltage circuits.

The small digital meter is handy when taking low-resistance measurements, continuity, and ac power line voltages (Fig. 34-17). The DMM is ideal when taking low-resistance continuity tests of resistors, diodes, transistors, and ICs on the control board. The ac voltage tests can be made with the DMM across the primary winding of power transformer triacs and oven relays. Never use the small meter to take voltages in the high-voltage circuits. Some VOM or DMM meters might go up to 3500 V, but they should never be used for high-voltage tests.

Remember, the oven voltage on larger ovens might exceed the maximum voltage (4.8 kV) range of an analog VOM or DMM and

■ **34-17** *The portable DMM is handy to take resistance and continuity measurements on the oven circuits.*

destroy the meter. When checking the HV diode, if it goes open or if you forget to return the diode to common ground, the ac voltage is now applied across the meter terminals. Usually, these meters are damaged beyond repair and might explode in your face. Use the small meter for resistance, continuity, and ac power line voltage tests in the microwave oven.

The magnameter

The magnameter is a specialized instrument especially designed to speed up servicing the microwave oven. You can quickly take high-voltage and current measurements without possible danger of high-voltage shock and injuries. Magnameter model 20-226 provides current and HV tests with one set-up. No changing of test clips is needed; you only need to read the voltage and flip the switch for current measurements (Fig. 34-18).

Most domestic ovens have a positive ground in the high-voltage circuit. The magnameter was designed to test both positive and negative ground ovens. There is no possible shock or arcover with test leads that are rated for up to 10 kV. No metal knobs or switches are exposed, preventing possible shock.

The meter can measure up to 10 kV in high-voltage measurements. Most domestic ovens operate between 1.5 and 3 kV, and commercial ovens operate from 2 to 4 kV. The toggle switch is

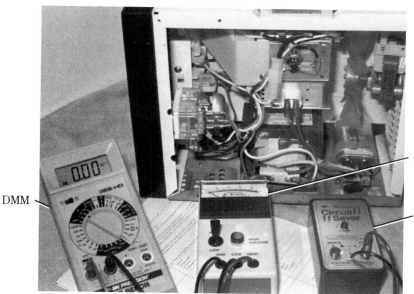

■ **34-18** *The DMM, magnameter, and the Circuit Saver will quickly service the microwave oven.*

flipped to high for voltage tests. Flip the toggle switch to the low position to measure plate current of the magnetron. You can easily locate a magnetron pulling heavy current, determine if the HV diode is leaky, if the HV capacitor is open or leaky, and if the magnetron tube is weak (Fig. 34-19). This meter will indicate if the high-voltage circuits are normal or defective.

Circuit saver

The circuit saver is designed to take the place of a blown fuse and is handy to have around, but it isn't necessary. You can save a lot of money on buying many 15-A fuses in intermittent and blowing fuse cases. These chemical fuses get quite expensive when servicing microwave ovens. The circuit saver is ideal when the oven intermittently blows the main fuse. Simply reset the circuit breaker and proceed again (Fig. 34-20).

Besides replacing the fuse, this tester can be used to substitute an HV diode within the oven. Just clip the circuit saver across the HV diode terminals. A red voltage warning light indicates that high voltage is present. When using the circuit saver for diode substitution, if the red light does not appear after a few seconds, shut down the oven. Check for leakage or other defective oven circuits.

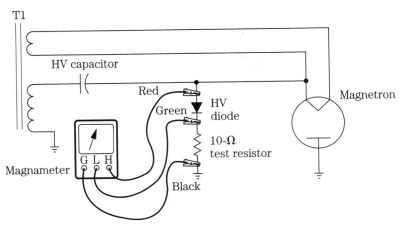

■ **34-19** *Connect the magnameter to the high-voltage diode and across the 10-Ω test resistor for current and HV measurements.*

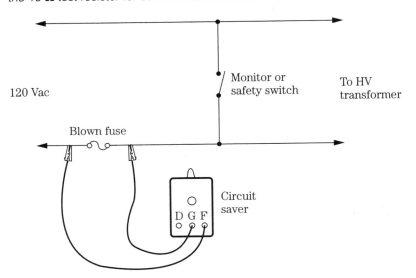

■ **34-20** *The Circuit Saver will save a lot of those expensive chemical fuses when servicing intermittent blowing of fuses.*

Before connecting the circuit saver into the oven circuits, discharge the high-voltage capacitor (Fig. 34-21).

Servicing the low-voltage circuits

Although with several interlock switches and relays, the microwave oven might look quite difficult to service. You can use

■ **34-21** *Simply push the red button on the Circuit Saver to reset the unit if an overload occurs.*

components in the oven circuits to tell what circuits are working. Use the fan blower rotation or the oven light to indicate low voltage (120 Vac) has reached this far (Fig. 34-22). If the oven light comes on, it indicates that voltage is present and that the interlock switches, fuse, thermal cutout, timer, and the oven switch are functioning.

If the fan blower is rotating and there is no oven action, suspect either a defective triac assembly, an open primary transformer winding, thermal cutout on the magnetron, or a defective oven relay. Monitor the low-voltage circuits by connecting a pigtail light or low-voltage VOM or DMM (120 Vac) across the primary winding of the power transformer (T1). If the indicator light or voltmeter does not light or show a power-line voltage, the triac, power relay, or control board is defective (Fig. 34-23).

Discharge the HV capacitor and then connect the various test instruments. It's best to clip the voltmeter across the primary winding and use the test light to indicate defective components. Clip the test light across the triac assembly. If the light comes on, the triac or no gate voltage from the control board is applied to the triac assembly. Suspect a leaky triac assembly if the oven comes on and will stay on and cannot be shut off until the ac cord is

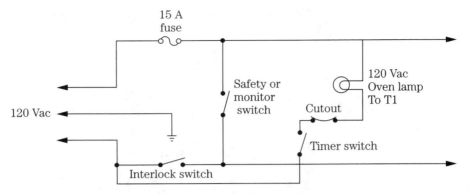

34-22 *The electronic technician can use the oven lamp and blower fan to indicate how far the 120 V travels in the low-voltage circuits.*

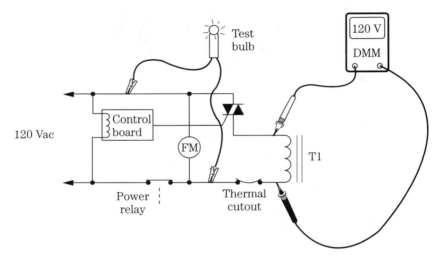

34-23 *Monitor the low-voltage circuits with a pigtail test light and DMM.*

pulled. The open triac should have no resistance measurement between MT1 and MT2.

Check the continuity between any triac terminals with VOM, DMM, digital VOM, or with the diode test of a digital meter (DMM). Switch the meter to the diode test and read the resistance between G and MT1. This measurement should be around 0.554 Ω in either direction with a normal triac. No reading is found in forward or reverse measurements between MT1 and MT2 (Fig. 34-24). A low-resistance measurement in both directions of any two terminals indicates a leaky triac assembly. The gate voltage might be somewhere between 5 and 10 V.

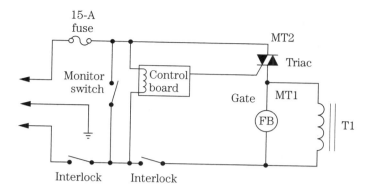

■ **34-24** *The triac component can be checked with resistance measurements to determine if there is any leakage between any two elements.*

Control board panel

The front pushbutton panel might control the auto defrost, timer, cooking, power, and start and stop operations. Cooking control is a feature for automatic food temperature control. It is used when cooking, reheating, or warming up food. If the temperature-control sensor is plugged in, the oven will operate until the food reaches a certain temperature. The control board might control time and power-level operation, temperature, and memory operation.

The control board is powered with a low-voltage transformer. The oven operations are controlled with a microcomputer, which in turn controls the voltage to the triac or to an oven relay. Besides the transformer, the microcomputer might control a pulse oscillator, buzzer, cooking display, variable-power relay control, power relay or triac, and membrane switch keyboard.

The panel control board is acceptable for lighting or power outage, which might destroy the critical components on the board. Suspect a defective control board with no display, flashing display, improper numbers and incomplete segments, improper countdown, time counting down and no cooking operation, oven stopping automatically, no action on time and temperature buttons, faint display, no temperature probe action, and oven stopping after the start button is pushed. Usually, the control board is replaced, not serviced. Most manufacturers request that control boards be sent in for repair. Always return the defective control board if it is within the warranty period. The low-voltage circuits can be checked in the control-board circuits, but it is a waste of time to proceed further.

High-voltage problems

Suspect a weak magnetron, improper heater terminals, and no stirrer fan operation if it takes a long time to cook the food. Discharge the capacitor. Connect the magnameter to the magnetron tube and the HV diode. Flip the switch to low and check for a poor current reading. Suspect a defective magnetron if the current is low and the high voltage is normal (Fig. 34-25). If the high voltage is low, check the HV diode and capacitor for leakage. An open HV capacitor will kill the voltage-doubler circuits. Inspect the heater terminals for a poor terminal connection. Clean the contacts and solder heater terminals to the transformer cables.

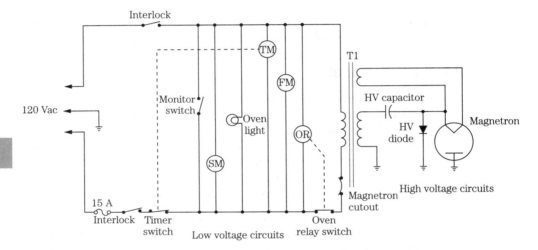

■ **34-25** *If low voltage is at the primary winding of T1, suspect a defective component in the high-voltage circuits if there is no oven cooking action.*

If the HV ceases and there is no magnetron current after the oven has been operating for several minutes, suspect that the thermal cutout is defective or that the fan blower is not operating. The magnetron might be overheating and kicking out the thermal unit mounted on the body of the magnetron. Usually, when the magnetron is overheating, excessive current will show on the magnameter.

The high-voltage circuits and current of the magnetron might be checked by inserting a 10-Ω resistor in series with the HV diode. Simply remove the ground chassis screw of the diode and clip the 10-Ω resistor to the diode and the chassis ground. A normal oven might operate at 1800 V and pull 300 mA of current, resulting in a 2.75-V measurement across the resistor (Fig. 34-26). Often, if the measured voltage across the resistor measures 2.5 to 5 V, the magnetron is normal. If the voltage is quite high across the resistor, the magnetron is pulling heavy current. Discharge the high-voltage capacitor before connecting the resistor or voltmeter in the oven circuits.

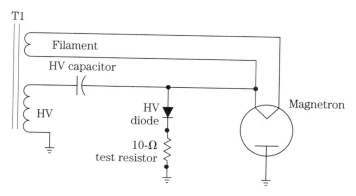

34-26 *Insert a 10-Ω 10-W resistor in series with the HV diode to measure the voltage and current of magnetron.*

Intermittent cooking

Monitor the power line voltage at the power transformer primary winding with an ac voltmeter. Connect the magnameter into the high-voltage circuit. The magnameter will indicate if high voltage and current are drawn in the high-voltage circuits. If the oven cuts out after being on for 10 to 15 minutes, suspect a defective thermal unit. Connect the pigtail light across the thermal cutout terminals (Fig. 34-27).

Notice if the light comes on when the oven shuts down. This indicates a defective thermal switch or the magnetron is overheating. Check the current and voltage measurements on the magnameter. If the current goes high when the oven shuts down, suspect a leaky magnetron. Check the intake air vents for plugged areas. The magnetron might be overheating with no fan motion. If it is defective, replace the thermal cutout and take another test. If the oven is still intermittent, replace the magnetron.

Critical resistance measurement

Discharge the HV capacitor. Measure the resistance from one side of the heater to ground or across the HV diode. No measurement should be found here unless a 10-MΩ bleeder resistor is across the

34-27 *Check the intermittent thermal unit with a pigtail light clipped across the terminals.*

HV diode. If the measurement is 10 MΩ with the bleeder, the HV diode and magnetron are normal. If a low measurement is found, remove the ground end screw of the diode and take another test. A low-resistance measurement (under 10 kΩ) indicates a leaky magnetron. Remove the heater terminals from the circuit and check resistance from the heater terminals to the metal (anode) of the magnetron or ground. Replace the magnetron if any signs of leakage are measured.

Oven leakage tests

According to Federal laws, the electronics technician should take an oven leakage test after all oven repairs. The leakage tester should be a certified meter, such as the Holaday Model H1-1500, Narda 8200, and the Simpson M380 model (Fig. 34-28). For a microwave oven in the customers' home, the U.S. government leakage standard is 5 mW/cm2. If the reading is greater, the oven should not be used and should be reported to the oven manufacturer.

Take microwave leakage tests around the front door and gap areas. Keep the probe perpendicular to the surface and slowly move it along the door area. Pull slightly on the door handle while the oven is operating and take another leakage reading. Oven door leakage might result from an improper seal or choke area, a warped door, or a loose door hinge. Take a leakage test around the door's window area. Doublecheck all air inlet openings and vented areas. List the oven leakage readings on all oven warranty repairs.

Conclusion

Discharge the HV capacitor before attaching instruments or working on the oven. Do not place tools in the oven while it is operat-

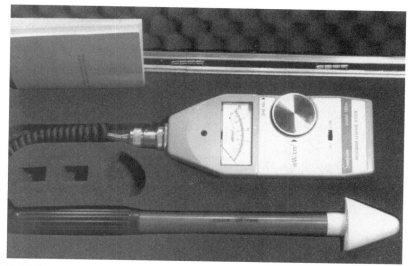

■ **34-28** *After repairing the oven, take a critical leakage test around the oven doors with a certified leakage tester (Simpson M380).*

■ **34-29** *Check the oven for possible voltage leakage and make sure that the oven is grounded with a three-prong plug to the metal cabinet.*

ing. Play it safe while servicing the microwave oven. Check for oven leakage between the metal cabinet or screen and the ground screw on the outlet receptacle. Check the center ground terminal of the cord to the metal oven cabinet for zero resistance measurement (Fig. 34-29). Take the leakage test along the front of the door seams and window area after the oven is repaired.

List of microwave oven manufacturers

Amana Refrigeration, Inc.
Amana, IA 52204
(319) 622-5511

Caloric Corp.
Topton, PA 19562
(215) 682-4211

Crosley Corp.
P.O. Box 1959
Winston-Salem, NC 27102

Emerson Radio
One Emerson Ln.
North Bergen, NJ 07047

Frigidaire Div. WCI
300 Phillips Rd.
Columbus, OH 43228
(614) 272-4100

General Electric Co.
Appliance Pk.
Louisville, KY 40221
(502) 452-4311

Goldstar Electronics Inc.
1050 Wall St. W.
Lindhurst, NJ 07071
(201) 480-8870

Hardwick Stove Co.
790 King Edward Ave.
Cleveland, TN 37311
(615) 478-4610

Hitachi Sales Corp. of America
401 West Artesia Blvd.
Compton, CA 90220
(213) 537-8383

Hotpoint
General Electric Co.
Appliance Park

Louisville, KY 40227
(502) 452-4311

J. C. Penney Co.
(Check at local store.)

Jenn-Air Corp.
3025 Shoddard Ave.
Indianapolis, IN 46226
(317) 545-2271

Kitchen Aid
701 Main St.
St. Joseph, MI 64501
(616) 982-4500

Litton Microwave Cooking Products
4450 Mendenhall Road S.
Memphis, TN 38101
(901) 366-3000

Magic Chef
740 King Edward Ave.
Cleveland, TN 37311
(615) 472-3371

Maycor (Division of Maytag)
240 Edwards St.
Cleveland, TN 37311

Maytag
One Dependability Sq.
Newton, IA 50208
(515) 792-7000

Modern Maid
Topton, PA 19562
(215) 682-4211

Montgomery Ward
(Check with local store.)

Norelco American Phillips
High Ridge Pk.
Stanford, CT 06903
(203) 329-5700

Panasonic Co.
One Panasonic Way
Secaucus, NJ 07094
(201) 348-7185

Quasar
1325 Pratt Blvd.
Elk Grove Village, IL 60007
(312) 228-6366

Roper Sales Co.
1507 Broomtown Rd.
Lafayette, GA 30728
(404) 638-5100

Royal Chef
Gary & Dudley
2300 Clinton Rd.
Nashville, TN 37209

Sampo of America
5550 Peachstreet 2nd Blvd.
Norcross, GA 30728
(404) 449-6220

Samsung Electronic Corp. America
One Samsung Pl.
Ledgewood, NJ 07852
(201) 691-6200

Samsung Electronics
301 Mayhill Street
Saddlebrook, NJ 07662
(201) 587-9600

Sanyo/Fisher, Inc.
21350 Lassen St.
Chatsworth, CA 91311
(818) 996-7322

Sharp Electronics Corp.
Sharp Plaza
Mahwah, NJ 07430
(201) 265-5600

Tappan Appliance Division
300 Phillips Rd.
Columbus, OH 43228

Tatung of America
2850 El Presidio St.
Long Beach, CA 90810
(213) 637-2105

Thermador/Waste King
5119 District of Bondward
Los Angeles, CA 90040
(213) 562-1133

Toshiba America, Inc.
Home Appliance Division
82 Otawa Rd.
Wayne, NJ 07470
(201) 628-8000

Whirlpool Corp.
2000 M-63 North
Benton Harbor, MI 49022
(616) 926-5000

White Westinghouse
Div. WCI
300 Phillips Rd.
Columbus, OH 43228
(614) 271-4100

Magnameter™ and Circuit Saver™ obtained from:
Electronic Systems, Inc.
624 Cedar Street
Rockford, IL 61102

35
Troubleshooting and repairing CD players

HOME, AUTO, PORTABLE, AND BOOM-BOX CD PLAYERS ARE available in the consumers electronic field. The auto CD player might be located in the trunk of the automobile. The home CD player operates from the ac power line and in the latest models might contain a CD changer, where five to eight CDs are stacked and loaded (somewhat like the phonograph). These home CD players might be top and front loaded, but the auto player is front loaded.

Portable CD players might operate from batteries or a power adapter. The boom-box might include an AM/FM/MPX radio, cassette, and CD player. Most portable and boom-box CD players are top loaded. The portable CD player might be listened to with headphones or via a line connection to the external audio amplifier.

Safety precautions

The service technician should avoid looking directly at the laser beam. Keep a CD on the disc platform at all times or keep your eyes at least 25 inches from the optical laser beam. Replace the critical parts with originals. Place a conductive mat under the test equipment and player while you are servicing it. Wear a wrist strap to leak off the body charges to the chassis (Fig. 35-1). Do not forget to remove shorting or interlock devices after repairing the CD player. Take critical leakage tests.

Remove the power cord from the power outlet and take a leakage test from the ac plug to the metal chassis screw. Short both power cord prongs together in this test. Connect one test lead of the ohmmeter to the ac plug and touch the other probe to the front metal screw cover or chassis. The required safety resistance should be less than 1.0 MΩ. If not, check for leakage of switch, wiring, ac capacitors, and power transformer.

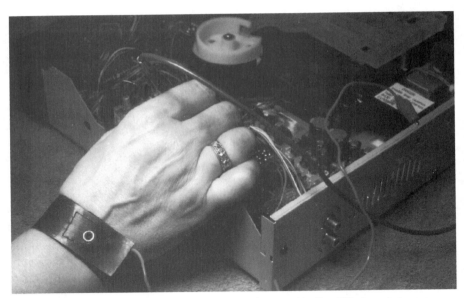

■ **35-1** *Wear an antistatic wrist strap and ground to the CD chassis to protect the delicate optical assembly and IC, microprocessor, and LSI from static electricity damage.*

A hot leakage test can be made between CD player and earth ground. You can use a water pipe, ac conduit, or metal junction box for this test. Check the current between the metal (earth) ground and the metal screw of the CD player with a current measurement. Plug in the ac cord and turn on the switch. The current should not exceed 0.5 mA.

Laser beam protection

Usually, interlock or safety switches are designed to disable the laser beam in the CD player operation. A safety switch is engaged and the player will not rotate unless the top plastic cover is closed in a portable CD player. You can look directly down on the laser lens assembly before loading the disc in portable players. In other CD players, the unit will not operate until a disc is loaded. Here, the LED light shines on a phototransistor and shuts down the laser beam without the disc loaded (Fig. 35-2). Remember to keep your eyes away from the laser beam while servicing the CD player.

Check the interlock switch on the portable CD player when the laser light and the disc platform rotate with the lid open. If the disc

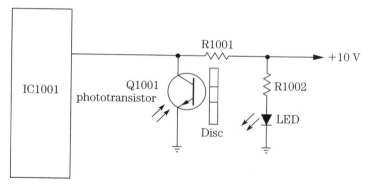

■ **35-2** *The interlock will shut off voltage to the optical pickup assembly if the disc is not loaded.*

motor starts to rotate without a CD loaded, suspect a defective LED or phototransistor circuit. Measure the voltage applied to the LED. Check both the LED and phototransistor with a diode-junction test of the DMM. No reading indicates an open LED or a phototransistor with reverse test leads. Leakage in both directions on the phototransistor indicates a leaky transistor.

Required test equipment

Besides hand tools and test equipment on the service bench, you might need several different pieces of test equipment.

- ☐ Optical power meter
- ☐ Dual-trace oscilloscope
- ☐ Digital multimeter (DMM)
- ☐ Signal generator
- ☐ AF oscillator or combination function generator
- ☐ Capacitance meter
- ☐ Frequency counter (Fig. 35-3)
- ☐ Test discs
- ☐ Schematics
- ☐ Special test tools, test jigs, wrist strap, etc.

You might have all of the test equipment listed, except for the power meter, special jigs, and alignment tools. Special test jigs can be purchased from the manufacturer that you do warranty CD repair for. Several test discs might be listed in the service literature; the most common ones are the YEDS7, YEDS18, and SZZP104E (Fig. 35-4).

■ **35-3** *The frequency counter might be used to check frequencies and alignment procedures in the CD player.*

■ **35-4** *The test disc is used in alignment tests. It should be returned to the original plastic package after the tests are completed.*

Compact disc cleaning

The compact disc should be handled at the edges or at the center hole with your thumb and finger. Handle it carefully so that you don't scratch the rainbow side, opposite of the CD label. The audio information is found on the plastic side and is placed downward with the labels upward when loading. The compact disc is made up of three layers of different material. A clear plastic material contains the audio information with tiny pits and islands of digital information. A reflective aluminum coating is placed over this. Acrylic resin is applied over the reflecting coating for protection.

Keep the CD and lens assembly clean of dirt and dust. Make sure that it is returned to the plastic sleeve or carton. Wipe the CD clean with a soft cloth starting from the center out. Do not wipe the CD in a circular motion. Many different CD cleaning kits are on the market. Some are wet and others are dry. These kits might include fluid in a spray bottle, a lamb skin wiper, cleaning cloth, and a brush. The automatic cleaner can be used with a wet or dry cleaning method. Some of these automatic cleaners are operated by hand or with several batteries. By cleaning the CD and lens assembly, skipping or audio fallout can be eliminated.

Surface-mounted components

Like the TV chassis and cassette players, the compact disc player has many surface-mounted devices placed on the PC wiring. You might find PC wiring on both sides of the main board in some players, such as the portable or compact disc player. Surface-mounted components found in the CD player might consist of ICs, microprocessors, LSIs, transistors, diodes, LEDs, capacitors, and resistors (Fig. 35-5). Some CD manufacturers might use only ICs and LSIs on the PC wiring side and regular parts on the top side. Often, the outline of the CD or LSI is found on the top side of the board with terminal numbers.

Dual op-amp components might be found to operate the motor circuits. One large op-amp IC might be used to drive the signal of several motors, focus and tracking coils. The signal- and system-control IC or LSI might have over 80 separate terminals. Check chapter 6 for more CD surface-mounted devices.

Laser pickup assembly

The laser optical assembly is one of the most critical and delicate assemblies in the CD player. The optical lens assembly might consist of objective lens, collimating lens, beam splitter, photodetectors, focus-tracking coils, monitor and laser diode. A basic optical assembly might have four photodetectors (ABCD), two tracking photodetectors (E and F), the focus tracking coils, monitor diode

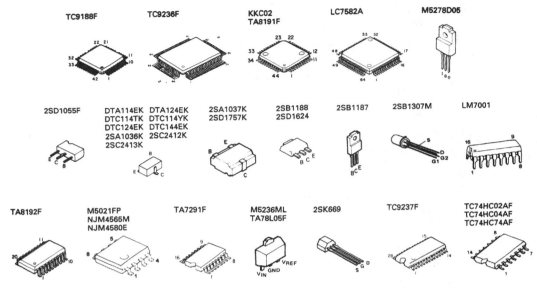

■ 35-5 *SMD solid-state devices are used throughout the PC board of a CD player, mounted on the wiring side.*

(MD), and laser diode (LD) (Fig. 35-6). You might find the automatic power control (APC) located in some pickup assemblies.

The photodetector diodes pick up the digital information from the rotating disc (EFM signal) and feed it into an RF amplifier. Because the pickup signal is very weak, an RF amplifier IC or transistors serve as preamplifiers. Besides supplying an EFM signal, the photodetector diodes (ABCD) provide tracking error signal. Sometimes the diodes are called *HF sensors*.

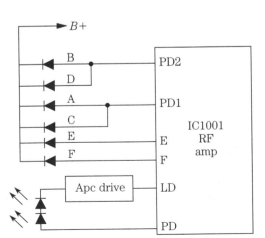

■ 35-6 *The photodetectors pickup the digital information from the disc, lens, and optical assembly, and they are amplified by RF amp IC1001.*

Troubleshooting and repairing CD players

The photodetector diodes (E and F) produce a tracking error signal for the error control circuits to keep the laser beam on track. If the laser beam moves away from the pits, the tracking error signal generates an error signal so that the beam spot correctly tracks the line of pits. In some players, the tracking photodetectors are called *tracking sensors*.

The laser diode emits a beam of light directly aimed at the pits on the disc surface. You cannot see this beam of light. Do not look directly at the laser lens assembly while servicing the CD player. Use extreme care while cleaning the lens assembly and pickup optical assembly. A laser optical diode is also beamed on a monitor diode to keep the lower output power at a constant level. The laser assembly might be located under a flapper assembly in some CD players (Fig. 35-7).

The laser assembly picks up encoded signals as a series of tiny pits with a spot of 1.6 micrometers in diameter. When the disc rotates and shines the beam on the pits, the photodetectors detect the absence of the pits within a fixed period of time. Changes in the re-

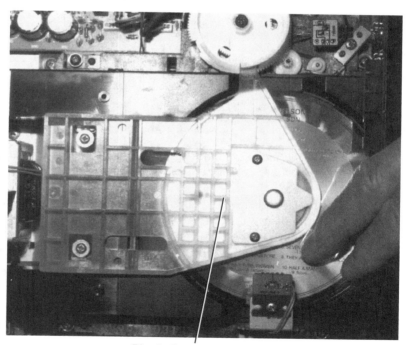

Plastic flapper assembly

■ **35-7** *A clapper assembly might be used to hold the disc in position after being loaded in the table-top CD player.*

flected light correspond to the recorded signals. The laser diode is a semiconductor diode made up of gallium-arsenide material, with a wavelength of 780 nM.

Focus and tracking coils

The focus and tracking coils are located in the optical assembly (Fig. 35-8). These coils keep the laser beam focused on the pits and precisely tracking as the disc rotates. These two coils will move and hunt when the CD player is first turned on. The beam from the pickup must remain focused on the disc surface to accurately read the pit information. If the focus on the pits drifts off, the focus servo circuits move the objective lens up or down to correct the focus. If the laser beam is reflected from the disc, it is directed to another lens and prism to the photodetector sensors, where it is split into four and forms a perfect circle.

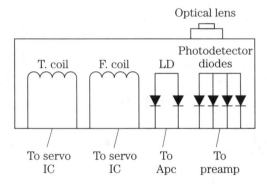

■ **35-8** *The focus and tracking coils are used in the optical assembly, along with the laser diode and photodetector diodes.*

The focus and tracking coils can be checked with the low-resistance measurement of the DMM. Around 30 Ω should be measured on the focus coil and 10 Ω for the tracking coil. If the focus coil resistance drops below 10 Ω and the tracking coil below 1 Ω, suspect a defective coil assembly. You might notice that if the ohmmeter leads are attached to either coil, the coil might shift or move, indicating normal operation. Remember that the tracking coil moves horizontally while the focus coil moves up and down.

The block diagram

The block diagram can be used to locate a possible defective circuit or stage as in the schematic diagram. You can trace out the signal paths from the RF amp to the audio line output circuits (Fig. 35-9). The tracking and focus error circuits can be traced from the RF amp to the servo and system control circuits. Although the block diagram might not point out the defective component, it will help to locate the defective section. Besides illustrating how the sections are tied together, the block diagram can be used to tell how the circuits operate in the CD player.

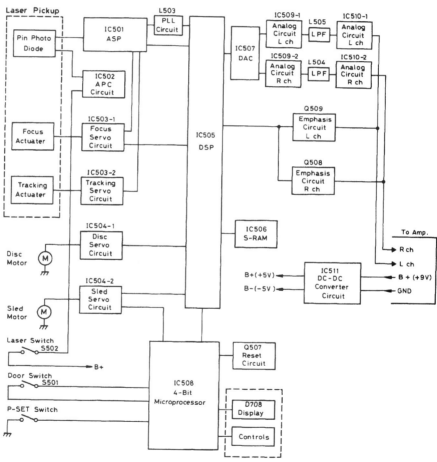

■ 35-9 *A block diagram of a portable CD player.*

Signal circuits

The signal circuits consist of an RF amp, signal processor, sample and hold, and a digital-to-analog (D/A) audio output circuit. You might find the RF amp and servo circuits in one large IC or LSI. The signal from the photodetector diodes is very weak and must be amplified by the RF amp IC before it is sent to the signal processor and servo IC. The RF amp might consist of several transistors in the early models; ICs are used in the present chassis. The RF amp might include track error, focus error, mirror hold, APC, FOK amp, FOX comparator, and EFM comparator (Fig. 35-10).

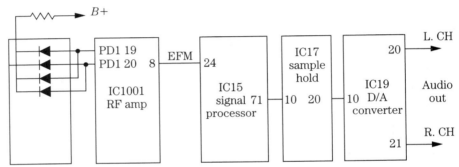

■ **35-10** *The RF amp amplifies the weak photodetector signal and feeds the EFM to pin 24 of the signal processor, and on to sample/hold and D/A converter.*

The 8-to-14 modulation (EFM) signal at pin 8 of the RF amp (IC1001) is fed to pin 24 of the signal processor (IC15). This EFM waveform can be found with the scope at pin 8 or a test point found on most RF amps. If the EFM signal is missing, suspect that the optical laser assembly or RF amp IC1001 is defective, or that there is an improper low voltage at the RF IC. If the EFM signal is not found at the servo IC, the CD player might automatically shut down. The EFM waveform is one of the most important signs in troubleshooting the CD player (Fig. 35-11). An EFM waveform at this point will indicate that everything is normal.

The EFM output is fed to pin 24 of a sample and hold IC17 and out at pin 20 to the D/A IC19. IC19 changes the digital signal to analog audio. The audio signal might be found at the line output jacks and fed to the earphone amplifier in some models. Audio from the line stereo output jacks might be cabled to the external audio amplifier. Muted transistors might be found at each stereo channel.

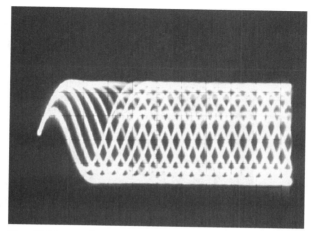

■ **35-11** *The EFM waveform or eye pattern must be used at the signal processor and servo system IC or the player will shut down.*

Signal processor circuits

The eye pattern or EFM waveform is sent to the signal processor IC that contains the clock generator, data latch, data concealment mute, digital filter, timing controls, EFM modulator, error correction, CLV servo, and the servo control. The interleaving and EFM signals are processed with signal processor or modulator. The interleaving data is memorized inside of the RAM IC and the data called in at exactly the same sequence as the original signals. Interleaving occurs when another program is included in segments so that both can be executed at the same time.

The PLL circuit is tied to the signal processor and consists of a VCO (voltage-controlled oscillator). The RAM and PLL or VCO circuits are in a separate IC tied to the signal processor. A test point at the VCO oscillator might indicate that the EFM signal or signal processor IC is operating in the signal path with a scope waveform (Fig. 35-12).

Auto focus and tracking error signals

Besides the EFM signal from the RF amp IC, the auto focus and tracking error signals are fed to the servo circuits to keep the beam in focus with accurate tracking. The beam from the laser pickup must remain focused on the disc surface to accurately read the information. If the focus strays off the pits, the focus servo moves the object lens up or down to correct the focus. When a beam is irradiated through a combination of cylindrical and con-

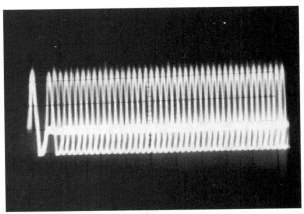

■ **35-12** *The PLL-VCO waveform is at the signal processor; it indicates that the signal is normal at this stage.*

vex lenses, the beam is elongated and then becomes a perfect circle. The focus error circuit is designed to detect changes in the distance to the disc and thereby ensures that the laser beam spot is kept in proper focus on the reflecting surface of the disc. The focus error (FE) signal developed in the RF amp is found at pin 2 and is fed to the servo circuits (Fig. 35-13).

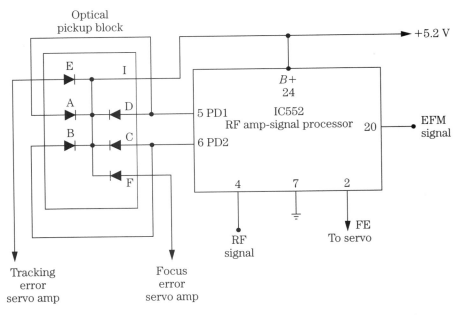

■ **35-13** *Besides the RF and EFM signal, the RF amp supplies a signal for the servo IC, focus error, and tracking error circuits.*

The tracking error (TE) circuit generates an error signal when the laser beam spot moves away from the center of the pits. TE signal is used to ensure that the beam correctly tracks the line of pits. This tracking error circuit is found inside of the RF amp or signal processor IC. The output TE signal is fed to pin 1 to the servo circuits for correction.

Servo circuits

The servo IC or surface-mounted LSI provides signal and power to the focus and tracking coil driver, sled or slide motor drive, spindle servo, tray motor drive, track jump, and search circuits. In some CD circuits, the tray or loading motor might be controlled by a system control IC. You might find that some of the servo control circuits are combined within the signal processor IC. The focus gain, focus offset, focus balance, tracking gain, tracking offset, and tracking balance adjustments are located in the servo system (Fig. 35-14).

CD adjustment controls

■ **35-14** *Check for focus gain, focus offset, focus balance, tracking gain, tracking offset, and tracking balance adjustments on the top side of the chassis.*

The TE and FE signals from RF IC are fed to pins 11 and 20 of IC102. Here, the sled motor is driven by Q206 and Q207 from pin 24 of IC102 (Fig. 35-15). The focus coil has driver Q201 and Q202 fed from an FEO signal at pin 20. TAO signal at pin 27 is fed to driver Q204 and Q205 to drive the tracking coil. In some chassis, the spindle or disc motor is driven by a common driver IC from the control system. A combination of transistors and IC drivers might be used in the various focus and tracking coil circuits.

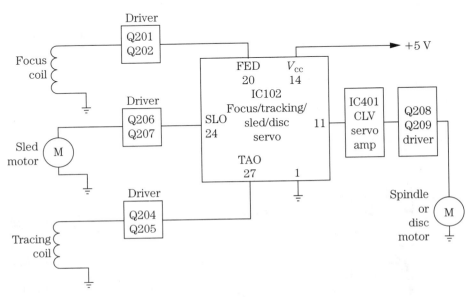

■ 35-15 *Servo IC102 drives the transistors and the IC driver components to the focus and tracking coil, sled, and spindle motor.*

CD motor circuits

Most CD players have three basic motors, loading, sled or slide, and disc motor. Within the boom-box and portable CD player the loading motor is eliminated with top loading. Portable CD players might have a disc and spindle motor. The auto and table-top CD changers might have four or more motors operating at different times (Fig. 35-16). You might find a slide, disc, up/down, magazine, and loading motor within the table-top changer.

Loading motor

The loading motor might be referred to as a *tray motor*. The tray or loading motor moves the loading tray in and out for loading and unloading the disc. The loading motor drives a plastic gear assem-

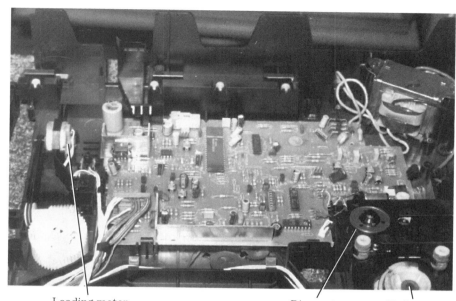

■ **35-16** *You might find four or more motors in the auto or table-top CD changer.*

bly that moves the tray assembly. This same plastic gear might raise and lower the clamper assembly. If the tray moves outward, the clamper raises and after loading, it adds pressure on the loaded CD and holds it in position.

The tray open and close switch is found on the front panel of the CD player and is fed into the servo control IC to move the loading motor. The output from the servo is fed to a driver IC or transistor that provides a voltage to operate the loading motor (Fig. 35-17). A positive and negative voltage from driver IC205 is fed to the motor, to open or close the tray, or change the direction of or the rotation of the motor. The defective loading motor can be checked with voltage or low-resistance tests.

Disc and SLED motor circuits

The disc motor might also be called a *spindle* or *turntable motor*. This disc motor starts to rotate after the disc is loaded. At the end of the disc motor shaft, a small platform is mounted for the disc to sit on. The disc starts out at 500 rpm and slows down as the laser pickup moves toward the outside disc (200 rpm). IC107 drives the disc motor with the signal from the servo IC105 (Fig. 35-18).

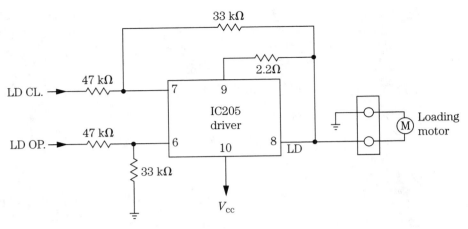

■ **35-17** *The signal from the system-control IC might have an IC driver to the loading motor.*

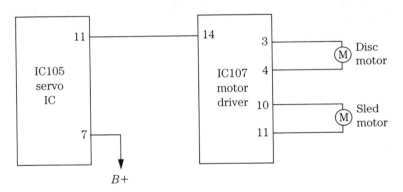

■ **35-18** *The disc and sled motor are driven by a common motor driver IC (IC107).*

You might find that the sled, slide, feed, or carriage motor moves the optical assembly across the disc area, from the center to the outside edge. Usually, this motor is gear driven to a rotating gear that moves the laser down two sliding bars or rods. The feed motor might operate from a transistor, IC, or from the same IC as the disc motor within the portable CD player. The whole boom-box PC board and CD mechanism mounts at the top of the cabinet (Fig. 35-19).

Check the disc and slide motors with low-resistance continuity and voltage tests. Check the driver IC with voltage tests if the motors appear to be normal. These small motors can be checked with a C battery by removing motor leads from the circuit.

CD   Rails
PC board

■ **35-19** *The CD chassis and optical laser assembly are mounted at the top of a boom-box combination CD player.*

Other motors

Besides the slide and disc motors, you might find the up/down, magazine, and loading motor in the table-top CD changer (Fig. 35-20). The up/down and magazine motors assist in loading and playing the disc while the magazine motor rotates the turntable (carousel) or changes the different discs for playing. The magazine motor might be called the *turntable motor*. Both of these motors might operate directly from the dc voltage source or from the motor-driven IC. In several carousel CD players, the carousel, loading, and chucking motors operate from a system-control microprocessor.

Low-voltage power supply

Check the low-voltage power-supply source if any circuit or motor fails in the CD player. Improper or no voltage source can make each circuit fail. Most electronics technicians check for dc voltages before proceeding to the suspected circuit. A quick voltage test across the large filter capacitor will indicate if the dc circuits are active. Then, check the voltage supply pin (V_{CC}) of the IC, transistor, or motor ter-

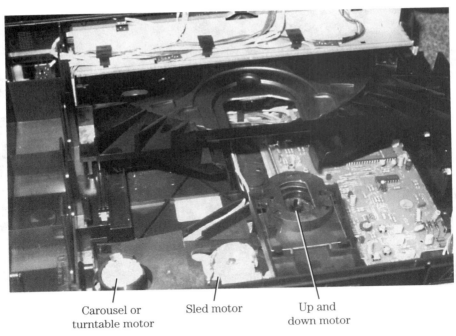

■ 35-20 The CD turntable or changer might have an up/down, magazine, and chucking motor besides having the regular disc and loading motor.

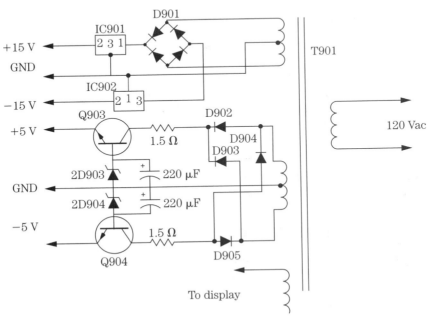

■ 35-21 The low-voltage power supply might have full-wave, bridge, and half-wave rectification with transistor, IC, and zener diode regulation.

Troubleshooting and repairing CD players

minals (Fig. 35-21). Check the motor voltage at the PC board terminal connections instead of trying to measure it on the motor terminals, in case the motor is buried or is under the main PC board.

The low-voltage power-supply circuits might consist of many circuits with several different dc output sources. Transistor, IC, and zener diode regulation is found in the low-voltage circuits. You might find bridge, full-wave, and half-wave rectifier circuits in the low-voltage power supply. Improper voltage sources might result from leaky regulators, electrolytic capacitors, resistors, overloads, and diode rectifiers. Suspect a defective silicon diode or leaky filter capacitor if the fuse keeps blowing. Doublecheck the voltage regulator transistors for leakage and burned zener diodes. Do not overlook damaged resistors in the regulator circuits.

Check the supply voltage at the suspected IC or transistor and then check the voltage source. Notice if an IC or transistor is used as the voltage regulator (Fig. 35-22). Leaky voltage regulators will have low output voltage; if the regulator is open, they will have no output voltage. Often, zener diodes will overheat and become leaky or shorted. Replace both transistor regulator and zener diode if the diode shows signs of overheating or leakage. Dried-up

■ **35-22** *Transistor and zener diode voltage regulators are used on the top side in a CD player.*

electrolytic filter capacitors will result in a very low voltage source. Check capacitors with a capacitor tester.

Audio circuits

The audio circuits might consist of a D/A converter, audio amp, line output, and a headphone jack. A digital signal is fed into the digital-analog (D/A) IC, where it is converted and separated to audio stereo channels. Each audio channel is filtered and muted in most CD players. The audio (when muted) is cut off in changing different operating modes and if no audio is heard in the channel. Separate audio amplifiers might be provided for headphone operation (Fig. 35-23). Headphone amps are usually used in portable and boom-box CD players.

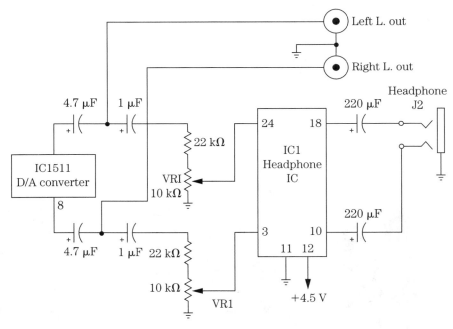

■ **35-23** *A separate IC might contain both stereo channels in the headphone audio circuit of CD player.*

The headphone amplifier is taken from the line output signal, where it is amplified and controlled to a headphone jack. The IC amp output might have a muted transistor in each channel. In the boom-box player, the audio output signal is taken from the line output and is switched into the regular radio and cassette amplifier. You might find a muting system for both line output and headphone operation. In some CD players, muting is automatic when the disc stops, when it is turned on, when it is accessing operations, and when it is paused (Fig. 35-24). The muting voltage to turn off and on the mute transistors might come from the muting or system-control IC. Audio can be signaltraced from the D/A output through the line and headphone circuits with the external audio amp.

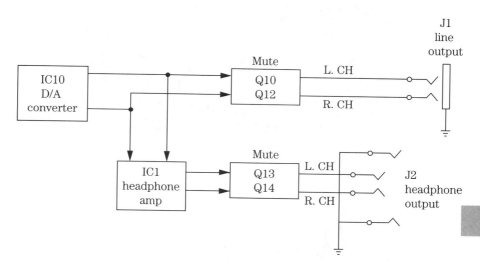

■ **35-24** *You might find transistor muting in both the line and headphone output circuits.*

CD player adjustments

Besides the DMM, the oscilloscope, AF oscillator, frequency counter, test discs, and various scope bandpass filter test circuits are needed for complete CD electronic adjustments (Fig. 35-25). Choose a laser power meter to ensure that the laser diode operates properly. By taking a laser diode radiation test with a power meter, you know that the laser diode is normal. If the diode wavelength wattage readings are low, the whole optical assembly must be replaced, resulting in a very expensive repair. Usually, if the laser diode is less than 0.1 mW, the service life of the diode has expired. Immediately take a laser power measurement to determine if the CD player can be serviced (Fig. 35-26).

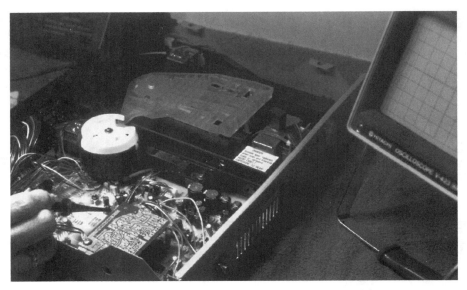

■ **35-25** *The oscilloscope and frequency counter are used in compact disc alignment procedures.*

■ **35-26** *Determine if the compact disc player optical assembly is operating if there is a critical power meter test.*

RF/HF/EFM tests

Adjust the laser gain control until the EFM signal is at least 1.3 V p-p on the oscilloscope. Check each manufacturer's literature for correct adjustment. If the radio frequency (RF), high frequency

(HF), and EFM signal is not found at the output of the RF amp, the CD player will shut down. Sometimes the EFM signal or waveform can be seen before the unit shuts down, indicating that the optical assembly and RF amplifier are operating properly. If the EFM signal is not found at the output of that RF amp, suspect a defective transistor, IC, or optical assembly, or suspect that the supply voltage is improper. The HF or eye pattern should be clean and distinct (Fig. 35-27).

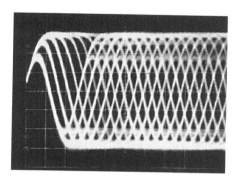

■ **35-27** *The HF or EFM eye pattern should be clear and distinct.*

PLL/VCO waveform

The PLL/VCO circuit consists of an 8.6456-MHz voltage-controlled oscillator and can be adjusted with the frequency counter. The PLL/VCO waveform indicates that the signal processor is normal. Check with each manufacturer for the correct PLL/VCO adjustment. Some manufacturers use a scope instead of a frequency counter for this adjustment.

Focus and tracking offset adjustments

The focus offset adjustment might be the same as the RF or EFM adjustment in the CD player. This adjustment might be called the *jitter* or *eye pattern adjustment*. Adjust the focus offset adjustment for less jitter and a clear-cut diamond-shaped opening. Use a test tape when making these adjustments.

The tracking and focus error adjustments are taken across each coil with the oscilloscope. For the focus gain adjustment, connect scope terminals across the focus coil. Play a test disc. Adjust the focus gain between 500 and 600 mV p-p or adjust it with the manufacturer's recommendations (Fig. 35-28).

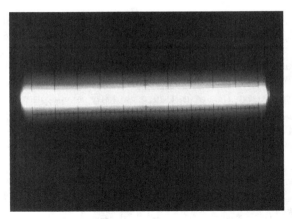

■ **35-28** *The waveform of a focus error signal at the focus coil winding.*

For tracking error or gain adjustments, connect the scope across tracking coil terminals. Play the test disc. Adjust the tracking gain control for a 1.8 to 2.2 V p-p across the tracking coil (Fig. 35-29). Check each manufacturer's correct tracking error adjustments.

If the sound skips when the player is jolted or bumped, the tracking gain might be set too close or be too small. If a small scratch appears on the test or play disc, suspect that the tracking gain is set too high.

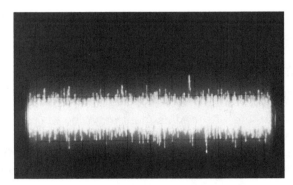

■ **35-29** *The waveform taken of the tracking error signal at the tracking coil winding.*

Troubleshooting CD players

Take critical waveforms and voltage measurements when servicing the CD player. Remember that no EFM signal might cause the player to shut down. Try to obtain service literature and a schematic from a local CD distributor, service depot, or manufacturer. Follow the signal circuits with waveforms to solve the many CD symptoms. In fact, the servo waveforms can be traced right to the focus and tracking coils, driver IC, or transistors of nonoperating motors, and audio can be signaltraced from the D/A converter to the line output or headphone jacks (Table 35-1). Keep a file or mark the symptoms and actual defects on the schematic to solve the many CD symptoms. Servicing the compact disc player is just another electronic product that the electronics technician is required to repair.

■ **Table 35-1 CD troubleshooting chart.**

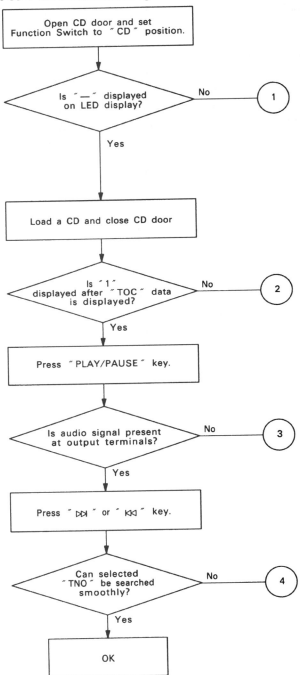

Troubleshooting and repairing CD players

[Repair Item 1]

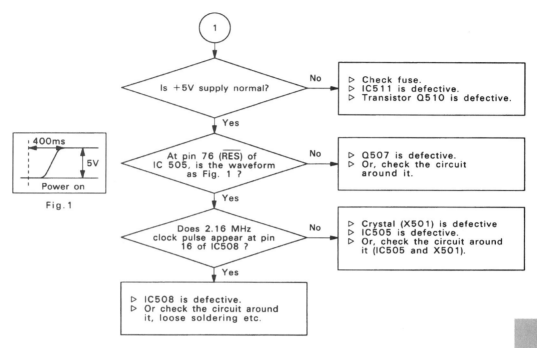

[Repair Item 2]

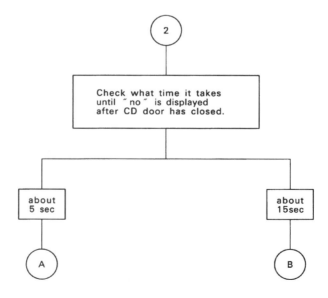

Troubleshooting CD players

■ Table 35-1 Continued

[Repair Item 2-A]

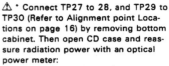

⚠ * Connect TP27 to 28, and TP29 to TP30 (Refer to Alignment point Locations on page 16) by removing bottom cabinet. Then open CD case and reassure radiation power with an optical power meter:
If an optical power meter is not available watch the radiation through the lense of pickup unit with distance more than 30 cm from Pickup Unit.

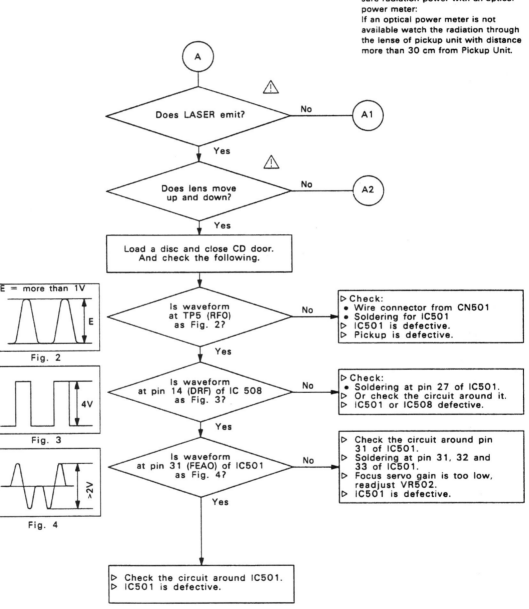

[Repair Item 2-A1]

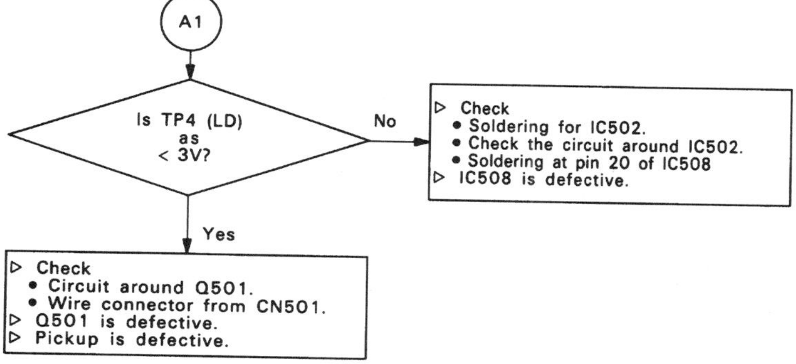

```
        (A1)
         │
         ▼
   ┌───────────┐
   │ Is TP4 (LD)│   No    ▷ Check
   │    as     ├──────►     • Soldering for IC502.
   │   < 3V?   │            • Check the circuit around IC502.
   └─────┬─────┘            • Soldering at pin 20 of IC508
        Yes                ▷ IC508 is defective.
         ▼
   ▷ Check
     • Circuit around Q501.
     • Wire connector from CN501.
   ▷ Q501 is defective.
   ▷ Pickup is defective.
```

[Repair Item 2-A2]

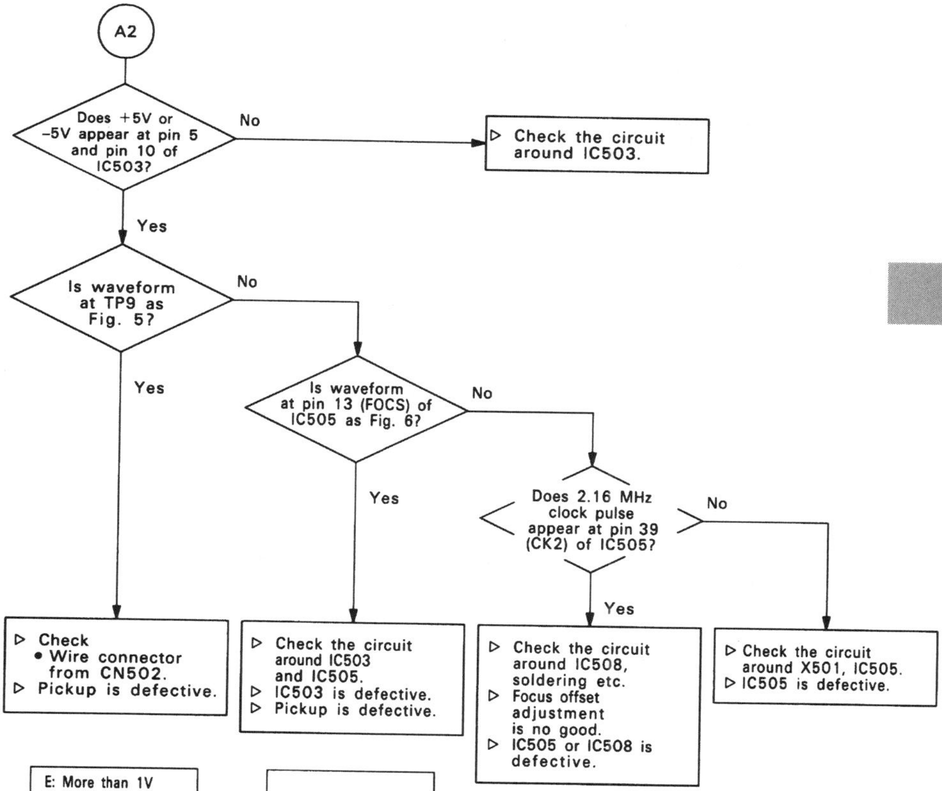

```
          (A2)
           │
           ▼
   ┌──────────────┐
   │ Does +5V or  │  No    ▷ Check the circuit
   │ −5V appear at├─────►    around IC503.
   │ pin 5 and    │
   │ pin 10 of    │
   │ IC503?       │
   └──────┬───────┘
         Yes
          ▼
   ┌──────────────┐
   │ Is waveform  │  No
   │  at TP9 as   ├─────┐
   │   Fig. 5?    │     │
   └──────┬───────┘     │
         Yes            ▼
                 ┌──────────────┐
                 │ Is waveform  │ No
                 │ at pin 13    ├──────┐
                 │ (FOCS) of    │      │
                 │ IC505 as     │      ▼
                 │  Fig. 6?     │  ┌──────────────┐
                 └──────┬───────┘  │ Does 2.16 MHz│ No
                       Yes         │ clock pulse  ├─────┐
                                   │ appear at    │     │
                                   │ pin 39       │     │
                                   │ (CK2) of     │     │
                                   │ IC505?       │     │
                                   └──────┬───────┘     │
                                         Yes            │
```

▷ Check
  • Wire connector from CN502.
▷ Pickup is defective.

▷ Check the circuit around IC503 and IC505.
▷ IC503 is defective.
▷ Pickup is defective.

▷ Check the circuit around IC508, soldering etc.
▷ Focus offset adjustment is no good.
▷ IC505 or IC508 is defective.

▷ Check the circuit around X501, IC505.
▷ IC505 is defective.

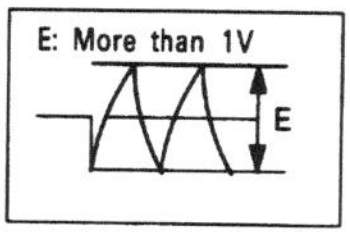

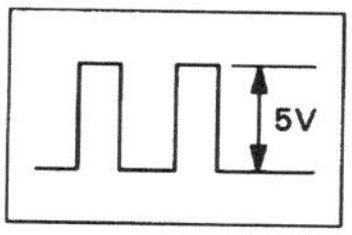

Fig. 5 — E: More than 1V

Fig. 6 — 5V

Troubleshooting CD players

■ **Table 35-1 Continued**

[Repair Item 2-B]

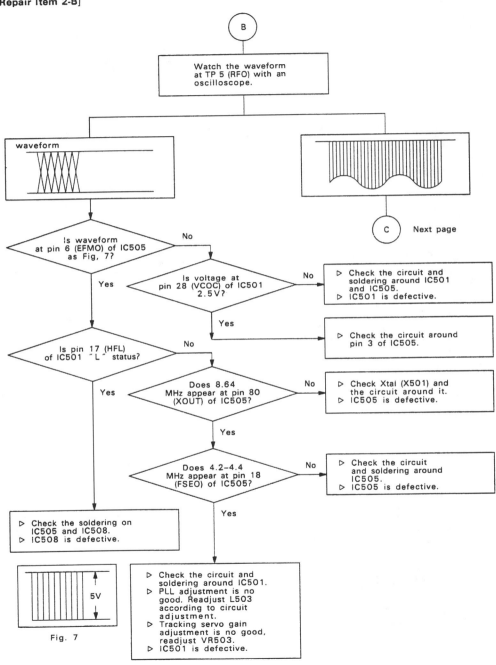

Troubleshooting and repairing CD players

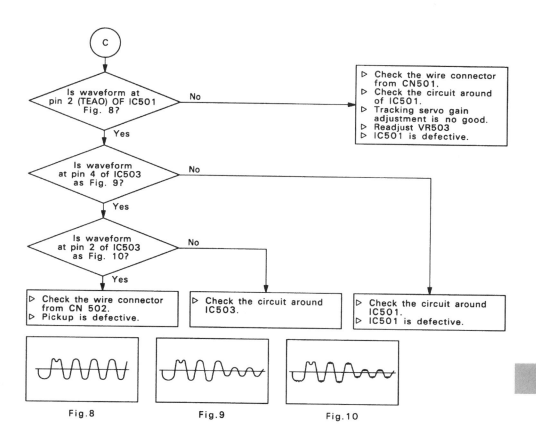

■ Table 35-1 Continued

[Repair Item 3]

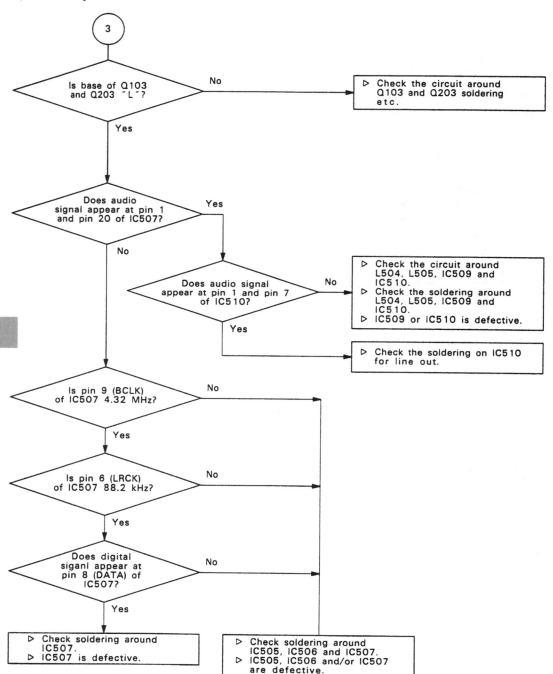

Troubleshooting and repairing CD players

[Repair Item 4]

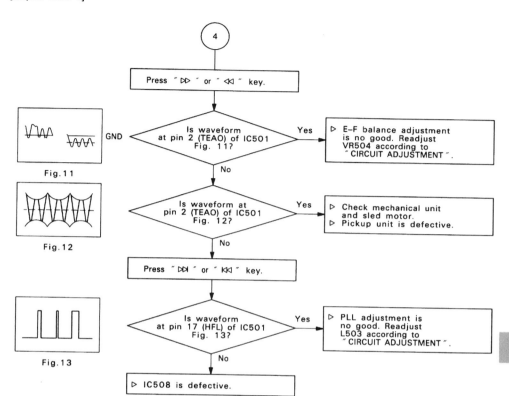

36

VCR and camcorder repair

THE RECORDING AND PLAYBACK CIRCUITS ARE SIMILAR IN the VCR and camcorder, except that the camcorder has a camera section. The VCR picks up the TV signal with a varactor tuner, processes the signal and records the program on the tape. While the camcorder lens and CCD imager takes a picture, the signal is processed and the picture and sound are recorded on the tape (Fig. 36-1). A VCR plays back the recorded tape through the TV set or monitor, and the camcorder can play back through the TV.

■ **36-1** *The VHS camcorder mechanism operates in the same manner as the VCR—only on a smaller scale.*

VCR and camcorder block diagram

The overall block diagram of the VCR may consist of an RF converter, tuner, on-screen display, VIF/SIF/AV, video in select, video signal processing, video preamp, audio in select, audio R/P amp, display, IR receiver, reset, microcomputer, digital servo, ML drive, reel sensor, and video heads (Fig. 36-2).

The 8-mm camcorder block diagram might consist of a CCD imager, timing generator, sync generator, process auto, matrix, white balance switch, and encoder. The VTR or VCR section consists of a servo, drum driver, capstan driver, loading motor driver, FG amp, pre-emphasis, audio amp, timing phase, rec/PB amp, flying erase head, drum and heads, microphone, and system control (Fig. 36-3).

VCR VHS and Beta formats

The VHS and VHS-C formats use the same size tape for record and playback operations. The video home system (VHS) and compact cassette (VHS-C) play in the same VCR. The VHS cassettes are low in price and can be played back in camcorders and VCRs. In standard play (SP), you can get two hours of recording on a standard T-120 cassette. You can get up to 6 hours of recording in extended play (EP) mode. The super and HQ recordings provide more lines of resolutions, which means that the pictures are more detailed.

The Beta VCR uses Beta recording tape. The Beta cassette is smaller and can play from 15 minutes up to 5 hours of recording. The L-750 cassette is slightly smaller than the standard VHS cassette. The Beta cassette will not fit into the VHS or VHS-C machines or vice versa (Fig. 36-4).

Camcorder VHS, VHS-C, and 8-mm formats

A VHS-C cassette uses the same size tape width as VHS, but in a smaller container. The VHS-C cassette is inserted into the VHS adapter and then plugged into the VHS camcorder or VCR. VHS and VHS-C both record and play back in VHS camcorders and VCRs. The same playing time is found in both VHS VCR and camcorders. The compact VHS-C camcorder is small in size, light to carry, and easier to take on vacations or outside for viewing. The VHS and VHS-C tape is 12.6 mm wide.

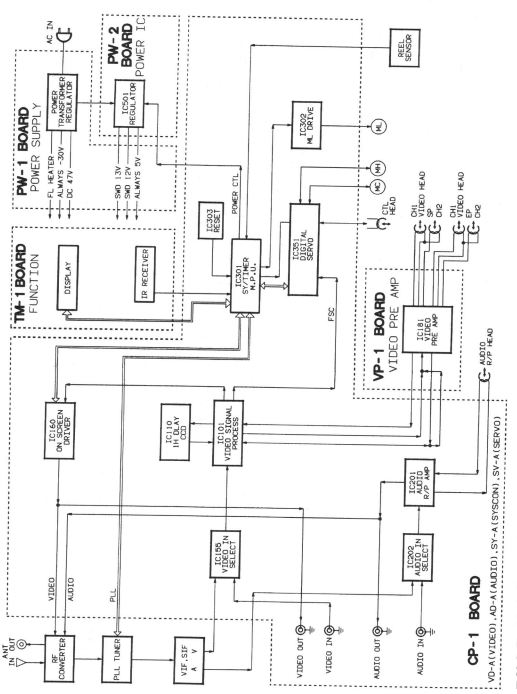

36-2 The block diagram of a double-azimuth 4-head VHS VCR.

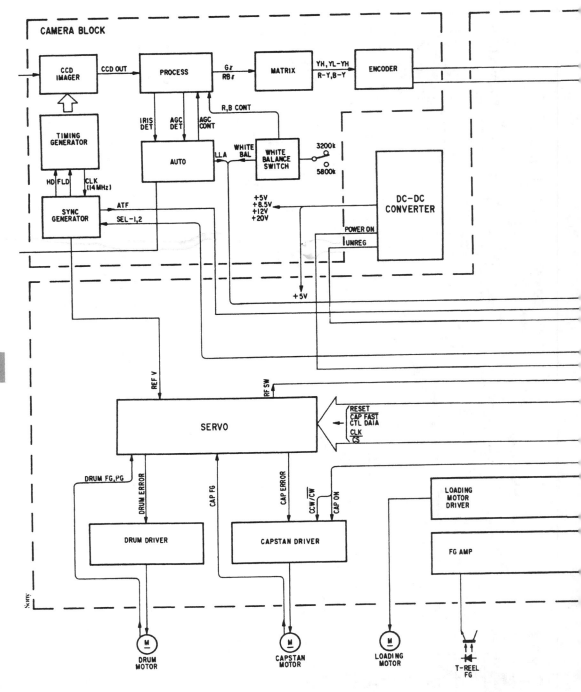

36-3 The block diagram of an 8-mm camcorder.

VCR and camcorder repair

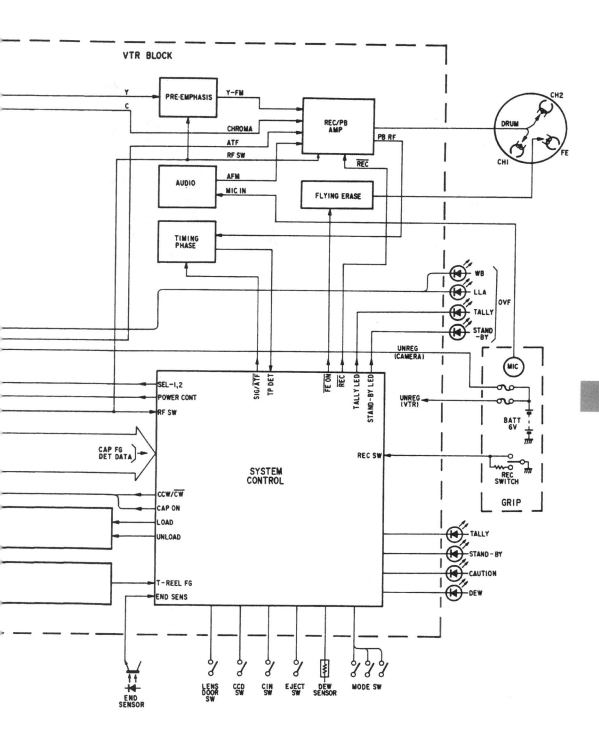

Camcorder VHS, VHS-C, and 8-mm formats

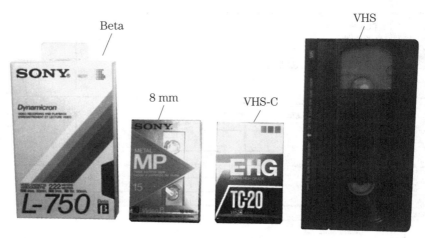

■ **36-4** *Here are 8-mm, Beta, VHS-C, and VHS cassettes.*

The 8-mm cassette operates with a small lightweight format and thinner tape. The 8-mm videocassette operates 15 minutes up to 2 hours playing time. You can play the 8-mm camcorder back through the camcorder viewfinder and through the TV set. This 8-mm cassette cannot be played through a VHS or VHS-C camcorder or VCR. The 8-mm cassette tape is only 8-mm wide.

VCR tape heads

The Beta VCR has a minimum of two heads at 180° apart. The drum or head rotates counterclockwise at a speed of 30 revolutions per second. The tape wrap-around is somewhat greater than 180°. The tape moves the same direction as the video heads. The video head consists of a top and bottom drum or cylinder assembly (Fig. 36-5).

The VHS cylinder head consists of a top-moving cylinder or drum and a stationary (stator) lower cylinder. The lower cylinder contains the motor part of the tape head. The upper cylinder or drum in the VHS and VHS-C video recorders and camcorders have a 41-mm diameter with four heads (Fig. 36-6). The tape wrap for this system is 270° and it rotates at a speed of 2700 rpm in the camcorder. Head switching is done in the record and playback modes of the VHS and VHS-C models.

The camcorder 8-mm drum or cylinder is smaller in diameter and contains two channels with a flying erase head (FE) circuit. The FM audio signal is recorded along with the video, rather than at the edge of the VHS tape (Fig. 36-7).

■ **36-5** *The video head in the VCR consists of an upper drum or cylinder and a lower drum. The top half rotates.*

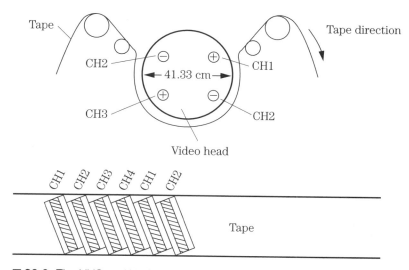

■ **36-6** *The VHS and VHS-C cylinder video head configuration in the camcorder.*

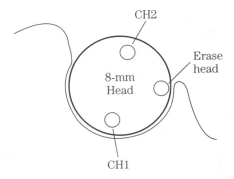

■ **36-7** *Two video heads with a "flying" erase head are used in an 8-mm camcorder.*

Tape head cleanup

Several methods are used to keep tape oxide dust from packing on the tape head, from a cleaning stick and solvent to cleaning cassettes. A video head cleaning fluid can be used with a cleaning cloth to keep the drum clean (Fig. 36-8). If the dirt on the video head is too stubborn to be removed by a cleaning cassette or cartridge, use a cleaning stick and solvent. Some VCRs have an automatic tape head cleaner built inside to keep oxide from building up on the heads and the tape.

■ **36-8** *The video and tape heads can be cleaned with chamois, cleaning spray, cloth, and methyl alcohol.*

Some manufacturers recommend a chamois cleaning stick with solvent. Move the videotape drum or cylinder to a position at which the video heads are easy to clean. Wipe each head several times with the head cleaning stick. When pressing your finger on the head cleaning stick to clean the heads, do not press too hard (Fig. 36-9). Do not move the cleaning stick up and down against the video heads and cause damage. Other VCR manufacturers recommend that you should clean the complete cylinder and video heads with a leather chamois (moistened with methyl alcohol) wrapped around your finger. Move the chamois leather to the left and right several times horizontally, to clean the video heads.

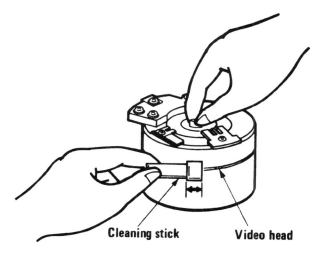

■ **36-9** *Cleaning the video heads with a cleaning stick while turning head with the other hand.*

Automatic head cleaner

In some of the latest VCRs, the auto head cleaning mechanism is a lever cleaner assembly. A roller cleaner assembly is attached to the top of the lever cleaner assembly so that it cleans the rotary head by pressing against the drum (cylinder) when it is loading and unloading (Fig. 36-10). The lever cleaning assembly is always pressed against the circumference of the main cam by the torsion-coiled spring.

VHS tape-loading paths

After insertion of the cassette, tape is normally wound to the head drum based on a full-loading system in the VHS VCR machine. The

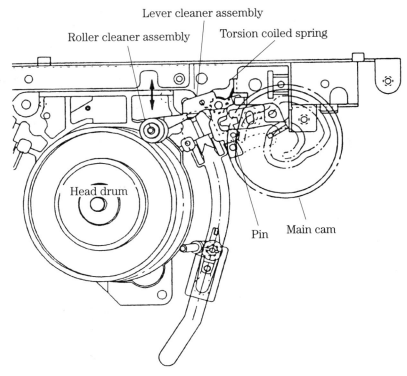

36-10 *The automatic head cleaner rolls against the video heads on the drum or cylinder to the clean heads in loading and unloading mode.*

VHS tape head path is rotated by the capstan and pinch roller assembly past the AC head, roller and leading guides, head drum, roller and leading guides, impedance roller, and back to the cassette take-up reel assembly (Fig. 36-11). In the early loading systems, the tape was wrapped around the tape head in only the play and F/R search modes, but never in the machine's full-loading mechanism. In every mode, the tape is kept wound onto the head drum. Tape tension applied to the head drum or cylinder when stopping is lowered by optimizing the balance between the left to right, and brake forces act in the left and right reel table. This prevents the V-head from scratching the tape when stopping in the FF/REW-Stop modes.

In FF/REW running, to ensure highly stable tape running in any condition, design is given to the foundation brake torque (soft brake) to the reel table and the tape winding angle, with respect to each tape guide. Some units are designed so that the supply-side

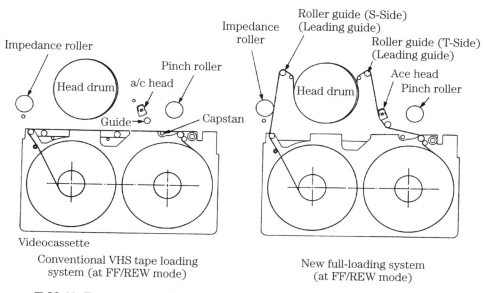

36-11 *The conventional and full-loading system of the VHS VCR unit.*

fixing guide and the FTE (full-track erase) head do not come in contact with the tape at the time of FF/REW operation.

Beta VCR loading path

In the early Beta VCRs the direction of the tape travel path and tape threading path rotates against the drum or top cylinder (Fig. 36-12). Two motors were used, one that provides belt drive to the capstan, head drum, and tape reels in the cassette. A small dc motor is used to drive the threading ring, and a belt drives a small toothed wheel that engages the rim of the threading.

Camcorder tape transport

The drive system consists of those parts that run the tape. The capstan motor pulls the tape across the tape heads with a pinch roller and capstan shaft. Most capstan motors are belt driven with a belt pulley attached. The capstan motor might be belt or gear driven to the various mechanical assemblies, providing tape movement in play, record, rewind, fast forward, and search modes (Fig. 36-13). Because the capstan motor travels at different speeds in the various modes, the dc voltage must be controlled by a servo, capstan speed, and phase-control system.

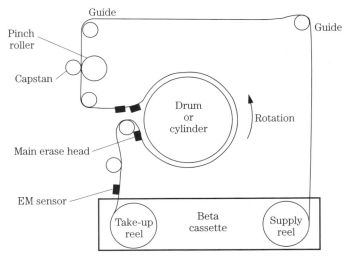

■ **36-12** *The Beta VCR loading tape path.*

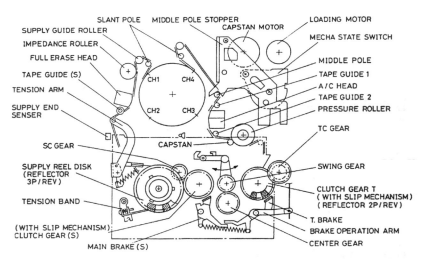

■ **36-13** *The main chassis of a VHS-C camcorder with the layout of the top-mounted part.*

Periodic check and lubrication

To keep the VCR and camcorder in tip-top shape, a maintenance schedule should be followed. The following maintenance and periodic checking procedures are recommended to ensure proper operation, and to protect the tape from dirt and damage. Check Table 36-1 for VHS VCR maintenance and Table 36-2 for camcorder checkups.

Table 36-1 VHS VCR periodic check items.

○ Cleaning ◎ Confirmation △ Lubrication

Part		Usage Time (hours) 500	1000	1500	2000	2500	3000	3500	4000	4500	5000	Remarks
Tape Guide System	Tape running surfaces	○	○	○	○	○	○	○	○	○	○	
	ACE head	○	○	○	○	○	○	○	○			
	Drum (cylinder)	○	○	○	○	○	○	○	○	○		Head life depends heavily on operating conditions.
Drive System	Loading belt	◎	◎	◎	◎	◎	◎	◎	◎	◎	◎	
	Reel drive belt	◎	◎	◎	◎	◎	◎	◎	◎	◎	◎	
	Intermediate gear, pulley axles		△		△		△		△		△	Absolutely avoid oil on tape running surfaces.
	Capstan axles		△		△		△		△		△	
	Loading motor		◎		◎		◎		◎		◎	
Performance Check	Back tension torque		◎		◎		◎		◎		◎	50 ± 10 g-cm
	Brake system		◎		◎		◎		◎		◎	
	FF, REW, REV, PLAY torque		◎		◎		◎		◎		◎	FWD: 600 g-cm and over REW: 750 g-cm and over PLAY: 100 to 160 g-cm REV: 150 to 240 g-cm

Table 36-2 Camcorder periodic checking and lubrication chart.

Parts maintained every	1000 hrs.	2000 hrs.	3000 hrs.	4000 hrs.	5000 hrs.
Video head	R C	R C	R C	R C	R C
Audio/control (A/C) head	C	C	C	C	R C
Full erase (FE) head	C	C	C	C	R C
Capstan flywheel	C	C	C	C	C
Center pulley	C	C	C	C	C
Supply guide pole	C	C	C	C	C
Take-up guide pole	C	C	C	C	C
Tension band	C	R C	C	R C	C
Supply reel disk	C	C	C	C	C
Supply gear	C	C	C	C	C
Take-up gear	C	C	C	C	C
Impedance roller	C	C	C	C	C
Pressure roller	C	R C	C	R C	C
Pulley belt	C	R C	C	R C	C
Capstan belt	C	R C	C	R C	C
Capstan motor		R		R	
Loading motor		R		R	
Cylinder motor		R		R	
Between both guide roller bases and gutters on chassis		H		H	
Pressure roller arm		H		H	
Loading gear		H		H	
Catcher block		H		H	

Periodic check and lubrication

■ Table 36-2 Continued.

Parts maintained every	1000 hrs.	2000 hrs.	3000 hrs.	4000 hrs.	5000 hrs.
Shaft of cassette holder switch		H		H	
Gear of supply loading ring		H		H	
Gear of take-up loading ring		H		H	
Shaft of FE head	S	S	S	S	S
Shaft of supply reel disk	S	S	S	S	S
Shaft of supply sub gear	S	S	S	S	S
Shaft of supply gear	S	S	S	S	S
Shaft of take-up gear	S	S	S	S	S
Shaft of take-up reel gear	S	S	S	S	S
Shaft of pressure roller	S	S	S	S	S
Shaft of pressure roller arm	S	S	S	S	S
Shaft of cam gear	S	S	S	S	S
Shaft of A/C head base	S	S	S	S	S
Shaft of loading gear (1)	S	S	S	S	S
Shaft of loading gear (2)	S	S	S	S	S
Shaft of center pulley	S	S	S	S	S
Shaft of ring idler gear (1)	S	S	S	S	S
Shaft of ring idler gear (2)	S	S	S	S	S
Shaft of loading ring spacers	S	S	S	S	S

R . . . Parts replacement
C . . . Cleaning
H . . . Grease Hitazal or Froil
S . . . Pan motor oil or Slidas oil

Besides cleaning the tape heads, clean the entire drive system (Fig. 36-14). Clean the loading belt or replace, reel drive belt, pulleys, supply wheel tables, guide assemblies, capstan, pinch roller, impedance roller, FTE head, AC head, tension pole, fixed guides, and angle poles. In fact, clean all parts that the tape touches in the various operations. Moisten a soft cloth with methyl alcohol and wipe off the drive system parts.

VCR reel table drive mechanism

The direct-drive capstan motor uses a mechanism system driven by a reel belt. The capstan and pinch roller are press fitted to allow the tape to run at a constant speed in play, F search, and R search modes. A friction gear between the capstan motor and reel table takes up the tape. In fast forward/rewind, tape high-speed, and take-up mode, the capstan and pinch roller are pulled away from each other, and the capstan motor and the reel table are linked directly by the clutch mechanism to carry out the tape take-up (Fig. 36-15). Power transmission to the pulley

■ **36-14** *Clean all tape path components and the video heads with alcohol and a cleaning cloth.*

of the clutch mechanism is transferred from a capstan motor via a rubber belt.

While in the reverse tape running mode, the winding load has to be taken in by the take-up reel table. In special playback operation, a mechanism for varying the take-up torque of the reel table is required to be changed, depending on the running direction of the tape. In this mechanism, the torque selected is made by varying the reduction speed ratio up to the reel table from the capstan motor on both the forward side and the reverse edge of the tape.

The rotation speed and torque needed for each mode of operation are converted by the clutch assembly of the drive power to the idler assembly. By using friction force, the left and right moving idler gear assembly is automatically rotated to the left and right, according to the direction of the capstan motor rotation. The mechanism is designed to drive the side of the reel table that the tape is transported by the capstan and pinch roller. The forward and reverse constant speed (play, record, F/R search modes) is carried out in the state in which the friction mechanism is being operated.

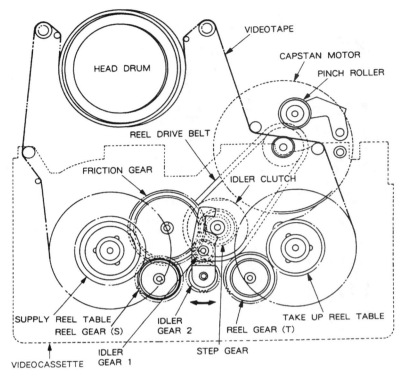

■ **36-15** *The take-up reel is operated by the friction drive of the reel gear, idler gear, and by the capstan reel drive belt.*

Rubber belt problems

A loose or enlarged belt might prevent VCR tape movement and basic capstan, friction gear, and reel take-up or supply reel operation. Oil on the capstan drive belt might cause slower tape movement or intermittent operation. Worn, cracked, or broken drive belts can cause improper speeds in the VCR. Check all belts for shiny surfaces, cracks, or broken areas. Check the drive belt friction and the idler clutch wheels for erratic or intermittent movements (Fig. 36-16).

Replace all drive belts with belts that have the original part numbers. When originals are not available, use universal belts and drive wheels from local or mail-order firms. Several VCR belt kits from different manufacturers offer convenience, economy, and a wide range of replacement parts. Pick up a belt gauge for a quick, accurate method of measuring the thickness, width, and inside circumference of broken or worn belts.

■ **36-16** *Check all rubber belts for cracked, loose, or worn areas, and clean the belt if the operation is erratic or intermittent.*

Camcorder camera sections

The basic camcorder circuit can be broken down into the camera and VCR or VTR sections. The VHS and VHS-C camera circuits might contain a lens assembly, automatic focus, iris or AIC control, and zoom motor assembly. The image sensor might be a CCD (capacity-charged device) or MOS (metal-oxide silicon) pick-up device. The CCD image sensor might contain over 250,000 pixels (picture elements). The drive pulse generator and preamplifier circuits are contained in the CCD or MOS circuits. Other VHS and VHS-C camera circuits are the sync generator, sample and hold, signal processor, automatic white balance, encoder, and power distribution systems (Fig. 36-17). The charge-coupled device (CCD) is used in most camcorders, although the metal-oxide silicon (MOS) image pickup device is used in Hitachi, RCA, Realistic, Kyocera, Pentex, and Sears models.

8-mm camera circuits

The 8-mm camera lens assembly might consist of a focus, lens, automatic focus (AF), iris control, and zoom lens assembly. The CCD image sensor might contain a color filter, CCD driver, and pulse generator in the CCD drive circuits. Luminance processing circuits might contain the iris control, AGC, YH LPF, YL LPF, vertical and horizontal aperture correction, mixer, BPF, chroma detector, white balance, R-YL/B-YL SW, R-YL modulator, B-YL modulator, and mixer (Fig. 36-18).

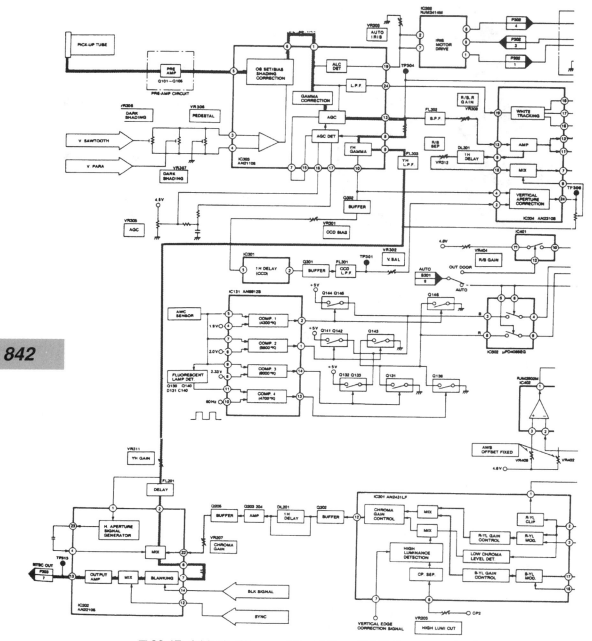

36-17 *A block diagram of the VHS and VHS-C camera circuits.*

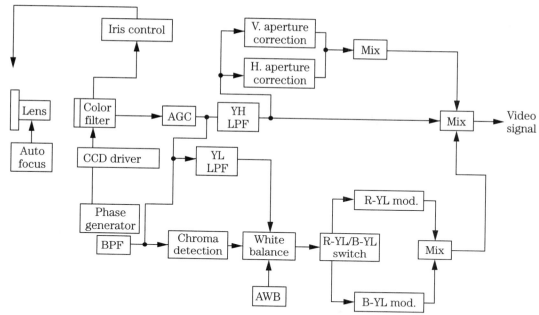

■ **36-18** *A block diagram of the 8-mm camera section.*

Electronic viewfinder

The electronic viewfinder (EVF) permits monitoring the picture or image being shot or played back. The EVF circuits resemble those from the small B&W TV chassis. The electronic viewfinder consists of a miniature picture tube with horizontal and vertical deflection circuits. A flyback transformer provides high voltage to the CRT anode terminal (Fig. 36-19). Vertical and horizontal sync circuits are generated and fed to the EVF deflection and VCR system-control circuits. A small amplifier and sync separator circuit add to the EVF circuits.

The video signal passing through the low-pass filter (LPF), which removes spurious high-frequency components, is applied to the sync separator circuits. These circuits separate the vertical and horizontal sync signals. The vertical sync separator separates the vertical sync signal from the composite sync signal. The vertical oscillator generates a sawtooth waveform to the vertical driver and to the vertical deflection coils. The synchronized horizontal drive pulse is applied to the driver, which drives the horizontal deflection coils to the small CRT or view area (Fig. 36-20).

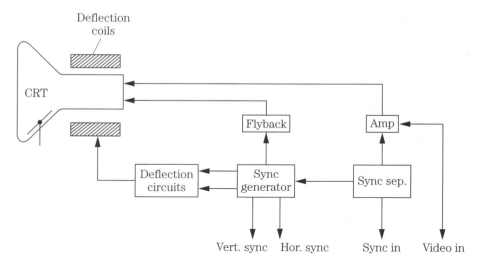

■ **36-19** *A block diagram of the electronic viewfinder circuits.*

■ **36-20** *The electronic viewfinder is fixed on a Hitachi camcorder; it is adjustable in the RCA 8-mm camcorder.*

Camcorder video circuits

The camcorder video circuits consist of the video in and out, headswitching, operations in the play and record mode, luminance signal recording and playback circuits, and the color signal recording and playback circuits. The video input circuits are switched into the input—either for camera or external devices connected to the AV connector. The video output signal is supplied to the AV connector.

The headswitching circuits in the camcorder are often controlled by an IC (Fig. 36-21). The headswitching circuits connect the video heads in the record or playback mode with several control signals. Record, inhibit, SW 30 Hz, ASBL/PB, REC signal, squelch, and head SW signals are used in the headswitching signals of 8-mm circuits.

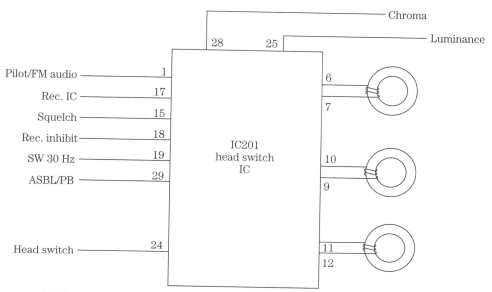

■ **36-21** *A block diagram of the head switching circuits in the 8-mm video circuits.*

When recording, the video signal is fed in at the line-in jack or from the camera video signal to the switching IC. The output signal is fed to the character mix signal of the luminance REC/PB process IC to the AGC amp. The REC switch turns the signal to the sub amp inside of the luma IC. The video signal might be switched to the EE amp to the character mix transistors or IC, video amp to the video output jack and to the electronic viewfinder (EVF). After the color and video signals are mixed, the signal goes to the REC amp, where it is switched to the video heads (Fig. 36-22).

Flying erase head

The flying erase head (FE) prevents the color "rainbow" effect when recording. It is mounted in the same drum or cylinder as the video heads. It erases CH1 and CH2 video tracks simultaneously 90° before the CH1 video head during recording. An erase head oscillator supplies the erase head with erase current (Fig. 36-23). It is controlled by the FE CONT signal that is supplied from the system-control micro compressor. When the camcorder is set to the record mode, the FE CONT signal becomes high (Hi) and the oscillator generates the erase current at about 7 MHz. The flying erase head is found in most 8-mm camcorders (Fig. 36-24).

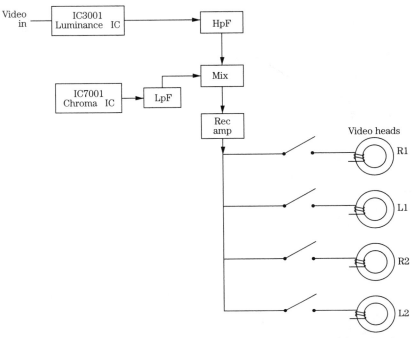

36-22 *A block diagram of the color and luminance mixing video circuits.*

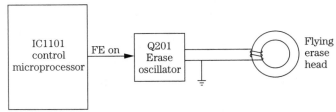

36-23 *A block diagram of the flying erase head circuits.*

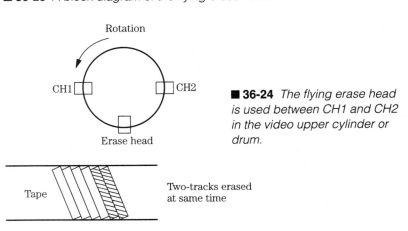

36-24 *The flying erase head is used between CH1 and CH2 in the video upper cylinder or drum.*

VCR servo and control systems

The VCR servo system must keep the cylinder or drum head, videotape, and capstan moving at a designated speed. If dark horizontal bands roll up the screen, suspect a defective servo control system. Intermittent or incorrect speeds might indicate a defective motor, a worn or loose belt, or an improper motor voltage. Check all belts for cracks and worn areas. Keep the upper cylinder or drum clean to prevent poor tracking and distorted sound (Fig. 36-25).

The capstan servo circuits control the speed of the tape moving past the video heads at the correct speed, according to the speed of the tape. You might find that the audio sounds too fast or slow, which indicates problems in the servo system circuits. When the sound is okay with a poor picture, suspect a defective cylinder system or circuits.

The capstan motor is controlled by the speed and phase-control circuits with a feedback frequency generator signal to the speed-control circuit. The control head amp and speed-select circuits feed to the phase control, as well as a 30-Hz reference signal. The upper head cylinder motor is controlled with a speed control and phase-control circuits. A pulse generator and a 30-Hz reference signal are fed to the phase-control circuits.

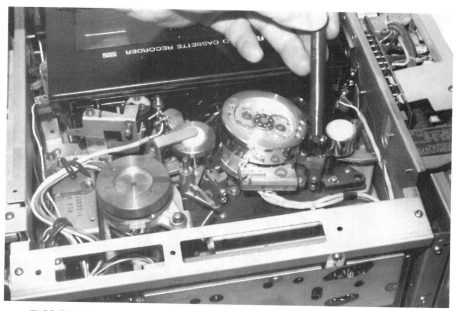

■ **36-25** *Clean the video heads and the audio head to prevent poor tracking and distorted sound.*

Camcorder servo and system control

The system control microprocessor or microcomputer IC monitors the supply-end sensor, take-up reel sensor, dew sensor, supply reel sensor, safety tab switch, cassette switch, and mechanism state mode switch (Fig. 36-26). When the tape reaches the end while moving in the forward direction, the light shows through the clear leader of the tape and allows light to shine on the surface of the supply-end photosensor and shuts off the tape movement. The tape might pull out, unwind, or clog up the mechanism if the take-up reel does not operate. The tape might be jammed if the reel stops during the unloading period and cannot eject the cassette. Loose tape must be removed from the VTR mechanism.

The dew sensor detects moisture within the VCR section and keeps the unit shut down. The resistance of the dew sensor increases as moisture increases, and the voltage is applied to the control IC and all VCR operation stops (Fig. 36-27).

The purpose of the drum or cylinder lock circuit is to inform the operator that recording is prohibited. Drum servo unlocking covers abnormal recordings. The speed-phase control IC compares

■ **36-26** *The top view of a VHS-C camcorder, showing the various components on the main chassis.*

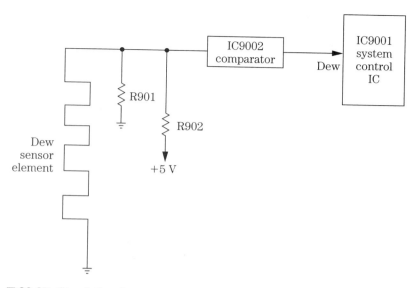

■ **36-27** Check the dew sensor circuits if the camcorder will not begin to operate.

the cylinder or drum speed circuit voltage with the cylinder lock voltage. The cylinder lock IC detects a pulse width of SW 30 Hz, obtained by the servo circuit for detecting a drop in the cylinder motor speed.

During recording, the servo circuits control the tape speed. The tape speed might be 1800 rpm for 8-mm and 2700 for the VHS-C camcorder servo circuit. During playback, it ensures the same accurate tape speeds, aligning the video track with the scanning of the video heads. The speed and phase control of the capstan and drum or cylinder motors are in the servo circuits. The tape speed control is to keep the speed of the video head track constant, and the phase control is performed by the tracking control system.

Problems within VCRs and camcorders

Besides cleaning the video tape heads to prevent a loss of video signals and dropouts, audio distortion might develop from a clogged audio tape head. Clean up all tape heads, guides, rubber pinch rollers, and threading mechanisms that might contain oxide, before attempting to service the camcorder or VCR mechanical or electronic problems. Most problems in the VCR and camcorder are defective spindles, loose screws, oxide on the capstan, slow speeds, intermittent operation, and poor loading.

Pulling of tape

Check for dirty tape heads, sticky rollers, or spindles when the unit is eating the tape. Pulling out of the tape from cassette might be caused by erratic or intermittent operation of the take-up reel, mechanical center bracket assembly, clutch assembly, friction gear, and function gear assemblies. Chewing or eating tape can be caused with a defective control IC, base assembly, clutch assembly, and transmitting arm unit. Clean all tape paths and heads to prevent tape eating or pulling (Fig. 36-28).

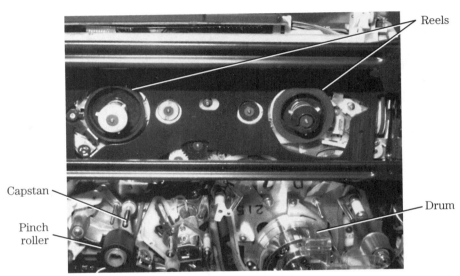

■ 36-28 *Clean the tape paths of reels, heads, spindles, rollers, and the capstan to prevent the tape from being pulled out.*

Cannot remove the cassette

When the cassette cannot be removed, suspect unraveled tape around the capstan and guide assemblies. Try to unload the tape by hand. Turn the capstan motor pulley by hand from the bottom side to remove the tape slack. If the tape or cassette will not come loose, the whole main mechanism must be removed. Sometimes the cassette mechanism and motor brackets might have to be removed (Fig. 36-29). Turn the main cam clockwise with your hand from the bottom side to unload the tape. Stop when the roller guide gets into the cassette lower part. Now turn the capstan motor in reverse direction to remove the tape slack. Keep the left side of the mode selector key pressed to perform front unloading. After the tape has been ejected, remove the cassette.

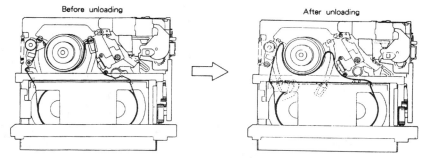

Perform unloading using the mode selector.

First rotate the capstan motor by hand from the bottom to take-up the tape slack, and perform front-unloading using the mode selector.

■ **36-29** *Rotate the capstan motor by hand in opposite directions if the cassette cannot be removed.*

Improper loading

In jammed VCRs or camcorders where a motor controls the loading mechanism, suspect that gum wrappers, candy, crayons, pins, pencils, etc. have been pushed through the cassette loading door area. The motor drive belt might be jammed, broken, or binding and will not rotate the loading mechanism. Check for foreign material down inside of the front-loading gear assembly, preventing rotation of the loading assembly (Fig. 36-30). Take continuity and voltage measurements on the motor and check the loading motor drive IC or check transistors for leakage or no supply voltage (Fig. 36-31). Suspect a regulator zener diode or transistor if the motor voltage is improper.

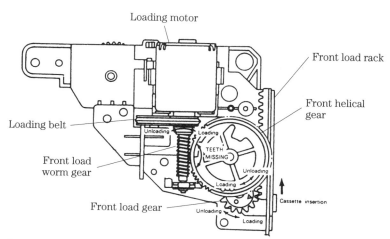

■ **36-30** *Improper loading might result from foreign objects inside the loading motor belt or worm gear assembly.*

■ **36-31** *Suspect a defective loading motor with voltage applied and no motor rotation.*

Shutdown after loading

Check the capstan driver IC and control system when the unit starts to play and shuts down. Suspect a defective drum or cylinder motor or driver and control IC when the cylinder loads and then shuts down. Suspect mechanical failure of a bracket assembly, a defective gear drive assembly, or a malfunctioning switch if the VCR begins to play then shuts down. Check the voltages on the capstan motor leads and check the cylinder or drive IC for the correct voltage source.

Capstan speed problems

The capstan motor provides tape motion in play, record, and fast-forward modes. No tape motion can be caused by a defective capstan motor, drive belt, driver, or servo IC. The capstan motor is driven by a speed control and capstan control IC in most VCRs and camcorders (Fig. 36-32). Check for the correct supply at both the driver and control IC. Take continuity and voltage tests on the capstan motor. Check for a worn or loose drive belt if the motor is operating. Replace the intermittent or erratic capstan motor.

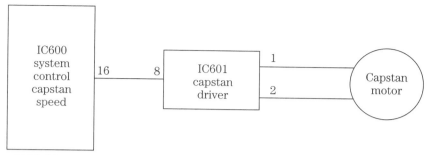

■ **36-32** *The capstan motor is controlled by a system-control system and by a servo IC to the capstan IC or transistor driver circuit.*

Erratic camcorder zoom operation

The zoom motor is located on the lens assembly (Fig. 36-33). The zoom motor might be controlled by the system control IC. Erratic or intermittent operation of the zoom motor can be caused by a bent motor assembly or by a clogged gear assembly. Dropping the camcorder or accidentally bumping the lens assembly against a tree or building can cause damage to the zoom motor assembly. Check the voltage applied to the motor terminals if the wide or tele buttons are pressed and nothing happens. Take a quick continuity test of the motor leads with the low-resistance range of the ohmmeter (Fig. 36-34). Replace the defective motor rather than trying to repair it.

■ **36-33** *Locate the zoom motor used on the lens assembly of the camcorder.*

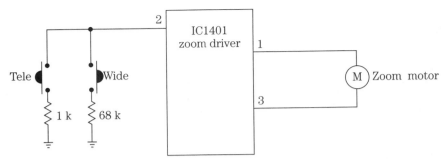

■ **36-34** *Check the continuity and voltage on the zoom motor. Test the switches and the zoom driver IC if there is no zoom motor operation.*

No camcorder auto focus

Check the auto focus motor and motor driver IC if the focus assembly will not respond (Fig. 36-35). Take continuity and voltage tests on the motor terminals. Measure the dc supply voltage at the motor driver IC. Check the infrared diodes with an ohmmeter and the preamp sensor circuits. Take voltage measurements on the infrared processor IC for poor auto focus or no auto focus motor operation. The focus lens assembly should move in and out when the camcorder pans to another scene.

■ **36-35** *The auto focus motor rotates the lens assembly to automatically focus the camcorder.*

Dead VCR or camcorder

Go directly to the low-voltage power supply if the camcorder or VCR is not functioning. Check for an open fuse (Fig. 36-36). The

■ **36-36** *Check the low-voltage fuses in the dead camcorder or VCR. Check out low-voltage power supply.*

power supply in the VCR can consist of a power transformer, full-wave and half-wave rectifiers, filter capacitors, IC and transistor regulators, and many different voltage sources. Check each voltage source after replacing the open fuse if one or more functions will not operate. Check for defective IC and transistor regulators (Fig. 36-37).

Suspect an open regulator transistor if no output voltage is feeding a special voltage source or circuit. Regulator transistors have a tendency to go open or become leaky. Test the regulator in the circuit and then test it again after removing it (Fig. 36-38). Intermittent circuits can result from an intermittent transistor or IC regulator. Do not overlook the possibility that zener diodes in the base terminal of a leaky regulator transistor are burned or damaged.

Poor VCR recording

Clean the audio, video, and erase heads if the VCR is recording poorly. Keep the video heads clean. Notice if the video and audio are both poor. In Beta and 8-mm circuits, both audio and video signals are picked up by the video heads. Check the safety switch, record sub assembly, and lower drum assembly if there is no video recording. If the video circuits do not record, suspect the luma record/playback process IC, transistor record switches, video

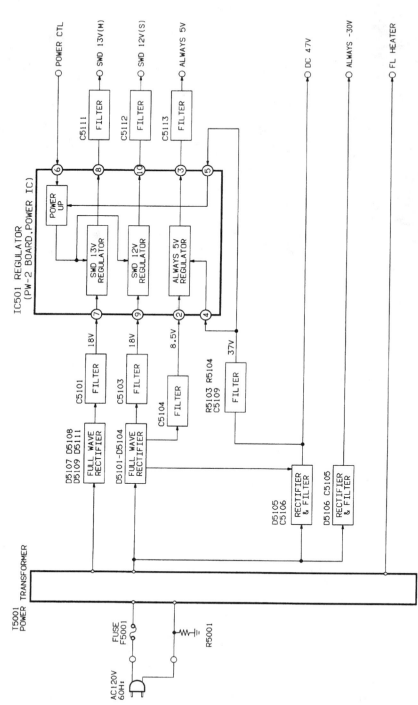

36-37 The block diagram of the low-voltage power supply in a VHS VCR.

VCR and camcorder repair

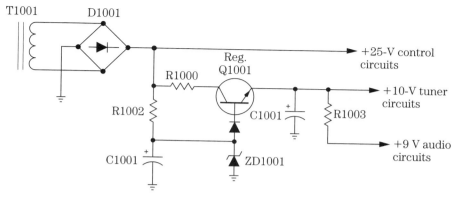

■ **36-38** *No audio can be caused by an open regulator transistor (Q1001) in the low-voltage power source.*

record amp, and the power supply. Make sure that the local oscillator circuits are functioning by taking a scope test at the video heads for FM/audio FM Beta recording (Fig. 36-39).

Check the VIF/SIF audio IC if there is no audio recording. Scope the video head amplifier IC with noise in the video playback circuits. Noise in the playback mode can be caused by improper tape contact. Check spindle posts 3 and 4 for correct tape contact (Fig. 36-40).

AUDIO FM RECORDING CIRCUIT

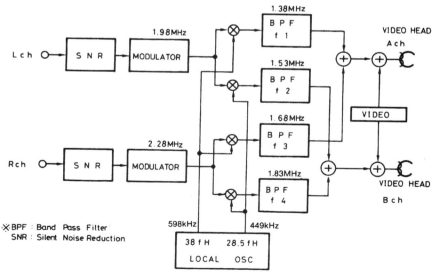

■ **36-39** *A block diagram of Beta audio FM recording circuits.*

Poor VCR recording

[Normal]

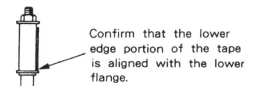

Confirm that the lower edge portion of the tape is aligned with the lower flange.

[Too low tape running position]

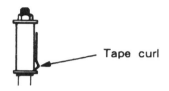

Tape curl

[Too high tape running position]

Tape curl

■ 36-40 *Poor sound and picture quality can result from an improper tape running on fixed spindles.*

Poor or no head erase

The audio erase and flying erase head is often excited from a bias oscillator circuit, and it is used to erase the audio and video portion from the tape. The full erase head erases the entire tape path. Clean both erase heads when cleaning the video heads. If crosstalk or poor picture quality is noted, scope the bias oscillator signal applied to both erase heads (Fig. 36-41). Check for improper alignment of the erase heads in respect to the running of the tape.

No audio in playback

Check the main audio IC that is connected to the play/REC head assembly in the playback mode. Locate the audio tape head on the tape path. In Beta and 8-mm machines, the same head picks up the video signal and also the audio signal. The VHS recorder's audio signal is picked up off of the tape with separate audio heads and rotary transformer windings. When Beta and 8-mm video and sound are affected, check the video heads and the related circuits.

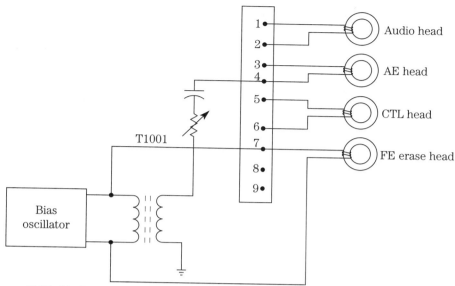

■ 36-41 *Suspect a defective bias oscillator circuit if the recordings have poor audio and picture.*

Troubleshoot the audio circuits with a signal generator and a scope. Scope the audio playback circuits. Clean the playback/record switch assembly.

Poor VCR picture in playback

Clean all video heads with alcohol and chamois. Snow and noise in the picture or a faint picture can be caused by a defective cylinder or drum amp IC. Resolder the pins on the head amp board and check the VHF block for a fuzzy or noisy picture. Check the tuner-demodular circuits if there is only half of a picture. Replacing the ICs within the video circuits can solve most playback audio and picture problems. Check for a missing control pulse if there is noise in the playback mode. A poor picture in the playback mode can be caused by a pre-record amp IC or luma process IC.

Check the supply voltages to all ICs and transistors within the playback circuits. Check the voltages on all IC and transistors to determine if a component is leaky. Inspect the PC board for burned or open resistors. Look for a leaky or open regulator transistor supply voltage in the playback circuits if there is no picture.

Camcorder motor circuits

Most loading, capstan, drum, auto focus, and zoom motors have an IC or have transistors as motor drivers (Fig. 36-42). When any one of the motor functions does not operate, check the continuity of the motor winding. Test the voltage applied across the motor terminals from the driver IC. Suspect that a driver IC or transistors are leaky if there is no voltage across the motor. Check both positive and negative voltages at the driver IC. If there are low voltages or no voltages, go directly to the low-voltage source in the low-voltage power supply. Replace the defective motor if it is erratic or if it has an open winding.

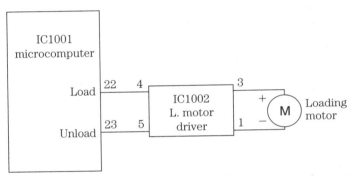

■ **36-42** *Check the voltage and continuity of the various motors used in the camcorder if the motor will not rotate.*

Conclusion

Only the highlights of camcorder and VCR repair is included in this chapter; entire books have been written about this topic. There is just not enough room to list all of the problems and various circuits. Troubleshoot the VCR and VTR mechanical and electronic circuits in the same manner. You can determine what section and components are defective with correct alignment and adjustment procedures. A good cleanup of video, erase, and audio heads have solved many video and audio problems within the camcorder and VCR circuits.

Servicing TV low-voltage and regulator circuits

SERVICING THE LOW-VOLTAGE POWER SUPPLIES IN THE TV chassis can be fun and quite easy to repair. Low-voltage circuits are the heart of any electronic product; no voltage source equals no operation. Next to the deflection circuits, the low-voltage power circuits cause the most problems within the TV chassis. Always use an isolation transformer when servicing the low-voltage line circuits in the TV set (Fig. 37-1). The power-line voltage supply should be respected.

■ 37-1 *Use the isolation power transformer when servicing electronic products.*

Line voltage circuits

When the first TV chassis was placed within the consumer electronic field, a power transformer supplied ac voltage to the low-voltage circuits. Today, most TV chassis use low-voltage circuits from the ac power line. Half-wave, full-wave, voltage-doubler, and bridge rectifier circuits are used in the power-line circuits. The 120-Vac power line is fused with a 3- to 5-A fuse, switched on by a simple SPST switch, a remote-control circuit, or with a relay. The half-wave rectifier removes one half of the ac cycle and is filtered by large-capacity input capacitor and voltage regulators (Fig. 37-2).

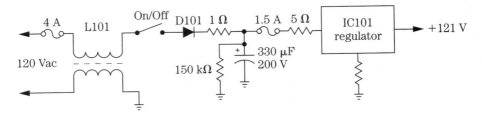

■ **37-2** *The half-wave rectifier might be used in the low-voltage circuit with a line-operated IC regulator.*

Often, the half-wave silicon diode is used in the imported and low-priced TV chassis. After the line voltage is rectified, the filter capacitor smooths out the ripple with voltage-regulator circuits. You might find a small 1.2- to 1.5-A fuse between the raw dc power circuit and the line-voltage IC regulator. Lower voltage circuits might be regulated with transistors, ICs, or zener diodes. Because the ac power supply might have a hot chassis, extreme care should be used while servicing the voltage sources.

Most problems with the line-voltage circuits are caused by blown fuses, leaky silicon diodes, voltage regulators, and defective filter capacitors. The line-voltage regulators have a fixed output voltage that feeds directly to the horizontal output and driver transistor circuits. Fixed output IC voltages might vary between 115 and 140 V.

Voltage-doubler power circuits

You might still find a few TV chassis that need repair in the voltage-doubler power supply. The ac line voltage connects to a voltage-doubling capacitor and hot chassis ground. Two silicon diodes are connected in the half-wave voltage-doubler circuits. A fairly high voltage can be developed with little expense by adding another electrolytic capacitor (Fig. 37-3). Here the output voltage is

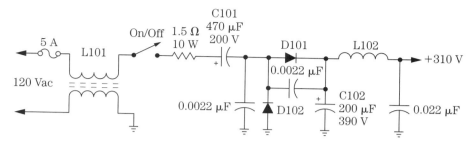

■ **37-3** *The voltage-doubler circuit uses an electrolytic capacitor at the input and it doubles the power-line voltage.*

over 300 V, which provides the supply voltage to the driver, output transistor, and boost circuits. Half-wave rectification provides raw $B+$ ripple voltage output to these circuits. Several scan-derived flyback voltage circuits might power the rest of the TV circuits.

Notice that the voltage-doubling capacitor (C101) is rated at 470 µF, 200 WVdc. A 200-µF electrolytic capacitor (C102) filters out the remaining dc source. A surge-limiting resistor is connected to the negative or ground terminal of doubler capacitor C101. Replace this capacitor with one that has a high working voltage rating (390 to 450 V). When either capacitor becomes defective or dries out, the output voltage will decrease.

Bridge rectifier circuits

The full-wave bridge rectifier circuits consist of four silicon diodes that appear directly across the line circuits. The four diodes are connected in a bridge configuration with two diodes grounded at the anode terminals and the cathodes connected to the voltage output source. A pair of diodes rectify the half-cycle in opposite sides of the bridge circuit and are actually wired in a series circuit. This bridge circuit is used in the majority of the low-voltage power-supply circuits in the present TV chassis. Bypass capacitors (0.01 µF) across the silicon diodes prevent diode oscillation into line.

Full-wave rectification provides less ripple effect for greater filtering applied to the dc output circuits. A large IC fixed-voltage regulator provides voltage to the driver and horizontal output circuits. The capacitor input voltage provides adequate filtering applied to the IC regulator (Fig. 37-4). Notice the 5-W surge resistor before the large filter capacitor and also at the input terminal of the IC regulator.

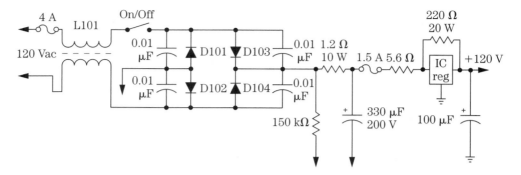

37-4 *A large electrolytic capacitor with higher working voltage is used to filter the dc ripple after the rectifiers.*

Another low-amperage (1 to 1.5 A) fuse is used to protect the horizontal and IC regulator circuits. If the fuse opens, the leaky component might be in the horizontal or IC regulator circuits. The IC line regulator might appear as a regular output transistor or in a flat pack with a heatsink (Fig. 37-5).

37-5 *The flat IC regulator is mounted on a large heatsink in the ac power chassis.*

You might find other transistor and zener diode regulators with a dropping resistor from the low-voltage circuit. Other low-voltage sources are taken from separate secondary windings of the horizontal output transformer. Both types of low-voltage sources might appear in the latest TV chassis.

Low-voltage transformer supply

Lower dc voltage supply might be used in B&W, color, and battery-operated TV chassis. The low-voltage combination battery and ac-operated TV chassis might be switched into the circuit with the four-prong ac cord. Low-voltage transformer circuits are used in the voltage-doubler or line-regulator circuits. Lower voltages might feed to the horizontal and vertical deflection IC, and the sync and sound input circuits.

Usually, the small power transformer is connected to the line circuits after the on/off switch in voltage-doubler or half-wave rectifier circuits. The secondary ac winding has a silicon diode as half-wave rectification, filtering the capacitor input (Fig. 37-6). Several voltage sources might connect to the very low-voltage output to supply voltage to the low-voltage sources. A surge or isolation resistor is used between the silicon diode and the transformer winding.

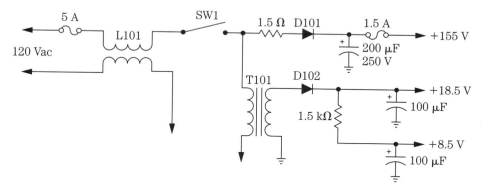

■ **37-6** *A low-voltage power transformer is used to lower the voltage applied to the horizontal deflection IC in several low-voltage power-supply circuits.*

In the latest TV chassis, you might find a pair of bridge-rectifier circuits in the line-voltage source and the low-voltage transformer regulator sources. In fact, some TV chassis might have two low-voltage sources powered by two separate power transformers. One of the power transformer power supplies has full-wave bridge rectification with a large input filter capacitor. The output voltage sources might have a transistor and a zener diode regulator. The 19- or 20-V source might feed the vertical output circuits (Fig. 37-7). Notice the large filter capacitor C101 in this transformer circuit. The 19-V source might power the audio output IC and the vertical output IC.

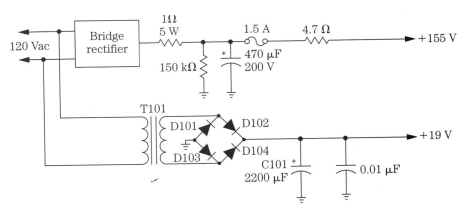

■ 37-7 *The low-voltage power transformer provides a low dc voltage to feed the vertical IC output circuits.*

You might find a transformer low-voltage circuit with half-wave rectification in the same TV chassis. Separate zener and transistor regulators might be used in the output voltage sources. This power source is fed to the microcomputer used in the control system. A large 1000-µF electrolytic capacitor provided filtering with a zener diode regulator at the 13-V source (Fig. 37-8).

SCR switching low-voltage circuits

The SCR switching regulator circuit might be used after the flyback winding and the bridge rectifier. The output voltage fed to the horizontal driver and output transistor comes from the cathode (K) terminal of SCR101. The input voltage is fed to the anode terminals of the SCR. The SCR101 gate voltage is fed from the zener diode and the voltage-dropping resistor in the gate circuit (Fig. 37-9). A pulse waveform is used at a power IC regulator with a 120-V B+ adjustment at pin 1. A horizontal waveform is fed back from the flyback circuits to pins 1 and 2 of the regulator.

By checking the dc output voltages in SCR101, you can determine what component is defective in the low-voltage power supply. Check the voltage at the cathode (K) terminal of the SCR (121 Vdc). If

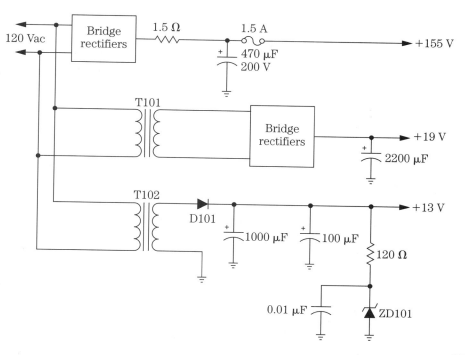

■ **37-8** *In low-voltage sources, a high-capacitance electrolytic capacitor smooths out the ripples in a 13-V power source.*

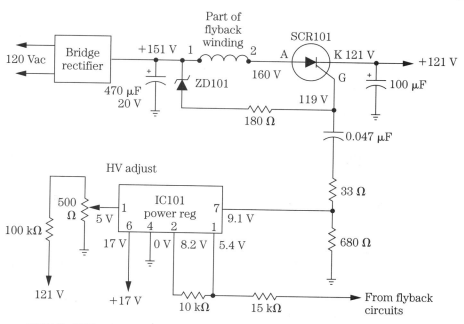

■ **37-9** *SCR101 is a switching rectifier and regulator circuit.*

SCR switching low-voltage circuits

there is no voltage, check at the anode (A) for 155 to 160 V. If there is no voltage here, there is a defective silicon rectifier circuit, an open flyback winding, bad winding connections at pins 1 and 2, or a leaky filter capacitor. If SCR101 is not conducting, check the gate voltage, ZD101, and the 180-Ω resistor for 119 V at the gate terminal. Check SCR701 for leakage and open conditions. If it's still missing, check the voltage, waveforms, and components tied to SCR701 and IC101.

SCR tests

The silicon-controlled rectifier (SCR) is a semiconductor that operates like a thyratron. The anode (A), cathode (K), and gate (G) elements are like the anode, cathode, and grid electrodes of the thyratron. The gate element controls the switching action of the SCR.

Check the suspected SCR with a resistance tester (Fig. 37-10). A low resistance between the gate and cathode are normal. If it is lower than 50 Ω, suspect that an SCR is leaky. Infinite measurement should be made from anode (A) to cathode (K). Replace the SCR with measurements under 3 MΩ between A and K elements. SCRs can break down, become intermittent, and leaky under load. Replace any SCR that you suspect is faulty.

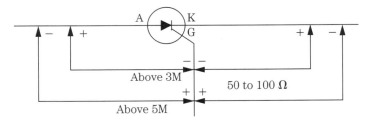

■ 37-10 *Resistance tests on the terminals of SCR might indicate a leaky or open semiconductor.*

B&W TV low-voltage supply

In battery and line operation of the portable B&W chassis, a small power transformer provides a low ac voltage to the silicon bridge rectifiers. A dc voltage jack might switch the batteries in and ac out with an external jack or ac cord. A dc voltage switching might occur at the dc switch or at the grounded circuit of the two silicon diodes. Notice that the transformer is in the circuit at all times with the on/off switch in the dc source (Fig. 37-11).

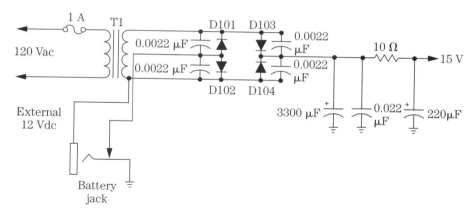

■ **37-11** *The low-voltage source in the black-and-white portable TV might consist of a small power transformer or it might accept a dc battery input.*

The secondary ac voltage (13.5 Vac) is applied to the bridge silicon diodes D101 through D104. Filtering of the bridge rectifiers is accomplished with a 3300-µF electrolytic and 220-µF capacitor. The 15-Vdc source is fed to the horizontal output transistor. You might find voltage regulator circuits connected to the 15.5-V source with transistor and zener diode regulators, providing lower voltages to other receiver circuits.

Different voltage sources

The raw line-filtered dc voltage (115 to 135 V) might be fed directly to the horizontal driver and output transistors or be voltage regulated with an IC power regulator. Different voltage sources might be taken from this regulated source, lowered, filtered, and regulated, to make up the various voltage sources that feed the many TV circuits. Lower dc voltages are required to operate transistor and IC circuits (Fig. 37-12).

In some TV chassis, these lower dc voltages are derived from the secondary windings of the flyback transformer. These scan-derived voltages are illustrated in Chapter 39. Separate voltage sources might be used in the low-voltage power supply of a TV chassis. These different voltage sources feed the critical stages and circuits with transistor, IC, and zener diode regulators, or with all three.

Besides providing voltages to the horizontal output transistor, several different voltage sources feed the other circuits in the TV chassis. A simple low-voltage source might include two other volt-

■ **37-12** *Critical voltage measurements in the low-voltage sources might help you to determine that a voltage source is defective or that components are overloaded.*

age sources, B and C (Fig. 37-13). Voltage source A supplies the voltage to the horizontal output IC. Voltage source B feeds the horizontal oscillator circuits, video IC, video amp, sync, and AGC circuits. Voltage source C supplies voltage to the IF, AFC, and tuner circuits. The normal flyback voltage sources provide for boost, CRT, video output, and vertical output circuits.

Line-voltage power-supply regulators

The dc line regulators are usually ICs, although some might resemble power output transistors. These IC regulator's output voltage might be ranged somewhere between 115 and 135 V. You might find a few new chassis that go up to 140 Vdc. Often, the output voltage feeds directly to the horizontal driver transistor and horizontal output transistor via the primary winding of the flyback. Most of these high voltage IC regulators have a fixed dc voltage at the output pin (Fig. 37-14).

Dual-transistor high-voltage regulators might have two transistorized devices in a parallel resistance circuit that regulates the line

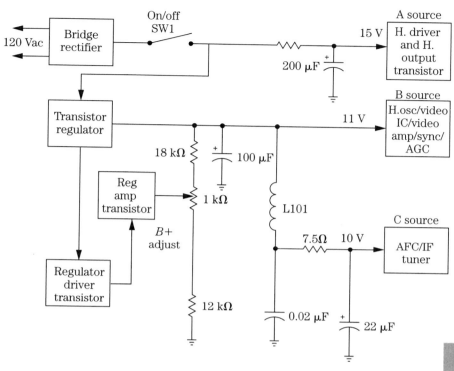

37-13 Notice the various circuits that receive voltage from sources A, B, and C.

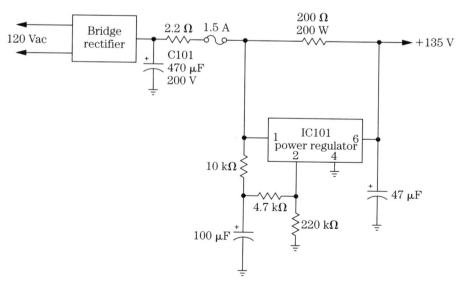

37-14 Most line-operated power regulators have a fixed output voltage.

Line-voltage power-supply regulators

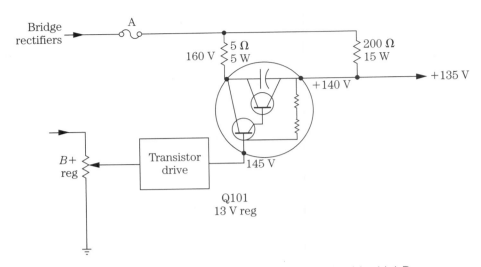

■ **37-15** *The dual-transistor in one voltage regulator provides high B+ regulation for horizontal driver and output transistors.*

voltage output up to 140 Vdc. You will find a driver transistor and a B+ adjustment in one leg of the 140-V regulator. The output voltage of the dual-packaged transistor regulator can be adjusted for the correct voltage on the anode terminal of the picture tube (Fig. 37-15).

Line-voltage regulators might become leaky, open, provide a low output voltage, and shut down the TV chassis. If no or low voltage is used on the horizontal output circuits, check the low voltage at the large electrolytic filter capacitor. Some TV chassis might have a test point on the PC board after the silicon bridge rectifier. If the voltage is normal, proceed to the voltage input pin of the IC regulator. A normal voltage at the input terminal and an improper voltage at the output terminal might indicate that an IC regulator is defective. Check the value of each resistor and capacitor tied to the IC pin terminals before replacing the defective part.

Transistor and zener diode regulators

For lower voltage regulation, transistors and zener diodes are used throughout the various dc sources. These types of voltage regulators are used in the low-voltage power supply and scan-derived flyback circuits. You might find a separate zener diode as a regulator or placed in the base circuit of the regulator transistor and connected to common ground (Fig. 37-16).

■ **37-16** *The voltage regulators might be mounted on a large heatsink.*

Transistor regulators have a tendency to go open or become leaky, resulting in abnormal output voltage. No voltage is the result of an open transistor, but the leaky regulator transistor has a lower output voltage. Often, if the regulator transistor is shorted or leaky, the zener diode in the base circuit overheats and might be damaged. Automatically replace the zener diode with a shorted regulator transistor. The leaky regulator transistor might prevent a $B+$ adjustment and also cause hum bars in the raster. Notice that the 6-V source (3) has a single zener diode as a regulator (Fig. 37-17).

Besides replacing the leaky transistor and zener diode, doublecheck the voltage-dropping resistors in the regulation circuits. A small electrolytic capacitor might become leaky or open and produce abnormal output voltages. Critical voltage, resistance, transistor, and diode tests, and electrolytic capacitors tested in the capacity tester solve most dc regulation circuits.

Usually zener diodes will overheat and become leaky. Very seldom do zener diodes open, unless they also burn. Check for white or burned marks on the body of the zener diode for overheating. A

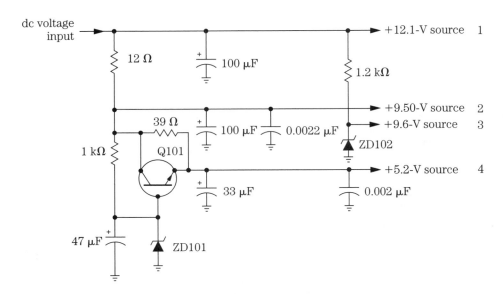

■ **37-17** *The 6-V output source (3) has a zener diode (ZD102) as a voltage regulator.*

critical resistance test from the cathode terminal to ground might uncover a leaky diode.

Keeps blowing the fuse

At least one line fuse is used in the low-voltage input circuits and in some chassis, another is used at the input of the IC regulator. In some chassis, the second fuse (1 to 2 A) is used to protect the horizontal circuits. The line fuse (4 to 5 A) might blow or open with a leaky silicon rectifier, a leaky or shorted filter capacitor, a leaky IC regulator, and leaky or shorted components in the horizontal circuits (Fig. 37-18). If both fuses open, suspect a leaky horizontal output transistor, driver, damper diode, or an insufficient base drive voltage to the output transistor.

A shorted horizontal output transistor and damper diode might destroy both fuses, isolation and surge resistors that are in series with bridge rectifiers, and the IC regulator. The leaky filter capacitor, IC regulator, and bridge rectifiers might blow the main and fuse, which protect the IC regulator. Fuses that have dark inside areas indicate a heavy overload or shorted component ahead of the blown fuse. A defective picture tube might cause the line fuse to blow.

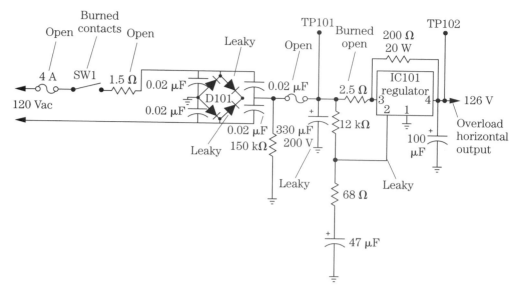

■ **37-18** *The various components in the low-voltage power supply might blow the various fuses.*

After replacing blown fuses, with burned, cracked, or open isolation and surge resistors, remove the horizontal output transistor from a possible overloaded circuit. Check the IC regulator for leakage. Take a critical resistance test from the electrolytic filter capacitor to ground. Check each silicon diode for leakage. Replace all defective components and test the low-voltage circuits at the IC regulator. Without a load, the output voltage might be higher than normal. Now test the horizontal output transistor, damper diode, and drive voltage on the horizontal circuits for overloaded conditions.

Check by the numbers

Check the IC regulator circuits by the number to locate the defective component in Fig. 37-19. After replacing the blown fuses, check the resistance at number 1 to common ground. If the resistance is below 500 Ω, remove the horizontal output transistor from the circuit. Check the voltage at number 1. If there is no output voltage or a low output voltage, go to test point number 2. Suspect a defective regulator IC if the voltage is normal here and there is no voltage or a low voltage at the output (or check point number 1).

Go to test point number 3 if no dc voltage is at test point 2. Check for a damaged isolation resistor (2.2 Ω) with dc voltage at test point 3. Suspect a blown fuse if the dc voltage is used at test point

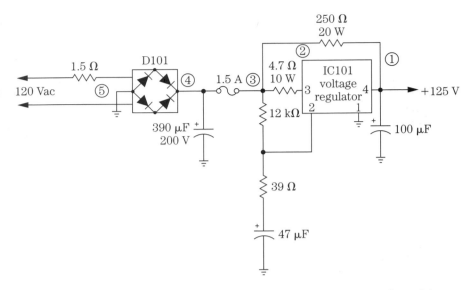

■ **37-19** *Service the low-voltage power supply by the numbers for quick, efficient repair.*

4 and not at test point 3. This dc voltage on test point 4 should be somewhere between 150 and 165 V. Take a critical resistance test across the main filter capacitor terminals (Fig. 37-20). Resistance below 1000 Ω indicates a leaky filter capacitor. A low dc voltage at test point 4 might result from an open main filter capacitor.

Suspect one or two burned or open diodes in the bridge rectifier circuits if there is no voltage or a low voltage at test point 4. Again, check the main fuse and surge resistor (1.5 Ω) for open conditions. Measure the 120-Vac voltage supplied to the bridge rectifiers. Check for an open or shorted silicon diode if there is an ac line voltage, but no dc output voltage. Test across each diode for leakage or open conditions. A leaky diode will have low resistance in both directions, but the open diode will have an infinite measurement in both directions.

Filter capacitor problems

A defective filter capacitor might blow fuses, create a low power supply voltage, insufficient width of raster and picture, hum, and black bars in the picture. A low voltage measured across the electrolytic filter capacitor might result from low dc input voltage or a dried-up and open capacitor. Often, the dried-up filter capacitor loses its capacitance and might produce insufficient width or

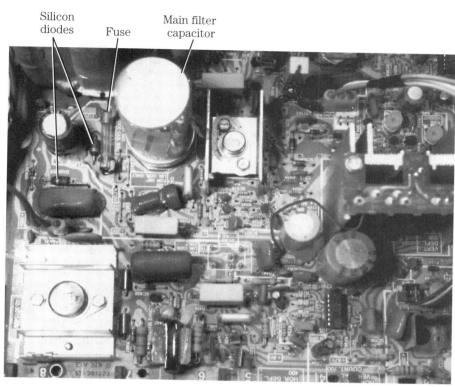

■ **37-20** *Locate the main filter capacitor and take critical resistance tests after discharging it.*

curved outlines, like that of a model (Fig. 37-21). You might find one or two large black bars in the raster if the TV has a defective filter capacitor.

The leaky electrolytic capacitor will have a low resistance across the two terminals if the positive terminal is removed from the circuit. Discharge the filter capacitor before taking resistance measurements. Place a screwdriver blade across the ground and positive terminals to discharge the capacitor. Take another resistance measurement for a leakage test. A normal or good electrolytic capacitor will charge up with the VOM meter hand going upward and slowly discharging as the hand returns to zero. The numbers on the DMM will change rapidly as the capacitor charges and discharges.

A defective electrolytic capacitor might have a white or black substance oozing out around the terminal connections. Sometimes the positive terminal might break right inside the capacitor, resulting in an open capacitor. Notice that the main line-filter capacitor has a

■ **37-21** *Insufficient width and a raster with curves indicates that the electrolytic filter capacitor is dried up.*

higher working voltage. Shunt another electrolytic capacitor across the suspected one to determine if it is defective. Observe the correct terminal polarity, required working voltage, and shut down the chassis before clipping another capacitor across the suspected one.

Black hum bars

One or two black bars appearing horizontally across the picture or raster might be 60- or 120-Hz hum bars (Fig. 37-22). A dark bar moving up the raster represents 60 Hz, but two hum bars are 120 Hz. The picture in the background might be pulling and tearing as the hum bars roll by. A hum might be heard in the speaker.

Leaky regulation transistors in the low-voltage sources and improper $B+$ adjustment might produce hum bars. Readjust the $B+$ control to eliminate the dark bars. Check the regulation transistor for leakage. Most regulator transistors show leakage between the collector and emitter terminals. If in doubt, remove the suspected transistor from the circuit and make another leakage test. You might have a low-resistance measurement in one direction with the diode test of the DMM and some leakage with the reverse test leads. The normal regulator transistor will show a low normal measurement in only one direction.

Do not overlook the possibility that a zener diode might be leaky if there are hum bars in the raster of a low-voltage voltage-regulator circuit. Always replace the regulation diode if it is leaky or if it was

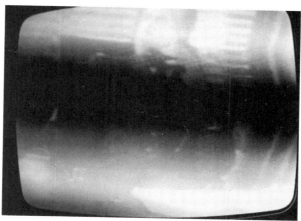

■ **37-22** *Black hum bars in the raster indicates that the filter capacitor and regulation network are defective.*

used with a leaky transistor regulator. Readjust the $B+$ or the high-voltage control after servicing a defective component in the regulation circuits.

If lightning strikes

If lightning or line outage strikes the TV set, the low-voltage power supply and tuner might be damaged. A slight lightning charge might only blow the line fuse, damage one or two silicon diodes, and the power cord connections. High voltages might damage the line-operated IC, plus the resistors and capacitors that are tied to the voltage regulator. If lightning has spread to other parts in the chassis, removing sections or burning off the PC wiring, the whole chassis might be totaled. It's just too expensive to repair (Fig. 37-23).

The low-voltage power supply must be repaired before other components and sections can be tested after lightning or power outage damage. Take a close look at the components tied to the IC line-operated regulator. If resistors and capacitors have been burned or are cracked, each damaged component must be replaced. Usually burned or smoked areas and blown-apart components around the ac line terminals indicate that the unit has been damaged by lightning.

On slightly damaged low-voltage power circuits, replacing leaky silicon diodes, line fuses, open isolation and surge resistors might only be required. If a black bar appears on the raster and the $B+$ control will not charge the low-voltage applied to the horizontal

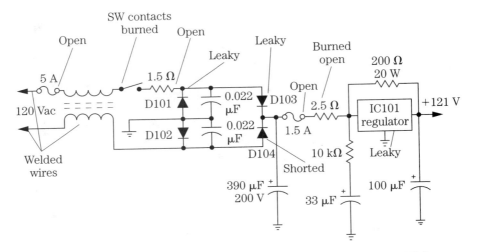

■ **37-23** Check the various components that might be damaged if lightning strikes the outside power line.

output transistor, suspect a leaky or damaged regulator IC. Inspect each small electrolytic capacitor tied to the IC circuits. Replace any damaged connecting resistors. Recheck the raster and properly adjust the low-voltage regulated output of the $B+$ or the high-voltage control. Replace the damaged IC regulator if the output voltage is low, if there is a black hum bar, or if no $B+$ control is used in the line-operated IC regulator.

Critical B+ adjustments

Improper adjustment of the low voltage that is applied to the horizontal output transistor might cause hum bars and high-voltage shutdown. If there is no adjustment of the $B+$ or if high-voltage control has no effect on the output of the low-voltage source, suspect component breakdown in the power supply. If the horizontal output transistor has been removed from the chassis with no load on the regulation circuit, the $B+$ control might not have any effect on the output voltage (Fig. 37-24).

Tune in a picture and set the brightness, color, and contrast controls at minimum. Connect the dc voltmeter (VOM or DMM) to the output terminal of the output regulator or at a fuse between the primary winding of the flyback and the low-voltage supply. Some TV chassis have a test point at the line-operation regulator output voltage source. Adjust the $B+$ control with the exact voltage applied to the horizontal output transistor. If in doubt, check the TV

■ **37-24** *Locate the B+ adjustment on the TV chassis and adjust it for the correct voltage applied to the horizontal output transistor.*

schematic or check the manufacturer's service literature for correct high-voltage adjustments. Monitor the HV with a high-voltage probe at the anode terminal of the CRT, to keep the high voltage at the manufacturer's specifications.

Chassis shutdown

Excessive line voltage, an improper voltage source, and defective safety or hold-down capacitors within the horizontal output circuits might cause a high-voltage shutdown. An improper setting of the $B+$ control might cause high-voltage shutdown. A defective high-voltage shutdown circuit might shut the chassis down. Check the low-voltage source and compare it to the schematic to determine if the voltage is too high. Some line voltages might provide higher line voltage (128 Vac) and cause high-voltage shutdown.

Repairing the vertical circuits

TROUBLESHOOTING MODERN VERTICAL CIRCUITS IS COME quite easy compared to repairing those from years past. The early transistor and tube circuits consisted of a multivibrator and relaxation vertical oscillator circuits. Then came the entire transistor chassis with a transistor sync, oscillator, and two output vertical transistors. Later came the vertical and horizontal deflection IC. You can locate the vertical output transistors and ICs without a schematic with the output IC on a separate heatsink (Fig. 38-1).

■ **38-1** *Locate the vertical section on the TV chassis by finding the vertical output IC on the large metal heatsink.*

In the beginning

Although you might service vertical circuits with transistorized outputs, this chapter deals entirely with vertical IC circuits. The new vertical circuits might consist of a vertical deflection or countdown IC. Besides the vertical oscillator or countdown circuits, a separate vertical amp might be included in the same IC. The new vertical circuits consist of a vertical output IC instead of two separate transistors. Again, the vertical power output IC is mounted on a heatsink with a fairly high operating voltage.

The vertical screen symptoms are the same in tube, transistor, or IC chassis. No vertical sweep is indicated by a horizontal white line (Fig. 38-2). Insufficient vertical sweep might appear from 1 to 10 inches on the raster. Intermittent vertical sweep might result in improper sweep or a white horizontal line. The raster might operate for several hours and then collapse. Vertical crawling and rolling can be caused by improper vertical sync and filtering action. Vertical foldover might result at either the top and bottom area of the picture.

■ **38-2** *A white horizontal line indicates that there is no vertical sweep.*

Just about any component within the vertical oscillator or output circuits can cause a horizontal white line. Improper height can be caused by a low supply voltage, improper vertical drive pulse, or a defective output IC. The intermittent can be caused by a defective coupling capacitor, intermittent supply voltage, poor connections, or intermittent ICs. Vertical bunching lines at the top of the picture might result from a defective output IC, electrolytic coupling capacitor, or open or leaky components tied to the vertical output IC circuits (Fig. 38-3). Check the vertical output IC, measure for improper voltages, and test the pincushion circuits for vertical foldover.

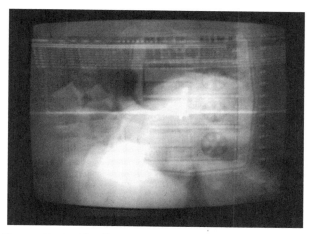

■ **38-3** *The vertical white line bunched at the top of the raster might be caused by vertical transistors, ICs, and leaky coupling capacitors.*

Vertical supply sources

Without the correct supply voltage, the vertical output IC might indicate insufficient sweep or a horizontal white line. Today, the vertical output IC might operate from the scan-derived flyback voltage source. The supply voltage might vary from 12 to 50 Vdc. The vertical output IC might be directly coupled to the deflection yoke winding with a return resistor and capacitor to common ground. Dc voltages are found on all IC terminals, except those that are connected to ground (Fig. 38-4).

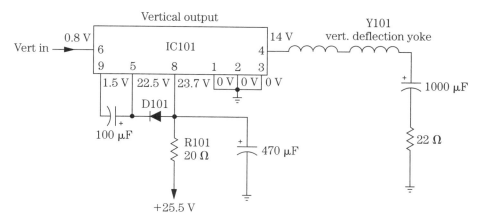

■ **38-4** *Check all voltages on each pin terminal of the vertical output IC to find a defective IC or power source.*

Check the supply voltage source on the vertical output IC. If the voltages are normal, scope the input and output terminals of the IC. With normal supply voltages and vertical sweep waveforms at the input terminals, suspect that an IC or the components connected to the IC are defective if there is an improper output sweep connected to the deflection yoke and pincushion circuits. Check the output drive at the countdown or vertical deflection IC with improper input waveform or no input waveform at the output IC.

Then there were none

Vertical circuits can easily be serviced without a schematic. More time might be needed, but with only two IC vertical components involved, the circuits are fairly simple. Locate the vertical deflection and output ICs. The vertical oscillator and amp stages might be found in a large IC, which includes the video, sync, hold-down, and deflection circuits. Locate the vertical output IC on a large heatsink (Fig. 38-5).

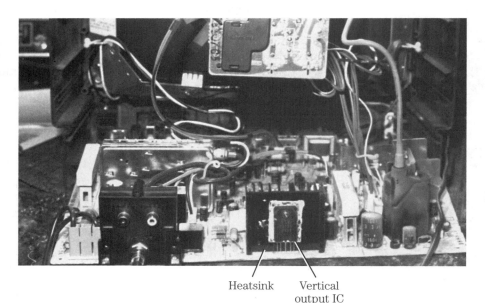

■ **38-5** *The vertical IC is usually mounted on a heatsink.*

Start at the vertical output IC with improper or no vertical sweep, when a schematic is not handy. In fact, you can start at the output and work toward the front-end vertical circuits—especially because the vertical circuits might be difficult to locate in an IC with

20 or more terminals. Take critical voltage tests on each terminal. The highest voltage should indicate the correct voltage supply pin.

With normal high voltage and horizontal circuits, you should be able to measure the dc voltage to the vertical output pin. Low voltage or no supply voltage can be traced back to the flyback circuits. Often, a single silicon diode, filter capacitor, and isolation resistor might connect to a flyback winding (Fig. 38-6). Check for open resistors and leaky diodes. Open or dried-up electrolytic capacitors (C411) might produce low supply voltage. Check the semiconductor or numbers found on the defective output IC for replacement.

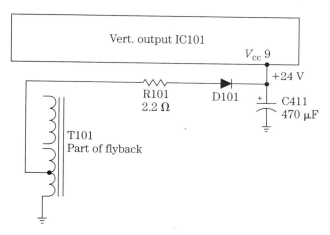

■ **38-6** *The low-voltage source feeding the vertical output IC might be from the flyback scan-derived voltage sources.*

Important vertical waveforms

The vertical output waveforms at the output terminal have a flickering sawtooth pulse that connects to the vertical yoke winding (Fig. 38-7). The waveform should have a drive waveform from 50 to 60 V, with an average at around 52 Vp-p. The vertical input voltage at the input terminal of the output IC should measure somewhere between 0.5 to 3 Vp-p. The drive voltage is greater with a vertical amp stage found in the deflection IC, which indicates a high or low drive voltage.

The critical vertical waveform taken at the output of the deflection IC, to the input of the output IC, and at the output of the power vertical IC will indicate what stage contains the loss of vertical sweep. If a normal 50-V vertical waveform is applied to the vertical yoke winding and there is only a white line, suspect an open yoke

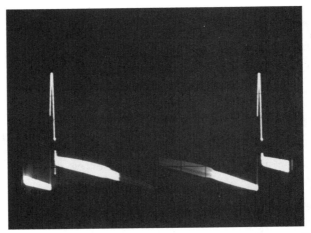

■ **38-7** *The vertical waveform fed from the output transistors or ICs to the yoke winding.*

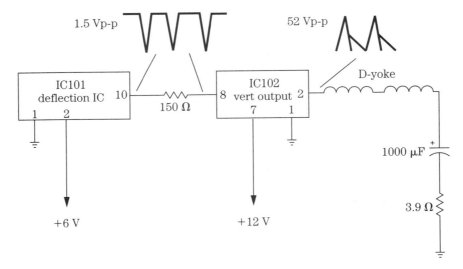

■ **38-8** *Check the vertical circuits with the scope for correct waveform and drive voltages on the deflection and output IC.*

winding, poor terminal connections at the yoke, open capacitor, or an open return resistor (Fig. 38-8). The vertical circuits can easily be serviced with a DMM and an oscilloscope.

Vertical countdown circuits

The TV signal processor and deflection IC might have a vertical oscillator, sync separator, sync trigger, and a vertical amp inside of

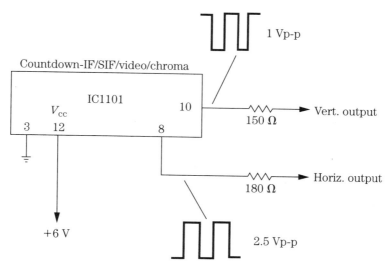

■ **38-9** *The vertical countdown IC might have an internal vertical amp, providing from 1 V to 3.5 Vp-p to the vertical output transistor or IC.*

one large IC (Fig. 38-9). The vertical output amp signal is then applied to the vertical output IC. In some chassis, you might find transistors doing the same operations with a vertical power output IC.

A vertical countdown circuit consists of a divider system of taking a horizontal and vertical signal from the countdown circuits to each deflection system. The system might be crystal controlled in some chassis. The vertical and horizontal sync circuits are developed within the same large IC. Checking waveforms with the scope is one of the best methods for troubleshooting vertical ICs. Sometimes in transistor vertical circuits, the scope can only be used to indicate if the vertical signal is present or missing because of failure in the vertical feedback circuits.

Scope the vertical output terminal of the countdown or deflection IC to determine if the vertical drive signal is present. Suspect a defective vertical IC or an improper supply voltage at the supply terminal of the countdown IC. Usually, if you have horizontal sweep in the raster, the IC is operating. Remember that the supply voltage for the vertical circuits might originate in the flyback derived-scan circuits. Do not forget to check components that are tied to the vertical countdown IC terminals for no vertical drive waveform.

By the numbers

Check the vertical circuits by the numbers (Fig. 38-10). Start at the vertical output signal from the countdown or deflection IC. This waveform can be a sawtooth or a sharp pulse with a peak-to-peak voltage 0.5 to 3.5 V. The shape and amplitude of the waveform depends if a vertical amp stage is found inside of the deflection IC. Take a scope waveform at point number 1. Notice if the vertical sweep waveform has the correct drive voltage.

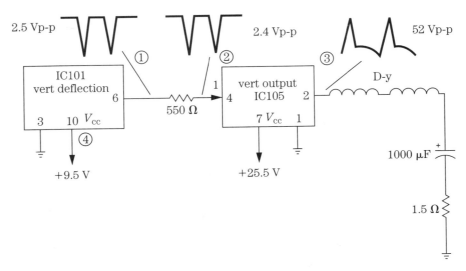

■ **38-10** *Check the vertical circuits by the numbers.*

Proceed to point 2, at the input terminal of the vertical output IC. This p-p signal is a little less than found at number 1 because the signal must pass through a small isolation resistor. Check the output signal at point 3. Here, the scan waveform voltage is fairly high (around 50 V p-p). The output waveform from the vertical output IC sweeps the vertical winding of the deflection yoke. Suspect that the output IC is defective if the vertical input drive signal is at the input terminal.

Vertical sync problems

The vertical input sync signal is developed from the composite video signal at the input of the countdown or deflection IC. The vertical sync circuits might consist of a vertical trigger and sync separator fed to the vertical oscillator circuits. Most sync circuits are found inside of the same IC. Tied to these vertical pin terminals are the vertical sync-shaping resistors, diodes, and capacitors. Check the IC internal sync functions by looking up the IC part number in a universal semiconductor manual.

Most TV manufacturers do not list the internal circuits on the schematic diagram. They are usually only listed in a boxed IC outline. This means that you do not know what terminal pins of the IC contain the vertical sync circuits and components. Very few sync problems are found in these circuits. If the picture is pulling or rolling up or down without a vertical hold control, suspect the vertical circuits within the deflection or countdown IC (Fig. 38-11). Take critical voltage measurements on the vertical sync and drive output pin terminals.

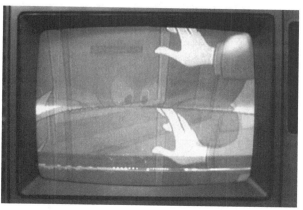

■ 38-11 *Suspect the vertical deflection IC and circuits if the picture rolls up or down without a vertical hold control.*

Intermittent vertical sweep

If the screen symptoms are an intermittent horizontal white line or an insufficient vertical sweep, try to isolate the problem within the vertical deflection or output IC (Fig. 38-12). Monitor the waveform at the drive pin terminal of the deflection IC with the oscilloscope. If the raster shrinks, notice a change in amplitude at the vertical drive signal. If the vertical amplitude drive waveform changes or lowers, suspect a deflection IC, components in the vertical oscillator circuits, or poorly soldered IC connections. First solder all pin terminals of the deflection or countdown IC (Fig. 38-13).

Proceed to the vertical output IC input terminal and monitor the drive waveform. Monitor the supply voltage output terminal of the IC with the voltmeter. If the drive waveform is normal with an in-

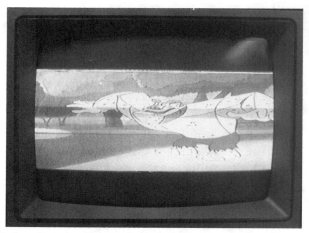

■ **38-12** *Insufficient vertical sweep results with not enough sweep to fill out the CRT raster.*

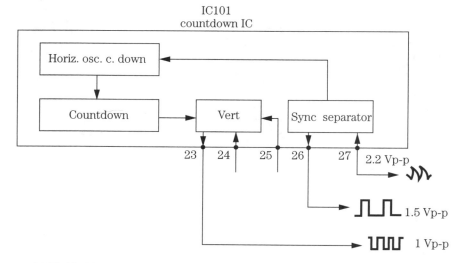

■ **38-13** *Most schematics do not show the internal vertical circuits. Check the universal replacement manual for correct pins and solder all terminals if there are intermittent vertical problems.*

termittent raster, monitor the output waveform with the scope. Notice a change in voltage if the raster seems intermittent. No waveform at the output pin might indicate that the output IC is defective or that the components tied to the IC terminals are defective. Suspect an open yoke winding, poor socket and board connections, and return capacitor and resistor if the waveform is fairly normal on the output IC terminal.

Supply pin terminal tests

Check the voltage at the supply pin terminal of the output IC if the vertical raster becomes intermittent. If the voltage rises rapidly or lowers, suspect a leaky or open vertical output IC. Do not overlook the possibility that a power supply source is defective. Remove the supply pin from the PC board with solder wick and a soldering iron. Now, monitor the voltage supply pin source. Suspect a defective IC if the voltage remains constant. Spray the IC with coolant and see if the voltage changes.

You can make a quick test of the suspected IC by removing it from the PC board. Compare the resistance measurements from the voltage supply pin of the suspected IC to the other terminals with the new replacement. Mark down the resistance measurements from each pin to the voltage supply pin (Fig. 38-14). Then, spray the IC with several coats of coolant. If a lower or an infinite resistance measurement is found compared to the same terminal, you can assume that the IC is defective. Replace it.

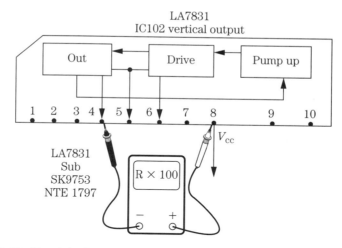

■ **38-14** *Check both replacement and removed output IC vertical terminals for leakage between the supply pin and between each pin terminal.*

Vertical voltage injection

Both vertical countdown and output ICs can be checked for intermittent or failure if the chassis shuts down before any voltage or scope measurements can be made (Fig. 38-15). Often, you can see the horizontal white line before the chassis shuts down, indicating

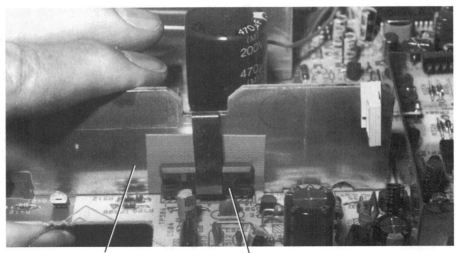

Metal shield heatsink   Vertical output IC

■ **38-15** *Locate the vertical output IC to input the external supply voltage to determine if the vertical circuits are normal.*

problems in the vertical circuits. If the vertical supply voltage is provided by the flyback scan-derived circuits, inject the external supply voltage to both the countdown and the output IC.

You can inject an internal voltage source to both vertical ICs, like the vertical transistorized circuits (Fig. 38-16). Locate the supply voltage pin (V_{CC}) on both vertical deflection and output ICs. The supply voltage pin is number 18 on the vertical deflection IC (IC201). Inject a supply voltage of 25 V to pin 8 of the vertical output IC (IC301). Connect the negative voltage clip to common ground. Here, you must have a variable power supply with two different voltage sources. Of course, several batteries can be connected in series to produce the same voltage.

Now, place the scope probe at pin 6 of the vertical deflection IC and notice if IC201 is functioning. No waveform or an improper drive waveform might indicate that a defective IC or that defective components are tied to the vertical circuits. Likewise, check the signal at the input and output terminals of IC301 at pins 6 and 4, respectively. If the input signal is normal at pin 6 and no waveform is at pin 4, suspect that IC301 is defective. Remember, external voltage injection is accomplished with the ac cord pulled from the receptacle.

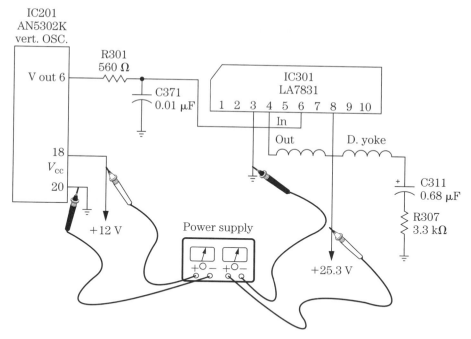

38-16 *Use a dual-variable power supply to inject the external dc voltage to the output and deflection IC and check the waveforms with an oscilloscope.*

Vertical foldover

Check for a leaky vertical output IC if vertical foldover occurs. Improper bias or leaky bypass capacitors tied to the vertical IC circuit can produce vertical foldover (Fig. 38-17). An improper supply voltage at the supply pin terminal might cause foldover. Vertical foldover can occur at the top and bottom of the raster. Often vertical foldover occurs with insufficient height problems.

Vertical linearity or foldover problems can be caused by vertical feedback and bias circuits. Check capacitors C451, C452, C453, C454, C455, and IC450 for foldover problems (Fig. 38-18). C445 is the vertical sweep return capacitor through resistor R450 (3 Ω). Do not overlook the possibility that a change in resistance in the vertical feedback circuits caused the foldover. Look outside of the main vertical circuits for foldover conditions if the voltage and the components in the vertical output circuits appear to be normal. Check for burned resistors, poor pincushion coil and transformer terminals, and poorly soldered board connections in the pincushion circuits for outside foldover problems.

■ 38-17 *Vertical foldover usually occurs at the top of the raster with defective components in the output circuits.*

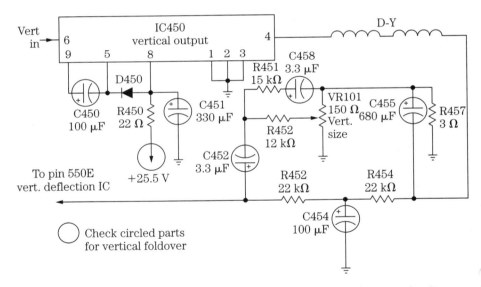

■ 38-18 *Check the components for foldover in the vertical output circuits.*

No work/no play

One of the biggest disappointments in electronic servicing is to find that the chassis will not function properly after a new IC has been installed. Check all soldered connections for the new transistor, IC, or diode with a magnifying glass. Make sure that all terminals are soldered well and clean. Check for sloppy solder running into another terminal or circuit. Clean all soldered points between each IC terminal pin or transistor element. Use a low-wattage soldering iron or a battery-operated iron on IC and transistor connections.

After installing a new IC, check each soldered terminal. Inspect the PC wiring going to each IC pin. With the RX1 scale of the ohmmeter, check from each pin terminal to the first component terminal on the same PC wiring. Sometimes a small piece of PC wiring might pull off with removal of the old IC. Naturally, the soldered terminal does not connect to the circuit. By checking each IC terminal pin to ground and also to a corresponding PC wiring, you can prevent a chassis from failing (Fig. 38-19).

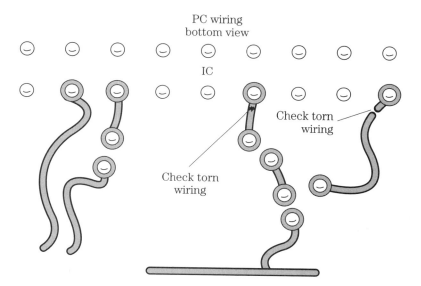

■ **38-19** *Doublecheck each IC pin terminal to a connected component on the PC board for possible cracked wiring at the IC terminals.*

Inside outside

Improper vertical sweep or no vertical sweep can be caused by defective components outside of the immediate vertical circuits. Check for an open yoke winding, electrolytic coupling capacitor, and sweep return resistor. A leaky or shorted yoke winding might provide improper vertical sweep and blow the main fuse. Inspect the yoke terminal connections or socket on the PC board (Fig. 38-20).

In larger TV screens, check the vertical pincushion transistors, pincushion transformer, and burned components for improper vertical sweep. Inspect all terminals of the pincushion transformer, or just solder each transformer board connection. Measure

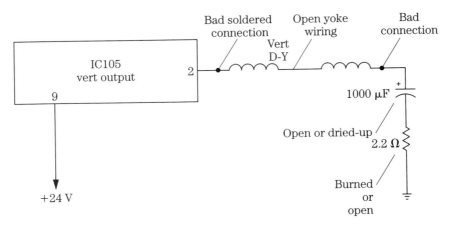

38-20 Look outside of the vertical circuits for a defective component that causes the improper vertical sweep.

the resistance of the primary and the secondary winding for open conditions at the pincushion transformer or coils. Sometimes a change in resistance of the primary winding will cause insufficient raster (Fig. 38-21). Check critical diodes and resistors for burned or open conditions. Rotate the width and linearity controls to locate a burned or open control in the pincushion circuits. Improper adjustments of the vertical height and linearity controls might produce vertical foldover.

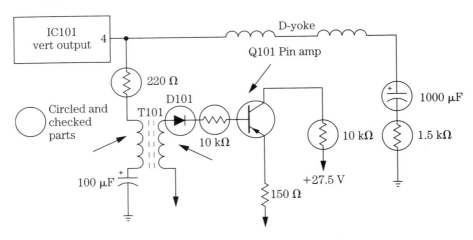

38-21 Check for burned resistors, pincushion transistors, diodes, and open or a change of resistance in pincushion transformer windings for vertical foldover.

An unusual vertical problem

Insufficient vertical raster was found in a Radio Shack TC-2700 27-inch color TV. The vertical output signal was checked at terminal 6 and it seemed fairly normal. The same waveform was found at the vertical output amp (IC301) on pin 6. If the scope probe was touched at output pin 4, about 27.5 Vp-p was found. Because a 50-Vp-p waveform should be found at output pin, IC301 was the suspected defective component (Fig. 38-22).

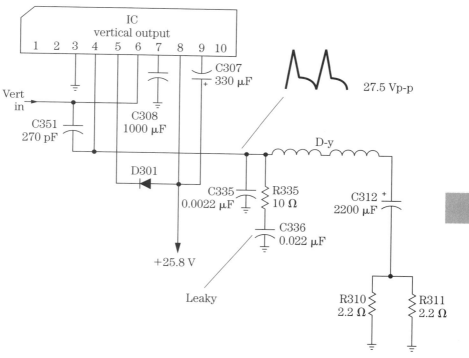

■ **38-22** *Insufficient height in a Radio Shack 27-inch TV set was caused by a leaky capacitor (C336) off of the yoke winding.*

Before replacing the vertical output IC, components tied to all terminals were checked. One end of diode D301 was removed from the circuit and it tested normal. C351, C308, and C307 were tested with a capacitor tester and were within the required limits. IC301 was replaced with a universal RCA SK9753. The results were the same, insufficient vertical raster.

Again, the waveform on pin 6 of IC301 was around 28 Vp-p. Perhaps the output yoke circuits were loading down the vertical output IC. Continuity measurements on the yoke assembly indicated normal resistance. Return capacitor C312 was checked with one lead removed and it tested 5 µF higher than the original capacitor (2200 µF). R310 and R311 measured little over 1 Ω, with both resistors in parallel.

Voltage measurements on supply pin 8 (V_{CC} 25.8 V) measured 25.2 V. No voltage measurements were listed on the schematic diagram for IC301.

Resistance measurements were made from each pin terminal of IC301 to common ground. Pins 3 and 4 showed leakage. Pin 3 was grounded and pin 4 was the output terminal. C312 was the likely suspect for leakage, but it checked normal. A closer look at the components tied to pin 4 and the vertical yoke winding indicated that C335 and C356 were tied to ground. Because R336 (10 Ω) should show signs of overheating, one end of C336 (0.022 µF) was removed from the circuit. C336 was found leaky, causing insufficient vertical sweep.

39
Troubleshooting the TV horizontal circuits

THE ELECTRONICS TECHNICIAN HAS MADE MORE MONEY servicing the horizontal circuits in the TV chassis than any other circuits. They just break down more often. New service techniques must be used to ensure quick and profitable servicing methods. Today, the horizontal circuits supply various voltages to other circuits in the chassis that before were provided by the low-voltage power supply. The horizontal circuits must operate before other circuits can function. Troubleshooting the horizontal circuits can be fun—and very frustrating at times (Fig. 39-1).

These horizontal circuits can be serviced rather quickly with a DMM and an oscilloscope. With critical voltages and resistance measurements, you can locate the defective component. Scope waveforms at various stages might indicate if the horizontal stages are performing. Manufacturers' schematics or Howard Sam Photofacts of

■ **39-1** *The electronic technician taking critical voltage measurements in the portable TV chassis.*

the different TV chassis help tremendously in speeding up the service process (Fig. 39-2). Although many of the regular horizontal oscillator and output symptoms are the same, several new circuits with the same symptoms are used in today's solid-state chassis.

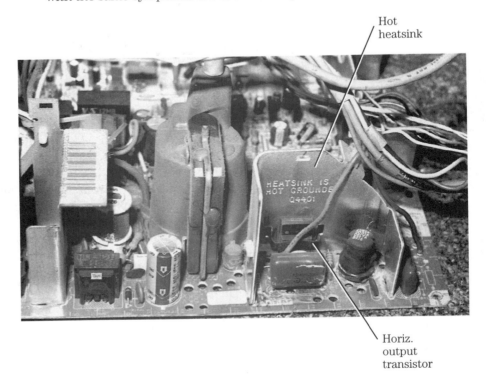

■ **39-2** *The horizontal driver transformer mounts close to the horizontal output transistor and the flyback transformer.*

To locate the intermittent, high-voltage shutdown, and chassis shutdown symptoms, the variable power line transformer is a necessity. The old TV tube chassis contained a power transformer that separated the hot TV chassis from power line and ground. In the present 120-Vac power-line chassis, just connecting a test instrument to it might damage the chassis and the test instrument. Play it safe by plugging the TV chassis into an isolation power transformer (Fig. 39-3).

Countdown and deflection circuits

In the early TV chassis, tubes were used to provide horizontal sweep to the yoke and picture tube. Along came the transistor os-

■ **39-3** *Plug the TV chassis into the isolation transformer before attempting to service the color TV chassis.*

cillator and drive circuits. Today, the horizontal oscillator and drive amplifier might be contained in one large deflection or countdown IC. The early deflection IC contained only the horizontal and vertical oscillator and driver circuits. Now, these same circuits are contained in a single IF/SIF/chroma/vert/horiz/AFT IC (Fig. 39-4).

Drive signals developed for the vertical and horizontal scan systems are derived from a single oscillator housed in the deflection or countdown IC. The master oscillator might operate at eight times the nominal horizontal frequency. This signal is applied to a horizontal divider stage and a vertical countdown stage inside of the IC. The output from the horizontal divider stage might be applied to a burst gate state to develop a burst signal, and to the horizontal amp stage to provide a horizontal drive signal. Some horizontal circuits use a frequency divider or window system to develop the horizontal and vertical drive signal. The horizontal oscillator frequency (15,734 Hz) might be crystal controlled (Fig. 39-5).

Horizontal driver circuits

The electronic driver circuits should not be confused with the golfer's drive club, as one provides a lift or drive signal to the horizontal output transistor. The horizontal drive pulse is provided by the deflection or countdown IC and is applied to the base terminal

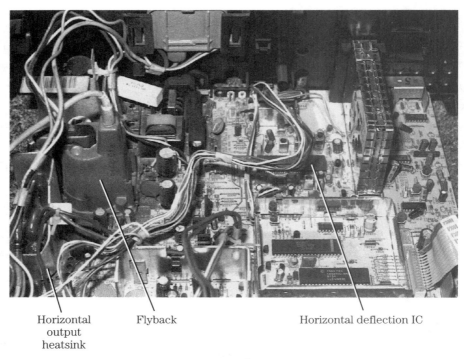

■ **39-4** *Locate the horizontal deflection IC, which is located in a large IC with other circuit features.*

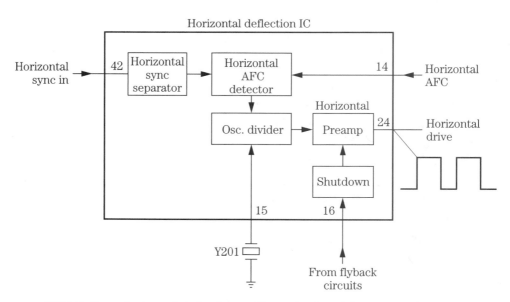

■ **39-5** *Some horizontal deflection oscillator circuits might be crystal controlled.*

Troubleshooting the TV horizontal circuits

of the driver transistor. The square-wave drive pulse might have an amplitude voltage of 0.82 to 3.5 V p-p. This drive pulse is greatly amplified by the horizontal driver transistor, from 150 to 250 Vp-p (Fig. 39-6).

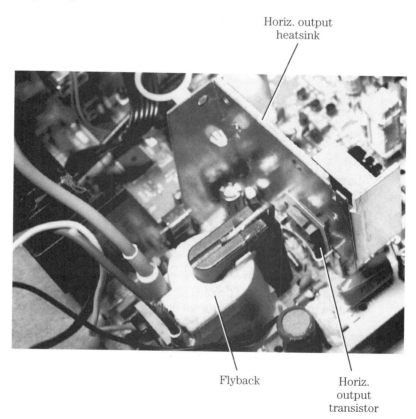

■ **39-6** *The integrated horizontal output transformer (flyback) might have internal HV diodes and capacitors, scan-derived low voltages, and HV circuits.*

The horizontal drive transistor operates at a much higher voltage than most transistors and from the same voltage source as the horizontal output transistor. In a line-operated power supply, the voltage source might be from 100 to 140 V from a line-operated fixed IC regulator. The primary of the driver transformer has a higher average resistance (75 to 150 Ω) than the secondary (0.05 to 0.35 Ω). Poor connections on the driver transformer might cause poor start-up/shutdown symptoms (Fig. 39-7). Often, the horizontal driver transistor will run quite warm without a drive signal applied to the base terminal. No output drive signal is used with a very low collector voltage.

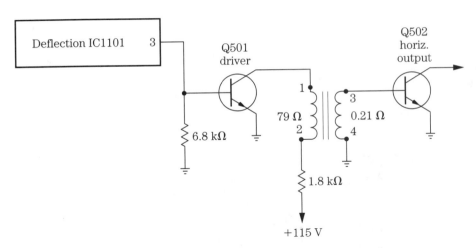

■ **39-7** *Check for poorly soldered connections on the driver transformer for chassis shutdown or improper start-up.*

Horizontal output circuits

The horizontal drive pulse is supplied to the base of the horizontal output transistor. The output of the horizontal output transistor is used to power the horizontal deflection yoke and the flyback. Today, an integrated output transformer (IH VT) develops the high voltage, a low voltage B+ source for the CRT driver transistor, the focus CRT filament voltage, and the screen grid voltages. Scan-derived low-voltage sources might also be used in the secondary circuits of the flyback (Fig. 39-8).

The horizontal output transistor (HOT) might have a damper diode connected to the collector terminal. You might find the damper diode located inside the horizontal output transistor in many circuits. The damper diode prevents oscillations and ringing in the TV power supply. These damper diodes are rated from 1200 to 1800 V. In some horizontal chassis, the driver transformer, output transistor, and flyback transformer might be eliminated from ground or it might have its own so-called hot chassis (Fig. 39-9).

The new horizontal silicon output transistor might have a T-048 mounting instead of the regular T-040 or T-042 case. Many of these output transistors operate within a 1500-V and 4- or 5-A range. The average input drive pulse might vary from 10 to 20 Vp-p and the output at collector terminal, 400 to 1000 Vp-p. Horizontal output transistors are replaced more often than any other transistors on

39-8 *Internal arcing inside of the flyback transformer might destroy the flyback transformer and horizontal output transistor.*

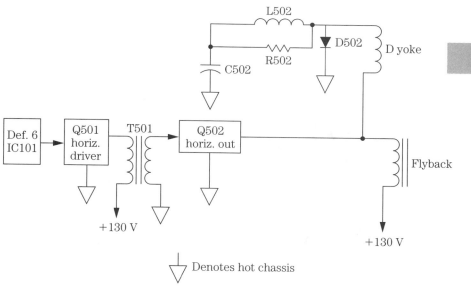

39-9 *The horizontal circuits might be above ground or called a hot ground in some TV chassis.*

the TV chassis. A defective output transistor can be located with a resistance diode test from collector (body) terminal to chassis ground. A low-resistance measurement (below 10 Ω) might indicate a leaky or shorted transistor and damper diode (Fig. 39-10).

Horizontal output circuits

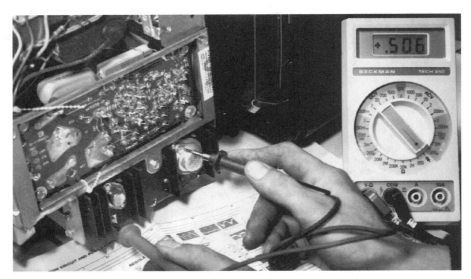

■ **39-10** *Check the suspected horizontal output transistor from the metal body (collector) to the ground with the diode test of the DMM for a leaky transistor or damper diode.*

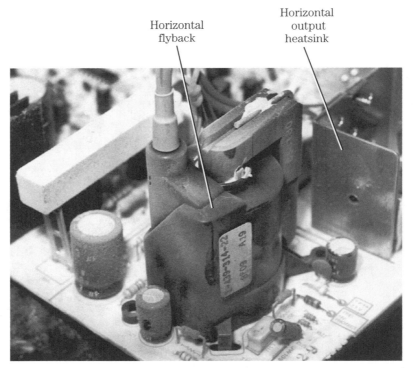

■ **39-11** *The defective flyback might arcover internally, short turns, and produce a low voltage or no high voltage.*

Troubleshooting the TV horizontal circuits

High-voltage problems

Low or improper high voltages might result from a defective drive signal, output transistor, yoke, hold-down capacitor, or flyback transformer (Fig. 39-11). High voltage might increase with open safety or hold-down capacitors, higher low supply voltage, or an improper setting of the HV or $B+$ adjust control. High voltage might arcover at this high-voltage cable, or at the flyback or anode of the picture tube. A failure of the disable or HV shutdown circuits might result in high-voltage arcover. Loud popping at the anode terminal or spark gaps indicates that a very high voltage is present. High-voltage capacitors inside the flyback might arcover and destroy the horizontal output transistor and transformer (Fig. 39-12).

Damper diodes

Damper diodes usually become leaky and destroy the horizontal output transistor and the line fuse. If the damper diode appears open or is disconnected from the circuit, the output transistor will overheat and be destroyed. If the damper diode becomes defective inside of the horizontal output transistor, replace the output transistor (Fig. 39-13). The early damper diode was connected to the collector circuit of the horizontal output transistor, alongside of the safety capacitors. You might find two damper diodes in a series-to-ground circuit of some horizontal output circuits. Replace the damper diode with a 1200- to 2000-V operating voltage.

High-voltage hold-down problems

If the high voltage cracks and pops, then shuts down, suspect open safety or hold-down capacitors. These capacitors are located in the collector circuit of the horizontal output transistor and the primary winding of the flyback transformer. Replace the safety capacitors with ones that have the exact capacity and voltage in these hold-down circuits. Never use a 450- or 1000-V capacitor in the collector safety circuits. Often, these capacitors go open or one lead comes loose from the circuit, letting the high voltage increase tremendously above the safety range of the TV (Fig. 39-14). Notice that these components have a star next to them, indicating that they are safety replacements.

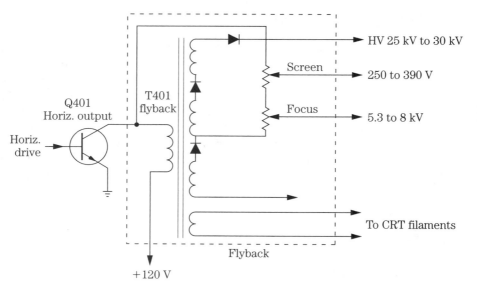

■ **39-12** *High-voltage capacitors inside of the IHVT might arcover and destroy the output transformer.*

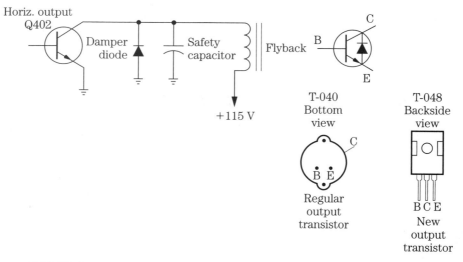

■ **39-13** *Replace the horizontal output transistor with a leaky damper diode, if the diode is mounted inside of the transistor.*

Troubleshooting the TV horizontal circuits

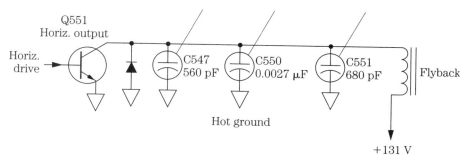

■ **39-14** *Open or defective safety capacitor in the collector circuit of horizontal output transistor might produce excessive high voltage at the CRT.*

Flyback operation and problems

The collector voltage for the horizontal output transistor is drawn from the $B+$ supply through the IHVT primary winding. Current is drawn through the primary winding and the yoke winding to the common ground. The horizontal output transistor turns off as the beam reaches the right edge of the picture tube, starting the beginning of the horizontal retrace. After the horizontal transistor turns off, current flows through the horizontal yoke winding and the IHVT primary winding into the retrace capacitor, where energy from the system is temporarily stored (Fig. 39-15).

If the retrace capacitor is fully charged, the current flow reverses and begins flowing from the retrace capacitor back into the primary winding of the IHVT and the horizontal yoke windings. Then, the electron beam from the right edge of the picture is deflected back to the left edge to begin the next time, completing the retrace.

The IHVT primary decreases toward zero when the first half of the horizontal scan continues or when the current flows through the damper diode to the yoke. If the current reaches zero (the electronic beam in the center of the picture), the horizontal output transistor begins to conduct. This method reverses the current flow and begins drawing current from the IHVT primary winding and from the horizontal yoke winding to the ground, completing the second half of the scan.

The defective flyback might short turns inside of the plastic coating, overheat and run warm, contain leaky HV diodes in the secondary winding, and destroy the horizontal output transistor (Fig. 39-16). A ticking noise heard close to the transformer might indi-

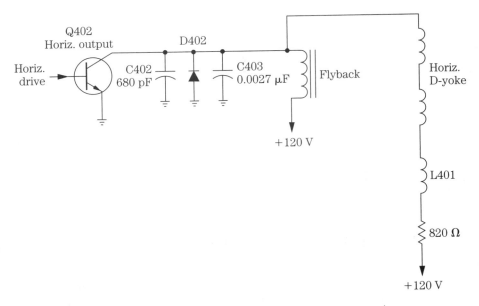

■ 39-15 *Current flow through the yoke winding with retrace action.*

■ 39-16 *The focus and screen controls are tapped from the high-voltage winding and the focus voltage might vary from 5.3 to 8 kV.*

cate that the flyback transformer is defective, loading down of the secondary voltage circuits or producing an improper drive voltage. The defective flyback might cause high-voltage shutdown. If the primary winding is open, no voltage is measured at the collector terminal of the horizontal output transistor.

Scan-derived screen and focus voltage

Scan-derived secondary voltage for the screen and focus voltage applied to the picture tube elements are connected to the HV winding. The focus voltage might vary from 5 to 8 kV while the screen voltage might vary from 250 to 390 V. Both voltages are tapped from the HV diode to the focus and screen controls (Fig. 39-17).

Both focus and screen voltages might be included in one ceramic component, connected directly to the flyback. In early chassis, the screen and focus ceramic assembly were a separate unit wired to the flyback. Today, the focus and screen assembly might be mounted with the horizontal output transformer. If an arcing or defective focus and screen assembly is used, the whole flyback

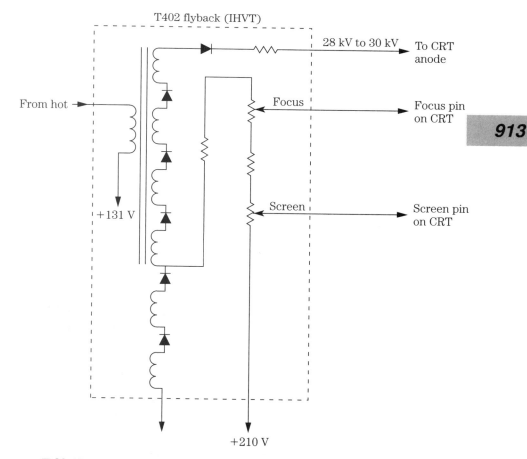

■ 39-17 *In today's color TV chassis, the screen and focus assembly are part of the flyback assembly.*

Critical waveforms

component must be replaced. The defective assembly might contain poor ceramic resistance control and arcing inside, which produces thin horizontal white lines in the picture.

Critical waveforms

Scope the horizontal circuits with the oscilloscope at each stage to determine what section ceases to function. Start at the horizontal output pin terminal of the deflection IC with a square waveform with average 2.5 Vp-p (Fig. 39-18). Waveform (1) is used at pin 64 of the deflection IC included with many other circuits in IC1101 (Fig. 39-19). The collector output terminal of driver transistor

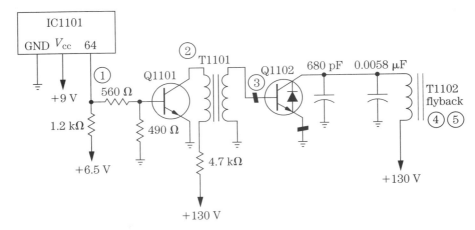

■ **39-18** *Scope the various numbers to locate the defective horizontal output stage.*

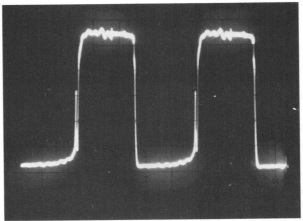

■ **39-19** *The waveform used at number 1 of the output of the deflection IC.*

Q1101 contains an amplified square wave (2) with an average around 185 Vp-p (Fig. 39-20).

The secondary winding of T1101 provides base drive voltage for output transistor Q1102 (Fig. 39-21). If the base drive voltage is not present, the horizontal output transistor is soon damaged (3). Hold the scope probe close to the flyback transformer to indicate that this stage is functioning (Fig. 39-22) (4). The horizontal output waveform can be seen connected to the metal body of the collector terminal through a 0.0058-μF 2-kV capacitor (Fig. 39-23) (5). This waveform should be clean cut; if oscillation and ringing lines appear on the waveform, suspect a defective flyback or yoke assembly. Distorted waveform or retrace spikes might indicate that the flyback transformer has a cracked core. The base wave-

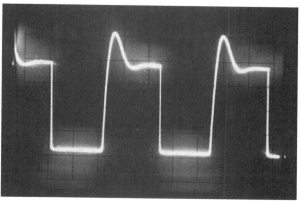

■ **39-20** *The waveform used at number 2, at the collector terminal of the driver transformer.*

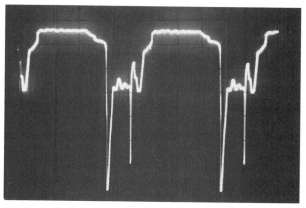

■ **39-21** *The drive waveform at the base terminal of the horizontal output transformer, number 3.*

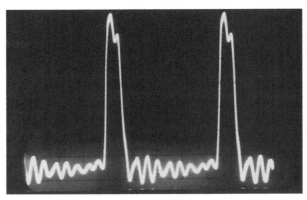

■ **39-22** *The waveform used with the scope probe held close to the flyback transformer.*

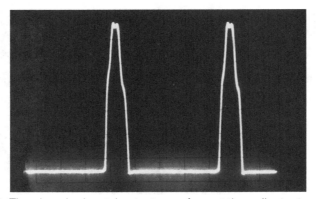

■ **39-23** *The clean horizontal output waveform at the collector terminal of output transistor number 5.*

form of the horizontal output transistor will be quite different with the transistor out of the circuit.

If improper waveforms or no waveforms are used at each number checkpoint, suspect a defective component in the preceding stage. In scan-derived voltages, the voltage applied to the deflection or countdown IC will not be there if the horizontal circuits are not functioning. A high $B+$ voltage might be at the horizontal output transistor collector terminal. Low collector voltage is at the driver transistor if there is no input drive pulse.

Pulled in on each side

Insufficient width might result from low $B+$ voltage, defective high-voltage regulator circuits, SCRs, or horizontal output transistors. Poorly soldered connections of the pincushion coil or transformer, regulator, and a burned driver transformer can produce poor width (Fig. 39-24). Improper adjustment of the HV or $B+$ control might cause insufficient width. A burned or open isolation resistor in the $B+$ supply to the primary winding of the flyback might occur in some TV chassis. Check the bypass capacitors in the collector circuits of the horizontal output transistor and the deflection yoke return winding.

■ **39-24** *Sides pulled in on the picture or raster might indicate problems in the horizontal output circuits or low-voltage power source.*

Horizontal foldover

Check the horizontal output circuits for foldover symptoms. The leaky horizontal output transistor and flyback might produce horizontal foldover (Fig. 39-25). Open bypass capacitors in the high-voltage regulator or SCR horizontal circuits can cause some types of foldover. Check for poorly soldered terminals of reactors and regulation transformers. Check capacitors C415, C416, C417, and C418 for poor horizontal linearity and foldover (Fig. 39-26).

Deflection yoke problems

Shorts between the vertical and horizontal windings might blow the main fuse. Shorted turns in the horizontal windings might cause chassis shutdown. Sometimes the chassis might operate for a few minutes and then shut down. Open horizontal yoke winding or poor socket and board terminal connections might not show any signs of sweep (Fig. 39-27).

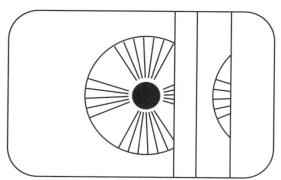

■ **39-25** *Horizontal foldover occurs in the horizontal output and flyback circuits.*

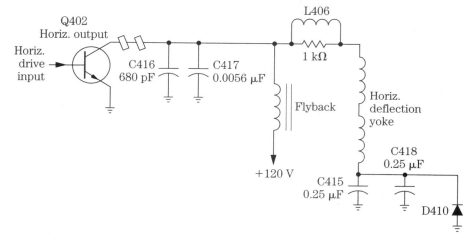

■ **39-26** *Check C415, C416, C417, and C418 for poor horizontal linearity or foldover in the output circuits.*

■ **39-27** *An open horizontal yoke winding and poor connections might cause no sweep on the raster.*

Troubleshooting the TV horizontal circuits

The defective yoke might load down the flyback and horizontal output transistor, indicated by lower high voltage at the anode terminal of the picture tube. The horizontal output transistor might run warm with an overloaded or a leaky yoke assembly. Remove the red lead from the yoke assembly. The low dc supply voltage at the horizontal output transistor will increase immediately when the yoke lead is removed. High voltage might only be one half of the original HV.

Check the yoke with the low-resistance scale of the ohmmeter. The total windings might be from 3 to 15 Ω. The larger the picture tube, the greater the wire diameter and the lower the resistance of the deflection yoke winding (Fig. 39-28).

■ **39-28** *The yoke assembly might have a resistance between 3 to 15 Ω in the latest TV chassis.*

HV shutdown

Excessive high voltage used at the anode terminal of the picture tube might cause high-voltage shutdown. Most TV chassis today have a horizontal disable circuit or some method to shut down the chassis when the high voltage reaches a certain level. A defective safety or hold-down capacitor circuit might cause excessive high voltage. An improper setting of the $B+$ or the high-voltage control might cause HV shutdown (Fig. 39-29). High line voltage might also cause HV shutdown.

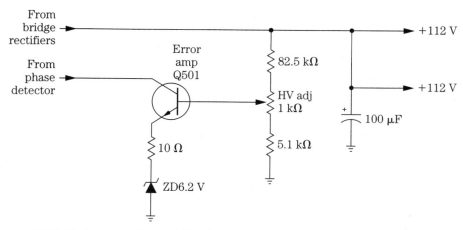

■ **39-29** *Improper B+ or HV adjustment might cause high-voltage shutdown.*

Use a universal power-line isolation transformer to slowly raise the line voltage. Monitor the HV at the CRT anode terminal with an HV voltmeter probe. Raise the voltage and notice at what line voltage and high voltage the TV chassis shuts down. If the chassis shuts down with 80 to 90 Vac applied, suspect a defective horizontal output circuit or flyback transformer. Check the horizontal output circuits for a defective component. Doublecheck all safety capacitors. High-voltage shutdown might result from a defective HV shutdown circuit.

Horizontal disable circuits

If the high voltage becomes too high, the shutdown circuit turns the chassis off. In early disable circuits, excessive high voltage triggered the horizontal oscillator transistor and made the chassis go into horizontal lines (Fig. 39-30). These horizontal lines could not be straightened up until the high-voltage shutdown problem was solved. The primary concern of the high-voltage shutdown circuits is to protect the operator from radiation at the picture tube because of excessive high voltage.

Horizontal disable circuits might consist of a silicon-controlled rectifier (SCR) to kill the horizontal oscillator circuit. A leaky high-voltage SCR can apply excessive high voltage to the output transistor. Other disable circuits can kill the drive pulse at the horizontal drive transistor. Today, the shutdown or disable circuit is fed back from the flyback transformer to a shutdown circuit in the deflection IC.

■ **39-30** *The horizontal oscillator might be disabled if the high voltage reaches an excessive level, producing horizontal lines in the picture.*

X-ray protection might be accomplished with a pulse from the winding of the high-voltage transformer (Fig. 39-31). The flyback pulse is rectified by D510 with the voltage sent to a 10-V zener diode. To activate the x-ray or high-voltage protection circuit, the voltage at Z530 must exceed this threshold voltage. During normal operation, the voltage at pin 16 of the deflection IC1101 is 0 V. If the high voltage increases, the rectified voltage of D510 will also increase and will trigger zener diode (Z530) into conduction, shutting down the TV chassis. The chassis can be reset; let it sit for a few seconds and turn it on again. Check the shutdown components if the chassis shuts down too soon or if it won't shut down the chassis at all.

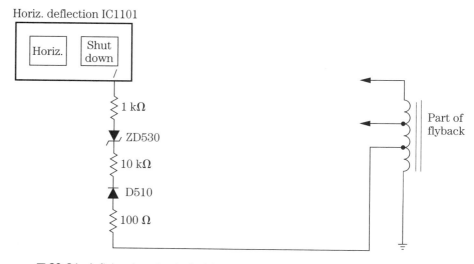

■ **39-31** *A flyback pulse is fed back to the shutdown circuit internally inside of the deflection IC.*

Horizontal disable circuits

High-voltage shutdown test

In some TV chassis, you can see if the horizontal disable circuits are operating by shorting a test point to ground or by momentarily shorting two different test point lugs (Fig. 39-32). If these two points are shorted, the chassis will automatically shut down the chassis. The chassis must be started up again to resume TV operation. If the set does not lose the raster and sound when it is shorted, suspect that an HV shutdown circuit is defective. Check all components within the shutdown circuits. Scope the waveform or pulse from the flyback transformer. Check for poor connections or open windings if there is no waveform.

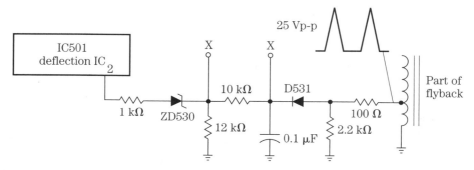

■ **39-32** *Test out the x-ray or HV shutdown circuits by shorting the (x) terminals.*

Keeps blowing the fuse

Check for a defective component in the horizontal circuits, such as leaky output and driver transistors, damper diodes, flyback transformers, tripler units, and yoke assemblies if the unit keeps blowing fuses. Clip a 100-W lightbulb across the fuse holder and start removing components, starting with the output transistor, which might be leaky or shorted. The light will go out if a shorted component is removed from the low-voltage and horizontal output circuits. Suspect the horizontal output circuits if both the line and $B+$ fuses are blown open (Fig. 39-33).

Suspect a leaky yoke, tripler, or flyback transformer if it takes several seconds before the fuse opens up. All three units might run warm before the fuse opens. If a flyback or output transistor is suspected to be leaky, use the variable isolation transformer to prevent damaging another new output transistor replacement. Isolate a possible shorted horizontal yoke assembly by removing the red lead from the PC board.

3-amp 250 V
Line fuse

■ **39-33** *Suspect a defective horizontal output transistor and flyback if the line fuse keeps blowing.*

Notice if it takes 4 or 10 seconds for the fuse to blow with a chassis having an HV tripler component. Disconnect the input lead from the tripler unit (Fig. 39-34). If a lightbulb is used as a fuse, the light will go out after removing the horizontal output transistor and the tripler input lead. The leaky HV tripler might arcover between the unit and chassis, and produce loud arcover noises.

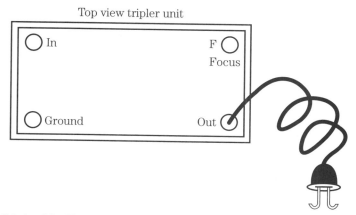

■ **39-34** *In older TV chassis, the tripler unit might become leaky and damage the fuse and output transistor.*

Keeps destroying the output transistor

Suspect a leaky output flyback transformer, open damper diode, open tuning and safety capacitors, or an improper sweep pulse if the output transistor is damaged repeatedly. Often, the output transistor indicates that there is a short between the collector and the emitter terminal (Fig. 39-35). Remove one end of the damper diode and safety capacitors, and test each one individually. Check the damper diode with the diode test of the DMM and check the safety or hold-down capacitors with the capacitance meter.

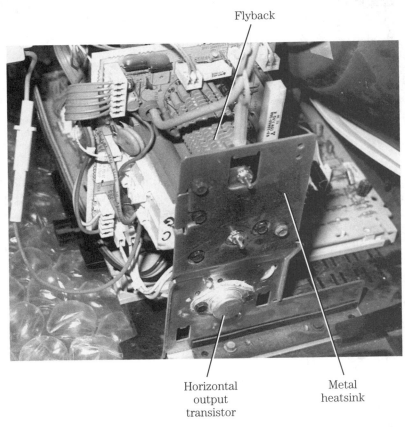

■ **39-35** *Suspect an open damper diode, shorted yoke assembly, flyback transformer, and overloading in the output circuits that destroys the output transistor.*

Most ceramic or molded safety capacitors have the value marked on the body of the capacitor. Dipped ceramic hold-down or bypass capacitors might have a color code or a value stamped in numbers. Capacitors that have several numbers stamped on it indicate that the first two digits designate the value of the capacitor, and the third number indicates the number of zeros to add at the right-hand side of the number. For small bypass capacitors marked in picofarads, multiply it by 10-2 (Fig. 39-36).

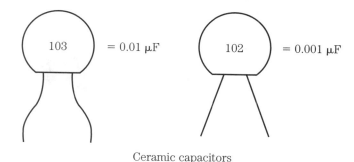

Ceramic capacitors

■ **39-36** *Small bypass ceramic capacitors might be marked with numbers to identify the correct value.*

For example, a capacitor marked 103 has a value of 10 followed by three zeros, or 10,000 pF or 0.01 µF. A bypass ceramic capacitor marked 102 has a value of 10 followed by two zeros, or 1000 pF or 0.001 µF. Some of these capacitors might be confusing and they provide no numbers or color codes. The best method is to test each capacitor on the capacitor tester.

Replacing hot transistor

After determining that the horizontal output transistor is defective, remove two metal screws and pull the transistor from the socket (Fig. 39-37). Remove one screw of the new T-048 output transistor and unsolder the terminal leads from the PC wiring to replace the leaky or shorted transistor. Check the back side of the transistor for a mica insulator. Often, a new mica insulator is included with the hot replacement. Test the new transistor in a beta tester or with the diode-transistor test of the DMM before installing it.

Silicone grease should be applied to the mounted side of the transistor and heatsink (Fig. 39-38). If a piece of mica insulation is used, place grease on both sides of the insulator. Place the transistor terminals into the correct holes and solder them. Replace the mounting screws. Make sure that the replacement has the same or higher voltage and current rating before replacing it.

■ **39-37** *Remove two metal screws so that you can replace the horizontal output transistor.*

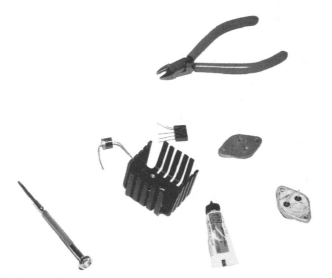

■ **39-38** *Place silicone grease on the output transistor and the heatsink to dissipate heat. Choose clear silicone grease as the white grease tends to mark up your clothes, hands, and tools.*

Intermittent horizontal problems

Monitor the TV chassis with a voltmeter and oscilloscope (Fig. 39-39). Place the scope probe next to the flyback transformer to determine when the waveform collapses. Place an HV meter probe at the anode terminal (under the rubber cup) to determine if the high voltage lowers when the chassis becomes intermittent. If the output waveform disappears in chassis shutdown, move the scope probe to the base terminal of the horizontal output transistor. Keep moving the scope probe and the voltmeter back toward the horizontal deflection IC output terminal if it is intermittent (Fig. 39-40).

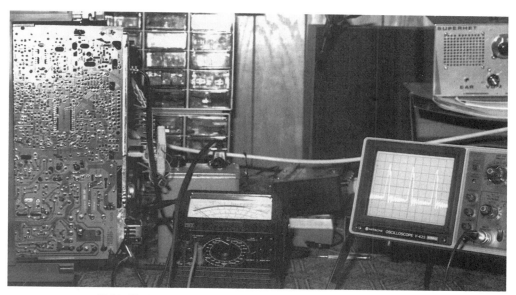

■ **39-39** *Monitor the horizontal output circuits by placing the scope probe near the flyback transformer.*

Lines in the raster

Fine white lines in the picture or raster might result from a defective focus and screen grid control assembly. Rotate the focus and screen controls and notice if the lines disappear. Firing inside of the IHVT flyback transformer might cause noisy lines in the picture. Sometimes you can hear the arcing and at other times you can't. Check the rubber HV socket if it isn't properly hooked into the anode button for arcing lines (Fig. 39-41). Make sure that the aqueduct ground wire is tight and grounded against the picture

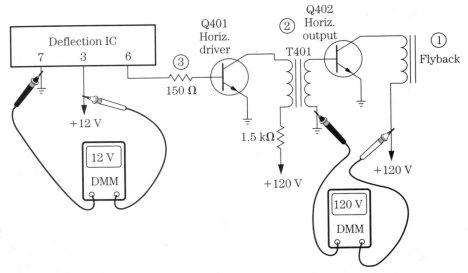

39-40 *Scope the horizontal output circuits by the number for the correct waveform.*

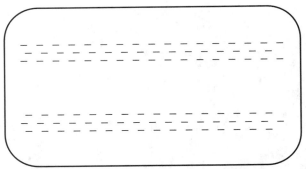

39-41 *Fine white and dark lines running horizontally across the picture might be caused by an arcing focus control and an ungrounded picture tube.*

tube bell area. Check for a broken or cracked core at the flyback transformer for noisy lines in the raster.

Remove the outside antenna lead or cable to determine if the noise is picked up or is inside of the TV set. Intermittent arcing of a tripler unit might cause dark firing lines in the picture. Vertical "jail-bar" dark lines to the left or right side of the raster are called *Barkhausen oscillations* and they might result from defective horizontal output circuits or from a defective flyback transformer. Several dark bars to the left side of the picture might be caused by dried-up or open filter capacitors in the primary winding of the flyback transformer (Fig. 39-42). Check for poor filtering in the horizontal oscillator circuits by shunting the small electrolytic capacitors.

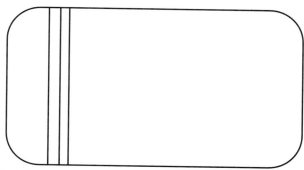

■ **39-42** *Jail bars to the left side of the raster might be caused with dried-up filter capacitors in the flyback supply source in some TV chassis.*

Repairing special TV circuits

SOME OF THE SPECIAL CIRCUITS IN THE TV CHASSIS include: color, secondary voltages, chopper, VIPUR, SMPS, comb and saw filter, computer, on-screen display, standby, and stereo sound circuits. Most of these circuits can be serviced with a good DVM and an oscilloscope (Fig. 40-1). The color, SMPS, and chopper circuits are more difficult to repair in the special TV circuits. Troubleshooting stereo sound and standby circuits might take up added service time although the comb and saw filter circuits seldom cause service problems.

■ **40-1** *Take critical voltage and waveforms to service the chopper, SMPS, on-screen display, and stereo sound circuits.*

Repairing the color circuits

In today's chassis, the color circuits are found in a large IC that might also include the video, luminance, hold-down, and deflection circuits (Fig. 40-2). The color problems include: poor color, intermittent color, color bars, and no color. Make sure that a good B&W picture is visible before attempting to service the color sections. Besides the oscilloscope, connect a color-dot bar generator to check the color waveforms (Fig. 40-3).

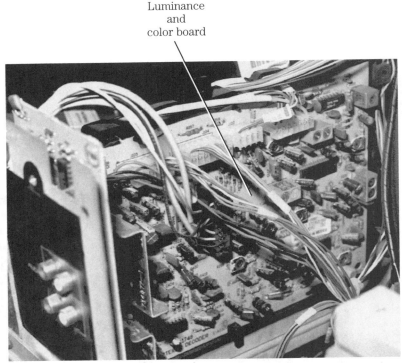

■ **40-2** Locate the color circuits on a special PC board or TV chassis by locating the 3.58-MHz crystal.

Connect the color generator to the TV antenna terminals for correct color waveforms. Most TV schematics have waveforms tested and taken with a color-bar generator. Knowing what terminals of the IC connect to the correct color pins is very important. Most color circuits can be serviced with an oscilloscope, DMM or DVM, and a color-bar generator.

■ **40-3** *Connect the color-bar generator to the antenna terminal to service the color waveform in the color circuits.*

Critical test points

You might find critical color test points in many TV schematics for easier servicing. Check for the critical voltage at the supply pin (V_{CC}) if the color IC has no color. No voltage or low voltage on pin 23 might indicate a defective power supply or leaky IC. If the voltage measurements on any pin are within 1 or 2 V of those shown on the schematic, go to the next pin terminal. A low-resistance measurement on the voltage supply pin might indicate that an IC is leaky (Fig. 40-4). Fairly normal voltage measurements might not indicate that the IC is open or defective.

Check the color waveform on color oscillator pin 13 (Fig. 40-5). An improper oscillator waveform indicates that there is no color in the picture. Locate the oscillator pin connected directly to the color crystal. You can locate the color circuits or correct the dc on the chassis with the color crystal nearby. Measure the voltage found on the oscillator pin. A defective crystal might cause intermittents, no color, or color bars.

Scope the input and output color waveforms on the chroma IC. Turn the color level control wide open, especially with weak or poor color symptoms. Remember that the color-bar generator should be connected for all color waveform tests. Check the color input waveform at pin 3 and check the output at pins 20, 21, and 22 (Fig. 40-6). A normal demodulator waveform at the chroma

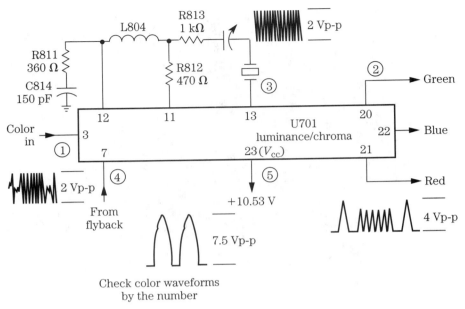

■ **40-4** *Check the voltage supply pin to determine if the chroma IC or the power source is defective.*

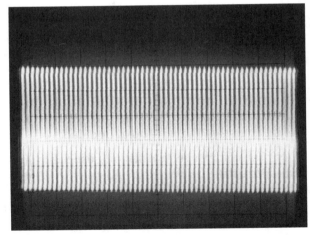

■ **40-5** *The color 3.58-MHz oscillator waveform is used on the output pin terminal of the chroma IC.*

output terminals indicates that the color circuits are normal. Critical voltage measurements on the output terminals indicate that the output terminals are normal and that all three voltages should be within 1 or 2 V of each other.

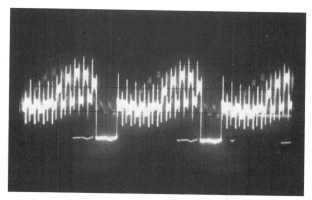

■ **40-6** *The color output scope waveforms at the color output terminals of chroma-luminance IC. These waveforms feed to the color output transistors.*

Check the color output transistors if one color is missing and if the output color waveforms are normal at the chroma IC. Take critical voltage measurements on each output transistor and compare it to the schematic. If blue is missing from the picture, check for improper voltage measurements on the blue output transistor. Remember, a defective color gun assembly of the picture tube of any color might cause the missing color. Most color output transistors are located on the CRT socket assembly (Fig. 40-7).

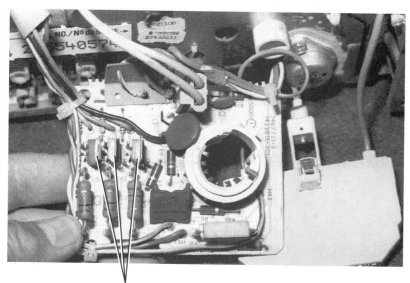

Color output transistors

■ **40-7** *The color output transistors are usually located on the CRT socket board.*

Last, but not least, check the waveform supplied by the flyback to pin 7 of the color IC. If this waveform is missing, no color will be found in the picture. Improper connections on the flyback terminals, burned resistors or a leaky diode in the blanking circuits, burst keying, and black-level clamping circuits might prevent the waveform or pulse from reaching the chroma IC (Fig. 40-8). This color waveform might change a little if a TV station is tuned in, instead of with the color generator.

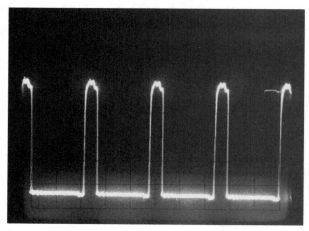

■ **40-8** *The pulse waveform used at the chroma IC from the flyback circuits.*

To receive the correct color waveforms, rotate the color bars into the picture. You might want several color bars in the picture if the generator has only three color bars (Fig. 40-9). By taking critical color waveforms and voltage measurements, most color problems can be solved.

Servicing RCA's chopper and VIPUR circuits

The main components of the RCA chopper power supply are the chopper regulator transistor, chopper transformer, regulator control transistor, x-ray latch, and over-current shutdown transformer (Fig. 40-10). A chopper-type regulator circuit is found in RCA's CTC131 and CTC132 chassis. The ac line voltage is applied to a full-wave bridge rectifier circuit and a standby power transformer (T101). The transformer is active at all times if ac power is applied. The bridge rectifier circuit supplies an unregulated 150 V to the chopper output circuit through the chopper transformer T105 (Fig. 40-11). The standby transformer (T101) supplies

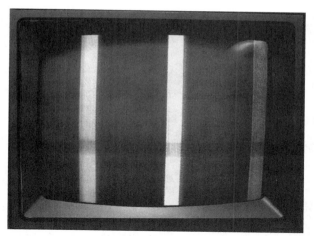

■ **40-9** *Rotate the color-dot bar generator to three or more color bars to take waveforms in the color circuits.*

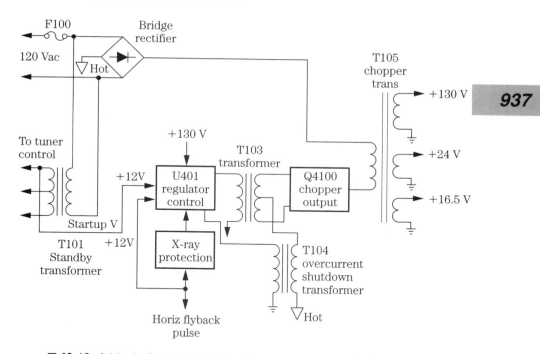

■ **40-10** *A block diagram of RCA's chopper power-supply circuits.*

power to the tuner control module and standby start-up power to the chopper regulator IC (U401).

The pulse-width-modulated (PWM) chopper-regulated power supply is similar to the horizontal deflection system that is used in

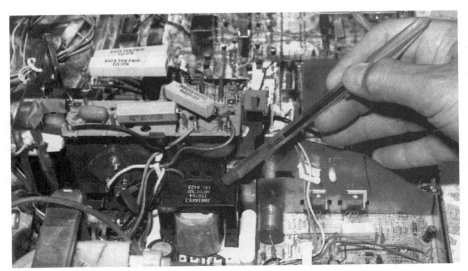

■ **40-11** *Locate the chopper transformer used on RCA's CTC131 chassis.*

many TV chassis. The regulator free-running oscillator frequency of approximately 15 kHz is triggered by a horizontal pulse that is derived from the flyback transformer, locking it to the horizontal scan frequency (15,734 Hz).

The output of the regulator circuit is applied to the regulator drive transformer (T103). The increased pulses of T103 control the on/off state of the chopper input circuit. The on/off state of the chopper output circuit causes a pulsating dc action to occur in the primary winding of the chopper transformer. Increased pulses in the secondary windings are rectified to produce a number of dc sources to power the TV.

Most problems in the chopper power supply are a leaky chopper transistor, open fuse, open isolation resistors, and open silicon diodes in the raw ac power supply (Fig. 40-12). The chopper transistor mounts on a small metal heat shield close to the flyback transformer. Check the regulator IC (U401) for poor voltage regulation. Take all voltage measurements from hot ground in the chopper circuits.

VIPUR power supply

The variable interval pulse regulator power supply is free running with a fixed time for the pulse-width modulator (PWM). Regulation is provided by varying the period of frequency of the PWM stage. A pulse is generated on regulator output transformer

■ **40-12** *The finger points down at the location of the chopper output transistor in RCA's CTC131 chassis.*

(T4100) pin 11, and is rectified to provide a +12-V run voltage. The +12-V source is applied to a voltage-divider network at pin 1 of the IC, which controls the period if the frequency of the PWM is providing regulation (Fig. 40-13).

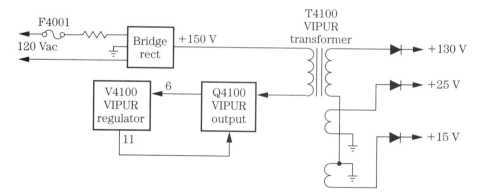

■ **40-13** *The block diagram of the RCA VIPUR regulation and T4100 transformer circuits.*

IC regulator output (U4100) is applied to the MOSFET output transistor Q4100. The energy developed in the primary winding is transferred to the secondary windings, which are rectified to produce several different voltage sources.

Protection is provided with overcurrent sensing and x-ray protection. If the MOSFET transistor begins to draw too much current, it is sensed at pin 11 of the regulator IC. This would trip the shutdown latch circuit inside of the IC and turn off the drive pulse to the VIPUR output transistor. If a defect occurs that would cause the high voltage to increase, Q4900 would turn off, applying voltage to pin 12 of regulator IC U4100, to a shutdown mode.

Most problems caused in the VIPUR power supply circuits are a leaky output transistor (Q4100), fuse, silicon diodes in the raw $B+$ supply, and poorly soldered connections around CR4701, CR4702, and CR4703. If CR4701, CR4702, and CR4703 are defective, they can destroy a different voltage source. Replace these diodes with 3-A silicon diodes.

Troubleshooting Sylvania's SMPS circuits

The switched-mode power supply (SMPS) used in some Sylvania's TV chassis has a self-oscillating transformer-coupled voltage regulator. A switch-mode transformer (T401) is driven by 155 Vdc. IC404 provides line isolation by the optoisolator IC. 155 Vdc is applied to the primary winding of the switch-mode regulator transistor (Q400) (Fig. 40-14).

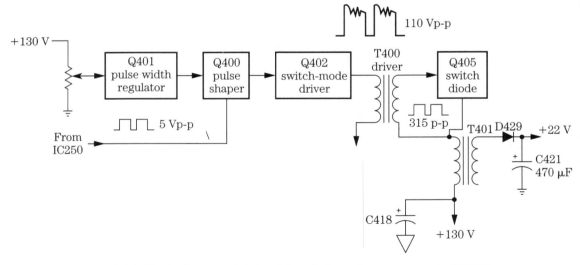

■ **40-14** Block diagram of Sylvania's switch-mode power supply (SMPS) circuits.

Magnetic energy is stored in the transformer during the on time of the transistor and is transferred to the secondary winding during the off time of Q400. The amount of energy transferred is controlled by the self-oscillating frequency and by the on/off time of transistor Q400. The self-oscillating frequency of the power supply under load conditions ranges from 20 to 40 kHz. Depending on the load, the frequency can range from 10 to 60 kHz.

Notice that on the primary side, the input circuits operate with a common hot chassis. Use an isolation transformer when servicing this chassis. A 100-W lightbulb can be connected to the 130-V source as a load. Remove the horizontal output transistor for this test to eliminate the rest of the output circuits. Take scope waveforms on the base and collector terminals of the pulse shaper, switch-mode driver, and switch-mode regulator.

Comb and saw filter circuits

The saw filter circuits are also called the *surface acoustic-wave saw-filter network* (Fig. 40-15). The saw filter takes the place of IF stages used in radio receivers and the TV chassis. A saw filter is made up of piezoelectric material with two pairs of transducer electrodes. One terminal is the input, another the output, and the cen-

■ **40-15** *The saw filter component located in the RCA CTC146 chassis.*

ter terminal grounded. Voltage applied to terminals causes distortion and mechanical waves. There are no adjustments on the saw filter, like the regular IF transformer. In the radio receiver, the saw filter is located between the converter and the second IF and between the second IF and the detector stage. You might find a saw filter between the tuner and the IF amplifier, and between the IF preamp and the sound IF amp stage in the TV chassis (Fig. 40-16).

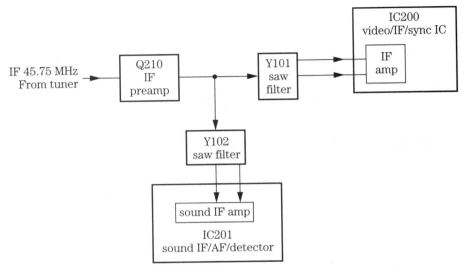

■ **40-16** *A saw filter network might be used between the IF preamp and the video IF IC, and between the IF preamp and the sound IF amp IC.*

Most saw filter networks do not cause much trouble. You can check them with a crystal checker or a DMM. Remove the saw filter unit from the circuit, because low-value resistors might be used in the input and output circuits. Check with the ohmmeter across both input and output terminals with infinite resistance in a normal saw filter component. Check the input resistance across terminals 4 and 5, and check the output resistance across terminals 2 and 3 (Fig. 40-17).

A resistance test does not indicate that the saw filter is normal. Check the saw filter in and out of the circuit with a crystal checker. A crystal tester will indicate if the saw filter is oscillating. Check both input and output terminals in the same manner. The crystal meter will indicate a low reading if it is tested in the circuit and it has a high measurement out of the circuit (Fig. 40-18). If an input signal is used at the input of the saw filter and not at the output, substitute a new saw filter network.

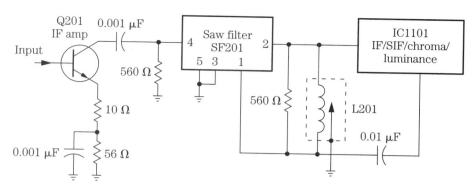

■ **40-17** *The saw filter might be located between IF amp Q201 and IC1101, the IF/SIF/chroma/luminance IC.*

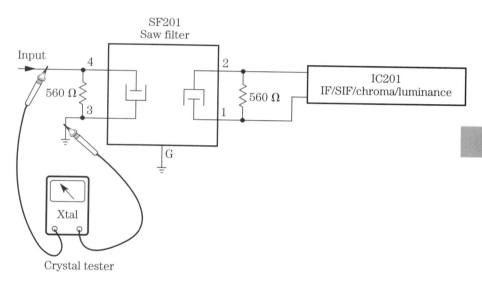

■ **40-18** *Check the saw filter by connecting across terminals 3 and 4, and 1 and 2 for a reading on the crystal tester.*

The comb filter is a selective device that will pass a series of frequencies within a band while rejecting other frequencies in between. The comb filter network response represents the teeth of a comb (Fig. 40-19). The purpose of the comb filter in the TV chassis is to separate the chrominance and luminance portions of the composite video signal prior to the low-level processing of these circuits. This eliminates cross-color, color bleeding, and large areas of color dots.

The comb filter in RCA's CTC140 chassis is composed of two primary parts, U2300, and delay line DL2600. DL2600 is a glass delay

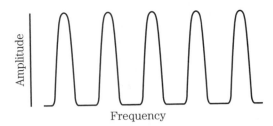

■ **40-19** *The comb filter frequency response represents the teeth of a comb for which it was named.*

line. U2600 is composed of a video buffer and a delayed signal amplifier. The comb filter is used between the IF video and the chroma-luminance circuits. U2600 separates the chroma and the luminance of the input video signal (Fig. 40-20).

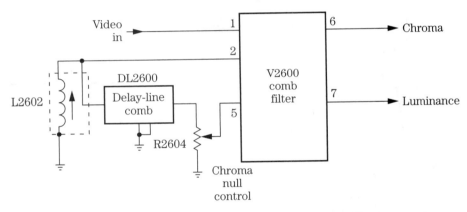

■ **40-20** *The comb filter separates the color or chroma and luminance signals of the video input signal.*

The comb filter circuits in a Sylvania TV chassis consist of the glass delay line of Y600, the delay line driver (Q600), the luminance inverter (Q605), the luminance buffer (Q610), the luminance equalizer (Q615), and the peaking buffer (Q620). The composite input video signal is fed into Q600, is amplified, and connected to pin 1 of the glass delay line of Y600. The luminance signal is in phase and added with R601 (Fig. 40-21).

The luminance signal is passed to the luminance buffer (Q610) for further processing. High-frequency components of the luminance signal are developed in the circuits of the luminance buffer (Q615) and Q620 to pin 10 of chroma-luminance IC640. The luminance inverter (Q605) and the chroma signal coming from the input at the base terminal of Q600 is fed to pin 3 of IC640.

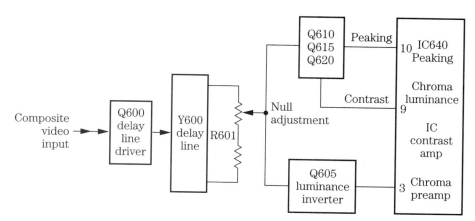

■ **40-21** *The block diagram of the comb filter circuits in Sylvania's C9 chassis.*

The comb filter adjustment in Radio Shack's 27-inch color TV chassis (TC-2700) consists of an oscilloscope, 12-Vdc power supply, and color-bar pattern generator. Adjust VR251 so that the waveform is minimum. Also adjust L204 so that waveform B is at minimum amplitude. Repeat both adjustments until waveform B is at minimum (Fig. 40-22). Check the manufacturer's service literature for any comb-filter adjustments.

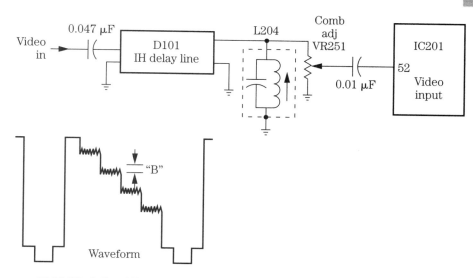

■ **40-22** *Adjust VR251 and L204 for a minimum "B" amplitude on the color stair-steps.*

Computer tuning circuits

You will find system-control tuning and microcomputer operation in the latest TV chassis. The control system consists of a system-control microcomputer IC, an analog interface IC, a band-switching IC with front panel controls, and a remote-control receiver. The system-control IC scans the front-panel keys and monitors for a key press if ac power is applied. IC3000 determines what key is pressed and starts the right program sequence for the key that was pressed.

A data bus line carries communications between the microcomputer and analog interface IC at pins 31 and 32. The timing clock is crystal controlled for the analog interface IC and microcomputer IC at pins 1 and 2 (Fig. 40-23). Information carried on the data bus lines can be local and remote functions, initial commands of the analog interface IC and the tuning commands.

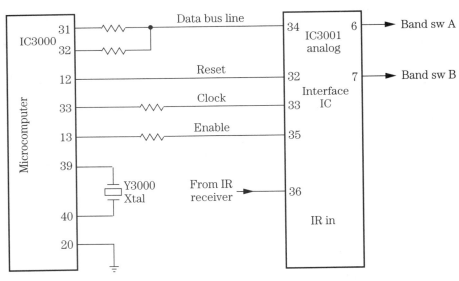

■ **40-23** *The data bus line, reset, clock, and enable circuits are connected between microcomputer IC3000 and analog interface IC3001.*

System-control microcomputer IC3000 has the different front-panel manual control functions, such as: power on/off, set-up, volume up and down, tuner, display, video, and channel up and down. All of these functions initiate in the microcomputer circuits.

Analog interface IC3001 initiates data in the bus line, band-switching signal, picture control, audio, and supplies pulses to be filtered to create dc voltages. These dc voltages are applied to control the audio and video circuits. IC3001 contains infrared remote receiver input at pin 36.

IC3001 also controls the on-screen display applied to the video processing circuits (Fig. 40-24). The video processing circuits control the signals to turn the picture tube guns on and off, resulting in an on-screen display. The on-screen display signal is mixed with the luminance signal. IC3000 controls the tuning operation by sending channel information via the serial data bus line to IC3001. The tuning signal is sent on to a band-switching IC that tunes to the correct channel.

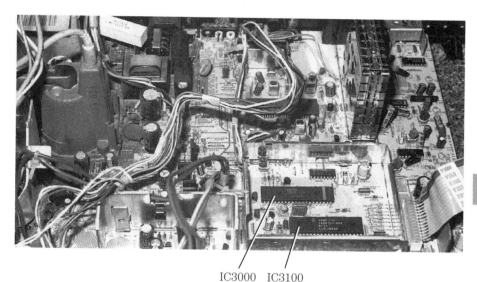

IC3000   IC3100

■ **40-24** *The microcomputer (IC3100 and IC3000) analog IC used on the PC board.*

The tuning system selects channels with remote control, channel scan, keyboard, and on-set channel scan. The keyboard might be connected by direct PC wiring on connections applied to a socket so that it can be easily removed. Band and reference switching controls the voltage applied to the varactor tuner.

AFT operation keeps the tuner on the correct frequency and will allow good color and sound. The AFT circuits monitor the tuner output frequency by applying a correction voltage if the frequency starts to shift or drift. If the tuner output frequency increases, the automatic fine-tuning (AFT) circuit produces a lower voltage to bring the tuning system back to the correct frequency. If the frequency decreases, the AFT voltage increases to bring the tuner back on frequency. Some TV chassis have an AFT test point to check the AFT voltage. Many different types of ICs are used in system and microcomputer controls, but all end up controlling most of the same circuits.

TV picture on-screen display

On-screen display data is generated by the analog interface IC, although the microcomputer IC might give instructions. The on-screen pictures are inserted in the chroma circuits; the black edge information is placed in the luminance circuits. The on-screen signal is fed from the analog interface IC to a color buffer and on-screen color driver, to the color gun in the picture tube. A black edge signal is sent from the analog interface IC to a black inverter stage, luma buffer transistor, and to the emitter circuit of each color output transistor (Fig. 40-25).

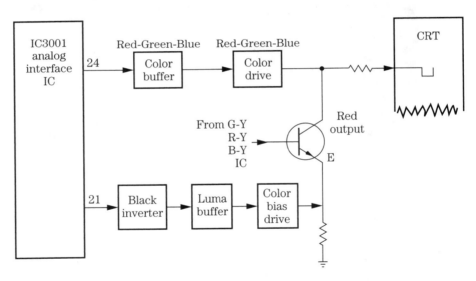

■ **40-25** *The block diagram for the on-screen display fed into each color output transistor.*

The display of numbers, letters, and dots or dashes of the various operation controls are placed on the screen in the on-screen display. On some TV screens, the contrast, color, tint, brightness, volume, treble, and bass bars or dashes, with correct functions are shown on the face of the picture tube. You can adjust these controls manually or with the remote control. The operation might be called *menus* or *characters* by several manufacturers.

By placing the function, such as treble and graduation of bar dashes on the screen, you can see the best level of color picture or sound (Fig. 40-26). This on-screen display information is applied over the color telecast picture.

■ **40-26** *The vertical lines at the bottom of the raster indicate how much treble is applied to the sound stages.*

In the RCA CTC159 chassis, the on-screen display control signal is fed from pin 24 of the analog interface IC (U3300) and fed to the on-screen display circuits and ending up at the R-Y, G-Y, and B-Y CRT bias/drive circuits. The on-screen display output is from U3300, but there is no black edge signal because it is provided in the on-screen display IC (U1001). This on-screen display signal is applied to pin 47 of U1001 (Fig. 40-27).

Problems used in the on-screen display might include scrambled on-screen display, no green character, no black edge, solid green

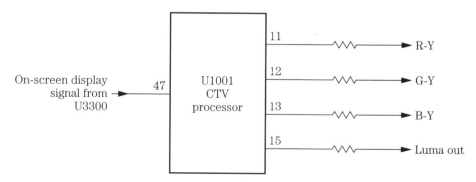

■ **40-27** *The on-screen display signal is applied to pin 47 of the CTV processor.*

screen, and no luminance with normal chroma. Check the on-screen display and the black output terminals on the analog interface IC, the control data from system control, and either the microcomputer or analog interface IC for scrambled on-screen display. For no green characters, check the on-screen display signal of the analog interface IC, green buffer, green drive transistor, and leaky diode.

Check the on-screen display black output from the analog interface IC, and the black inverter transistor and silicon diode between the black inverter and the luma buffer transistor for poor or no black edge. For a solid green screen, check the on-screen display signal from the analog interface IC, defective analog IC, shorted or leaky green buffer, and the corresponding silicon diode. Also check for a green drive or chroma processing system.

If color is in the picture, but there is no luminance, check the output of the on-screen display signal at the analog interface IC. If the signal is low, suspect the analog interface IC. Test the black inverter transistor for leakage. Suspect luminance processing or drive problems if the black inverter transistor is normal.

Mono audio circuits

The video signal from the luma/chroma processing IC is passed through a 4.5-MHz bandpass filter network (CF1100) to the sound IF limiter stage. This sound signal is detected by the sound IF detector and is coupled to the attenuator stage that also controls the volume control. The sound output is fed to the preamp audio circuit and is directly applied to the audio preamp and audio output circuits. A speaker is capacitively coupled to the output audio circuits (Fig. 40-28).

TV stereo audio circuits

The baseband audio signal is developed within the luma/chroma IC and fed to the input of a stereo demodulator or decoder IC. In some chassis the stereo decoder or sound might be contained in a module off from the regular TV chassis. The stereo audio signal might be amplified before or inside the matrix or decoder IC. The decoder IC might send the mono audio signal out to a stereo-mono mode switch. The amplified audio signal is applied to the decoder block, to the first phase detector, and to the loop detector. In some chassis, the amplified stereo signal is fed to a matrix and to a differential amplifier IC.

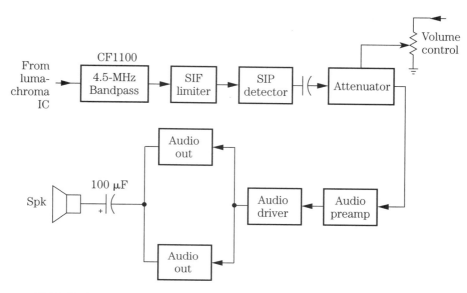

40-28 *The mono sound circuits might have two transistors in push-pull in the audio output circuits.*

The decoder block sends the same signal out of pins 4 and 5 of the decoder IC. The left channel audio is at pin 4 and right channel audio is at pin 5 (Fig. 40-29). This stereo signal is sent to the attenuator IC, where balance adjustment and attenuation occurs with the capacitively coupled audio output signal. Some stereo chassis have a logic switch that switched the external input audio directly to these stereo channels. A volume-control IC is used between the matrix and differential IC in some stereo circuits.

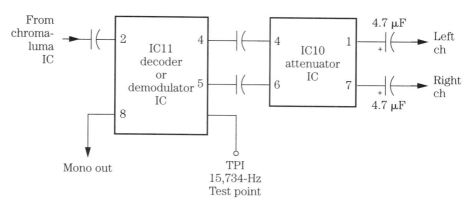

40-29 *The stereo decoder or demodulator provides a stereo signal to the attenuation IC (IC10).*

The left and right stereo channels, fed from the attenuator IC, are fed to the input sound circuits of the audio output IC. In some chassis, a monitor output jack can be used with separate transistors as monitor amplifiers. The power IC amp is capacitively coupled to the stereo right and left speakers (Fig. 40-30).

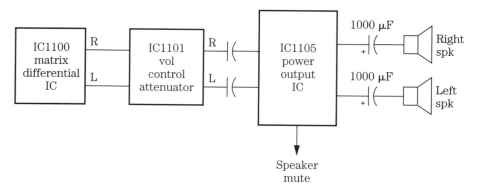

■ **40-30** *The block diagram of the stereo TV audio from the matrix IC (IC1100) through the power output IC (IC1105).*

Troubleshooting stereo decoder circuits

Make sure that audio is present from the baseband output of the stereo decoder or demodulator. This audio signal can be signal-traced with an external audio amp. Check the stereo audio signal at the output of the decoder IC, matrix-differential IC, logic switch IC, volume-control attenuation IC, and to the input and output audio terminals of the output sound IC. If the stereo sound is missing at both channels of a particular IC, if there is only one channel of audio from the IC, or if the audio is intermittent, suspect a defective IC. Test the 15,734-MHz divider frequency with the frequency counter on terminal 10 of the decoder IC.

Determine if the audio stereo systems are in mono or stereo. If the stereo LED is on, the audio stages are in stereo mode. If the light goes out, the audio circuits are in mono mode or there is no stereo reception. Remember, not all TV programs are broadcast in stereo.

Some stereo decoder units are serviced by replacing the entire module. Other circuits can be serviced with an external audio amp, oscilloscope, and DMM. Usually, there are no special requirements to adjust the stereo decoder.

Stereo headphone operation

In some TV chassis, stereo headphone operation might be required with a separate headphone jack. The stereo audio signal is tapped off after the attenuation IC and is applied to the op-amp headphone circuit. Here, two separate hi-fi line jacks can be connected to the external amp and for headphone operation (Fig. 40-31). Check for defective headphone jacks, op amps, and the supply source if there is no audio headphone operation.

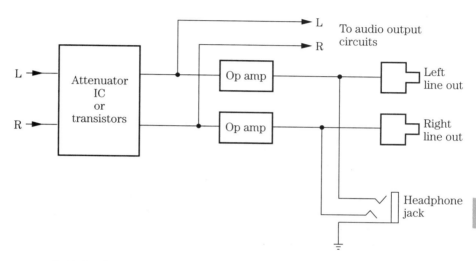

■ **40-31** *Some TVs might have stereo headphone jacks, which are connected to op-amp output amplifiers.*

Picture in a picture

You can see more than one picture on the TV screen with special picture-in-a-picture circuits. The regular large picture can be seen with several small pictures within the larger picture. These pictures can be rotated or changed in a position on the screen with zoomed, swapped, moved, or frozen inside of the large picture. Most picture-in-a-picture movements start in the right-hand corner (Fig. 40-32).

The picture can be moved with the move button and exchanged with the big picture by pressing the swap key or button. Press the freeze button to hold still the small or large pictures. The zoom-in button increases the size or close-up of each small picture each time that the zoom-in button is pressed. The zoom-out button decreases the size of the small picture. Several other features in-

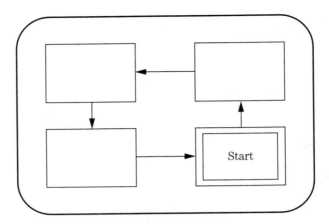

■ **40-32** *The small picture-in-a-picture can be moved around the screen area of the picture tube.*

cluding strobe, pan, multi-channel, and special-effect modes can be used in some picture-in-a-picture chassis.

The picture-in-a-picture operation might be contained in two or three different modules or separate units. Pressing the PIP button activates the feature. A small picture will appear on the screen and contain the same features as the large picture. The sound at the speaker belongs to the large picture. Press the PIP button again to turn off the picture in a picture.

TV and CAV video, y and c signals are fed to a signal-select switch IC. The R-Y and B-Y signal is applied to the picture-in-a-picture processor IC (Fig. 40-33). The burst oscillator provides a 3.58-MHz signal to lock in the big picture chroma signal to the encoder and the picture-in-a-picture processor IC. The horizontal and vertical sync outputs are applied to the picture-in-a-picture processor. The picture-in-a-picture has a voltage-controlled 20-MHz oscillator (VCO) to synchronize internal tuning with external horizontal sync signals.

The digital video information is stored in the RAM IC. This digital information can be operated by the signal processing circuits. Analog-to-digital conversion occurs in the picture-in-a-picture processor and converts analog luminance, B-Y and R-Y signals into digital information.

The R-Y and B-Y, and Y/C signals are fed through an encoder IC to the fast switch IC and applied to the TV chassis. The signal information is connected to the on-screen display or line drive circuits of the picture tube.

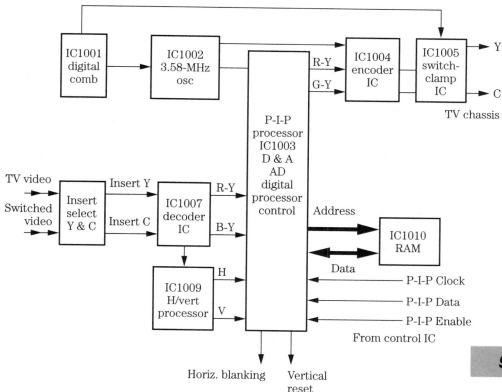

■ **40-33** *The block diagram of the picture-in-a-picture module, as applied to the Y and C circuits of the TV chassis.*

To troubleshoot the picture-in-a-picture circuits, the different modules might be interchanged with new modules or sent in for repair. Check the various input signals into the various jacks of the picture-in-a-picture module. Scope the vertical input pulse, oscillator frequency, clock, data, and enable lines on the control micro IC. Measure the voltages that feed the picture-in-a-picture module. If the voltage and waveform signals are normal, the trouble is likely in the main board. Check the manufacturer's troubleshooting techniques to service the main picture-in-a-picture circuits.

Standby circuits

Standby circuits are on all the time if the TV chassis is shut off so that the remote control can be operated. The standby power circuits provide voltage to the standby circuits with the TV off. Although these circuits do not develop many problems, check the standby circuits if the remote will not turn on the TV and the remote is okay. Use an isolation ac transformer when servicing TV standby circuits (Fig. 40-34).

■ **40-34** *The standby circuits are located close to the power supply circuits within the TV chassis.*

If the set does not respond when you press the on/off button on the remote-control transmitter, the remote control circuit might possibly be defective. Check to see if the TV chassis is dead by operating the on/off button at the set, or shunt around the standby circuits.

Sometimes, the standby circuits are not included on the TV schematic. You can spot a power relay of the standby circuits on the PC board. If not, trace the switch contacts back to the power line to determine which contacts must be shunted to apply ac to the TV chassis. You know that there are problems in the standby circuits if you bypass the relay or standby circuits and the chassis begins to operate.

The standby power supply circuits might be powered with a low-voltage ac transformer, chopper transformer and SMPS, or a switched power transformer (Fig. 40-35). Often, the low voltage is rectified by a bridge-rectifier circuit with transistor and zener diode regulation. The standby voltage source might vary from 5 to 12 V. You might find both a 5- and 12-V standby voltage source.

Problems within the standby circuits might be caused by dirty relay points or an open relay solenoid coil. Clean the contacts with

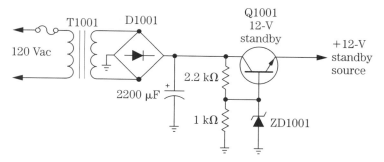

■ **40-35** *The standby power supply might consist of a separate power transformer, bridge rectifiers, transistor, and zener diode.*

cardboard, sandpaper, and cleaning fluid. If voltage is used across the solenoid winding, suspect that a coil is open. Take continuity measurements on the coil and also on the contact points.

Check the standby voltage output source with the voltmeter. No voltage might indicate that a fuse or that the primary winding of the standby transformer is open. If the dc voltage is at the large filter capacitor and there is no output at the regulator transistor, suspect that the transistor is open. A leaky regulator transistor and zener diode might keep blowing the line fuse. Low output dc voltage might result from transistor and zener diode regulation. Regulator transistors might appear leaky or open if the dc standby voltage is missing. An improper voltage from the standby voltage source might prevent the remote circuits from operating.

Pincushion problems

On small-screen sizes, the conventional linearity coil and pin-connected yoke provide adequate screen connection. In large TV screens, a pincushion connection circuit corrects distortion on the face of the large picture tube. The pincushion circuits are defective if the outside edge of the TV raster bows. Improper adjustment of the pincushion coils might bow the picture at the top and bottom of the raster.

The primary of the pincushion transformer is in series with the horizontal yoke assembly. By modulating the current through the pin transformer, most of the regulated $B+$ voltage appearing across the yoke winding changes, resulting in a change of horizontal width. This width will increase as more $B+$ voltage appears across the yoke winding (Fig. 40-36).

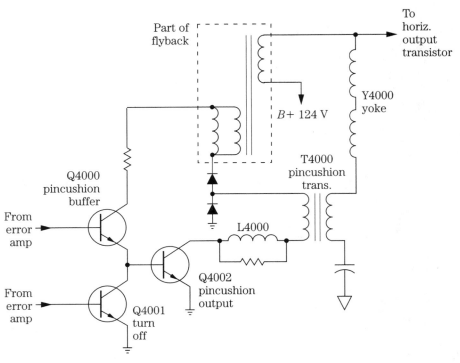

■ **40-36** *Check the pincushion output transistor if the picture bows at the outside edge.*

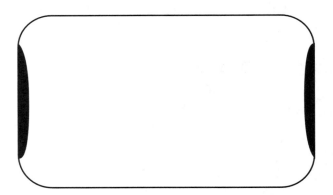

■ **40-37** *After replacing the leaky pincushion output transistor for a bow in the picture, make the proper pincushion adjustments.*

A correction circuit develops a parabola waveform that is applied to the pincushion transformer. Suspect a pincushion output transistor if the raster bows (Fig. 40-37). Suspect an intermittent power output pincushion transistor if the sides are bowed in and suddenly return to normal. Do not overlook the possibility that there are poorly soldered connections on the pincushion transformer (T4000) if there is poor width and raster bowing. The pincushion output transistor might run warm in normal operation. Excessive width can be adjusted with the width control.

41
Keep that test equipment in tip-top shape

NO MATTER HOW CAREFUL THE ELECTRONICS TECHNICIAN is with critical test instruments, accidents do occur, test instruments are knocked off benches, and constant use destroys the test leads and probes. During servicing procedures, the test instrument probe might disconnect and fall down into the line or high-voltage circuits. Accidentally applying a test instrument that was not designed for high-voltage circuits might destroy components within the tester (Fig. 41-1). The electronics technician must learn to repair the damaged test equipment or send them back to the factory for repair. Test instrument repairs can be made on those slow, snowed-in, or rainy days.

■ 41-1 *Expensive test instruments should be handled with extreme care.*

Simple maintenance

Replacing batteries and fuses within the battery-operated tester is quite easy, provided that you have the right fuse. Sometimes the manufacturer will provide a spare fuse alongside the fuse area or within the service package (Fig. 41-2). If not, these small fuses can be difficult to obtain. These small test instrument fuses might have a rating of 0.5 A at 250 V. It is wise to have a spare fuse for each

■ **41-2** *Most test instruments have an extra fuse alongside the fuse or battery compartment.*

test instrument. Of course, the test instrument breakdown occurs when the equipment is needed the most.

Battery replacement is easy because most shops carry all kinds of batteries for various consumer electronic products. Most batteries in the portable test instruments are the standard 9- and 1.5-V AA types. Some test instruments have a "LO BAT" indicator, indicating that it is time to change the batteries, while others have a battery test switch. The battery is getting weak in the DMM when the numbers appear light and are not a deep black color.

Test instrument schematics

All test instrument service operation manuals and literature should be kept in a safe spiral file. Besides showing the technician how to operate it, a schematic with a parts list might be included. These schematics are handy if the test instrument quits operating. Do not try to repair critical circuits without a schematic.

Fill out the registration card when the new test instrument arrives. Do not try to service or repair any test instrument that is still under warranty. Take or send the instrument back to the wholesaler from which it was purchased. You might find a printed slip, enclosed with the service literature, of the various U.S. authorized service centers. These authorized service centers will handle any parts or service problems, either in or out of warranty.

Where to obtain parts

Critical parts should be obtained from the manufacturer of the test instrument or from factory authorized service centers. Components that have a 1- or 5-percent tolerance should be purchased from the manufacturer outlets (Fig. 41-3). Critical trimmer capacitors, crystals, 1- or 5-percent resistors, ICs, and special components should be obtained from the manufacturer outlets. Look up the critical part numbers on the same sheet as the schematic. Regular 10- and 20-percent capacitors, resistors, and diodes can be taken from your own stockroom. Send or take back foreign-manufactured test instruments to the place it was purchased for warranty repairs.

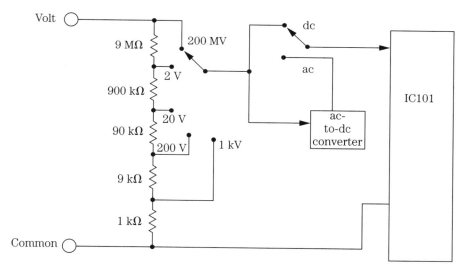

■ **41-3** *Replace critical 1% or 5% resistors in the voltage and resistance ranges of the test instrument.*

Instrument warnings

Treat your test instruments like diamonds. To avoid electric shock, disconnect the test leads before removing the battery or fuse. Open the cabinet and replace the battery and fuse. Do not touch any other area inside of the test instrument case. Replace the fuse with one of the exact type (Fig. 41-4). Higher current fuses might cause a fire or added damage to the test instrument. Some test instruments have a red ribbon under the fuse for easy removal.

Be careful when using the FET, DMM, and VOM meter when taking voltage measurements. There is always the possibility of danger-

■ **41-4** *Check the fuse before replacing it. It is wise to have a complete set of fuses for each test instrument on hand. Tape the fuse to the schematic diagram.*

ous voltages being present in any piece of electrical or electronic equipment. Always use extreme caution when making high-voltage measurements, which might appear at unexpected places in a specified defective circuit. Although some meters will measure up to 300 V, use extreme caution when measuring voltages of 150 V and above. Solder a piece of bare hookup wire to attach the test instrument clip so that it will not fall down into a higher voltage and destroy the test instrument. Solder a bare piece of hookup wire to the PC wiring and attach the instrument clip so that it will not fall down into high-voltage areas and damage the tester.

Always start with the highest voltage and current range available. Never attempt to measure a voltage when the function is set to resistance or current because it might burn out the meter movement or other circuitry (Fig. 41-5). Although some small instruments have diode protection, high voltages might damage the meter. Never attempt to measure current when the tester is set for resistance, and likewise never attempt to measure ac voltages or current with the tester set to a dc mode or the meter circuitry can be damaged.

Do not attempt to measure strong RF voltages with a small multimeter. Remember that in small FET or VOM multimeters voltage and resistance measurements are made with the meter connected in parallel. Current measurements are made with the meter connected in series. Do not use a multimeter to take high-voltage measurements in microwave ovens (Fig. 41-6).

■ **41-5** *Before taking voltage, resistance, and current measurements with FET, DMM, or VOM meters, make sure that the tester is set for that range.*

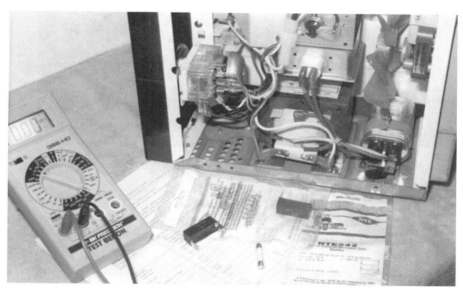

■ **41-6** *Do not use a VOM, DMM, or FET analog meter to check high voltages in a microwave oven. Use it only for continuity, low voltage, and resistance measurements.*

Never stand on a damp or wet floor when making voltage measurements. Stay away from metal work tables, metal water or gas pipes, metal posts, and metal electrical conduit. Accidental contact between grounded metal objects and high voltages can be dangerous to your health. When taking voltage measurements in the TV chassis, keep the other hand in your pocket or on your hip, and do not grasp the metal chassis. Although a small shock might not be dangerous, your body reaction might cause you to bump or fall against a higher voltage, knock off test equipment, and damage other chassis on the service bench (Fig. 41-7). Always plug the electronic product into a variable isolation transformer to prevent damage to test instruments that are ac operated.

■ **41-7** *Keep your hands off of the chassis when taking high-voltage measurements in the TV chassis.*

When the multimeter is not in use, turn it off—especially when leaving the bench at night. Always use well-insulated test leads. Never use test leads with frayed or broken insulation.

Repairing cords, plugs, and jacks

The ac power line cords can be damaged by placing sharp, heavy objects on them that cut through the insulation. Cracked or worn ac cords should be replaced. Instrument test cords or cables that have been cut or melted must be repaired or replaced. Dropped or instruments knocked off the bench might end up having damaged test leads replaced. Damaged test cords or cables can be replaced with regular flexible test lead cabling.

Because of heavy usage, the DMM or voltmeter test leads can be replaced with a test lead kit. Often, the banana plugs become flattened or worn and will not make good contact or stay plugged into the meter (Fig. 41-8). These banana plugs can be replaced separately or you can purchase an entire meter cable kit. Sometimes the set of cables might have to be replaced two or three times a year. Try to replace with the same test leads and some are rated up to 1200 V.

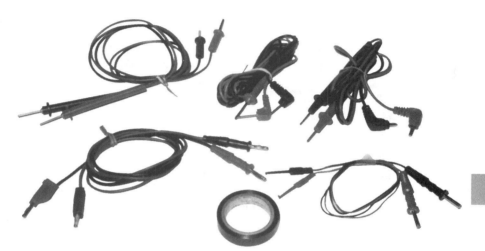

■ **41-8** *Banana plugs and standard plugs in test cables can be replaced if they are worn, flattened, or have poor contacts.*

High-voltage multiplier probe cable should be replaced with flexible test lead cable that will withstand 30 to 40 kV. Do not replace them with ordinary meter test cable or speaker wire.

Test equipment BNC and regular connectors and jacks are available at most wholesale houses or Radio Shack. Replace the connectors, plugs, and jacks with those that are the same type as the originals. Multimeter test jacks are usually mounted inside of the meter area for shock-proof protection. Replace them with the exact type of internal safety jacks and plugs (Fig. 41-9).

Internal VOM and DMM repairs

If the letters or numbers appear weak and not dark black on the DMM, replace the battery. Likewise, if the VOM or DMM will not

■ **41-9** *The various test jacks, plugs, and probe tips can easily be repaired by universal part replacements.*

show any measurements, replace the fuse. Sometimes when accidentally taking voltages with the FET or VOM set on resistance, circuit damage and a blown fuse might result. Check for blown resistors and burned diodes. Often, these are critical resistance in the ohmmeter and voltage ranges, and they should be replaced with parts that have the same tolerance (Fig. 41-10). This means getting components from the manufacturer or factory parts depot. Small-signal or power switches can be replaced locally, but selector switches must be replaced with components that have the exact part number.

Expect heavy damage if the DMM or VOM is used to measure high voltage in the microwave oven. Do not use these small meters for high-voltage measurements in the oven. Extreme damage might result and even the factory can't repair these multimeters satisfactorily.

Although the input and meter circuits are protected by a fuse, extreme high voltage might damage input resistors and diodes, switching transistors, FETs, and op amps in the FET analog multitester. Meter damage can occur in the VOM or FET testers although protection diodes are used for protection. When the analog meter itself is damaged, it might cost too much to have the meter replaced or to have the tester repaired by the factory. Simple meter repairs can be performed by the electronics technician (Fig. 41-11).

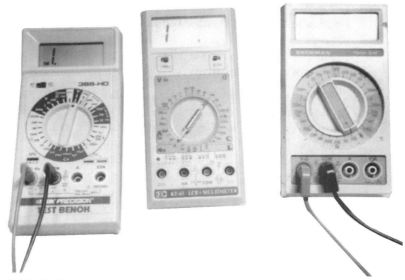

■ **41-10** Burned or damaged resistors used in the DMM, LCR, and VOM should be replaced with parts that have the correct resistance tolerance.

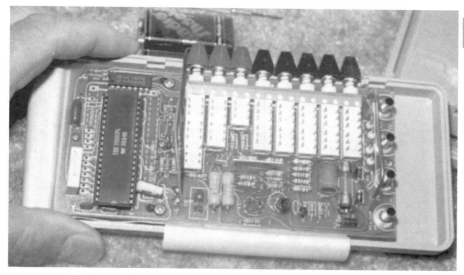

■ **41-11** Most VOM, DMM, and FET meters can be serviced by the electronics technician with parts from the shop.

Power supply repairs

Test instruments that operate from the ac power line have a low-voltage power supply, with bridge rectification, filter capacitor input, and regulation. Check the main fuse used in the ac line circuit or ac adapter. Leaky bridge rectifiers can be replaced with universal replacements. Most components in the low-voltage power supply can be replaced with electronic parts from local wholesale houses or mail-order firms.

To ensure circuit and overall operation capability, transistor, IC and zener diode regulators are used in sweep generators, frequency counters, oscilloscopes, distortion meters, power supplies, audio oscillators, RF generators, and universal test systems. Most of these regulators can be replaced with universal parts. Check the circuit diagram for the correct voltage drop of IC or zener diode regulation. Some meters have a positive and negative voltage source (Fig. 41-12).

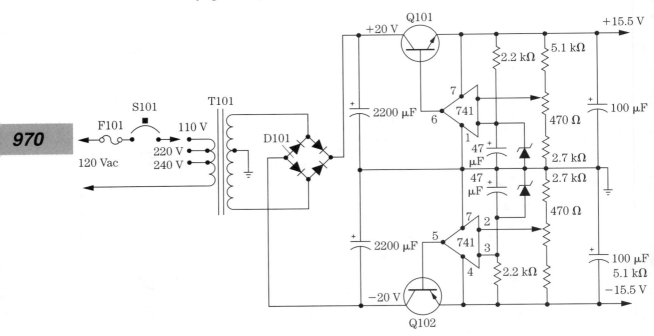

■ **41-12** *The function generator might have a positive and negative voltage source in the ac power-supply circuits.*

Besides the test instrument that operates from the power line, a battery-operated test instrument might have a voltage regulation circuit. Some of these units also have an external jack for a dc outside voltage source or so that they can be operated with AA batteries. Here, D1 and D2 are used for polarity protection with IC101 as the 5-V regulator (Fig. 41-13). The unregulated voltage might

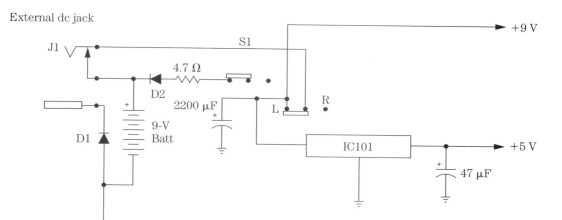

41-13 *You might find that the internal battery-operated circuits contain transistor, IC, and zener diode regulators.*

be connected (VD) to display circuits; the regulated 5 V applies to the input, amplifier, and electronic display IC.

External power supplies have power transformers, silicon diode rectifiers, large filter capacitors, and IC and transistor regulation stages. Most parts can be replaced by the electronics technician in small external power supplies (Fig. 41-14). Besides the on/off switch, fuse, and silicon diodes, voltage regulator transistors or ICs can be replaced with universal parts. Heavy-duty power supply

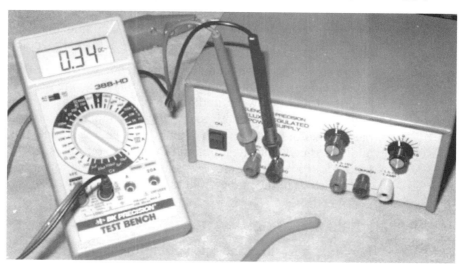

41-14 *Check the output voltage of the external power supply. Check diodes, transistors, and IC regulators for leakage or open conditions.*

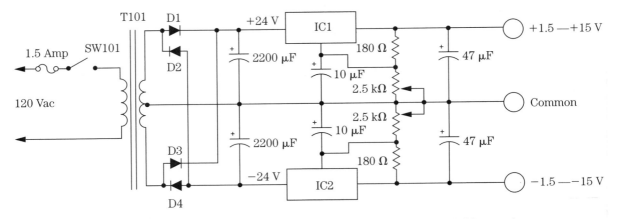

■ **41-15** *The external power supply might have full-wave and bridge rectifiers in a positive and negative power supply.*

regulators, transformers, and motors must be replaced by components that have the original part number (Fig. 41-15).

Servicing the external audio amp

The simple external audio amplifier might consist of an input and output IC driving a 4- or 5-inch speaker. Some experimental amps might have an FET transistor as an input and a simple LM386 IC as an output, driving up to an 8-inch speaker for mono audio troubleshooting (Fig. 41-16). These sound circuits can be repaired like any audio circuit. Simply test each transistor with the transistor tester or the diode junction test of the DMM. You might find an AF amp, driver, and two output transistors are in the early audio amps.

Scope each stage with an audio signal generator tied to the input cables. Go from stage to stage with the oscilloscope to locate the source of dead or weak audio. A weak signal or a loss of signal might result from defective electrolytic coupling capacitors, bypass capacitors, and an open or an increase in resistance. If normal audio is traced in at the input and there is no output, suspect a defective output IC or defective transistors. Check all bias diodes and resistors in the audio circuits.

Broken dials and knobs

Purchase new dials and knobs from the manufacturer. Regular small push-on or screw-type knobs can be used temporarily until regular knobs are available. Damaged knobs with calibrating

■ **41-16** *The external audio signaltracer might contain an FET transistor and LM386 IC to drive a 4- or 5-inch speaker.*

marks and plastic skirts should be replaced with original types of knobs. Of course, if the test instrument is 8 or 10 years old, they might not be available. Small push-on or clip-on knobs on a switch bank should be replaced with originals.

Generator repairs

Signal, audio, and function generators can easily be damaged when a high voltage comes in contact with the RF or audio test cables. Besides blowing a line fuse, damage to the front end or signal output might damage a coupling capacitor, transistor, or IC. It is best to place a 0.05-μF 450-V blocking capacitor in series with these generators' main signal probe or connection for added protection. Usually, low-value resistors might be used in the output circuits, which can be destroyed (Fig. 41-17).

Clean erratic hand switches or function switches in the generator by spraying cleaning fluid right down into the switch area. Rotate the switch back and forth to clean the dirty contacts. Likewise, clean the dirty output controls and trimmer pots if they become erratic in operation. Replace worn controls with universal replacements (Fig. 41-18).

Besides replacing the fuse, check the voltage at the largest filter capacitors in the low-voltage power supply, if it is ac operated.

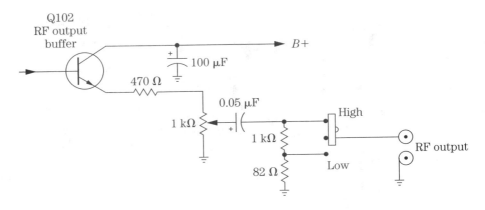

■ **41-17** *If a high voltage is applied in a circuit where the generator is attached, small input resistors can be damaged.*

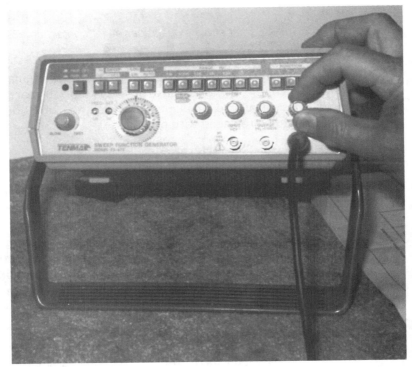

■ **41-18** *Replace a worn volume or gain control on test instruments if they are worn or erratic.*

Some of these generators might be operated from both batteries and the ac power line. Usually, the small power transformer has a line fuse with either a bridge or full-wave rectifier circuit. You might find zener diode and transistorized voltage regulation (Fig. 41-19). Low voltage might result from dried-up electrolytic capacitors, burned voltage-dropping resistors, and zener diodes or overloaded circuits. Suspect poor filtering if the signal generator adds 60 Hz to the electronic product being aligned or tested. Remove the generator from the product to make sure the 60- or 120-Hz hum bars are in the test instrument.

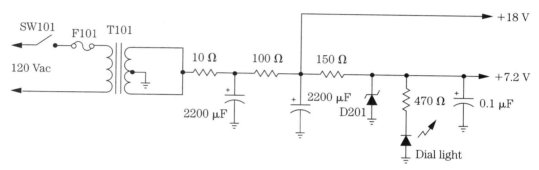

■ **41-19** *A simple low-voltage power supply or adapter used in the signal or function generator.*

Transistor tester repairs

The low-priced transistor-diode tester might consist of LEDs as indicators, instead of an analog large meter. Several transistors with an oscillator IC make up the heart of a transistor tester. Three of the LEDs indicate if the transistor is an NPN or PNP, and if it tests normally. In the diode test, two diodes indicate the diode polarity and that it tests OK (Fig. 41-20).

In larger transistor testers, the analog meter registers the ac beta reading on HI with a LO measurement of the leakage. This type of transistor tester actually provides a transistor beta test. The dynamic test of the beta and the NPN or PNP polarity tests are provided. Three separate test cables clip to the transistor for in- or out-of-circuit tests (Fig. 41-21). With this beta tester, transistors can be matched for critical matched transistor circuits.

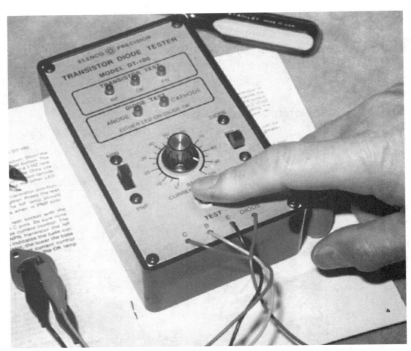

■ **41-20** *The simple diode and transistor tester might have LEDs that light, indicating leakages and correct PNP or NPN polarity.*

■ **41-21** *The beta ac transistor tester can be damaged if power is left in the cassette player with transistor leads clipped into the circuit.*

Keep that test equipment in tip-top shape

Capacitor tester repairs

Choose a capacitor tester that will test larger filter capacitors from 0.1 to 20,000 µF. You might find small capacitor testers included in the DMM ranges and they will only list up to 2000 µF. The extended-range capacitance meter will test the high-value filter capacitors that are used in large audio amplifiers. Most problems in the capacitor testers are dirty pushbutton controls and banana plugs. Clean the ganged pushbutton switches with cleaning fluid. Since the connecting meter leads are short, replace the banana plugs with new plugs if erratic measurements exist or if the plugs will not stay in the recessed jacks (Fig. 41-22). The extended capacitor tester might have a time-base crystal-oscillator circuit.

■ 41-22 *Clean erratic pushbutton switches if the capacitor meter readings change rapidly on the LCD display.*

FET meter repairs

Repair the FET analog meter like any VOM or DMM. Transistors can be tested with the transistor tester. Replace the batteries if the meter will not balance or reach the end of the scale. Check for burned resistors if either the voltage ohmmeter or the current range is erratic or will not function (Fig. 41-23). If the meter hand will not move, check for an open fuse or defective meter. If the meter has been dropped and the meter hand is damaged, purchase a new meter because meter repair can exceed over half the cost of a new FET meter.

■ **41-23** *Check for burned components in the FET analog meter if it is left on the wrong resistance scale when attempting to take voltage measurements.*

Send the FET meter in for repair if high voltages have damaged the measurements. Critical resistors must be replaced with components that have the same factory part numbers. If the meter is set to resistance and is taking voltage measurements, suspect that the meter or circuitry is damaged. Although the FET circuit might be protected by a small fuse, check the meter setting before attempting to make a voltage, resistance, or current measurement.

Portable color-bar generator repair

Check the color-bar generator battery if the color patterns are weak or will not function. Replace the 9-V alkaline battery. If the tester has been dropped, check for damage to the front slide switches. Clean the switch contacts if color-bar, dot, or crosshatch signals are intermittent or erratic (Fig. 41-24).

If the color functions are missing, check voltages on the ICs and transistors. Take in-circuit tests on each transistor. Look for broken boards or snapped component leads if the instrument is dropped. A circuit diagram is a must-have item if you are trying to repair generators or oscilloscopes. For difficult repairs or test instrument calibration, send the unit back for factory service. Do not

■ **41-24** *Erratic or intermittent color bars or crosshatch lines on the TV screen might result from dirty switch contacts.*

attempt to repair any test equipment that is still under warranty. Send or take it back for factory warranty service.

Cabinet repairs

Just cleaning up the front panel of equipment can improve the appearance considerably. Touching up the paint job could make it appear to be new again. Use auto enamel spray paint, which is available at hardware, auto, and department stores, to repaint metal.

Glue broken pieces back together with contact cement or epoxy. Small jagged holes can be filled with epoxy cement or an auto body repair kit. A small can of auto body repair kit can fix dents, holes, and scratches. Sand down all rough outside surfaces and apply one or two coats with auto body filler. Sand out all rough areas with fine sandpaper and spray on a couple coats of auto enamel finish paint. The soldering gun might be a little heavier, but it works and looks like a new one (Fig. 41-25).

Soldering iron repairs

Besides replacing soldering iron tips, replacement of the cleaning sponge and plugs cause the most service problems, within the solder iron stations. Burned or worn conical, long conical, screw-

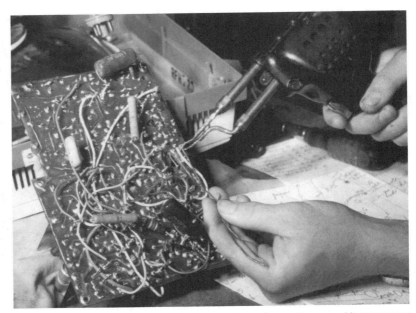

■ **41-25** *Repair the large soldering gun handle and case with epoxy or with an auto body kit if it is dropped and the pieces must be cemented together.*

driver, and large screwdriver tips are replaceable. Replace the whole soldering iron assembly if the iron becomes defective, if the cord becomes worn or frayed, or if the side mount plug is broken (Fig. 41-26).

If the battery in the cordless soldering iron needs to be charged all of the time or will not hold a charge, replace the nicad battery. Remember, this battery can be half the cost of a new battery-operated iron. Replacement of soldering iron kits and accidentally dropping the iron on the tip area are the common battery iron problems (Fig. 41-27). Erratic operation of the switch and light can be caused by a dirty hold switch. Clean it with sandpaper and cleaning fluid.

Factory instrument adjustments

Precision test instruments must be accurate in measurements. They should be sent into the factory for correct calibration every five years. A microwave-certified radiation meter should be sent in for recalibration every three or four years or if the probe is dropped or damaged. Frequency counters and digital function meters should be checked against a standard frequency or sent in for fac-

■ **41-26** *Suspect a broken cord plug, frayed, or cut cable, and an open iron element if the regulated soldering iron is intermittent or will not heat up.*

■ **41-27** *Besides replacing the broken or burned tips of the battery-powered soldering iron, clean the on/off switch and the battery contacts if charging and operation is intermittent.*

tory repairs and calibration. If a critical test instrument needs repair or calibration, send it back to the manufacturer.

Of course, get an estimate before the repairs are finished—especially with older test equipment. If the test equipment is too complicated to repair, just send it in. First, make arrangements with the wholesaler who sold you the equipment. You might find a factory-repair center only a few hundred miles away. Most electronics technicians can make small repairs without any difficulty. A good test instrument should not be catching dust on the top shelf, just because it needs attention. Place it back into operation by repairing it yourself or by sending it back to the factory for repair.

Taking care of business

42

EVERY ELECTRONICS TECHNICIAN STARTS OUT AS A NOVICE and works for years until he or she becomes a professional electronics technician. Some technicians might become specialists in repairing VCRs, computers, camcorders, medical instruments, satellites, marine radios, video equipment, and armed forces gear. The radio and TV technician require 5 to 7 years in the field, before entering into their own service business.

The CET test

The electronics technician can add a CET symbol at the end of his or her name, like the doctor, by passing the Certified Electronic Test (CET). To become fully certified by the International Society of Certified Electronic Technicians (ISCET), you must have at least four years of formal training and experience. In addition, he or she must pass both the 75-question Associate and the 75-question Journeyman test. The passing grade for each of the multiple-choice exams is 75 percent.

The Associate examination covers basic electronics fundamentals and each Journeyman option covers a certain or specialized field of electronics. The electronics technician or student can apply for Associate-level certification with less than four years of experience. The basic subjects are in Electronics Math, dc and ac Circuits, Semiconductors and Transistors, Electronic Components, Test instruments, Tests and Measurements, and Troubleshooting.

The individual Journeyman exams focus on audio, computers, communications, consumer electronics, medical electronics, industrial electronics, radar electronics, and video electronics. The fee for each exam is $25. If the technician fails any one exam, taking it again is free. For more additional information on becoming a CET Technician, contact ISCET at: 2708 West Berry St., Fort Worth, TX 76109, (817) 921-9101.

You can prepare for any of these exams by studying material and literature that is related to the CET test. ISCET offers study materials to prepare for each exam for $10. You can purchase CET books by Ron Crow, Dick Glass, and Sam Wilson. These books are published by McGraw-Hill, Blue Ridge Summit, PA 17294-0850, (717) 794-2191. Becoming a CET electronics technician is an added feature to your servicing career, but not a requirement for becoming an excellent consumer electronics technician.

Business ethics

Before entering the business of electronics, become the best electronics technician you can be. Be honest with yourself and your fellow man or customer. Treat your fellow electronics technician like you want to be treated. Never run down another business—even if you feel that he or she does lousy work. Know the electronics business, charge a fair price, collect for that repair, and you can have a successful business. Learn all the facets of business from experience in a well-managed service establishment.

Most electronics technicians have the electronic skills and experience in repairing electronic products, but are less knowledgeable about business practices. Understanding the business end might take more time. The basic principle of professional business records is one of the most essential parts of any business. You must know where to start, know where the business has been, where it is, and where it can go and grow to in the years ahead. Accurate bookkeeping is very important in any business.

Radio and TV sales and service

The electronics technician in the radio and TV servicing field must be able to repair radios, TVs, camcorders, boom-box and CD players, VCR, audio, and just about anything electronic. The old-timer repaired radios for a living until TV came along in the 1950s. Although the TV chassis does not break down as in past years, the electronics technician must combine other electronic products for service revenue.

Besides servicing and selling TVs, the small radio and TV establishment might also service and sell radios, CD players, boom-boxes, VCRs, camcorders, and cassette players. Service what you sell. Years ago, most radio and TV shops sold TVs to add to their income. Today, just about every type of business is selling TVs at rock-bottom prices. TV sales might be rough in some areas. Some electronics businesses only use sales as a method to gain new customers, a form of advertising.

The electronic service technician who knows his or her trade, makes the best repairs possible, in the shortest time, and does not have to worry about keeping busy. Today, the good, professional, knowledgeable, TV technician will be hunted down by customers who want the repair done at once with a fair price. The word-of-mouth advertising is the best.

Be honest with the customer. If the repair is too difficult and you cannot repair it, tell them so. Take it to a professional who specializes in the same brand of TV. Call the electronics distributor, manufacturer, or manufacturer's service depot for help. There is always someone out there who can service the most difficult problem.

Get into the factory warranty service programs of foreign and American manufacturers. Many of these manufacturers are looking for a qualified shop to do their warranty work. Like General Motors, the warranty charges for repairing defects in the car are sometimes greater than the garage would normally charge. You can make money with well-known brands of equipment. Besides getting paid for these repairs, you can quickly pick up new customers if you do a good service job.

Servicing microwave ovens is right down the TV alley, so to speak. Most of the tools and test equipment used in the TV repair business can be used to service most microwave ovens. The only extra test equipment that you might need is a magnameter for high voltage and current tests. Sign up for microwave oven warranty service. Usually, the repaired oven stays repaired with very few callbacks. Check Chapter 34 for microwave repair or purchase McGraw-Hill's *Microwave Oven Repair (3rd Edition)*.

TV repair and satellite business

Besides repairing TVs, the electronics technician might want to add satellite receiver and antenna installation and repair to the present TV business. Although satellite TV is a specialized business, TV antenna repair and installations have always been a part of the TV repair business. Large screen satellite antennas are found in people's yards, in the country, on the farms, and in the city. Servicing and repairing satellite receivers and antennas adds up to extra income for the busy electronics technician.

In the fall of 1994, America's first satellite broadcasting system is ready to beam 150 TV channels directly to the customer's home with excellent reception. The DirecTV company, a unit of General Motors defense subsidiary, GM Hughes Electronics, supplies the space signal. To watch DirecTV programs, consumers must purchase reception gear called *Digital Satellite System (DSS)* (Fig. 42-1).

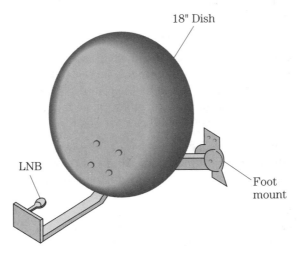

■ **42-1** *DBS can bring over 100 stations to the TV set in a den or living room.*

The basic RCA 18-inch dish with receiving equipment costs $699. A deluxe version that will operate two different TV receivers costs $899. The digital satellite system is another field in sales, service, and installation for the electronics technician. The DBS signal is sharp and clear, compared to the TV signal picked up from antenna or cable programs—especially on large 27- and 31-inch TV screens.

TV and VCR repair

Servicing and repairing VCRs and TVs go hand in hand. Besides recording, the VCR player might play back through the TV set or monitor. Just about every home has at least one videocassette recorder. Some homes have two or three. Cassette recordings of the latest movies can be rented for a few dollars and provide low-priced entertainment for the family. The electronics technician can become a VCR specialist with TV repair as a sideline (Fig. 42-2).

If the VCR is used constantly, the video heads must be cleaned. A good cleanup of the tape heads with cleaning cassette or cleaning supplies can add easy money to the cash register. A complete

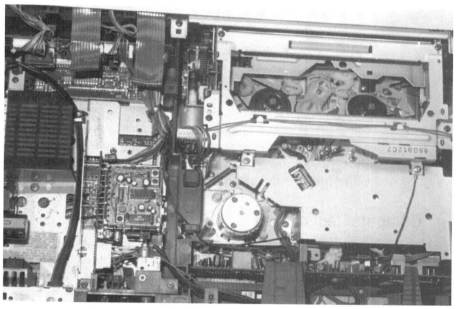

■ **42-2** *The VCR electronic technician can become a specialist or add service to the TV repair business.*

cleanup of tape heads, and tape path components can vary from $25 to $49. Of course, major VCR repairs might cost more than $100.

VCR repairs can be a very profitable electronics business. Besides servicing the consumer's machine, VCR warranty service can be provided by many foreign manufacturers. Line up several VCR manufacturers and electronics stores for added electronics business. Today, TV and VCR service specialists are in great demand. The technician who can service mechanical, electronic, control, and servo systems in the VCR can always have a job in the consumer electronics field.

Medical electronics technician

The electronics technician might want to enter the field of servicing medical electronic equipment. Many doctors, dentists, and other professionals in the medical industry might have to send back electronic equipment to the factory for repair. This takes a long time and some services might be interrupted. Hospitals have the most medical equipment in need of service. New medical equipment installation, service, and preventative maintenance requires the help of an electronics technician. It's possible to get linked up with manufacturers of medical equipment for warranty service or to work for an electronics firm that performs medical electronics repairs.

TV sets, monitors, heart rate meters, patient monitors, and biomedical equipment are only a few of the hospital units that constantly need service. Blood pressure monitors, X-ray machines, electrocardiogram amplifiers, recorders, electronic thermometers, and heart meters are also used in the hospital. If the factory must send out an electronics technician from several hundred miles away, service rates are very costly and it might take time before the repairs are completed. Servicing medical electronic equipment is just another field for the advanced electronics technician.

Computer technicians

One of the fastest-growing industries for electronic servicing is computers. Practically every business and household has a computer. Children are taught how to operate computers in grade school and high school. You can take a college or freshman course on how to operate and use the computer. Computer classes are held in the evenings for adults who want to learn more about computer operations (Fig. 42-3).

■ **42-3** *Soon you might find a computer in every home.*

Like any electronic and mechanical equipment, computers break down and need service. Today, technicians are needed to keep that computer, monitor, cable keyboard, and ink jet or laser printer in tip-top shape. The computer electronics technician is now in demand, for complete service and preventative maintenance.

Other electronics repair fields

There are many types of electronics to repair, such as home audio equipment, car stereo equipment, CB transceivers, home security systems, and communications equipment. Besides the special electronics fields, many electronic devices in the household need service, such as pocket pagers, garage door openers, microwave ovens, motor speed controls, bread baking machines, electronic games, battery chargers, photographic equipment, electronic corn planters, and electronic light dimmers. Electronics are everywhere (Fig. 42-4).

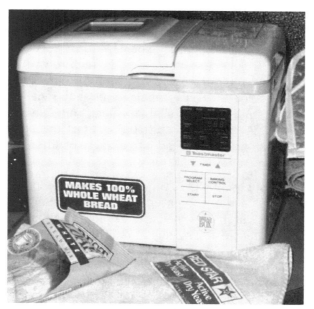

■ 42-4 Besides the TV set, there are many electronic products, such as a computer-controlled bread maker.

Getting into the business

In the 1950s, the electronics technician could start up in the radio repair business with $100. Today, the part-time TV technician should have at least $5000. To start a full-time TV service business today, you should have at least $15,000, plus a service vehicle and tools. This does not include the cost of a building to rent, utilities, electronic supplies, or operating expenses. If you want to start a part-time or full-time business in the home, you can get by on $10,000.

Many electronics technicians have started in business in their own homes and have been very successful. If you are not going into electronic sales, this is a good place to start. Of course, you must find out if a business permit can be obtained for your home residence or location. Check with city hall if a business permit is required for a new business. Some technicians have enlarged the double garage area or breezeway for an electronics service business. Usually, someone in the house can answer telephone calls while the technician is doing outside work. Do not choose the basement area because it is too difficult to lug heavy TVs or electronic equipment up and down the stairs. Keep the overhead as low as possible, for greater profit.

Before getting started in the TV or electronics business in a business district or in a small mall area, you should have enough working capital to pay a whole year's rent for the building. Shop or business floor space is expensive in these areas. The ordinary TV or VCR service cannot exist in high-rent districts. Even if the building rents for $1000 per month, that's $12,000 a year, and it takes a lot of service repair to just pay rent for the building. You might find, even with TV and electronic sales, that this type of location is difficult to start up in. Start out small and after a few years of growth, spread into a larger building or location. Most TV technicians start out this way.

TV test equipment is very expensive. Of course, the small VOM or DMM is cheap compared to electronic analyzing test equipment. Do not purchase expensive test equipment and let it set on a shelf to collect dust. The digital multimeter (DMM) with all the many testing features can service many different electronic products. The following test equipment should be used on the TV service bench:

- ☐ Digital multimeter (DMM)
- ☐ VOM
- ☐ Transistor/semiconductor tester
- ☐ Oscilloscope with at least a 5-inch screen, covering 40 to 100 MHz
- ☐ Multi-sweep generator
- ☐ RF signal generator
- ☐ Frequency counter

- ☐ Capacitance meter
- ☐ Color pattern generator
- ☐ CRT tester and restorer
- ☐ Isolation transformer
- ☐ Variable power supply
- ☐ Tuner subber
- ☐ High-voltage probe
- ☐ Soldering iron station
- ☐ Check Chapter 15 for additional and add-on-later test equipment.

Bookkeeping

Most electronics technicians excel at servicing electronic products, but fail to keep adequate books. Some prefer to have an outside bookkeeping firm keep all business records, but this is expensive when starting out. You might want to have someone in the bookkeeping business set up your books so that you can keep them up to date. In the beginning, you can do the books over the weekend or at night. If they are done each week, it only takes a few hours of your time.

Some electronics businesses like to set up a ledger to break down the labor charges away from the sales. The total sales and service charges are entered into the ledger as total income, with the sales in another column, labor in another, and sales tax in the final column. All sales from the cash register, with work sheets and pad charges are entered and described to what job or sale was income. At the end of the month, all columns are totaled, which includes total sales and labor, and the sales tax collected. These business ledger pages can be purchased from stationery, printing, and office supply stores. By keeping the labor separate, you can quickly see if you are charging enough or when to raise the labor rate.

At the end of the year, you might want a balance sheet and income statement for the year. The balance statement should include the total assets, cash, insurance, accounts receivable, and inventory, which will produce the total assets. Property and equipment should be listed under assets with separate land and building costs and equipment (less accumulated depreciation). The total assets should be listed in the left-hand column.

List the current liabilities in the right-hand column with accounts receivable, payroll payable, and total the current liabilities. Next, list the long-term debt, mortgage payments, and total the list of liabilities. List owners' equity in stock, paid-in capital, and retained earnings. Now total the owners' equity.

The year-end income statement should include the revenue of net sales and service, and other income, resulting in total revenue. The cost and expenses would include the cost of goods sold, insurance expenses, interest, and the total of cost and expenses. Next, list the earnings before income tax, income taxes, and the resulting total of the net income. It is best to have an outside tax firm do your state and federal income tax statements. If you have any trouble with the internal revenue service, the tax firm will help you explain the tax deductions to the government.

Business expenses should be kept on another separate sheet of the same ledger. This listing will include advertising, vehicle expenses, vehicle depreciation, total rent, depreciation of equipment, office wages, accounting, technicians' wages, owner wages, interest on investments, cost of parts, resulting in a total expense figure. All expenses paid out by cash should be listed. Taking the business expense away from the net sales and service equals the net profit of the business.

Advertising is quite expensive and comes off the top of your profit. But a business must advertise to keep its name in front of the public. Small ads in the newspaper (in the want ads, announcements, or on last pages) is the cheapest way to go. At first place, run Monday, Wednesday, and Friday. Only use radio advertising for sales of TVs and products. Often, the TV manufacturer will pay 50 percent of a co-op ad for radio or newspapers. TV advertising is too expensive for a TV-repair establishment.

A one-inch listing in the yellow pages of the telephone book helps after a year of operation. A small changeable sign in the front window on special sales or used TV sets adds up to quick sales. The best advertisement is your shop's name and telephone number painted on both sides of the panel or service truck. This keeps your name in front of the public at all times. Of course, word of mouth about reasonable service pays extra dividends. Remember every business loses 10 to 15 percent of customers each year because of people moving away, death, and some that you will never please, no matter how hard you try.

Sperry pricing book

You must make a profit or you will not stay in business very long. Some businesses figure that a 10 percent profit is reasonable. You can make more if you are a good businessperson. It all depends on what you charge for electronic service. When starting out, many technicians charge by what they think they know about electronics. Why not check around in the city, in another town, and see what each electronics business charge rates are? Do not take the lowest or highest price hourly rate, but stay in the middle. For instance, if one shop charges $10 per hour, another $15, and the highest is $20, take the $15 charge to start for the first year. Never price your service work below any one of your competitors. You can quickly start a price war and no one wins.

Some shops use flat-rate charges. This method might not be fair for you or for the customer. If you run into a tough problem each week, you have lost money. Giving estimates can be risky, for you must repair the TV set before you can give a correct estimate. Several states require estimates on various repair establishments, but explain that the estimate charges can vary 10 percent either way. Some electronics shops provide an estimate to just fix the product or a complete overhaul that is guaranteed 90 days. Most customers want the 90-day overhaul.

The Sperry-Tech System of charging is very accurate and covers every job in the electronics service establishment. The book gives accurate average times or minutes for every type of operation. When finished, you add up all the minutes. Check the minutes against the handy labor tables to get the total cost of service time. The Sperry-Tech System covers every small part replacement. From the various charts, you can pick out your hourly rate charge. You can get information for the new updated 6th edition tech guide in electronics pricing, by calling the toll free number: (800) 228-4338.

Sperry-Tech also has pricing order forms, listing the various replacement parts, technical charges, service calls, state tax, city tax, etc. Other electronics order forms are supplied by Tech Spray, O.W. Donnell, MM forms, and Ohelrich Publications. Check for service forms supplied by your local electronics distributor. Make sure that each form of service is itemized for the customer to see what service was performed and paid for.

Local and federal tax laws

Check city hall for a business permit and local city taxes. You might find that some cities require a simple test before a business permit is issued. Some cities have their own sales tax or one section of a large city might have a local sales tax. Inquire about antenna installation permits under city laws.

State sales tax is another tax form that must be filed each month or once every three months, according to state regulations. You must receive a sales tax permit from your state sales tax commission. Some states might collect sales tax on both labor and sales. Years ago, sales tax was only charged against sales. Remember the filing dates because you can be penalized with added interest rates if the sales tax report arrives late.

Every U.S. citizen and resident must file an estimate tax. If you work for some firm, income tax and FICA tax is withheld from your wages on a W-4 form. If in business for yourself, you must file an estimate tax and payable on April 15, June 15, September 15, and January 17th of the following year. Actually, the 1040 form is the estimated taxes paid on the amount of money you earned the same year. If you have a tax accounting firm make out your income tax forms, they can figure what amount the estimate tax should be on the quarterly forms for the following year.

Besides filing a government income estimate tax and FICA tax, you must pay a state income tax, if they have one. Usually, if you work for another electronics firm, they have a government or state tax schedule for the amount of salary on which you must pay taxes. On regular computation of net earnings from self employment, a Social Security Self-Employment tax form must be made out and paid. Check with your state internal revenue service if the state charges a separate income tax. Visit and set up your tax schedules with a local tax accounting firm. For government internal revenue forms, check the government forms that are located in your local post office.

Conclusion

Besides becoming a professional electronics technician, you must be a good businessman, if you want to succeed in business. Your business is opened for one reason and that is to make money. Do not become so busy that you lose money by failing to make out correct charges, overlook critical service parts, and forget altogether to make out a service order. You must collect on every repair or service, at least within 30 days. After a charge is not collected within 60 days, you can kiss it goodbye. Of course, you can turn it over to a collection agency (they get 50 percent) and sometimes you might get something in return.

Know at the end of each month how much money is netted, how much is lost, or on what month you come out of the red, and when the business starts making a profit.

Being your own boss is great, but there are also a lot of drawbacks. After a few months, you will find that you are spending several more hours working for yourself, than for someone else. You can take off when you want to, to a certain point, but the service work is still waiting until you return. Becoming an advanced or professional electronics technician is fun and exciting.

43

List of electronics manufacturers

THIS LIST CONTAINS 117 ELECTRONIC MANUFACTURERS IN the various electronics and entertainment fields. They are listed alphabetically with the type of equipment manufactured alongside of each listing:

Amplifiers	A	Speakers	SPK
Audiocassettes	AC	Tape decks	TD
Compact disc players	CD	Tuners	TN
Compact systems	CS	TVs	TV
Portable tape players	PT	Turntables	TT
Receivers	RC	VCRs	VCR

AAL Meter Group (SPK)
One Mitex Plaza
Winslow, IL 61089

Acoustat Corp. (A/SPK)
3101 SW First Terrace
Ft. Lauderdale, FL 33315

Adcom (SPK/TD/TM)
11 Elkins Rd.
East Brunswick, NJ 08816

Advent (SPK)
4138 N. United Pkwy.
Schiller Park, IL 60176

Afco Electronic Inc. (AC)
471 Roland Way
Oakland, CA 94621

Aiwa America, Inc. (AC/PT/SPK/TT) (Fig. 43-1)
35 Oxford Dr.
Moonachie, NJ 07074

■ **43-1** *Speakers might be contained in large columns, table-top cabinets, and mini-stereo speaker systems.*

Akar American, Ltd. (A/CD/PT/RC/SPK/TD/TT)
807 W. Artesia Blvd.
Compton, CA 90220

Alarm, Inc. (AC/CS/PT/TD/TT)
P.O. Box 550
Troy, MI 48097

All Channel Product (A)
4240 Bell Plaza
Bayside, NY 11361

Alpine Electronic of America (AC)
3102 Kashiwa St.
Torrance, CA 90505

Altec-Lansing International (SPK)
15155 Manchester Ave.
Anaheim, CA 92803

Amber Electronic (A)
218 Ridge St.
Charlottesville, VA 22901

American Acoustics, Inc. (AC)
12841 Western Ave.
Garden Grove, CA 92641

American Audio, Corp. (AC)
636 Forbes Blvd.
So. San Francisco, CA 94080

Analog & Digital System (A/SPK/TD/TN)
One Progress Way
Wilmington, MA 01887

Andante (A)
P.O. Box 5046
Berkeley, CA 94705

ARA Manufacturing Co. (AC)
606 Fountain Pkwy.
Grand Prairie, TX 75050

Audiovox Corp. (AC) (Fig. 43-2)
150 Marcus Blvd.
Hauppauge, NY 11727

■ **43-2** *The audiocassettes might be included in auto radio or cassette units.*

Bang & Olufsen of America (CS/RC/SPK/TD/TT)
1150 Freehamville Dr. (Fig. 43-3)
Mt. Prospect, IL 60056

Blaupunkt (AC)
2800 S. 25th Ave.
Broadnow, IL 60153

List of electronics manufacturers

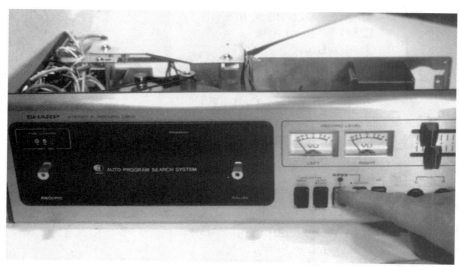

■ **43-3** *The compact systems might include a stereo recording deck.*

Bose Corp. (SPK)
The Mountain
Framingham, MA 01701

Bozach, Inc. (A)
68 Holmes Rd.
Newington, CT 06111

BSR USA (TT)
RR 303
Blauvelt, NY 10913

Carver Corp. (A)
19210 33rd Ave.
Lynwood, WA 98037

Clarion Corp. of America (AC)
661 W. Redondo Beach Blvd.
Gardena, CA 90247

Concord Systems Inc. (A/AC)
6025 Galanda Ave.
Tarzana, CA 91356

Craig Corporation (AC)
921 W. Artesia Blvd.
Compton, CA 90220

Curtis Mathes Service (TV)
1220 Champion Circle
Irving, TX 75006

Daewoo Electronics (AC/PT/TV)
100 Daewoo Pl.
Carlstadt, NJ 07072

D.B. Systems (A)
Main St.
Rindge, NH 03461

Denon America, Inc. (A/AC/CD/CS/RC/TN/TT)
P.O. Box 1139
W. Caldwell, NJ 07006

Electronic Industries (AC)
16940 Vincennes
S. Holland, IL 60473

Emerson Radio Corp. (AS/PT/TV/VCR)
North Bergen, NJ 07047

Fisher Corporation (A/CD/RC/SPK/TN/TT/VCR)
21314 Larsen St.
Chatsworth, CA 91311

Fortune Star Products (AC/CS/PT/SPK/TD/TT)
1200 23rd St.
New York, NY 10010

Fujitsu Corp. of America (AC)
19281 Pacific Gateway Dr.
Torrance, CA 90502

Arthur Fulmer Electronics Division (AC)
122 Gagoso
Memphis, TN 38103

General Electric Consumer Electronics
(A/CD/PT/RC/SPK/TD/TN/TT/TV/VCR) (Fig. 43-4)
Portsmouth, VA 23705

Goldstar Electronics (PT/TV/VCR)
1050 Wall St.
W. Lyndhurst, NJ 07071

Grundy/GR Electronics (AC/PT)
Glenpointe Center E.
Teaneck, NJ 07666

Hamimex, Inc. (CS/PT)
3125 Commercial Ave.
Northbrook, IL 60062

■ **43-4** *The boom-box might have a dual-cassette deck.*

Hanabasheya, LTD (AC/PT)
39 W. 28th St.
New York, NY 10001

Hitachi Sales Corp. (A/CD/CS/PT/RC/SPK/TD/TN/TT/TV/VCR)
401 Artesia Blvd.
Compton, CA 90220

Infinity Systems (SPK)
9409 Owensmouth
Chatsworth, CA 91311

JBL, Inc. (SPK)
8500 Balboa Blvd.
Box 2200
Northridge, CA 91329

Jensen, Inc. (AC/SPK)
4136 N. United Pkwy.
Schiller Park, IL 60176

JICL, LA Corp. (AC)
17120 Edwards Rd.
Cerritos, CA 90703

JR Loudspeaker (A/SPK/TT)
165 Broadway
Hudson, NY 10706

Jullette Electronics (CS/PT)
4615 NW 77th Ave.
Miami, FL 33160

JVC America (A/AC/CD/PT/RC/SPK/TD/TN/TT/VCR)
41 States Dr.
Elmwood Park, NJ 07407

Kenwood Electronics, Inc. (A/AC/CD/PT/RC/SPK/TD/TT)
1315 E. Watson Center Rd. (Fig. 43-5)
Carson, CA 90745

■ **43-5** *The high-powered AM/FM/MPX receiver might connect to a set of speakers or to the separate amplifier.*

Koss Corp. (PT/SPK)
4129 N. Port Washington Ave.
Milwaukee, WI 53212

Kraco Enterprises, Inc. (AC)
505 E. Euchal Ave.
Compton, CA 90224

Kyocera International (A/CD/RC/SPK/TD/TN/TT)
7 Powder Horn Dr.
Warren, NJ 07060

Linear Power, Inc. (A)
11545 D. Ave.
Auburn, CA 95603

Lloyds Electronics (CS/PT/VCR)
180 Raritan Center Pkwy.
Edison, NJ 08818

Luxman Division (CD/RC/TD/TT)
3102 Kashiwa St.
Torrance, CA 90505

Lyman Division (A)
3102 Kashiwa St.
Torrance, CA 90505

Magnavox (NAP) (CD/CS/PT/TV/VCR)
Box 6950
I/40 & Straw Plain Pike
Knoxville, TN 37914

Marantz Co. (A/AC/CD/CS/PT/RC/SPK/TD/TN/TT)
20525 Nordhoff St. (Fig. 43-6)
Chatsworth, CA 91311

■ **43-6** *The turntable might be belt driven in single-play component systems.*

MCI (AC/PT)
28 NW 8th Ave.
Hollandale, FL 03309

Midland International Corp. (AC)
1090 N. Topping
Kansas City, MO 64120

Mitsubishi (AC/CD/CS/RC/SPK/TD/TN/TV/VCR)
3030 E. Victoria
Rancho Dominguez, CA 90221

Montgomery & Wards (TV/VCR)
Montgomery & Wards Plaza
Chicago, IL 60071

Motorola, Inc. (SPK/TV/VCR)
4800 Alameda Blvd. N.E.
Albuquerque, NM 87113

NAD (USA) Inc. (A/CD/RC/SPK/TD/TN/TT)
675 Canton St.
Norwood, MA 02062

Nakamichi USA Corp. (A/AC/TD/TN/TT)
19201 S. Vermont Ave.
Torrance, CA 90502

NEC Home Electronics (CD/VCR)
1401 W. Ave.
Elk Grove Village, IL 60007

Nikko Audio (A/RC/TD/TN/TT)
5830 S. Triangle Dr.
Commerce, CA 90040

Olympus Corp. (PT/VCR)
Crossways Pk.
Woodbury, NY 11717

Onkyo USA Corp. (A/CD/RC/SPK/TD/TN/TT)
200 Williams Dr.
Ramsey, NJ 07446

Panasonic Consumer Electronics
(A/AC/CD/CS/PT/RC/SPK/TD/TN/TT/TV/VCR)
One Panasonic Way
Secaucus, NJ 07094

J.C. Penney Co. (TV/VCR)
6840 Barton Rd.
Morrow, GA 30260

Philco Consumer Electronics (CS/TV/VCR)
Box 6950
Knoxville, TN 37914

Phillips Auto Radio Division (AC)
230 Duffy Ave.
Hicksville, NY 11802

Phillips Consumer Electronics (TV/VCR)
Product Service
P.O. Box 967
Greenville, TN 37744/0967

Pioneer Electronics (A/AC/CD/CS/PT/RC/SPK/TD/TN/TT)
5000 Airport Plaza Dr. (Fig. 43-7)
Box 1540
Long Beach, CA 90801

List of electronics manufacturers

■ **43-7** *The high-powered stereo amplifier might drive four or more speakers.*

Polk Audio (SPK)
1915 Annapolis Rd.
Baltimore, MD 21230

Pyle Industries (SPK)
501 Center St.
Huntington, IN 46750

Quad Electrocoustics LTD (SPK/TN)
695 Oak Grove Ave.
Menlo Park, CA 94026

Quasar Co. (CD/CS/PT/SPK/TV/VCR)
Same as Phillips Electronics

Radio Shack (AC/CD/CS/PT/RC/SPK/TD/TN/TT/TV/VCR)
900 E. Northside Dr. (Fig. 43-8)
Ft. Worth, TX 76106

■ **43-8** *The stereo cassette tape deck might provide speaker level meters for recording.*

RCA Consumer Electronics (A/CD/SPK/TD/TN/TT/TV/VCR)
600 N. Sherman Dr.
Indianapolis, IN 46201

SAE (A/CD/RC/TD/TN)
1734 Gage Road
Montebello, CA 90640

Sampo (TV/VCR)
5550 Peachstreet 2nd Blvd.
Norcross, GA 30071

Samsung Corp. (PT/TD/TN/TV/VCR)
1250 Valley Brook Ave.
Lyndhurst, NJ 07071

Sansui Electronics (A/AC/CD/PT/RC/SPK/TD/TN/TT/VCR)
1250 Valley Brook Ave.
Lyndhurst, NJ 07071

Sanyo Electronics, Inc. (AC/CD/CS/PT/RC/TD/TT/TV/VCR)
1200 W. Artesia Blvd.
Compton, CA 90220

HH Scott, Inc. (A/CD/RC/SPK/TD/TN/TT)
20 Commerce Way
Woburn, MA 01808

Sharp Electronics (CD/PT/RC/SPK/TD/TT/TV/VCR)
10 Sharp Plaza (Fig. 43-9)
Paramus, NJ 07652

■ **43-9** *The portable CD player might include an AM/FM/MPX receiver, cassette player, and top-side CD operation in a boom-box player.*

Sony Corp. of America
(A/AC/CD/PT/RC/SPK/TD/TN/TT/TV/VCR)
Sony Dr.
Park Ridge, NJ 07074

Soundcraftsmen, Inc. (A/TN)
2200 S. Richey
Santa Ana, CA 92075

Soundesign (PT/RC/SPK/TT)
34 Exchange Pl.
Jersey City, NJ 07302

Sparkomatic Corp. (AC)
Rt. 6 & 209
Milford, PA 18337

Speco Division (A)
1172 Rt. 109
Box 624
Lyndhurst, NJ 11757

Studer Rivex America (A/AC/RC/SPK/TD/TN/TT)
1425 Elm Hill Pike
Nashville, TN 37210

Sylvania Corp.
(See Phillips Consumer Electronics)

Symphonic Electronics (PT/VCR)
1825 Arcadia Blvd.
Compton, CA 90220

Tandberg of America (A/TD/TN)
Labriola Ct.
Armonk, NY 10504

Tatung Co. of America (PT/TV)
2850 El Presidio St. (Fig. 43-10)
Long Beach, CA 90810

Teac Corp. (TD/VCR)
7733 Telegraph Rd.
Montebello, CA 90640

Technics (A/CD/RC/SPK/TD/TN/TT)
One Panasonic Way
Secaucus, NJ 07094

Teknika Electronics Corp. (TV/VCR)
353 Rt. 46 West
Fairfield, NJ 07006

Thomson Consumer Electronics (See GE and RCA)
P.O. Box 1976
Indianapolis, IN 46206

■ **43-10** *The portable TV set might contain a varactor tuner with a system-control IC.*

Threshold Corp. (A)
1832 Tribute Rd.
Suite E
Sacramento, CA 95815

Toshiba America (CD/CS/PT/TD/TV/VCR)
82 Totowa Rd.
Wayne, NJ 07470

TZL International (AC/PT/TD/TN)
1523 NW 79th Ave.
Miami, FL 33126

Ultimate Sound (AC)
138 University
Pomona, CA 91768

UCM Corp. (CS/PT)
3843 Carson St.
Torrance, CA 90503

Wald Sound (PT/SPK)
11131 Dora St.
Sun Valley, CA 91352

Winegard Co. (A)
3000 Kurwood St.
Burlington, IA 52601

Yamaha Electronic Corp. (A/AC/CD/PT/RC/SPK/TD/TN/TT)
6660 Orangethorpe Ave.
Buena Park, CA 90260

York Electronics (CS/PT)
405 Minnisink Rd.
Totowa, NJ 07512

Zeff Advanced Products (A)
2135 Stone Ave.
Modesto, CA 95351

Zenith Corp. (TV/VCR)
1900 Austin Ave.
Chicago, IL 60639

Electronics book publishers

Banner Technical Books
1203 Grant Ave.
Rockford, IL 61103

McGraw-Hill
1221 Avenue of the Americas
New York, NY 10020

Howard W. Sams & Co.
2647 Waterfront Parkway E. Drive
Indianapolis, IN 46214

WEKA Publishing, Inc.
P.O. Box 4510
Greenwich, CT 06830

Electronics magazines

Electronic Handbook
P.O. Box 5148
North Branch, NJ 08876

Electronic Servicing & Technology
76 N. Broadway
Hicksville, NY 11801

Electronics Now!
500-B B1-County Blvd.
Farmingdale, NY 11735

Popular Electronics
500-B B1-County Blvd.
Farmingdale, NY 11735

Glossary

ac (alternating current) The type of electricity that is used in the home, business, and in industry.

ACC (automatic color control) A circuit like the AGC that is used to maintain a constant color signal level in color circuits.

ac hum A low-pitch sound heard in defective amplifiers and audio circuits. The common ac hum heard with defective filter capacitors is 60 and 120 Hz.

acoustic feedback A sound or vibration from the speaker that is fed back to the amplifier or turntable. The sound might be hum, rumble, howling, or distortion in the speaker.

ADC (analog-to-digital converter) A device used to convert analog information (usually sound) from digital information.

AF (auto focus) The focus servo that moves the lens assembly to correct the focus of the beam.

AFC (automatic frequency control) A method of maintaining the frequency in FM receivers, TVs, and VCRs from drifting off channel.

AFM (audio frequency modulation) The process by which both Beta and VHS systems encode the audio track recording.

AFT (automatic fine tuning) In the TV and VCR circuits, the AFT compensates for frequency drifting. No fine tuning control is used.

AGC (automatic gain control) Adjusts the radio tuner sensitivity according to the strength of the signal in the receiver. It maintains an overall constant picture level in the luminance circuits of TVs and VCRs.

air suspension Another name for an acoustic-suspension speaker.

alternate channel selectivity The ability of a tuner, in decibels (dB), to filter out the alternate station alongside the one that is already tuned in.

AM (amplitude modulation) With the AM process, the program information is imposed on a carrier signal of constant frequency to vary its amplitude level in proportion to the level of audio. The standard broadcast station band uses amplitude modulation.

amp Abbreviation for amplifier or ampere.

ampere The standard unit of current. Current is measured by the rate at which charge flows.

amplification The process by which the amplitude of a signal is increased.

amplifier A tube, transistor, or IC combination (with other components) to increase the amplitude of either voltage or current in sound circuits.

amplitude The strength of a waveform, expressed in height. It can either be positive or negative.

analog filter A filter system in the CD player that reduces and cancels out noise. The analog filter is used before and after the D/A converter in some players.

analog meter A meter that has a pointer that deflects with current flow, like the VOM.

anode The positive (+) element of a vacuum tube or semiconductor. In the TV chassis, the anode is the highest voltage terminal of the picture tube. The anode element is grounded in the microwave oven. It is the positive electrode of a two-terminal semiconductor device.

antenna Any device to collect a broadcast signal for the TV or radio receiver.

anti skating A force applied to the pivoted tone arm to counteract the tendency of the arm to be pulled toward the center of the record or disc.

APC (automatic phase control) The circuit that automatically maintains an overall constant picture level in the VCR luminance circuits.

APC (automatic power control) The circuit that keeps the laser diode optical output at a constant level in a CD player.

audio frequency Any sound that is capable of being heard (20 to 20,000 Hz).

auto eject The tape player feature that automatically ejects the cassette at the end of the playing time.

auto record level Circuits that automatically control the recording level.

auto reverse The ability of a cassette player to automatically reverse directions to play the other side of the tape.

auto turntable A record turntable that automatically plays each loaded record and shuts off without help from the operator.

azimuth The angle at which the tape meets the moving tape. The head azimuth adjustment is used alongside of the tape head.

azimuth control A control to adjust the angle of the tape head to correct misalignment in the stereo tape player.

back EMF Sometimes called *counter EMF*. A force produced by an inductor that opposes that which produced it.

baffle The board on which the speaker is mounted.

balance The control in the stereo amp equalizes the output audio in each channel.

balanced modulator A circuit designed to output the frequency differences or sum of its two signals in the VCR.

balun coil Baluns are used to a balanced input to an unbalanced antenna.

bandpass amplifier The bandpass amplifier in the color circuits passes a certain band of frequencies.

bandwidth A frequency limit between a device or circuit that transmits ac energy with low loss. See *bandpass amplifier*.

Barkhausen The effect of when two vertical lines appear in the left side of the picture.

base One of three leads of a bipolar transistor.

bass reflex A bass reflex system vents the backward sound waves through a tuned vent or port, thus improving the bass response.

battery A battery consists of two or more cells that store and provide a dc voltage.

BCD (binary code decimal) The number system used by compact discs.

bell or buzzer An electronic or mechanical device that is used as a sound indicator.

Beta A tape head system developed by Sony for VCR recorders.

bias A high-frequency current applied to the tape head winding to prevent low distortion and noise while recording on the tape. A predetermined level to establish a threshold or operating point. The battery might provide bias voltage. A bias resistor in a tube or emitter circuit of a transistor, where a bias voltage is developed by current flowing through that resistor.

bipolar-junction transistor A two-junction transistor—either PNP or NPN. A bipolar transistor is current operated and uses both electron and hole conduction.

bit A binary number or digit. A group of bits makes a word. There are 16 bits in most CD players.

block diagram A drawing or schematic that shows the different sections or stages of a system.

blower motor The blower motor keeps the magnetron cool in the microwave oven.

booster amp A separate amplifier connected between the main unit and the speakers in an auto stereo installation.

BPI (bits per inch) The number of bits placed in a linear inch of disk space.

breadboarding The assembling of electronic circuits.

bridge rectifier Four diodes are wired in a series circuit to provide full-wave rectification of a two-lead power transformer. The TV chassis might have a bridge rectifier after the line fuse.

bridging Combining both stereo channels of the amp to produce a monaural signal with almost twice the normal power rating in a car stereo system.

brightness Refers to both the amount of illumination on the TV screen and the control that is used to adjust the brightness lead.

burst In color TV circuits, the burst is a precise timing signal. The burst keeps the color oscillator in the TV locked to the TV broadcast signal.

burst oscillator The precision 3.58-MHz oscillator is vital to color reception. It is kept in step (sync) by the burst.

buzz Sometimes called *intercarrier buzz*, a raspy version of ac hum, usually caused by an improper adjustment of the sound IF circuits.

B-y The blue component of a color picture minus the monochrome in the TV color circuits.

byte A set of eight bits. A byte is approximately equivalent to a character.

cabinet The box that contains the electronic equipment.

capacitance The electronic "size" of a capacitor. Sometimes called capacity. The basic unit of capacitance is a farad.

capacitive filter A capacitor used in the filter supply circuit, which is connected in parallel with the circuit connected.

capacitor A capacitor permits storage of electricity between two insulated conductors. The capacitor stores electric energy, blocks the flow of direct current, and permits the flow of ac current. The capacitor is available in many different sizes and shapes.

capstan The shaft that rotates against the tape and pinch roller, moving the tape past the tape heads in cassette decks and VCRs.

carrier The radio signal that carries the sound or picture information from the transmitter to a receiver.

cartridge A component that holds the stylus or needle and follows the record groove on the phonograph.

cassette A tape-based recording and playback medium. "Compact" cassettes are used for audio recording and VHS cassettes are used for video recording.

cathode The cathode element provides the source of electrons. The cathode element is used in a tube, and in the magnetron tube, it has a very high voltage.

center tap A center connection of a power, audio, and RF transformer.

ceramic capacitor A ceramic dielectric material is used between the capacitor plates. It is used for RF, bypass, and coupling applications.

CH The abbreviation for channel. The stereo audio circuits have a right and left audio channel. A quadraphonic amp has four channels.

channel separation The degree of isolation between left and right channels, often expressed in decibels (dB). The higher the decibels, the greater the separation.

chassis The physical framework that holds the working parts in the amplifier, tuner, radio, cassette, CD player, TV, or VCR. The chassis can be metal, plastic, or a PC board.

chip A chip is a miniature component that is soldered directly onto the PC wiring. They can consist of resistors, capacitors, diodes, and transistors.

choke An inductance used in a circuit to present high impedance to frequencies without limiting the flow of direct current.

chroma Another term for color. Color amplifiers are often called chroma amplifiers. Chroma is the color portion of a color signal. The color quality is called *hue*.

chrominance The color portion of a video signal.

circuit A path in which electrical current can flow.

circuit diagram A circuit with electronic symbols or parts.

clamper A tube or semiconductor that restricts a wave at a certain dc level.

clamper assembly The clamper assembly fits over the compact disc after it is loaded. Remove the clamper assembly to get at the laser pickup assembly in the CD player.

clear button The button that erases the last digit entered or the complete program if it is pressed immediately after the band button on some CD players.

clipping Removing or cutting off the signal of a waveform that contains distortion, which can be seen on an oscilloscope. A clipper circuit in the TV chassis separates the sync (timing) signals from the picture information.

clock An electronic circuit that provides correct timing in a CD player.

closed-loop drive A tape transport system that drives both incoming and outgoing tapes. It is controlled by contact with the capstan at each end of the head assembly.

CLV servo (constant linear velocity) An electronic system that provides CLV during disc playback. The CLV circuits control the spindle, turntable, and disc motors.

coaxial cable A shielded wire around an insulated center wire connecting the antenna to the receiver.

coaxial speaker A speaker with two drivers mounted on the same frame. Usually coaxial speakers are used in home and auto radio-player systems.

collector One of the three leads of a bipolar transistor.

color bar generator The color bar generator provides patterns for color alignment and adjustments with date, crosshatch, and color bars.

color burst A sensor in a VCR that senses when color burst signals are present.

color killer A special circuit whose function is to turn off the color amplifier circuits when a black-and-white signal is received in the TV receiver.

comb filter The comb filter circuit separates the luminance (brightness) and the chroma (color) video information, eliminating cross-color that might occur in a TV.

compact disc (CD) The compact disc provides noiseless high-fidelity music on one side of the rainbow surface of the disc.

compliance The ease with which the stylus can be deflected by the groove wall. A high-compliance stylus will yield readily to the forces exerted on it by the record groove.

component Refers to small electronic parts.

connector An electrical path of two or more components tying them together for easy disconnection.

contrast The depth of difference between light and dark portions of a TV scene. The name given to the control for adjusting the contrast level.

control board The control board in the microwave oven controls the time and operation of the oven cooking process.

convergence A system that brings the three electron beams together in a picture tube so that they pass through the same hole in the shadow mask and strike the correct dots on the screen.

converter A stage in the tuner of a TV or radio receiver that converts an incoming signal to an IF signal.

corona Similar to an electric arc of a much higher voltage (thousands). Corona occurs as a continuous path through the air between two points.

CPV A computer-type processor used in the master and mechanism circuits.

crossover A filter that divides the signal to the speaker into two more frequency ranges. The highs go to the tweeter and the lows to the woofer.

crosstalk An improper adjustment of the tape head might cause crossover between the two different tracks. Crosstalk is leakage between one channel to another in tape players and VCRs.

CRT (cathode ray tube) Another name for the color picture tube.

crystal A quartz of synthetic material-like slab or wafer having the property of vibrating at a precise rate or frequency. Each crystal is cut and ground to vibrate at the desired frequency.

cue To scan the playback at a faster speed than in the forward direction in a cassette player and VCR.

cueing lever A method of raising or lowering the tone arm using a lever.

current The flow of electricity. Current is measured in amps or milliamperes.

cutoff A transistor or tube so biased, only a small amount of current flows.

cycle See hertz (Hz).

cylindrical lens A special lens in the optical system of the laser pickup assembly in the CD player.

D/A converter A special stage that separates the audio signal from the digital signal in the CD player.

DAC Digital-to-analog converter.

damper A diode, tube, or semiconductor used in horizontal amplifier circuits to suppress certain electrical activity in the TV set. The damper motor in the microwave oven opens and closes to let out hot air.

DAT Digital audio tape.

DBX The noise reduction system in which the program is compressed before being recorded and expanded in the playback mode.

dc Abbreviation for direct current.

DDC The direct-drive cylinder is used in VHS VCR players or camcorders.

decibel (dB) A measure of gain, the ratio of the output power voltage with respect with the input, expressed in log units.

de-emphasis A form of equalization in FM tuners to improve the overall S/N (signal-to-noise) ratio while maintaining the uniform frequency response. A de-emphasis stage follows the D/A (digital-to-analog) converter in CD players.

deflection The orderly movement of the electron beam in a picture tube. Horizontal deflection pertains to left-right movement, and vertical deflection is the up-down movement of the beam. The deflection coil is mounted in the bell and neck of the CRT.

deflection IC Both the vertical and horizontal oscillator and amplifier circuits are located in one IC. Many circuits might be contained within the same deflection IC.

deflection plates Plates are elements of a CRT tube that operate like a yoke assembly.

defrost The defrost circuit in the microwave oven removes ice from frozen or cold food for quick microwave cooking.

degaussing Demagnetizing. In the color TV set, the degaussing circuit demagnetizes the front of the TV screen each time the TV is turned on.

delta circuit A three-phase electrical circuit with no common ground.

demodulator A demodulator separates or extracts the desired signal, such as sound or picture information.

detector Same as a demodulator. A detector in the AM radio detects audio from an IF frequency. Silicon detectors are used in many consumer electronic products.

deviation The swing of the FM carrier when it is modulated in a VCR.

dew A warning light that might come on in a VCR or camcorder, indicating that there is too much moisture. Some players will remain in shutdown if too much moisture occurs.

dielectric A material that serves as an insulator. A dielectric material is used between the metal foils of the capacitor.

dielectric constant A number that indicates the relative quality of the dielectric material.

digital Information expressed in binary terms. Signals representing numbers or characters are of discrete, rather than of continuously variable values. Digital tuning is used in auto and home radio receivers.

digital filter A low-pass filtering network.

digital multimeter (DMM) The digital multimeter measures voltages, current, resistance, and tests diodes. Today, the DMM might also measure capacitance and frequency, and test transistors with an LCD display.

diode A two-element electronic device—a tube or semiconductor. The most common application is converting ac to dc (rectification).

DIP A dual in-line package.

dipole A type of antenna that uses two horizontal quarterwave elements. A folded dipole is a common element in FM and TV antennas.

direct current (dc) Current that flows only in one direction. A battery is a source of direct current.

direct drive A direct-drive turntable has a motor shaft connected to the turntable without any coupling or belts. The capstan motor might be direct-driven in VCRs, camcorders, or cassette players.

disc The CD disc is inserted into a CD player that contains music, messages, or information.

disc holder The disc holder or turntable that sits directly on top of the motor shaft in a CD player.

discriminator An audio detector in FM or TV sound circuits.

dispersion The spread of high frequencies from a speaker, measured in degrees (the angle in which the speaker radiates sound).

distortion In a simple sine-wave signal, distortion appears as multiples (harmonics) of the input frequency. A type of distortion is the clipping of the audio signal in audio amplifiers. Distortion can be traced with an oscilloscope.

dither A very low-level noise filter added to a digitized signal to reduce high distortion caused by quantizing the low-level signals used in the CD player.

dl Delay line used in a TV set or VCR.

DMM Abbreviation of digital multimeter.

DNR (dynamic noise reduction) A noise reduction system that reduces the high frequencies.

Dolby noise reduction A type of noise reduction that operates by increasing the high-frequency sound during recording and decreasing them during playback, thus restoring the signal to the original level by eliminating tape hiss. The three types of Dolby noise-reduction systems are Dolby A, B, and C.

door assembly The front door assembly opens so that food can be cooked in the microwave oven.

DOS Disk operating system. Programs that direct operation of a disk-based computing system.

drain One of the three leads of a field-effect transistor.

driver In a speaker system, each separate speaker is sometimes called a *driver*. The driver in the audio amp might drive the audio signal to the audio output transistors or ICs.

drive system The motor, belts, and gears that rotate the turntable. No belts or pulleys are in a direct-drive system.

dropout Dropouts are caused by dirt, scratches, and foreign material in a CD disc. The optical system cannot read the digital information with this type of substance. Dropout might result from oxide or coating of dust on the tape or video heads in a camcorder or VCR.

dry cell A primary cell made up of zinc and carbon with an electrolyte.

dual capstan Dual capstans and dual flywheels are used in auto-reverse cassette players and they can play in both directions. In some tape decks, a second capstan might be added, creating a closed loop method with stable tape motion and better performance.

dynamic A dynamic speaker has a voice coil that carries the signal current with a fixed field (permanent magnet) and moves the coil and cone. This same principle applies to headphones.

dynamic range The ratio between the maximum signal level and the minimum level expressed in decibels (dB). The full dynamic range of the human ear can be recorded on the compact disc.

E-E (electronics to electronics) The picture viewed on a TV screen when a recording is made on the camcorder.

E-F balance After changing the optical-laser pickup assembly, the balance of the E-F diodes must be adjusted for tracking-error detection in the CD player.

efficiency The percentage of the electrical input power to a given speaker that is connected to audio energy. High efficiency might mean less amplifier power is required to reproduce the music for listening.

EFM signal (8- to 14-bit modulation). A very complex encoding scheme used to transform the digital data to a form that can be placed on the disk. This information is modulated by the EFM in the CD player.

EIA Electronic Industries Association.

electricity Flow of current of electrons along a conductor.

electrolytic capacitor A type of polarized capacitor.

electromagnet A magnetic field with electric current through a coil of wire.

electromotive force (EMF) Term of voltage or a force that moves electrons.

electron The smallest electrical charge. A negatively charged particle.

electronic component An individual part in electronic circuits.

electronic speed control An electronic method of controlling the motor speed.

electronic voltage Same as RMS voltage; 0.7071 times the peak voltage.

electronics The physical action of electrons in electronic devices, such as vacuum tubes, ICs, transistors, etc.

electrostatic An electrostatic speaker, headphone, or meter phone uses a thin diaphragm with voltage applied to it. The electronic field is varied by the signal voltage.

element A part of an electronic device, such as the terminal of a transistor, tube, or diode.

emitter One of the three legs of a bipolar transistor.

emphasis A process of boosting the high-frequency portions of the video signal.

equalization (eq) Alteration of the frequency response so that the frequency balance of the output equals the frequency balance of the input in tape players. Equalization is also used to correct response deficiencies in speakers.

equalizer A device to change the volume of certain frequencies in relation to the rest of the frequency range. Sliding controls might be used in auto radio and cassette player equalizers.

erase head A magnetic component with applied voltage or current to remove previous recording or noise in the tape. Some erase heads operate with only a magnet as a head.

extended play (EP) Refers to the six hours of playing time obtainable with a T120 VHS cassette played in a VCR machine.

eye pattern The RF signal waveform at the RF amplifier in the CD player. The waveform is adjusted so that the diamond shapes in the eye pattern are clear and distinct.

face plate The front assembly of a picture tube. In a color tube, it includes the tricolor phosphor and the aperture mask.

fader A control in an auto receiver or cassette player that is used to control the volume balance between the front and rear speakers.

farad The unit of a capacitance. Broken down in microfarads and picofarads.

ferrite A compressed, powdered iron material.

FET Field-effect transistor.

FET multimeter A meter using an FET transistor with a high-impedance input.

FF Fast forward.

FG Frequency generator is used in servo circuits of VCR machines.

field One scanning of the scene on the face of the picture tube, in which an alternate line is temporarily left blank. The scan duration of a field is 1/60 second. A field consists of 262.5 horizontal scanning lines across the picture tube.

filter The electrolytic filter capacitor is used in the low-voltage power supply. A circuit that selectively attenuates certain frequencies, but not others, such as those in its pass-band frequency.

fixed capacitor A capacitor with a fixed value.

flag waving Flag waving can cause a bending or flapping from side to side of the top part of the screen.

floating ground A common ground or chassis above the auto ground.

fluorescent display A display that shows the functions of the electronic product.

flutter A change in the speed of the disc or record on a turntable. Practically no flutter exists in the CD player.

flux 1. Flux is a material used in solder. 2. The magnetic field of a coil.

flux density The number of flux transitions that are written along a given length of disk surface.

flyback, retrace The trace or return movement of the electron beam in a picture tube after completing each line and each field. These retrace lines are blacked or blanked out in TV chassis.

flyback transformer Another name for the horizontal output transformer. The flyback transformer provides sweep and builds up high voltage for the anode voltage of a picture tube. The flyback can provide scan-derived voltage sources for other circuits in a TV chassis.

focus In the TV set, the picture is cleared up and is sharp with a focus adjustment. Focus in the camcorder has a motor that brings the picture into focus.

focus coil A coil at the end of the focus drive transistor or IC that moves the object or actuator assembly up and down to maintain focus on the disc of a CD player.

focus error (FE) The output from four optosensing elements is supplied to the error signal amp and a zero output is produced.

The error amp corrects the signal voltage and goes to the servo IC to correct the focus in the CD player.

focus offset The offset adjustment mode at the RF amp with a test disc. A good eye pattern means that the diamond shape in the center of the waveform is clear on the oscilloscope.

focus OK circuit (FOK) The circuit that generates a signal to determine when the laser spot is in the reflecting surface of the disc. The FOK signal is high when the laser is in focus of the CD player.

focus servo The servo IC controls the correct focus applied to the focus-driving circuit. The focus servo IC detects the errors and corrects them.

folded horn speaker The system that forces the sound of the driver to take a different path to the listener. One of the most effective types of speakers available.

forward bias The biasing of a transistor or diode that controls conduction or no conduction.

frame One complete picture in television or video.

frequency The number of recurring alternations in an electrical wave, such as ac, radio waves, etc. Frequency is specified by the number of alternations occurring during 1 second and given in hertz (cycles per second), kilohertz (1000 cycles), and megahertz (million cycles).

frequency counter The frequency counter test instrument counts the frequency of various circuits. The frequency range might vary by 2 Hz up to 100 MHz.

frequency response The range of frequencies that a given piece of equipment can pass to the listener. The ideal frequency response of a given amplifier is 20 Hz to 20 kHz. It is often defined with decibel variation over a flat specified frequency range.

front-to-back ratio A ratio of forward resistance to reverse the resistance of a diode.

full-range speaker A speaker system with only one driver, which reproduces the normal frequency range without help from another speaker.

full-wave rectifier Uses two diodes to change ac to the dc voltage.

fuse A safety device with a low-melting-point metal. If the current passing through exceeds the safe level of the fuse, the metal melts or opens the circuit.

fuse resistor A service that senses an increase in current of the transformer secondary and opens the primary winding circuit.

FZC circuit The circuit that detects when the FE signal reaches 0 V. It is used together with the FOK circuit to determine the focus-adjustment timing in the CD player.

gain The amplification of an electronic signal. Gain can be given in decibels (dB).

gain control A control to adjust the amount of boost in the signal.

gap The critical distance between the pole pieces of a tape head. The gap area of a cassette head is quite small compared to a camcorder or VCR tape head.

gas Refers to undesirable presence of a trace of gas inside a vacuum tube.

gate A circuit that delivers an output with a specific combination of the input. The gate is used in analog and digital circuits. One of the three legs of a field-effect transistor.

generator An electromechanical force that produces electrical power.

ghost A double-exposure scene on the TV screen. Usually a faint picture is used around the regular image, caused by two different signals from the same station; one signal is delayed in time.

glitch A form of audio or video noise or distortion that suddenly appears and disappears during VCR operation.

graphic equalizer An equalizer with a series of slides that provides a visual graphics display.

ground A point of zero voltage or the common voltage return for the components within a circuit, sometimes referred to as the *earth ground*. The common ground can be a metal chassis in the amplifier or receiver. All test equipment should be connected to common ground when servicing electronic products.

grounding A common ground.

guard band The space between the video tracks on the video tape when in SP mode.

G-y The green color signal minus the monochrome.

HAD The horizontal drive signal in a VCR.

half-wave rectification A single diode converts ac to dc voltage.

Hall-effect IC An external magnet causes current to flow in the IC.

harmonic A multiple of a given frequency.

harmonic distortion The additions of harmonics not present in the original recording. Harmonic distortion should be less than 1 percent.

head A magnetic component with a gap area that picks up signals from the moving tape.

head cylinder A cylindrical piece of metal that houses the video head. Sometimes called a *drum*.

head switching Switching the video heads in and out of the video recording circuits of a VCR.

head switching pulse The signal that is applied to the head amplifier to perform head switching.

head unit The central control of a car stereo system, which contains the radio, cassette deck, CD player, and amplifier.

heatsink A clamp to radiate heat in soldering; heatsink to dissipate heat from the mounted part on a heatsink.

helical scan system The diagonal stripe system by which very high video frequencies are placed on the tape in both Beta and VHS format.

henry A unit measurement of inductance.

hertz (Hz) Cycles per second (cps), the unit of frequency.

high voltage Generally refers to anode voltage applied to the picture tube.

high-voltage probe The high-voltage probe is a test instrument that measures the high voltage on the picture tube anode terminal.

hiss The annoying high-frequency background noise found in tape decks, CD players, and record players.

HOT Horizontal output transformer, which steps up low voltages, usually with driver and horizontal output transistors. It is sometimes referred to as a *flyback transformer*.

hue In color TV, the basic color that distinguishes red from green and blue, etc.

hum A type of noise that originates from the 60-Hz power line, caused mainly by poor filtering in the low-voltage power supply. Improper grounding at the input of the amplifier can cause hum.

hybrid circuit A combination of tubes and transistors or of PC wiring and hand wiring.

hysteresis A magnetic loss in a transformer or coil that is caused by residual magnetism.

IC (integrated circuit) A structure similar to a module, in which a number of parts required for the performance of a complete function are prewired and sealed. It is not repairable.

idler wheel The wheel in the phonograph that rotates the turntable. The idler wheel in the cassette player helps to determine the speed of the capstan flywheel.

IDM (intermodulation distortion) Distortion at frequencies that are the sounds and differences of multiples of the input frequencies.

IF (intermediate frequency) IF transformers are used in radio receivers and TV sets. The TV chassis might have an IF stage that contains video and audio signals, and it also has a sound IF (SIF) transformer.

IHVT The integrated horizontal or flyback has HV diodes and capacitors molded inside of the flyback area. IHVT transformers might provide several different voltage sources for other TV circuits.

impedance A combination of resistance and reactance, expressed in ohms. A speaker's impedance might be 2, 4, 6, 8, 10, 16, or 32 Ω.

impedance matching When the impedance of a device is matched to the impedance of another device (such as between a TV and its antenna) to allow maximum transfer of energy.

index The subdivision of a CD-ROM track.

index search When using a disc with index coding, press track (skip forward) or track (reverse) until the desired number appears in the display. If the track or index does not exist on the disc, the player will search to the end of the disc and stop the CD player.

inductance The property of a circuit that causes a magnetic field to be induced.

inductive filter A filter circuit with an inductor connected in series with a voltage source or load.

inductor A device that is used to oppose any change of current.

infinite baffle A completely sealed box that encloses the speakers.

in-line picture tube A recent development in color picture tube construction that produces three basic colors in adjacent strips or bars, instead of the earlier types, which produce a three-dot or triad group. Improved quality, simplified design, and maintenance is claimed in this design.

input Where signals enter a circuit.

interlock A safety device in the CD player that activates when loading the disc. The interlock might lower or cut off the power to the optical assembly when it is loading. Also, the interlock is a switch in a microwave oven.

intermodulation distortion (IMD) The presence of unwanted frequencies that are the sum and differences of the test signals. Blurring or smearing of sound is the result of IMD. These levels should be below 0.1%.

ion trap See *trap*.

ips (inches per second) The measurement of tape speed.

isolation transformer The isolation transformer might be of the variable type that raises or lowers the power line voltage to the TV set or other ac products. Always use an isolation transformer in today's electronic products.

jack A convenient, nonpermanent connection post, such as the headphone jack of the portable CD player. The DMM has jacks for the different measurements with a black and red cable with probes.

jitter The playback picture might have jitter with too much flutter or wow. The picture shakes.

key-off eject The key-off eject is used to prevent damage to the tape and capstan of the auto cassette player. The cassette is ejected when the ignition key is turned off.

kHz (kilohertz) 1000 Hz.

kinne This is often referred to as the *picture tube*.

laser A narrow intense beam of single-wavelength light.

laser assembly The assembly that contains the laser diodes, and the focus and tracking coils. The optical assembly might be known as a *laser assembly*.

laser current The laser current indicates if the laser diodes are defective in the CD player. The laser current can be tested with voltage or with a laser power meter.

laser diodes The diodes that pick up the coded information from the disc along with the optical pickup assembly in a CD player.

latch switch An interlock switch is located and controlled by the oven door in a microwave oven.

leakage Undesired current flow through a component.

leakage tester A government-approved survey instrument to measure radiation leakage of the front door, vents, and waveguide areas in microwave ovens.

LED (light-emitting diode) The low-power diodes used for optical readouts and displays in electronic equipment. LEDs are available in many colors.

level The strength of the signal; also the alignment of the tape head with the tape. The turntable or disc motors should be level in some CD players.

light switch Controls the oven light, depending if the door is closed or open in a microwave oven.

linearity Picture symmetry. Vertical linearity refers to symmetry between upper and lower halves of the picture, while horizontal linearity refers to the right and left side of the picture.

line filter A device that is sometimes used between the ac wall outlet and a radio or TV set to reduce or eliminate electrical noises.

line output Line input or output jacks are used in amplifiers, cassettes, and CD players. The line signal is usually a high-level signal.

line, transmission The antenna lead-in wire or cable.

loading motor The loading motor loads the disc in a CD player and a cassette in a cassette player, camcorder, or VCR.

loading tray The tray that holds the small compact disc during the loading of a CD player.

local DX (or LO/DX) switch A switch that increases sensitivity of the tuner for distant (DX) stations.

lock, horizontal An adjustment in some TV sets for setting the automatic frequency operation on the horizontal sweep oscillator.

long play (LP) A speed on the VCR that provides four hours of recording on a 120-minute VHS cassette.

loss Usually refers to the amount of signal loss in a TV antenna lead-in line.

loudness The volume of sound. Loudness compensation is controlled by the loudness control of an amplifier.

loudness circuit A switch to change the frequency response contour of the sound by increasing the bass at low volumes.

loudness compensation A switch that boosts a low sound level to compensate for the natural loss of sound at the human ear when the sound level is reduced.

low-voltage regulator Low-voltage regulators are formed in the low-voltage power supply. The regulators can be transistors, ICs, or zener diodes.

LSI (large-scale integration) Many components built into one large chip. Includes the processors, ICs, and CPUs in compact disc players, TVs, and VCRs.

luminance A portion of the video signal that contains sync, and black-and-white information.

magnet A component that attracts magnetic materials.

magnetic The magnetic phono cartridge has a moving vane between the coil assembly. The output voltage is very low and must have a preamplifier stage ahead of it, unlike the crystal cartridge.

magnetic field The force that surrounds the magnet and produces magnetic lines.

magnetism Ability to attract iron, cobalt, and nickel.

magnetron A vacuum tube in the microwave oven that controls the speed of electrons with a magnet and electrical field. The frequency of these waves is 2450 MHz.

mega The prefix to designate 1,000,000.

memory The program memory of a CD player. Some program memories can play up to 15 selected tracks.

memory counter A system that allows the tape to be rewound to any point.

memory preset An instant recall feature in computerized timers.

metal tape The high-frequency response and maximum output level are greatly improved with metal tape. The pure metal particles on the tape make the cassettes more expensive than the normal oxide cassettes.

MHz One million hertz.

micro The prefix to designate 0.000001 (10^{-6}).

microprocessor An IC that serves many different functions. The microprocessor might be used in the control and mechanism system of a VCR and camcorder.

milli Prefix used to designate 0.001 or (10^{-3}).

MIR (mirror detector circuit) The circuits used to detect the mirror portion of the disc between the tracks and outside of the lead-out track and also in the detection of disc flaws.

modular chassis A TV chassis that is made up entirely of separate modules for each circuit in the TV set.

modulation The way in which one signal modifies or controls another signal for the purpose of carrying information.

module A subassembly of a number of parts, usually including transistors, ICs, and diodes.

molecule The smallest quantity of a compound of elements.

monitor To compare signals. To monitor voltage or waveform in an intermittent circuit. A monitor light might indicate that the chassis is functioning. An ordinary TV hooked up to the VCR can be used to view or monitor the tape.

monitor diode A special photodiode to monitor the light output of the laser to a connection circuit.

motor-boating A "putt-putt" sound caused in the audio sound input and output circuits. Motor-boating might be caused by poorly filtered circuits.

MPU Another term for CPU. A microprocessor unit.

multimeter A test instrument to measure voltage, current, and resistance.

music search A feature that finds the beginning of each song automatically.

mute switch An electronic device that controls the muting of the output line audio circuits in the compact disc player. The mute button in the remote-control transmitter temporarily shuts off the sound of a TV set.

mutual inductance Measured in henrys. Inductance that results in interaction of an adjacent inductor.

noise Any unwanted signal that is unrelated to the desired signal. Noise can be generated during the rewind or playback of tape and disc.

noise suppressor A filter to reduce background noise.

NPN transistor Transistor made of N and P materials arranged in an NPN sequence.

NR Abbreviation for noise reduction.

NTSC Abbreviation for National Television Systems Committee. Identifies the U.S. color television standard.

open circuit No flow of current. A switch opens or connects the circuit.

optical lens The lens assembly in the laser pickup. The optical lens should be cleaned by wiping it with a lint-free cleaning paper or with camera lens cleaning products.

oscillator Generator of a signal, such as a 3.58-MHz color subcarrier signal (the RF oscillator in the TV).

oscilloscope A test instrument that can share exact waveforms throughout the TV chassis to help you to troubleshoot circuits and locate defective components.

output power The output of an amplifier, rated in watts. To double the volume a 10-fold increase in power is required.

oven light Furnishes light to the oven cavity. Some ovens have two lights.

oversampling When the digital data derived from the disc is sampled at a rate higher than normal.

oxide The magnetic coating compound of the recording tape. After many hours of operation, the oxide dust might flake off and coat the tape head, resulting in weak, distorted music or erratic pictures.

parallel A method of connecting components where all components involved connect to a common point so that each component is independent of the other components.

passive component A component that does not change value as a result of the power applied to the circuit.

passive radiator A second woofer cone that is added, but without a voice coil. The efficient second speaker cone is driven by the inside pressure movement of other speakers.

pause control A feature to stop the tape movement without switching off the machine. Pause is used in cassette decks, VCRs, camcorders, and CD players.

PC board Printed circuit board. A subassembly of various parts, on a phenolic or fiberglass board, on which the interconnections are printed on the metal veins or paths.

peak The maximum level of a signal. A peak indicator light shows that the signal levels are exceeding the recorder's ability to handle peaks without distortion.

peak to peak (p-p) The highest positive and negative levels during a cycle of alternating current.

permanent magnet The magnet retains its magnetism even after a force has been removed.

permeability The ease with which magnetic lines pass through an object.

PG A pulse generator is used in VCR servo circuits.

phase Sound waves that are in sync with one another. The positive terminals of a speaker are wired together so that the cones will go in and out together or in phase.

phono plug A particular type of small plug that transports the cartridge signal to the amplifier. Also known as an *RCA plug*.

phosphor The coating on the interior of the face plate of a picture tube, which gives off light when it is struck by an electron beam. The chemical composition of the phosphor determines the color of the light it will emit.

photodiode A light-sensitive device that is used in the laser assembly to bring the RF signal to the preamp stages of the CD player.

pickup The laser or optical assembly might be called the *pickup assembly* in a CD player.

pickup motor The motor used to move the pickup assembly toward the outer edge of the disc. Also called *sled*, *slide*, or *feed motor*.

picture projection Three small projection color tubes are used to project the TV image in the front or rear of a large TV screen. The projection tubes are used inside of the cabinet of a rear-projection color set.

picture tube The picture tube receives the video color signal that displays the picture on the CRT raster.

PI filter A filter circuit with the capacitor before the inductor and one after in series with the output load.

pit A microscopic depression in the surface of a compact disc.

pitch control A control that slightly changes the speed of the turntable motor.

playback head The head that often includes both playback and record in one assembly. In some tape players, the single-play head is also known as a *monitor tape head*.

PLL (phase-locked loop) An accurate system with an FM multiplex decoder based on feedback control of the oscillator. A PLL VCO circuit might also be found in the digital control processor of the CD player.

p-material The transistor material that has a deficiency of electrons.

PN junction In a semiconductor diode or transistor object where the P and N materials are joined.

PNP transistor A transistor using P and N material in a PNP order to form the semiconductor device.

polarity A voltage source has a negative and positive polarity like a battery.

positive ion The deficiency of an atom of one electron is called a positive ion.

potentiometer Variable resistor with the center terminal varying the resistance.

power $E \times I = P$, or voltage times current equals watts (W).

power rating The amount of heat that a resistor can dissipate.

power transformer A transformer that provides power for electronic equipment.

preamp The preamp amplifies weak RF signals from the photodiodes in the pickup head. The preamp provides EFM, FE, and TE signals. The first amplifier.

preamp output Low-level output signal from the preamplifier to be connected to the main power amp.

preventive maintenance Inspections and corrections to keep the electronic chassis operating.

primary The input terminals of a power transformer.

printed-circuit board (PC board) Copper lines and traces on a phenolic or fiberglass board.

proton The positively charged particle of an atom's nucleus.

pulse A single signal of very short duration that is used for timing and sync purposes. Sync pulses are the best example of this type of signal. Pulses occur in precisely measured bursts.

purity, color The display of the various true colors without any accidental or unwanted contamination of one color by any of the others. Color purity largely depends on correct convergence adjustments.

Q The figure of merit of a capacitor, inductor, or of an LC circuit. Can be the symbol for selectivity.

quantization The number of possible values available to represent various levels of amplitude of a digital audio system. The resolution of quantization is 16 bits in the compact disc player. Quantization occurs within the RAM and the control system processor.

quartz lock The quartz crystal speed detector that locks in the precise turntable speed for the motor.

radial arm The radial arm consists of a motor and pickup head assembly that moves the head in an arc across the disc in a CD player.

radial tone arm A linear-tracking tone arm that moves along a track that is parallel to the record radius in a phonograph.

radiation leakage RF microwave leakage that might occur around the door and vent areas in a microwave oven.

radio frequency (RF) Frequencies above dc and below light.

RAM (random access memory) Used to store information and audio data in the CD player.

rated power bandwidth The frequency range over which the amplifier supplies a certain minimum power factor, usually from 20 to 20,000 Hz.

recording level meter The meter that indicates how much signal is being recorded on the tape. The recording meter might be a vane type, an LED, or a fluorescent panel. It is useful for preventing overloads during normal recording.

regulator A transistor, IC, or zener diode that regulates the voltage for a given circuit in power supplies.

reject lever A lever that rejects or deletes a given track in a cassette or a record on the record changer.

relay A relay is energized by a voltage within the solenoid and it pulls the switch contacts closed. An oven relay provides power to the HV transformer and magnetron. The power oven relay might be controlled by a digital programmer circuit. The ac relay in a TV chassis applies power to the power supply or it turns on the set.

reluctance Opposition to the flow of magnetic lines.

remote control A hand-held transmitter that controls functions of a TV set, VCR, cassette player, radio receiver, and CD player. Today, most remote controls are infrared operated.

repeat button The button that replays the same track of music in the CD player.

residual magnetism The magnetism that is left after the magnetizing force has been removed.

resistance Unit of ohms with the symbol of *R*. A device, component, or circuit, that has opposition to current flow.

resistor A device that limits the flow of current, providing a voltage drop. The symbol of a resistor is R.

resonance The condition of a circuit when capacitive reactance equals inductive reactance.

retrace The return movement of the scanning electron beam from the extreme right to extreme left and from the bottom to the top of the raster in a TV.

retrace blanking The extinction or darkening of the light on the face of the picture tube during retrace time in order to make these lines invisible. If the retrace blanking fails, white lines sloping downward from right to left would be seen on the screen.

review To rewind the tape.

RF Abbreviation for radio frequency.

ribbon speaker A high-frequency speaker that uses a light ribbon material suspended in a magnetic field to generate audio when current is passed through it.

ripple Spikes or pulses that ride on top of direct current when ac is rectified to dc.

ripple voltage Peak-to-peak (p-p) voltage variation produced by a rectifier circuit.

rotary transformer A device used to magnetically couple RF signals to and from the spinning video heads in a VCR.

R-y The red color of the overall color picture signal minus the monochrome.

sample/hold (SH) The circuits in each stereo channel that sample incoming data and hold it momentarily.

sampling frequency The rate or the numbers of times that a signal is sampled in digital audio. The sampling rate of a CD player is 44.1 kHz.

sandcastle The sandcastle operation is a three-level signal pulse that includes horizontal and vertical blanking and burst keying pulses.

saturation The inability to receive further information. The recording tape is saturated when it cannot hold any more magnetic signals. Saturation can occur in a magnetic tape head.

scan The ability of a tuner to search the band and sample each station for a few seconds.

scanning lines The horizontal lines in the picture or raster. The scanning lines make up the picture from left to right (looking at the front of the TV screen).

schematic A drawing with symbols of an electrical or electronic circuit.

SCR (silicon-controlled rectifier) Is used in the low- and high-voltage regulator power-supply circuits.

search To scan the playback picture forward at a faster rate of speed in a VCR or camcorder.

secondary The output side of a transformer.

self-erase A degrading or partial erasure of information on the magnetic tape.

self-powered speakers A speaker with a built-in amplifier.

semiautomatic In turntables, semiautomatic operation requires that the tone arm be placed manually on the record to start. The turntable will shut off and set the tone arm down automatically.

semiconductor A general name given to transistors, diodes, and similar devices that use semiconducting materials.

sensitivity 1. Ability of a receiver to pick up the weakest RF signal. 2. A measure of how much signal is needed to provide a combined noise and distortion level in the audio output from 30 to 50 dB below the output signal level. The sensitivity of a speaker is the measured output of the speaker (in dB) compared to the input.

separation The complete separation between the stereo audio channels.

series A connection between a number of components or tubes in chain fashion, one component following another. If any one component opens or burns out, it breaks the series circuit.

series circuit Circuit arrangement of components when total current flows through each one.

servo Refers to the servo control or tracking circuits that keep the laser pickup in the grooves at all times. Most servo components in the CD players and VCRs are IC processors.

servo control Refers to the servo control IC that controls the focus and tracking in CD players and controls the motor speeds in VCRs.

shadow mask Same as an *aperture mask*.

shield A metallic enclosure or container surrounding a component, tube, cable, etc.

shielded cable A wire with a metal casing on the outside to prevent unwanted electrical energy from reaching the inner conductor.

short circuit Low resistance that bypasses the flow of current.

shunting Placing a component in parallel with another component or part.

signal Any form of detectable electronics information.

signal processing In CD players, converting the laser beam signals to audio signals with audio processors.

signal-to-noise (S/N) The ratio of the loudest signal to that of hiss or noise. The higher the signal-to-noise ratio, the better the sound.

skating If the stylus is worn or if the pressure is wrong, the stylus might skate across the record.

skew The actual change of size or shape of the video tracks on the tape from the time of recording to the time of playback in the VCR. Skewing is the distortion or bending of the upper part of a VCR picture.

skip Certain compact discs have index points that allow different movements of parts of pieces of music, such as symphonies, to be selected. To set the index number, press the skip or track index number button.

slide See *pickup motor*.

solenoid A switch consisting of an electric coil with an iron-core plunger, which is pulled inside the coil by the magnetic field. The solenoid switch is often used on cassette, tape, and CD players instead of mechanical switches.

solid-state A term indicating that equipment uses semiconductors and not vacuum tubes.

sound bars Thick horizontal bars that appear on the TV screen because unwanted sound energy is reaching the picture tube. Sound bars are caused by a misadjusted circuit.

source One of the three leads of an FET transistor.

spindle motor The spindle motor can be called the *turntable motor* or *disc motor*. The spindle motor rotates the disc at a variable speed from the beginning to the end of the disc.

SS Slow and still picture modes of a VCR.

standard play (SP) The speed at which a two-hour (T120) VHS cassette plays in a VCR machine.

standing waves A wave created by bouncing or reflecting back the original wave. Standing waves can either cancel themselves or cause distortion.

static electricity Electricity generated by friction that provides a path to arcover.

stator Fixed plates in a variable capacitor.

stirrer motor A motor or blade that circulates the microwaves within the oven cavity. The stirrer blade might be driven by a fan or blower motor by a long drive belt in the microwave oven.

stylus A jewel or diamond tip needle that rides grooves of the record.

subcarrier The color picture information carrier. It is called a *subcarrier* because it is a secondary carrier in the particular channel. The color subcarrier frequency is 3.58 MHz.

super tweeter A tweeter that handles very high frequencies.

surface-mounted components Surface-mounted parts are soldered into the circuit on the same side as the PC wiring. Surface-mounted components are now used in practically all electronics products.

sweep marker A generator used by the electronics technician for TV receiver alignment.

switch A device that connects or disconnects a circuit.

symptoms Characteristics of a circuit that indicate a malfunction.

sync An abbreviation for a synchronizing signal. It is a timing signal or series of pulses sent by the transmitter and used by the receiver to stay in precise steps with the transmitter in the TV chassis.

sync clipper See *clipping*.

synchronous Where signals are coordinated through the use of a master clock circuit.

sync separator A circuit in a TV receiver that separates the sync from the picture information or the vertical sync pulses from the horizontal sync pulses.

temperature probe When inserted into meat or other food in the oven cavity, the probe determines the temperature and cooking time.

tension error Refers to skew.

test cassette Recorded signals on a test cassette for alignment and adjustment procedures on the cassette player or VCR.

test disc A compact disc used to make alignments and adjustments to the CD player.

timer A device to control the amount of cooking time.

tolerance The positive and negative error between the actual and indicated value of a part.

total harmonic distortion (THD) A percentage of harmonic distortion in components. To measure how accurate the amp is, a signal is fed in, and the harmonics are measured at the output.

track One lane of spiral pits on the surface of a compact disc.

tracking The tracking force is how much downward pressure is needed for the tone arm to track along in the groove. The spinning action of the video heads during playback when accurately tracking the video RF information laid down during recording in the VCR.

tracking coil A coil located with the tracking driver transistor or ICs that moves the lens back and forth across the disc for accurate tracking in a CD player.

tracking servo The IC processor that keeps the laser beam in focus and tracking correctly.

track-kick circuit The circuit used when the laser beam is skipped to a relatively close pit track during accessing and cue/record mode operations. Skipping is achieved by applying kick and brake pulses to the tracking core with the tracking servo loop open.

track offset adjustment An adjustment made with a test disc and the oscilloscope. The test is of the tracking jump waveform on the RF board of a CD player.

transducer A device that converts energy into another form. The speaker changes electrical energy to sound energy. A cartridge converts mechanical movement to electrical energy.

transfer rate The speed at which a hard or floppy drive transfers information.

transformer An electrical or electronic device that can step up or step down voltages and matching impedance.

transistor Three-legged device with a base, emitter, and collector terminal.

tray The loading tray or drawer in which the CD disc is to be played or placed.

triac An electronic switch to apply ac voltage to the HV transformer. Usually, the gate voltage of the triac is controlled from the control board in the microwave oven.

triad The three-color, three-dot group (red, green, and blue) of which the color picture tube phosphor is made. Each group of three dots is a triad, and there are thousands of triads on a modern color tube screen.

troubleshooting The method of determining the failure of an electrical or electronic breakdown.

tube shield A metal sleeve that fits snugly over a glass tube and shields it from extraneous electrical impulses.

tuner The tuner picks up each broadcast TV signal and passes it to the IF circuits for amplification. The tuner might be operated manually or with remote control. Tuners are also used in radio receivers.

turns ratio The relationship of turns in the primary winding of a transformer to the number of turns in the secondary winding.

tweeter A high-frequency driver.

UHF Ultra-high-frequency radio and TV frequencies above 300 MHz. TV channels 14 to 83 are located in the UHF band.

vacuum tube An electronic tube with all air and gases removed.

variable capacitor A capacitor that can be rotated or adjusted for a different value.

variac The variac can raise or lower the power-line voltages to locate intermittent, defective flyback transformers and high-voltage shutdown problems.

VCO Voltage-controlled oscillator.

VCR Videocassette recorder.

vented speaker system Any speaker cabinet with a hole or port to let the back waves of the woofer speaker escape. A *bass reflex enclosure* is a type of vented speaker system.

vertical Pertaining to the circuits and functions associated with the up-down motion or deflection of the electron beam in a TV chassis.

vertical amplifier An amplifier that follows the vertical oscillator used to enlarge the vertical sweep signal.

VHF Very high frequencies. TV channels 2 through 13.

VHS (video home system) The system used in VCRs and camcorders.

video A term applied to picture signals or information (video, circuits, video amp, etc.) in TVs, VCRs, and camcorders.

video head The electromagnet used to develop magnetic flux, putting RF information on tape.

VIR Vertical interval reference. It is sometimes called the *broadcast-controlled color-correction system*. It is an automatic, station-controlled signal that is used to initiate color and luminance corrective action in the TV set, if it is so equipped.

voice coil The coil or wire wound over the end of the cone of the speaker, to which is connected to amplifier.

volt Unit measurement of voltage, which represents electrical pressure.

voltage divider An arrangement of resistors in a circuit to produce lower voltage, like that in a dc power source.

voltage-doubler circuit The high ac voltage from the secondary winding of the power transformer is applied to an electrolytic capacitor in the voltage-doubler circuits. A circuit that supplies dc voltage output that is almost twice the voltage input.

voltage drop The voltage remaining from the current flowing through a resistor.

voltage gain The ratio of output voltage to input voltage in a circuit.

VOM (volt-ohmmeter) The VOM measures resistance continuity, voltages, and current. The VOM utilizes a meter to display the measured readings.

VTR Video tape recorder.

v-v Video to video. A playback picture produced from a tape during playback.

watt The unit of electric power. A microwave oven might pull more than 200 watts.

wave The name given to each recurring variation in alternating electric energy, including radio and TV signals.

waveguide A metal enclosure for the conduction or transmission of microwaves. The waveguide in the microwave oven is between the magnetron and the oven cavity.

wavelength The distance between corresponding points of two successive ac waves.

W/CH Watts per channel.

wet cell A battery with a liquid electrolyte.

white clep A white clip circuit is used to cut overshoot spikes off, at an adjustable level.

width The width of a TV screen might be pulled in at each side, indicating problems within the horizontal deflection system. Poor width might be caused by HV regulator transistors, SCRs, zener diodes, and IC regulators.

wire-wound resistor A fixed resistor with resistance wire wrapped around a ceramic core.

woofer The largest speaker in a speaker system and the one that best reproduces the low frequencies. The low-frequency driver.

wow Variation in the speed of tape. *Wow* is a slow-speed fluctuation; fast-speed variation is called *flutter*.

x-demodulator The designation of the red (R-y) signal demodulator.

xtal Abbreviation for crystal.

yoke Deflection yoke. The deflection yoke is mounted on the neck of the picture tube in TVs and camcorders. "Electric" and "magnetic" means that the yoke imparts to the electronic beam the scanning of horizontal and vertical deflection.

Y-signal The black-and-white part of the video signal, containing black-and-white information. It is sometimes called the *brightness signal,* meaning the actual brightness and darkness (and all shades in between) of the picture. The Y-signal is used in TVs, VCRs, and camcorders.

Z-demodulator Same as the X demodulator, except for the blue (B-y) signal.

zener diode A special semiconductor diode with a reverse breakdown voltage. The zener diode provides a fixed voltage in the power circuits.

Index

A

AAL Meter Group, 997
ac voltage, 2
accounting, 991-992
Acoustat Corp., 997
Adcom, 997
adjustable shielded coil, **6**
Advent, 997
Afco Electronic Inc., 997
air core transformer, symbol, **30**
Aiwa America Inc., 997
Akar America Ltd., 998
Alarm Inc., 998
All Channel Product, 998
Alpine Electronic of America, 998
Altec-Lansing International, 998
AM circuits, 496
AM converter, 496, **497**
AM converter transistor, troubleshooting, 455-457, **456**, **457**, **458**
AM detector, 221-222, **223**
AM IF circuits, 497-500, **499**
AM radios, alignment, 515-519
AM receiver, 217, **218**
AM station, 455
AM/FM radios, troubleshooting, 512-515
AM/FM/MPX receiver, 493, **219**, **493**, **1003**
  alignment, 518
  block diagram of digitally controlled, **508**
  capacitors used in, 482, **483**

circuits/components, 730-731, **731**, **732**
different sections of on PC board, 205-206, **206**
portable, **1007**
stereo amplifier circuits, 211, **211**
troubleshooting, 238
  dead, 240, **241**
Amana Refrigeration Inc., 786
Amber Electronic, 998
American Acoustics Inc., 999
American Audio Corp., 999
amplification, 285, **285**
amplifiers, 205-221
  audio, 207-209, 972, **207**, **208**, **973**
  audio-frequency, 210-211, 285, 549-550, **210**, **285**
  Class-A, 286-287, **286**, **287**
  Class-B, 287
  color output, 220-221, **220**, **221**
  converter transistor stages, 216, **217**
  detector, 217
  IC, 217
  IC audio, 288, 290, **289**
  IF, 217, **218**
  mixer, 217
  noisy, 358-359
  operation/performance, 288, **288**
  operational, 21, **22**
  pre-, 209, 460-462, 547-549, 685-686, **209**, **548**, **685**

radio-frequency, 213-215, 217, **215**
stereo, 211-212, 290-293, 550, 552-553, **211**, **212**, **213**, **1006**
symbol, **30**
transistorized power, 213, **214**
troubleshooting intermittent problems, 675, **675**
TV radio-frequency, 216, **216**
video, 219-220, **220**
amplitude modulation (*see* AM)
Analog & Digital System, 999
Andante, 999
antennas
  safety precautions, 193
  symbol, **30**
Apollo Wholesale Electronics, 738
ARA Manufacturing Co., 999
Arthur Fulmer Electronics Division, 1001
audio
  dead channel, 295, **296**
  distorted, 296-297, **298**
  intermittent, 296, 676-677, **297**, **676**
  monophonic circuits, 511, **511**
  servicing
    power circuits, 301-303, **303**
    stereo circuits, 300-301, **302**

Illustrations are in **boldface**.

audio, *continued*
  table radio circuits, 300, **301**
  with DMM, 299-300, **299**
  signaltracing, 297-299, **298**
  stereo circuits, 370, **371**
  sound problems, 293-294, **294**
  waveform tests, 374-377, **374**, **375**, **376**
  weak sound, 294-295, **295**
audio amplifiers, 207-209, **207**, **208**
audio-frequency, 210-211, **210**
  IC, 288, 290, **289**
  preamplifiers, 209, **209**
  servicing external, 972, **973**
  signaltracing and, 357-359, **357**, **358**
  stereo, 211-212, 290-293, **211**, **212**, **213**
  troubleshooting, 675, **675**
audio circuits
  CD player, 810-811, **810**, **811**
  IC, 610-611, **610**
  mono, 950, **950**
  repairing directly coupled, 694, **694**
  signaltracing, 611
  television stereo, 950-952, **951**, **952**
  testing driver transistors, 609-610, **609**
  troubleshooting directly coupled in TV chassis, 695, **695**
audio driver transistor, testing, 609-610, **609**
audio frequency (AF) amplifier, 210-211, 285, 549-550, **210**, **285**
audio generator, 324-325, **324**
audio output transformer, 15
audio signal generator, 359-360, **360**

audio signaltracing, 359-362, **360**, **361**, **362**
Audio Video Parts Inc., 738
Audiovox Corp., 999
auto focus, 801-802, **802**
auto-stop circuits, 510, **510**
auto-transformer, 14
automatic iris circuit (AIC), 442

B

B-B&W Electronics, 738
Bang & Olufsen of America, 999
Banner Technical Books, 1010
bat wing switch, 11
batteries, 3
  camcorder, 487
  life expectancy in digital multimeter, 53
  mercury cells, 487
  nicad, 523
  replacing, 486-487, **487**, **488**
  symbol, **30**
  testing, 643, **643**
  types of, 486
bearings
  lubricating, 420-421, **421**
  squeaky, 419-420, **420**
beat frequency oscillator (BFO), 504
belts
  broken, 418-419
  camcorder, 417, **417**
  cracks or breaks in, 418, **419**
  drive, 425, 437, 840, **426**, **438**, **841**
  replacing, 484-486, **485**, **486**
bias oscillator, 545-546, **545**
bipolar transistor, 16
Blaupunkt, 999
boom box, **1002**, **1007**
  power supply for, 282-283, **283**
  replacing cassette motor in, 529-600

boost diode, 118, **118**
Bose Corp., 1000
Bozach Inc., 1000
bridge rectifiers, 259, 266, 863-865, **9**, **19**, **266**, **864**
  locating unknown, 412-413
  symbol, **476**
  testing with DMM, 69, **69**, **70**
  testing, 115-117, **116**, **117**
BSR USA, 1000
building (*see* kit and project building)
business
  accounting/bookkeeping, 991-992
  computer technician, 988-989
  ethics and morals, 984
  medical electronics technician, 987-988
  pricing your services, 992-993
  radio sales and service, 984-985
  satellite installation, 985-986
  starting a, 989-991
  tax laws, 993-994
  television repair, 985-987
  television sales and service, 984-985
  VCR repair, 986-987, **987**
bypass capacitor, 4, 91-92, 750, **91**, **92**, **750**
  leaky, 91-92,

C

Caloric Corp., 786
camcorders, 437-443, 841-846, **825**
  8-mm camera circuits, 841, **843**
  batteries, 487
  block diagram, 826, **828**, **829**
  camera sections, 841, **842**
  capstan motor, 463-464, **463**
  cleaning, 421, **422**

cracked/loose belts, 417, **417**
dew sensor circuits, 846, **847**
electronic viewfinder, 843-844, **844**
flying erase head, 845, **846**
focus motor, 441
formats, 826, 830, **830**
iris motor, 442
loading motor, 438
locating/replacing motor in, 537, 539, **538**
main chassis components, **848**
motor, 521, 860, **521**, **860**
preventive maintenance and lubrication, 836-837
servo and control systems, 848-849
supply and take-up reels, 442-443, **443**
tape transport, 835, **836**
troubleshooting,
  dead, 854-855, **855**
  erratic zoom operation, 853, **853**, **854**
  no audio in playback, 858-859
  no auto focus, 854, **854**
  no head erase, 858, **859**
  poor head erase, 858, **859**
video circuits, 844-845, **845**, **846**
zoom motor, 441, **442**
capacitance, unit of, 3
capacitance diode, 18
capacitance meter, 331-332, **331**
capacitor coupling, 686, **687**
capacitors, 3-5
  bypass, 4, 750, **750**
  ceramic, 751
  coupling, 4, 751
  defective filter, 4
  different types of, 4, **5**
  dip, 4
  discharging, 197

electrolytic, 4, 65, 100, 260, 268-269, 332, **6**, **65**, **66**, **100**, **269**, **332**, **864**
electrolytic film, 4, **4**
filter, 277
filter, 66, 260, 268-269, 277, 876-878, **66**, **260**, **261**, **268**, **269**, **877**, **878**
foreign chart of, 26
high-voltage, 764, 766-768, **764**, **767**
leaky, 91-92, 94, **91**, **92**, **94**
memory, 482
memory backup, 4
mounting types, 5, **6**
removing chip, 136-138, **137**
replacing, 482-484, **483**, **484**
silver mica, 4
surface-mounted, 5, 131-132, **132**
symbol, **29**, **30**, **31**
testing, 65-66, 598-599, **65**, **66**, **598**
types of, 482
variable, 504, **504**
capacitor tester, repairing, 977, **977**
capstan motor, 847
Carver Corp., 1000
cassette players, 424-428, 541-560, **1006**
  automobile, 531, **999**
  bias oscillator, 545-546, **545**
  block diagram of portable, **292**
  block diagram of stereo, 291, **291**
  boom box, 529
  capstan, 424-425, **425**
  circuit breakdown, 541-542
  cleaning with silicon spray, 752, **752**
  defective motor, 424, **424**
  drive belts, 425, **426**
  erase head, 546-547, **547**
  head azimuth adjustment, 560, **560**

head preamp circuits, 460-462, **461**
idler pulleys, 425
operation, 418
pinch roller, 426, **427**
portable, 528, 541, **541**
power supply, 280-281, **281**
recording, 553-557, **554**, **555**
replacing motor in, 527-531, **528**, **529**, **530**
soft-action door, 426
take-up and supply reels, 427-428, **428**
tape head, 427, 542-545, **543**, **544**
tape speed adjustment, 558
troubleshooting,
  erratic switching, 547
  defective playback, 559
  defective recording, 559
  intermittent recording, 752-755, **753**, **754**
  preamp recording circuits, 693, **693**
  types of, 541
cathode-ray tube (CRT) (*see also* picture tubes)
  circuits, 231-232, **232**, **233**
  components, 619, **620**
  repairing, 625-626, **626**, **627**
  replacing, 196-197, **196**
  tester and restorer, 335-336, 618, **335**
  testing, 624-625, **625**
CD players, 431, 434-436, 791-823
  adjustments to, 811-814
  audio circuits, 810-811, **810**, **811**
  auto focus, 801-802, **802**
  block diagram, 799, **799**
  cleaning, 794-795
  component layout, **385**
  defective loading circuit, 462-463
  focus adjustments, 813-814, **814**

1051

Index

CD players, *continued*
  focus and tracking coils, 798, **798**
  laser beam protection, 792-793
  laser eye damage, 200-201, **200**, **201**
  laser pickup assembly, 796-798, **796**, **797**
  locating/replacing motor in, 532-535, **533**, **534**, **535**
  low-voltage power supply, 807, 808-809, **807**, **809**
  motors, 521, 804-807, **805**
    disc, 435, 805-806, **806**
    loading, 434-435, 804-805, **434**, **435**
    SLED, 436, 805-806, **436**, **806**
    spindle, 805-806, **806**
    turntable, 805-807, **806**
  portable, **1007**
  RF/HF/EFM tests, 812-813, **813**
  safety precautions, 791-793
  servo circuits, 803-804, **803**, **804**
  signal circuits, 800, **800**, **801**
  surface-mounted components, 795, **796**
  test equipment, 793, 811, **794**
  tracking error signal, 802-803, **802**
  tracking offset adjustments, 813-814, **814**
  troubleshooting, 814-823
    tray motor rotating/no tray action, 435
    waveforms, 385
ceramic capacitor, 751
ceramic fuse, 481, **481**
certified electronic technician (CET), test, 983-984
chassis ground, symbol, **30**
chemical products, 720, 722
Chinese products (*see* import products)
chip capacitors, removing, 136-138, **137**
chokes, 5-6, 37
  filter, 269, **270**
  RF, 5
  symbol, **31**
chopper circuits
  RCA, 936-938, **937**, **938**, **939**
chopper power supply, 654, **655**
chroma IC, 713, **715**
circuit breakers, 191, 275, **191**, **275**
circuit diagrams (*see* schematic diagrams)
circuit saver, 777-778, **778**, **779**
circuits, integrated (*see* integrated circuits)
Clarion Corp. of America, 1000
Class-A amplifier, 286-287, **286**, **287**
Class-B amplifier, 287
clock radios, 500-502, **501**
  dial cord problems, 512, **512**
  imported, 727, **727**
coils, 5-6
  adjustable shielded, **6**
  ferrite core, **6**
  fixed, 37
  FM, 37
  focus and tracking, 798, **798**
  symbol, **30**, **31**
  voice, 14
color-bar generator, repairing, 978-979, **979**
color circuits, 932, **932**
color output amplifier, 220-221, **220**, **221**
color output transistor, 715, 718, 935, **720**, **935**
color pattern generator, 332-334, **333**
comb filters, 943-945, **944**, **945**
components (*see also* specific types of)
  comparison circuit, 413-414
  identifying, 2-24
  identifying and relacing unknown, 395-416
  identifying on PC board, 729-730
  intermittent problems with, 671-672, **672**
  locating, 712-715
  locating defective, 249-251, **250**, **251**
  locating replacement parts, 143-144
  manufacturer data, 414
  manufacturer parts depot, 414-415
  microwave oven, 764-774
  moving, 661-662, **661**, **662**
  part substitution, 141, **141**
  parts replacement, 469-491, 718-720, 728-729
    batteries, 486-487, **487**, **488**
    belts, 484-486, **485**, **486**
    capacitors, 482-484, **483**, **484**
    defective new part, 746-747, **747**
    diodes, 475-476, **476**
    fuses, 480-481, **481**, **482**
    ICs, 475
    imported, 737-740
    incorrect numbers, 491
    original, 470-471, **470**, **472**
    part numbers, 705-709
    RCA/GE guide, 728, **729**
    resistors, 477-478, 480, **478**, **479**
    SMDs, 476-477, **477**
    speakers, 488, 490, **489**
    switches, 487-488
    transistors, 472-475, **475**
    wholesale, 700
    universal, 259-260, 471, **259**, **471**, **473**

parts that cannot be replaced, 490-491, **490**
photographing, 415, **415**
purchasing at local electronics stores, 416
reasons for failure, 92-94, **93**, **94**
removing defective, 120-123, **121**, **122**
replacing defective, 256-258, **257**
solid-state, 15-19, 413, 727-728, **413**
surface-mounted, 125-144, **125**
10 service tests, 716-717
testing solid-state, 95-123
computer technician, 988-989
Concord Systems Inc., 1000
condenser microphone, 553
Consolidated Electronics Inc., 144, 738
continuity
  testing in motors, 524, **525**
  testing with DMM, 57-59, **58**, **59**
continuous wave (CW), 504
convergence
  adjusting, 633-634, **634**
  problems with, 634-635, **635**
  slotted shadow mask, 635-636, **636**
cook relay, 768, **768**
coolant, 720, 722, 744, 751, **744**
coupling
  capacitor, 686, **687**
  direct, 208, 683-702, **208**, **684**
  impedance, 208, 685, **208**
  RC, 208, 210, **208**, **210**
  transformer, 208, 684, **208**
coupling capacitor, 4, 751
Craig Corporation, 1000
Crosley Corp., 786
crosshatch, 333, **334**
crystal-controlled oscillator, 330-331, **331**

current, testing with DMM, 63-64, **64** (*see also* voltage)
Curtis Mathes Service, 1000

D

D.B. Systems, 1001
Daewoo Electronics, 1001
Dalbani Corp., 738
damper diodes, 908-910, **228**, **909**
  testing, 117-118, 602-603, **117**, **602**
dc motors, 522, **522**, **523**, **524**
dc ripple, 268
dc voltage, 6
defective filter capacitor, 4
deflection ICs, 604-605, **604**
deflection yoke, 622, 918-919, **622**, **918**, **919**
  testing, 607-608, **608**
defluxer spray, 720, 722
degaussing coil, 629-631, **629**
Denon America Inc., 1001
derived voltage, 230
detector amplifier, 217
detector diode
  symbol, **476**
  testing with DMM, 70, **70**
detectors, 221-222
  AM, 221-222, **223**
  FM, 221-222
dew sensor, 846, **847**
diagrams, schematic (*see* schematic diagrams)
diffused-junction transistor, 16
digital frequency counter, 324
digital multimeter (DMM), 51-73, 320, 775, **51**, **320**, **776**
  audio servicing, 299-300, **299**
  auto ranging, 56-57, **56**
  battery life, 53
  bridge rectifier tests, 69, **69**, **70**

capacitor tests, 65-66, **65**, **66**
checking 9-V battery with, 57, **57**
checking transistor function with diode tests, 70-71
continuity tests, 57-59, **58**, **59**
current tests, 63-64, **64**
detector diode tests, 70, **70**
diode tests, 67-69, 112-113, **68**, **113**, **114**
display readouts, 53
frequency tests, 66-66
inductance measurements, 73, **73**
operation, 52-54, **52**, **53**
precautions, 54-56, **54**, **55**
prices, 52
repairing, 967-968, **968**, **969**
resistance measurements, 79-82, **81**
resistance tests, 59-61, **59**, **60**, **61**
transistor checking in the circuit, 82-83, **82**
transistor tests, 71-73, **72**
voltage measurements, 62-63, 75-78, **62**, **63**, **75**, **76**, **77**
digital tuning, 507, 509-510, **508**, **509**
Digital-Key Corporation, 144
diodes, 16, 18-19, **17**
  boost, 118, **118**
  bridge, 412-413 (*see also* bridge rectifiers)
  capacitance, 18
  checking transistor function with diode tests, 70-71, **71**, **72**
  damper, 117-118, 602-603, 908-910, **117**, **228**, **602**, **909**
  detector, 70, **70**, **476**
  germanium, 7

diodes, *continued*
  high-voltage, 7, 118-120, 764, 766-768, **118**, **119**, **120**, **764**, **767**
  junction tests, 83, **84**
  laser, 797
  leaky, 105, **105**
  locating unknown, 411-412, **411**, **412**
  low-amp, 128
  photodetector, 797
  replacing, 475-476, **476**
  Schottky, 19
  silicon, 7, 9, 100, 112, 255, 712-713, **8**, **100**, **113**, **114**, **476**, **477**, **714**
  solid-state, 18, **19**
  surface-mounted, 18, 19, 127-129, **128**
  switching, 19
  symbol, **29**, **30**, **32**
  testing, 172
  testing with DMM, 67-69, 112-113, **68**, **113**, **114**
  transistor junction tests, 107-110, **108**, **109**, **110**
  varactor, 18, **20**
  voltage tests, 100-101, **100**, **101**
  zener, 18, 68-69, 101, 128, 255, 271-272, 277, 872-874, **20**, **101**, **271**, **272**, **277**, **476**, **873**
dip capacitor, 4
direct coupling, 208, 683-702, **208**, **684**
  horizontal circuits, 698-699, **698**, **699**
  ICs, 690-691, **690**, **691**
  NPN dc circuits, 686
  PNP dc circuits, 686
  preamp circuits, 685-686, **685**
  repairing audio circuits, 694, **694**
  transistor interaction, 687-690, **688**, **689**
  troubleshooting audio circuits in TV chassis, 695, **695**

  vertical circuits, 696-697, **697**
  video circuits, 699, **700**
disc motor, 805-806, **806**
distortion, 352, 375
distortion meter, 346-347
Diversified Parts, 738
double-pole double-throw (DPDT), 11
  symbol, **30**
drills, **526**, **527**
driver transformer, 15, **16**, **228**, **760**
driver transistor, **761**
  testing, 603-604, **603**
dual-in-line package (DIP), 20
dual-trace oscilloscope, 325, 373, **325**
duo-diode rectifier, testing, 114-115, **114**, **115**
dust-all spray, 720, 722

E

earth ground, symbol, **30**
East Coast Distributors, 738
education and training, CET test, 983-984 (*see also* business)
EFM signal, 800
Eiger Electronics, 738
electrolytic, symbol, **31**
electrolytic capacitor, 4, 260, 268-269, 332, **6**, **269**, **332**, **864**
  testing, 100, **100**
  testing with DMM, 65-66, **65**, **66**
electrolytic film capacitor, 4, **4**
Electronic Handbook, 1010
Electronic Servicing & Technology, 46, 1010
Electronic Industries, 1001
Electronic Systems Inc., 789
electronic viewfinder (EVF), 843-844
electronics, 1-26
  history, 1-2

identifying important components, 2-24
Electronics Now!, 1011
Electronics Warehouse Corp., 739
electrons, 765
electrostatic discharge, 135-136, 791, **137**
Emerson Radio Corp., 786, 1001
emitter-follower transistor, 76, **77**
eye pattern adjustment, 813

F

farad, 3
ferrite core coil, **6**
FET analog meter, repairing, 977-978
FET meter, 319, **319**
FET voltmeter, 85, 319
field coil speaker, 12
field effect transistor (FET), 16, 17
  metal oxide semiconductor, 17
  symbol, **29**
filter capacitors, 66, 260, 268-269, 277, **66**, **260**, **261**, **268**, **269**
  defective, 876-878, **877**, **878**
  leaky, 94, **94**
filter chokes, 269, **270**
filters
  comb, 943-945, **944**, **945**
  low-pass, 507, 843
  saw, 941-942, **942**, **943**
Fisher Corporation, 1001
fixed capacitor, symbol, **30**
flyback, 659, **659**
flyback transformer, 96, 705, 910-913, **97**, **706**, **911**, **912**
flying erase head, 845, **846**
FM converter, 497
FM detector, 221-222
FM IF circuits, 497-500, **498**, **499**

FM MPX circuits, 500, **500**
FM radios, alignment, 515-519
FM RF transistors, 459, 497, **460**
focus coil waveform, 389, **389**
focus controls, 913, **913**
focus error (FE) waveform, 388, **388**
foil pattern, 306
formats
  8-mm, 830, **830**
  Beta, 826, **830**
  VHS, 826, **830**
  VHS-C, 826, **830**
Fortune Star Products, 1001
Fox International LTD Inc., 144, 739
frequency, testing with DMM, 66-67
frequency counter, 328-331, **329**, **794**, **812**
frequency modulation (*see* FM)
Frigidaire, 786
Fujitsu Corporation of America, 1001
full-wave rectifier, 266, **9**, **266**
function generator, 970, **970**
fuses, 7, 273-275, **8**, **273**, **274**
  amperage, 480
  blowing, 874-876, 922-923, **875**, **923**
  ceramic, 481, **481**
  checking television, 595-596, **595**
  chemical microwave oven, 767, **767**
  low-amp, 7
  pigtail, 481
  plug-in ac line, **7**
  power-line, 188, **188**
  replacing, 480-481, **481**, **482**
  slow-blow, 274, 481
  speaker, 274
  symbol, **30**, **31**

test equipment, **962**
types of, 480

G

gas-filled regulator tubes, 272
General Electric Co., 786
General Electric Consumer Electronics, 1001
generators
  audio, 324-325, **324**
  audio signal, 359-360, **360**
  color-bar, 978-979, **979**
  color pattern, 332-334, **333**
  harmonic signal, 351, **351**
  noise, 354-356, **354**, **355**, **356**
  NTSC color, 334
  repairing, 973, 975, **974**
  RF signal, 326-328, 362-364, **327**
  sweep, 325-326, **326**
germanium diode, 7
germanium transistor, 16
Goldstar Electronics Inc., 786, 1001
grounding
  antistatic wrist strap, 135-136, 791, **137**, **792**
  hazardous work areas and, 193
  symbol, **30**
  test equipment, 192, **192**
  work bench, 190-191
Grundy/GR Electronics, 1001

H

half-wave rectifier, 266, **9**, **266**
Hamimex Inc., 1001
Hanabasheya Ltd., 999
Hardwick Stove Co., 786
harmonic signal generator, 351, **351**
HBF Electronics Inc., 739
headphones, 14, **14**
  PM, 14
  stereo, 953, **953**

symbol, **30**
heatsink, 260-261, 886, **262**, **886**
henries, 5
HH Scott Inc., 1007
high voltage
  discharging in microwave ovens, 774-775, **775**
  discharging in televisions, 623-624, **624**
  shutdown of in televisions, 919-920, **920**
  shutdown test, 922, **922**
high-voltage capacitor, 764, 766-768, **764**, **767**
high-voltage diode, 7, 118-120, 764, 766-768, **118**, **119**, **120**, **764**, **767**
high-voltage probe, 347, 349, **348**
high-voltage shutdown circuit, 758, **759**
high-voltage transformer, 764, 770-772, **764**, **772**
high-voltage TV circuits, 229-231, **230**, **231**
Hitachi Sales Corp. of America, 786, 1002
hold-down circuits, 910
horizontal circuits, 901-929
  intermittent problems, 927
  servicing directly coupled, 698-699, **698**, **699**
horizontal deflection IC, 710, 902, **710**, **903**
horizontal disable circuits, 920-921
horizontal driver circuits, 734, 904-905, **905**
horizontal foldover, 917, **917**, **918**
horizontal output circuits, 905-907, **906**, **907**
horizontal output transformer, **926**, **927**
horizontal output transistor, 226-227, 600-602, 659, 734, 757, **227**, **601**, **659**, **735**, **757**

1055

horizontal waveforms, 379, 381-382, **380**, **381**, **382**, **383**
Hotpoint, 786
Howard Sams Photofacts, 48-49, 414, 491, **48**, **414**

I

IC amplifier, 217
ICM International, 739
IF, 15, 455
IF amplifier, 217, **218**
IF circuits, 497-500, **498**, **499**
IF transformer, 15, 39
 symbol, **31**
impedance coupling, 208, 685, **208**
import products, 725-740
 distributors of, 738-740
 manufacturers/countries, 725
 radios, 730-732, **731**, **732**
 televisions, 732-735
 troubleshooting, 726-727
inductance
 measuring with DMM, 73, **73**
 unit of, 5
inductor, symbol, **30**
Infinity Systems, 1002
infrared remote tester, 644-645, **644**, **645**
integrated circuits (ICs), 16, 19-20, 659, **17**, **21**, **38**, **659**
 body part numbers, 396, **397**
 chroma, 713, **715**
 deflection, 604-605, **604**
 directly coupled, 690-691, **690**, **691**
 dual-in-line package, 20
 horizontal deflection, 710, 902, **710**, **903**
 leaky output, 92, **93**
 locating unknown part numbers, 404-405, **406**
 mounting, 154-156, **155**
 output, 550, **551**

recording, 556-557, **556**, **557**
removing, 138, **138**, **139**
replacing, 409, 475, **409**
resistance measurements, 105-107, **106**, **107**
sound output circuits, 610-611, **610**
surface-mounted, 129-130, **129**, **130**
symbol, **29**, **32**
testing, 89-90, **90**
testing before replacing, 80, **81**
vertical countdown, 888-889, **889**
voltage measurements, 101-102, **102**
voltage tests, 454-455, **454**
interlock switches, 768, **769**
iron-core inductor, symbol, **30**
isolation transformer, 15, 189-190, 264, 337, **189**, **190**, **264**, **337**, **902**

J

J.C. Penney Co., 1005
jacks, symbol, **30**, **31**
Japanese products (*see* import products)
JBL Inc., 1002
Jenn-Air Corp., 787
Jensen Inc., 1002
JICL LA Corp., 1002
jitter, 813
JR Loudspeaker, 1002
Jullette Electronics, 1002
JVC America, 1003

K

Kenwood Electronics Inc., 1003
Kevin Electronics, 144
key-type switch, 11
keypads
 microwave oven, 781
 servicing interface circuits, 651-652, **652**

troubleshooting, 652-653, **653**
kit and project building, 145-166
 assembling components, 153-154, **153**
 deluxe regulated power supply kit, 178-181, **179**
 from schematic to board mounting, 159-162, **160**
 IC mounting, 154-156, **155**
 inspecting parts, 151-152, **152**
 LCD meter kit, 175-178, **176**, **177**, **178**
 learning to solder, 146-148, **147**, **148**
 PC board layout, 157-159, **157**
 PC boards, 306-309, **307**
 reading directions, 150-151, **151**
 resistor color code, 156-157
 signal injector-tracer kit, 167-171, **168**, **169**
 test instrument kits, 167-183
 testing, 166
 tools, soldering, 148, 150, **149**, **150**
 transformer connections, 162-166, **162**, **163**
 transistor mounting, 154-156, **155**
 transistor test kit, 171-174, **172**, **173**
Kitchen Aid, 787
knife switch, 11
Koss Corp., 1003
Kraco Enterprises Inc., 1003
Kyocera International, 1003

L

large-scale integration (LSI), 16, 409-410, **21**, **410**
 surface-mounted, 129-130, **120**, **130**
laser diode, 797

laser pickup assembly, 796-798, **796**, **797**
laser power meter, 341-342, 811, **341**, **342**
laser remote tester, 645-646, **646**, **647**
lasers, eye damage from, 200-201, 792-793, **200**, **201**
LCD meter kit, 175-178, **176**, **177**, **178**
  parts checkoff, 177
  parts identification, 175-177
  testing procedures, 177-178
LCR meter, 344-345, **345**
leaf switch, 87, **88**
light-emitting diode (LED), symbol, **30**, **31**
line transformer, 15
line-voltage circuits, 862, **862**
line-voltage regulator, 870, 872, **871**
Linear Power Inc., 1003
Litton Microwave Cooking Products, 787
Lloyds Electronics, 1003
loading motor, 804-805
low-pass filter (LPF), 507, 843
low-voltage power supply B&W television, 868-869, **869**
  block diagram in VCR, **856**
low-voltage power transformer, 865-866, **865**, **866**
low-voltage shutdown circuit, 758, **759**
Luxman Division, 1003
Lyman Division, 1003

M

Magic Chef, 787
magnameter, 343-344, 776-777, **343**, **777**, **778**
Magnavox, 1004
magnetron tube, 764-765, 769-770, **764**, **765**, **766**
magnets, permanent, 14
Marantz Co., 1004
matching transformer, 15
Maycor, 787
Maytag, 787
McGraw-Hill, 1010
MCI, 1004
MCM Electronics, 144, 739
mechanical vibrator, 7
medical electronics technician, 987-988
memory backup capacitor, 4
memory capacitors, 482
metal-oxide resistor, 11
metal-oxide semiconductor (MOS), 17
metal-oxide semiconductor field effect transistor (MOSFET), 17
metal-oxide transistor, 16
  symbol, **29**
meter, symbol, **30**
meters (*see* test equipment)
microphones, condenser, 553
microprocessors, 16, 21, **17**, **21**, **22**
  removing, 138, **138**, 139
  surface-mounted, 129-130, **129**, **130**
  symbol, **29**
  voltage measurements, 101-102, **102**
microwave energy, 764-766
microwave ovens, 763-789
  chemical fuse, 767, **767**
  components, 764-774
  control board panel, 781
  cook relay, 768, **768**
  discharging high voltage, 774-775, **775**
  fan blower, 766, **766**
  HV circuits, 774, 782-783, **774**
  HV capacitor, 764, 766-768, **764**, **767**
  HV diode, 764, 766-768, **764**, **767**
  HV transformer, 764, 770-772, **764**, **772**
  interlock switches, 768, **769**
  low-voltage circuits, 778-780, **780**, **781**
  magnetron tube, 764-765, 769-770, **764**, **765**, **766**
  manufacturers, 786-789
  motor, 521
  piezo buzzer, 769
  separating high- and low-voltage circuits, 772
  shock hazards, 197, 199, **196**, **199**
  sizes/types, 763, **763**
  test equipment, 775-778
  testing, 343-344
  testing motor in, 535, 537, **536**
  thermal unit, 766, **766**, **770**, **771**
  triac, 768, **768**
  troubleshooting,
    intermittent cooking, 783
    radiation leakage, 784, **785**
  turntable motor, 770, **771**
Midland International Corp., 1004
mike, symbol, **30**
Mil Electronics Inc., 739
Mills Electronics Inc., 739
Mitsubishi, 1004
mixer amplifier, 217
Modern Maid, 787
monophonic audio circuits, 511, **511**
Montgomery Wards, 1004
Motorola Inc., 1004
motors
  camcorder, 521, 860, **521**, **860**
  capstan, 847
  capstan, 424-425, 440, 463-464, 847, **425**, **440**, **463**
  CD player, 521, 804-807, **805**
  continuity tests, 524, **525**
  dc, 522, **522**, **523**, 524

1057

motors, *continued*
  disc, 435, 805-806, **806**
  drum, 440, **441**
  erratic operation, 525-526
  flat armature, 526-527
  focus, 441
  intermittent, 524-525
  iris, 442
  loading, 434-435, 438, 804-805, **434**, **435**, **439**
  locating/replacing camcorder, 537, 539, **538**
  locating/replacing cassette, 527-531, **528**, **529**, **530**
  locating/replacing CD, 532-535, **533**, **534**, **535**
  locating/replacing phonograph, 531, **532**
  microwave, 521
  microwave turntable, 770, **771**
  noises in, 422, **423**
  overheated, 525
  phonograph, 432, 521, 563-565, **564**, **565**
  problems, 522-523
  replacing defective, 521-540
  SLED, 805-806, **806**
  slide/sled, 436, **436**
  spindle, 805-806, **806**
  squeaky bearings, 526
  tap test, 527
  test voltage supply, 539-540, **539**
  testing microwave oven, 535, 537, **536**
  tray, 435
  turntable, 805-807, **806**
  VCR, 521
  voltage tests, 523
  zoom, 441, **442**
Mouser Electronics, 144
multimeter, digital (*see* digital multimeter)

N

NAD USA Inc., 1005
Nakamichi USA Corp., 1005
NEC Home Electronics, 1005
Ness Electronics, 739
Nikko Audio, 1005
noise, 749-750, **749**, **750**
  intermittent, 742
  low frying/hissing, 750-751
  motor, 422, **423**
  tracing in amplifiers, 358-359
noise generator, 354-356, **354**, **355**, **356**
Norelco American Phillips, 787
NPN transistor, 17, 72-73, 76, 89, **18**, **76**, **89**
  directly coupled, 686
NTSC color generator, 334
NTSC vectorscope, 350

O

Ohm's law, 24
Olympus Corp., 1005
on-screen display, 948-950, **948**, **949**
Onkyo USA Corp., 1005
op amps, 21, **22**
  symbol, **30**, **32**
oscillators, 222-229
  AM/FM, 222, 224, **223**
  beat frequency, 504
  crystal-controlled, 330-331, **331**
  horizontal driver, 225, **226**
  horizontal output transistors, 226-227, 600-602, **227**, **601**
  horizontal/vertical, 224-225, **224**, **225**, **226**
  vertical output transistors, 227, 229, **228**
oscilloscope, 320-323, **322**, **323**, **812**
  dual-trace, 325, 373, **325**
  set-up chart to obtain a display, 323
  signaltracing with, 364, **365**
  troubleshooting radio circuits with, 377-378

output transistor, keeps destroying, 924-925, **924**

P

Pacific Coast Parts, 739
Panasonic Co., 787
Panasonic Consumer Electronics, 1005
Panson Electronics, 739
parts (*see* components)
Parts Express International, 144
PC board circuit analyzer, 350
PC boards, 305-316
  Audiovox automobile, 305, **305**
  broken, 669-670, **669**
  broken wiring, 668, **668**
  component layout diagram, 153, 157-159, **153**, **157**
  cracked and broken, 312-313, **313**
  double-sided wiring, 658, **658**
  identifying components on, 729-730
  identifying defective components, 90-92, **91**, **92**
  identifying parts on, 40-44, **41**, **42**, **43**
  intermittent connections, 141-142
  intermittent problems, 309-310, 312, **311**, **312**
  locating components on, 134-135, **135**
  locating/isolating problem area, 666-669, **667**, **668**
  locating parts on PC wiring, 315, **315**
  making your own, 306-309, **307**
  mounting components, 159-162, **160**, **161**
  numbered parts, 43-44, **44**
  one-piece television, **310**
  overheated terminals, 660, **660**
  replacing, 315-316

soldering all terminals, 670-671, **670**, **671**
surface-mounted devices, 42, 314-315, **314**
tie wires, 313-314
troubleshooting, intermittent problems, 662, 664, 666, **663**
VCR, 306, **306**
wiring layout chart, **308**
wiring problems, 309, **310**
pencil generator (*see* noise generator)
permanent magnet (PM), 14
  headphones, 14
  speakers, 12
permeability tuning, 505-506, **505**
phase-locked loop (PLL), 389-390, 813, **390**
  block diagram, **509**
Philco Consumer Electronics, 1005
Phillips Auto Radio Division, 1005
Phillips Consumer Electronics, 1005
phonograph pickup, 234, **234**
phonographs, 428-431, 560-565, **1004**
  installing new stylus/needle, 560-561, **561**
  locating/replacing motor in, 531, **532**
  motor, 521
  motor circuits, 563-565, **564**, **565**
  transistor circuits, 563-564, **563**, **564**
  troubleshooting,
    dead/no turntable movement, 428-429
    howling noise, 434
    miscellaneous problems, 431
    motor does not rotate, 432
    motor rotates/no turntable motion, 429

motor runs fast, 432
motor runs hot, 432
no muting, 434
pick-up arm won't move, 433
record arm keeper loose, 433-434
rumble noise, 434
slow rotation, 432
speed erratic, 433
speed problems, 429-430, **429**
speed won't change, 433
stereo magnetic cartridge hookup, 561, **561**
strobe speed indicator, 564, **564**
tone arm problems, 433
tone arm skips on record, 434
turntable runs slow, 432
turntable won't rotate, 432-433
two records drop together, 434
won't play last record, 431, 433
won't reject records, 430-431, **430**, **432**
photo transistor, symbol, **29**
photodetector diode, 797
picofarad, 3
picture tubes (*see also* cathode-ray tube)
  circuits, 620-622, **621**
  high-voltage problems, 618-619, **618**, **619**
  poor tube socket connections, 616-617, **616**, **617**
  removing, 626-629, **628**
  replacing, 615-637, 629-630, **629**
  safety factors, 622-624, **623**
  sizes, 620
  symptoms of failing, 615-616, **615**
  testing, 613, 618-619, **613**, **618**, **619**
piezo buzzer, 769

pigtail fuse, 481
pincushion transformer, 957-959, **958**
Pioneer Electronics, 1005
PNP transistor, 17, 72-73, 76, **18**
  directly coupled, 686
Polk Audio, 1006
Popular Electronics, 1011
potentiometers, 10
  symbol, **30**
power, electrical (*see* wattage)
power amplifier, transistorized, 213, **214**
power supplies, 3
  automobile radio, 281-282, **282**
  boom box, 282-283, **283**
  chopper, 654, **655**
  components, 265-267
  external, 700-702, **701**
  heavy-duty dc, 347, **348**
  line-voltage regulators, 870, 872, **871**
  line-voltage wall adapter, 280, **280**
  low-voltage, 263-284, 807-809, **263**, **807**, **809**, **856**
  full-wave, 265, **265**
  operation, 270-271, **270**
  portable radio/cassette player, 280-281, **281**
  RCA VIPUR, 938-940
  repairing, 969-972, **970**, **971**, **972**
  resistance tests, 276-280, **276**
  safety precautions, 264-265
  standby, 653-655, **654**, **655**
  switched-mode, 940-941, **940**
  testing voltage, 447-449, **448**
  troubleshooting television, 677-678
  variable, 337-338, **338**, **339**
  variable dc, 365-367, **365**, **366**

power supplies, *continued*
  voltage injection, 365-367, **365**, **366**
  voltage tests, 276-280, **276**
power transformers
  low-voltage, 865-866, **865**, **866**
  shorted, 96, **96**
  symbol, **30**
preamplifiers, 209, 460-462, 547-549, **209**, **548**
  directly coupled, 685-686, **685**
Premium Parts Electronics Company, 144
preset resistor, symbol, **30**
projects (*see* kit and project building)
pulse-width modulator (PWM), 938
purity, 631-633, **632**, **633**
  adjusting, 632-633
  adjusting on large screens, 633, **634**
  poor, 631-632
pushbutton switch, 11
Pyle Industries, 1006

Q

Quad Electrocoustics LTD, 1006
Quasar Co., 788, 1006

R

radiation, X-ray, 199-200
radio
  Audiovox automobile PC board, 305, **305**
  power supply for automobile, 281-282, **282**
  power supply for portable, 280-281, **281**
Radio Shack, 1006
radio-frequency (RF) amplifier, 213-215, 217, **215**
radios
  AM, 455-457, **456**, **457**, **458**, **459**
  AM circuits, 496
  AM converter, 496, **497**
  AM/FM, 512-515
  AM/FM alignment, 515-519
  auto stop circuits, 510, **510**
  circuits and components, 730-732, **731**
  clock, 500-502, 512, **501**, **512**
  dial cord problems, 512, **512**
  FM, 459, **460**
  monophonic audio circuits, 511, **511**
  portable, 494-495, **494**, **495**
  shortwave, 502-504, **503**
  signaltracing, 369-370, **369**
  troubleshooting, 493-519
    AM reception weak, 513
    AM/FM reception intermittent, 514
    amplifier problems, 675, **675**
    erratic FM noise, 514
    FM reception poor, 515
    FM reception weak, 514
    intermittent problems with, 672, 674-675, **673**
    loud tuning hum, 515
    no AM/FM normal, 513
    no FM/AM normal, 513
    no tuning in digital, 515
    using the oscilloscope, 377-378
  tuning,
    digital, 507, 509-510, **508**, **509**
    permeability, 505-506, **505**
    varactor, 506-507, **506**
    varicap, 506-507, **506**
  types of receivers, 493, **493**
raster
  insufficient vertical, 899-900, **899**
  troubleshooting, intermittent, 680-681, **680**
  vertical, 678-680, **679**
RC coupling, 208, 210, **208**, **210**
RCA Consumer Electronics, 1006
receivers
  infrared remote, 650-651, **651**
  RF remote, 648-649, **648**, **649**
recording
  intermittent, 752-755, **753**, **754**
  poor VCR, 855, 857, **857**, **858**
  recording circuits, 553-557, **554**, **555**
  IC, 556-557, **556**, **557**
  preamp, 693, **693**
  troubleshooting IC, 557-558
rectifier tube, 7
rectifiers, 7-9, 267 (*see also* diodes)
  bridge, 7, 115-117, 259, 266, 863-865, **9**, **19**, **116**, **117**, **266**, **476**, **864**
  duo-diode, 114-115, **114**, **115**
  full-wave, 266, **9**, **266**
  half-wave, 266, **9**, **266**
  selenium, 9, **9**
  silicon-controlled, 21-22, 866, 868, **23**, **867**
regulated power supply kit, 178-181, **179**, **180**, **181**
  component mounting, 179
  heatsink mounting, 179-180
  testing, 180-181
regulators, voltage, 599, **599**, **600**
relay, symbol, **30**
remote-control devices, 639-655, **639**, **640**
  batteries, 643, **643**
  infrared, 641, **642**
  remote tester, 644-645, **644**, **645**
  testing, 641, 643, 650-651, **651**

keypad interface circuits, 651-653, **652**, **653**
laser, remote tester, 645-646, **646**, **647**
receivers, 648-649, **648**, **649**
standby circuits, 653-655, **654**, **655**
transmitters, 640, **641**
testing, 649, **650**
universal, 646-648
resistance, 15
  hot and cold ac tests, 202, **203**
  IC measurements, 105-107, **106**, **107**
  measuring, 448-449, **449**
  measuring with DMM, 79-82, **81**
  tests and testing, 253-255, 692, 715, **255**, **716**
    in microwave ovens, 783-784
    solid-state components, 102-104, **103**, **104**
    with DMM, 59-61, **59**, **60**, **61**
  unit of, 10
resistance-capacitance coupling circuit, 207, **207**
resistors, 10-11, **10**, **11**, **17**
  burned, 277, **277**
  code, 25
  color code, 156-157
  flameproof, 11
  high-wattage wire-wound, 11
  metal-oxide, 11
  overheated and cracked, 93, **93**
  removing, 136-138, **137**
  replacing, 477-478, **478**, **479**
  surface-mounted, 130-131, **131**
  symbol, **29**, **30**, **31**
  universal carbon, 11
  variable, 10, 478, 480, **11**, **12**
RF choke, 5

RF signal generator, 326-328, 362-364, **327**
rocker switch, 11
Roper Sales Co., 788
rotary switch, 11
Royal Chef, 788

S

SAE, 1006
safety
  CD players and, 791-793
  circuit breakers, 191, **191**
  digital multimeter, 54-56, **54**, **55**
  discharging capacitors, 197
  electrical shocks in test equipment, 201
  electrostatic discharge, 135-136, **137**
  grounded test equipment, 192, **192**
  grounded work bench, 190-191
  hazardous work areas and, 193
  heat and burns, 193-194
  high-voltage precautions, 194, 196, **195**
  hot TV chassis, 187-188, **187**, **188**, **189**
  isolation transformer, 189-190, 264, **189**, **190**, **264**
  laser eye damage, 200-201, 792-793, **200**, **201**
  microwave oven shock hazards, 197, 199, **196**, **199**
  picture tubes and, 622-624, **623**
  power cords, 186
  power line problems (ac), 185-186, **186**, **187**
  power supplies and, 264-265
  protecting the customer, 202, **203**
  replacing the CRT, 196-197, **196**
  shock hazards, 185-203

test instrument warnings, 963-966, **964**, **965**, **966**
X-ray radiation, 199-200
Sampo of America, 788, 1007
Samsung Electronic Corp. of America, 788, 1007
Samsung Electronics, 788
Sansui Electronics, 1007
Sanyo Electronics Inc., 1007
Sanyo/Fisher Inc., 788
saw filters, 941-943, **942**, **943**
scan-derived secondary voltage, 913
schematic diagrams, 27-49, **27**
  block, 37
  electronic hobby and project, 45-46, **46**
  partial, 35, **36**
  pictorial, 35, 37, **36**
  sample, **2**
  servicing products without, 703-723
  symbols, 27-34
  terminal guides, 44-45, **45**
  test equipment, 962
  tie or terminal points, 37-40
  voltage source markings, 79, **79**
Schottky diode, 19
screen controls, 913, **913**
selenium rectifiers, 9, **9**
semiconductors, 7, 15-16
  locating defective with voltage tests, 97-99, **97**, **98**
servo circuits, 803-804, **803**, **804**
  VCR, 847
Sharp Electronics Corp., 788, 1007
shielding, symbol, **30**
shock hazards (*see* grounding; safety)
shortwave radios, 502-504, **503**
sideband (SSB), 504

1061

Index

signal generator
 audio, 359-360, **360**
 harmonic, 351, **351**
 RF, 326-328, 362-364, **327**
signal indicator, 352, **352**
signal injection, 352-353
signal injector/tracer, 338, **339**
signal injector-tracer kit, 167-171, **168**, **169**, **170**
 assembly instructions, 169-171
 operation, 168
 parts inspection, 169
 signaltracing, 171
signal processor, 800-801, **801**, **802**
signaltracing, 351-372, **353**
 audio, 297-299, 359-362, **298**, **360**, **361**, **362**
 audio circuits, 611
 audio signal generator, 359-360, **360**
 audio stereo circuits, 370, **371**
 external audio amplifier, 357-359, **357**, **358**
 harmonic signal generator, 351, **351**
 kit, 171
 noise generators, 354-356, **354**, **355**, **356**
 oscilloscopes, 364, **365**
 power supply voltage injection, 365-367, **365**, **366**
 radio circuits, 369-370, **369**
 random transistor tests, 368-369, **368**
 random voltage measurements, 367-368, **367**
 RF signal generator, 362-364
 signal indictors, 352, **352**
 signal injection, 352-353
silicon-controlled rectifier (SCR), 21-22, 866, 868, **23**, **867**
 tests, 868, **868**
silicon diodes, 7, 9, 712, 713, **8**, **477**, **714**

checking in televisions, 596-597, **597**
resistance tests, 255
symbol, **32**, **476**
testing voltage, 100, **100**
testing with DMM, 112, **113**, **114**
silicon transistor, 16
silver mica capacitor, 4
sine wave, 362
single-pole double-throw (SPDT), symbol, **30**, **31**
single-pole single-throw (SPST), 11
 symbol, **30**, **31**
SLED motor, 805-806, **806**
SLED motor waveform, 390-392, **391**, **392**
slide switch, 11
slow-blow fuse, 481
solder, removing excess, 120, **121**
soldering, 38, 146-148, 257, **147**, **148**, **257**
 components in TV chassis, 1
 PC board terminals, 670-671, **670**, **671**
 PC boards, 310, **311**, **312**
 tools, 148, 150, **149**, **150**
soldering iron, 86, 150, **87**, **149**
 battery-powered, **40**
 repairing, 979-980, **980**, **981**
solid-state components, 15-19
 imported, 727-728
 locating replacements, 413, **413**
 testing out of circuit, 120
 testing, 95-123, 445-455
 troubleshooting, 455-467
Sony Corp. of America, 1007
sound (see audio)
Soundcraftsmen Inc., 1008
Soundesign, 1008
sources, 143-144
 book publishers, 1010

electronics manufacturers, 997-1011
 import product distributors, 738-740
 magazines, 1010-1011
Sparkomatic Corp., 1008
speakers, 12, 14, **13**, **998**
 automobile, 490
 cone damage, 704-705, **705**
 cone material, 14
 field coil, 12
 PM, 12
 replacing, 488, 490, **489**
 symbol, **30**
 testing television, 608-609, **609**
 troubleshooting, 303-304, **304**
Speco Division, 1008
Sperry-Tech System, 992-993
spindle motor, 805-806, **806**
spindle motor waveform, 392, **393**
square wave, 361, 374, **361**, **374**
standby circuits, 653-655, 955-957, **654**, **655**, **956**
start-up transformer, 15, **16**
static electricity, 135-136, 791, **137**
step-down transformer, 15
step-up transformer, 15
stereo amplifiers, 211-212, **211**, **212**, **213**, 290-293, 550, 552-553
stereo decoder circuits, troubleshooting, 952
Studer Rivex America, 1008
supply pins, terminal tests, 893, **893**
supply voltages, testing, 78-79, **78**
surface acoustic-wave saw-filter network, 941, **941**
surface-mounted devices (SMD), 24, 42, 125-144, **24**, **125**
 capacitors, 5, 131-132, **132**
CD player, 795, **796**

diodes, 127-129, **128**
electrostatic-sensitive components, 135-136
ICs, 129-130, **120**, **130**
intermittent board connections, 141-142
locating on PC board, 134-135, **135**
locating replacement parts, 143-144
LSIs, 129-130, **120**, **130**
microprocessors, 129-130, **120**, **130**
part substitution, 141, **141**
parts layout, 142-143, **143**
PC board components, 314-315, **314**
removing, 136-138, **137**, **138**, **139**
replacing, 138, 140, 476-477, **139**, **142**, **477**
resistors, 130-131, **131**
symbols for various, **29**
testing components, 133-134, **134**
transistors, 126-127, **126**, **127**
sweep generator, 325-326, **326**
switched-mode power supply (SMPS), 940-941, **940**
switches, 11-12
  bat wing, 11
  defective, 11-12
  double-pole double-throw, 11
  function, 11
  interlock, 768, **769**
  key-type, 11
  knife, 11
  leaf, 87, **88**
  on/off, **12**
  pushbutton, 11
  replacing, 487-488
  rocker, 11
  rotary, 11
  single-pole single-throw, 11
  slide, 11
  symbol, **30**, **31**
  toggle, 11
  types of, 11
switching diode, 19
Sylvania Corp., 1008
symbols, 27-34
  air-core transformer, **30**
  air transformer, **30**
  amplifier, **30**
  antenna, **30**
  battery, **30**
  bridge rectifier, **476**
  capacitor, **30**, **31**
  chassis ground, **30**
  chokes, **31**
  coil, **30**, **31**
  detection diode, **476**
  diode, **29**, **30**, **32**
  dots and dotted lines, 33, **33**, **34**
  DPDT switches, **30**
  earth ground, **30**
  electrolytic, **31**
  field-effect transistor, **29**
  fixed capacitor, **30**
  fuse, **30**, **31**
  headphones, **30**
  IC, **29**, **32**
  IF transformer, **31**
  inductor, **30**
  iron-core inductor, **30**
  jack, **30**, **31**
  light-emitting diode, **30**, **31**
  metal-oxide transistor, **29**
  meter, **30**
  mike, **30**
  nonconducting lines, 34, **35**
  op amp, **30**, **32**
  phono jack, **30**
  photo transistor, **29**
  potentiometer, **30**
  power transformer, **30**
  preset resistor, **30**
  relay, **30**
  resistor, **30**, **31**
  shielding, **30**
  silicon diode, **32**, **476**
  SMD ICs, **29**
  SMD microprocessor, **29**
  SMD transistor, **29**
  SPDT switch, **30**, **31**
  speakers, **30**
  SPST switch, **30**, **31**
  stereo jack, **30**
  straight lines, 28, 32-33
  surface-mounted capacitor, **29**
  surface-mounted resistor, **29**
  tapped transformer, **30**
  transistor, **29**, **32**
  triac, **29**, **32**
  unijunction transistor, **29**
  varactor diode, **29**, **32**
  variable capacitor, **30**
  variable resistor, **30**
  variable transformer, **30**
  voltage regulator, **32**
  zener diode, **29**, **32**, **476**
Symphonic Electronics, 1008

T

Tandberg of America, 1008
tape head circuits, 232-234, **233**
tape player, defective, 87-88, **88**
Tappan Appliance Division, 788
tapped transformer, symbol, **30**
Tatung Company of America, 1008
Tatung of America, 788
Teac Corp., 1008
Technics, 1008
Teknika Electronics Corp., 1008
telephone product analyzer, 349
telephone product tester, 349
television frequency converter/modulator, 349
televisions, 567-591
  audio driver transistor, 609-610, **609**
  $B+$ adjustments, 880-881, **881**

televisions, *continued*
  basic circuit found in chassis, 205, **205**
  black-and-white, adjusting, 636-637
  block diagram, **569**
  cabinet repairs, 979
  capacitor tests, 598-599, **598**
  chassis shutdown problems, 757-761, **757**, **758**, **759**
  checking silicon diodes, 596-597, **597**
  checking the fuse, 595-596, **595**
  circuits/components that cause most problems, 567, **568**, **569**
  color circuits, 572, 932, **932**
  convergence,
    adjusting, 633-634, **634**
    problems, 634-635, **635**
    slotted shadow mask, 635-636, **636**
  CRT circuits, 231-232, **232**, **233**
  damper diode tests, 602-603, **602**
  deflection IC tests, 604-605, **604**
  deflection yoke tests, 607-608, **608**
  degaussing coil, 629-631, **629**
  different face/same chassis, 711, **711**
  different model/same chassis, 403-404, **404**
  discharging high voltage, 623-624, **624**
  driver transistor tests, 603-604, **603**
  high-voltage circuits, 229-231, **230**, **231**
  high-voltage precautions, 194, 196, **195**
  horizontal circuits, 570, **570**
  horizontal output transistor testing, 600-602, **601**
  IC sound output circuits, 610-611, **610**
  import circuits, 732-735
  lightning damage, 879-880, **741**, **880**
  locating various circuits, 712-715
  low-voltage power supply in B&W, 868-869, **869**
  low-voltage circuits, 571, **571**
  new chassis, 709-711
  on/off switch tests, 597
  one-piece PC board, **310**
  on-screen display, 948-950, **948**, **949**
  part numbers, 705-709
  parts mounted off the main chassis, 717-718
  parts replacement, 718-720
  picture in a picture, 953-955, **954**, **955**
  picture tube replacement, 615-637
  picture tube tests, 613, **613**
  portable, **1009**
  purity,
    adjusting, 632-633, **634**
    poor, 631-632
  radio-frequency amplifier, 216, **216**
  RCA chopper circuits, 936-938, **937**, **938**, **939**
  RCA VIPUR power supply, 938-940
  repairing, 593-613
  replacing the CRT, 196-197, **196**
  schematic comparison, 712
  service notes, 712
  servicing without a schematic, 703, **703**
  signaltracing audio circuits, 611
  sound circuits, 572
  speaker cone damage, 704-705, **705**
  speaker tests, 608-609, **609**
  stereo headphones for, 953, **953**
  sweep waveforms, 378-379, **379**
  Sylvania SMPS circuits, 940-941, **940**
  10 service tests, 716-717
  testing vertical output transistor, 237, **237**
  test points for color, 933-936, **934**, **935**
  troubleshooting,
    AGC problems, 582
    audio, 676-677, **676**
    black hum bars, 878-879, **879**
    black raster/no picture/no sound, 574, **574**
    bow in picture, 582
    bunching lines, 583, **583**
    chassis shutdown, 583-584, 881
    dead, 573
    diagonal horizontal dark bars, 581, **581**
    excessively bright raster, 576, **576**
    firing lines in the raster, 584, **584**
    fuses keep blowing, 588, 874-876, 922-923, **589**, **875**, **923**
    high-voltage arcover, 585-586, **585**
    high-voltage problems, 908, **909**
    horizontal foldover, 586
    horizontal white line, 465-467, 576-577, 594, 884, **466**, **577**, **594**, **884**
    hot chassis, 187-188, **187**, **188**, **189**
    hum bars in picture, 586, **587**
    HV shutdown, 586-588, 919-920, **587**, **920**
    insufficient height, 577, **578**

Index

intermittent horizontal problems, 927
intermittent picture/good sound, 575-576
intermittent problems, 658-661, 676-681, 744-746, **676**, **745**
multiple problems, 743
narrow picture, 578
no color, 579-580, **579**
no red in picture, 589
no sound, 580, **580**
no sound/no picture/no raster, 240, 242, **241**
picture rolls up/down, 890-891, **891**
poor focus, 578, **579**
poor width in chassis, 249, **249**
power supplies, 677-678
raster, 680-681, **680**
red color missing, 464-465, **464**
rolling pictures, 577
sides pulled in on each side of picture, 916-917
snowy picture/fair sound, 575, **575**
tuners, 677, 678
vertical foldover, 590, **590**
vertical raster, 678-680, **679**
weak color, 589
weak sound, 590
white lines in raster, 927-929, **928**, **929**
white raster/rush in sound, 575
tuners
  computers, 946-948, **946**, **947**
  testing, 611-612, **611**, **612**
VCRs built into, 725, **726**
vertical height, insufficient, 755-757, **755**, **756**
vertical output transistor replacement, 606-607, **607**
tests, 605-606, **605**, **606**
visual inspection, 704

voltage regulator tests, 599, **599**, **600**
waveform measurement, 657, **657**
television video analyzer, 349
terminals, 37-40
  fixed, 37
  guides, 44-45, **45**
  overheated, 660, **660**
  soldering, 670-671, **670**, **671**
  wire, **39**
test equipment, 317-350, 961-982, **317**
audio generator, 324-325, **324**
audio signal generator, 359-360, **360**
broken dials/knobs, 972-973
capacitance meter, 331-332, **331**
capacitor tester, 977, **977**
CD player, 793, **794**
circuit saver, 777-778, **778**, **779**
color pattern generator, 332-334, **333**
color-bar generator, 978-979, **979**
CRT tester, 618
CRT tester and restorer, 335-336, **335**
deluxe regulated power supply kit, 178-181, **179**, **180**
digital frequency counter, 324
digital multimeter, 51-73, 75-83, 112-113, 299-300, 320, 775, 967-968, **113**, **114**, **299**, **320**, **776**, **969**
distortion meter, 346-347
electrical shocks in, 201
factory instrument adjustments, 980-982
FET analog meter, 977-978
FET meter, 319, **319**
FET voltmeter, 85, 319

frequency counter, 328-331, **329**, **330**, **331**, **812**
fuses, **962**
harmonic signal generator, 351, **351**
heavy-duty dc power supply, 347, **348**
high-voltage probe, 347, 349, **348**
infrared remote, 644-645, **644**, **645**
instrument warnings, 963-966, **964**, **965**, **966**
isolation transformer, 15, 189-190, 264, 337, **189**, **190**, **264**, **337**, **902**
kits for building, 167-183
laser power meter, 341-342, 811, **341**, **342**
laser remote, 645-646, **646**, **647**
LCD meter kit, 175-178, **176**, **177**, **178**
LCR meter, 344-345, **345**
magnameter, 343-344, 776-777, **343**, **777**, **778**
maintenance, 961-962
microwave oven, 775-778
motor voltage supply, 539-540, **539**
noise generators, 354-356, **354**, **355**, **356**
NTSC color generator, 334
NTSC vectorscope, 350
obtaining parts for, 963
oscilloscope, 320-323, **322**, **323**, **812**
PC board circuit analyzer, 350
repairing cords/plugs/jacks, 966-967, **967**
required instruments for beginning technician, 85-86
RF signal generator, 326-328, 362-364, **327**
schematics, 962
signal indicator, 352, **352**
signal injector/tracer, 338, **339**

test equipment, *continued*
  signal injector-tracer kit, 167-171, **168**, **169**, **170**
  sweep generator, 325-326, **326**
  telephone product analyzer, 349
  telephone product tester, 349
  television frequency converter/modulator, 349
  television video analyzer, 349
  transistor-semiconductor tester, 320, **321**
  transistor test kit, 171-174, **172**, **173**, **174**
  transistor tester, 975, **976**
  tube tester, 336, **336**
  tuner-subber, 340, **340**
  universal video analyzer, 349
  universal video generator, 349
  vacuum-tube voltmeter (VTVM), 84-85, 319, **85**
  variable power supply, 337-338, **338**, **339**
  video head tester, 347
  video tracker, 349
  volt-ohm-millimeter, 83-84
  volt-ohmmeter, 318-319, 775, 967-968, **318**, **969**
  waveform and circuit analyzer, 350
  wow and flutter meter, 345-346, **346**
tests and testing
  after repairs have been made, 261-262
  batteries, 643, **643**
  bridge rectifiers, 69, 115-117, **69**, **70**, **116**, **117**
  capacitors, 65-66, 91-92, 598-599, **65**, **66**, **91**, **92**, **598**
  cathode-ray tube, 624-625, **625**
  color test points, 933-936, **934**, **935**

continuity in motors, 524, **525**
continuity with DMM, 57-59, **58**, **59**
current tests with DMM, 63-64, **64**
deflection yoke, 607-608, **608**
diode-junction, 83, **84**
diodes, 67-69, 100-101, 105, 112-113, **68**, **100**, **101**, **105**, **113**, **114**
  damper, 117-118, 602-603, **117**, **602**
  detector, 70, **70**
  high-voltage, 118-120, **118**, **119**, **120**
  zener, 68-69
duo-diode rectifier, 114-115, **114**, **115**
EFM, 812-813, **813**
frequency with DMM, 66-67
high voltage, 812-813, **813**
high-voltage shutdown, 922, **922**
ICs, 89-90, **90**
  deflection, 604-605, **604**
  resistance measurements, 105-107, **106**, **107**
  voltage measurements, 101-102, **102**
inductance, 73, **73**
inspecting chassis for defective parts, 95, **95**
microprocessor voltage measurements, 101-102, **102**
microwave ovens, 343-344, 535, 537, **536**
new parts, 746-747, **747**
on/off television switch, 597
picture tubes, 613, 618-619, **613**, **618**, **619**
power supplies, 276-280, **276**
  voltage, 447-449, **448**
  radiation leakage, 784, **785**

radio frequency, 812-813, **813**
remote controls,
  infrared, 641, 643, 650-651, **651**
remote transmitters, 649, **650**
required electronic, 75-94
resistance, 202, 253-255, 448-449, 692, 715, 783-784, **203**, **255**, **449**, **716**
  solid-state components, 102-104, **103**
  with DMM, 59-61, 79-82, **59**, **60**, **61**, **81**
semiconductors, 97-99, **97**, **98**, **99**
signaltracing, 351-372
silicon-controlled rectifiers, 868, **868**
SMD components, 133-134, **134**
solid-state circuits, 445-455
solid-state components out of circuit, 120
speakers, television, 608-609, **609**
supply pin terminal, 893, **893**
supply voltages, 78-79, **78**
10 service tests, 716-717
transistor junction-diodes, 107-110, **108**, **109**
transistorized circuits, 87-89
transistors, 97-99, **97**, **98**, **99**, 450, 691, **451**, **692**, **748**
  audio driver, 609-610, **609**
  driver, 603-604, **604**
  horizontal output, 600-602, **601**
  horizontal output with damper diode, 112, 174, **112**
  open, 110-111, **111**
  vertical output, 237, 605-606, **237**, **605**, **606**
  with DMM, 71-73, 82-83, **72**, **82**
tuners, 611-612, **611**, **612**

voltage, 446-447, 692, 715, **446**, **447**, **716**
 in motors, 523
 on ICs, 454-455, **454**
 with DMM, 75-78, **75**, **76**, **77**, **78**
voltage measurements, 62-63, 251, 253, **62**, **63**, **252**, **253**
voltage regulators, 599, **599**, **600**
Thermador/Waste King, 789
Thomson Consumer Electronics, 1008
 RCA/GE guide, 728, **729**
Threshold Corp., 1009
toggle switch, 11
tone controls, 672
tools, 258, **258**
 special electronic, 133, **133**
Toshiba America Inc., 789, 1009
tracking coil waveform, 387, **388**
tracking error signal, 387, 802-803, **387**, **802**
tracking sensors, 797
transformer coupling, 208, 684, **208**
transformers, 14-15, **15**
 audio output, 15
 auto, 14
 connecting, 162-166, **162**, **163**
 defective, 15
 driver, 15, **16**, **228**, **760**
 flyback, 96, 705, 910-913, **97**, **706**, **911**, **912**
 high-voltage, 764, 770-772, **764**, **772**
 horizontal output, **926**, **927**
 IF, **39**
 intermediate frequency, 15
 isolation, 15, 189-190, 264, 337, **189**, **190**, **264**, **337**, **902**
 line, 15
 low-voltage power, 865-866, **865**, **866**

matching, 15
pincushion, 957-959, **958**
power, 96, **96**
start-up, 15, **16**
step-down, 15
step-up, 15
symbol, **30**
transistor test kit, 171-174, **172**, **173**, **174**
 diode testing, 172
 mounting components, 172
 transistor testing, 174
transistor tester, repairing, 975, **976**
transistor-semiconductor tester, 320, **321**
transistorized power amplifier, 213, **214**
transistors, 16-17, **17**, 38
 AM converter, 455-457, **456**, **457**, **458**, **459**
 audio driver, 609-610, **609**
 bipolar, 16
 checking function of with diode tests, 70-71, **71**, **72**
 checking in the circuit, 82-83, **82**
 color output, 715, 718, 935, **720**, **935**
 defective, 18
 diffused-junction, 16
 directly coupled, 687-690, **688**, **689**
 driver, 603-604, **603**, **761**
 emitter-follower, 76, **77**
 field-effect, 16, 17
 FM RF, 459, 497, **460**
 germanium, 16
 horizontal output with damper diode, 112, **112**
 horizontal output, 226-227, 600-602, 659, 734, 757, **227**, **601**, **659**, **735**, **757**
 hot, 925, **926**
 junction-diode tests, 107-110, **108**, **109**, **110**
 leaky, 104, **104**
 locating unknown, 397-399, **397**, **398**, **399**

marking on PC board before removing, 401, **401**
metal-oxide, 16
metal-oxide semiconductor field-effect, 17
mounting, 18, 154-156, **155**
NPN, 17, 72-73, 76, 89, 686, **18**, **76**, **89**
open, 110-111, **111**
output, 924-925, **924**
phonograph, 563-564, **563**, **564**
PNP, 17, 72-73, 76, 686**18**
replacing defective, 405, 407, 472-475, **407**, **408**, **475**
silicon, 16
solid-state replacements, 413, **413**
surface-mounted, 17, 126-127, **126**, **127**
symbol, **29**, **32**
testing, 71-73, 97-99, 174, 450, 691, **72**, **97**, **98**, **99**, **451**, **692**
 bias voltage, 452, 454, **452**, **453**
 for signaltracing, 368-369, **368**
 unijunction, 16
 vertical output, 227, 229, 237, 605-607, **228**, **237**, **605**, **606**, **607**
 voltage measurements, 79, **80**
 voltage regulating, 872-874
 working voltages, 399-401
 of power output, 401-403, **402**, **403**
transmitters, remote-control, 640, 649, **641**, **650**
triac, 23, 768, **23**, **768**
 symbol, **29**, **32**
troubleshooting, 237-262
 AM converters, 455-457, **456**, **457**, **458**, **459**
 AM/FM/MPX receiver, dead, 240, **241**
 amplifiers, audio, 675, **675**
 audio, 293-294, **294**

troubleshooting, *continued*
  dead, 295, **296**
  distorted, 296-297, **298**
  intermittent, 296, **297**
  weak, 294-295, **295**
camcorders, 437-443, 463-464
  dead, 854-855, **855**
  erratic zoom operation, 853, **853**, **854**
  no audio in playback, 858-859
  no auto focus, 854, **854**
  no head erase, 858, **859**
  poor head erase, 858, **859**
cassette players, 424-428, 460-462
  defective playback, 559
  defective recording, 559
  erratic switching, 547
  IC recording circuits, 557-558
  preamp recording circuits, 693, **693**
  tape rotates/no audio, 242, **242**
CD players, 431, 434-436, 462-463, 814-823
  tray motor rotating/no tray action, 435
intermittent problems, 657-681
isolating different circuits, 243-244
keypads, 652-653, **653**
locating defective component, 249-251, **250**, **251**
mechanical defects, 417-443
microwave ovens, 763-789
  high-voltage circuits, 782-783
  intermittent cooking, 783
  radiation leakage, 784, **785**
motors, 521-540
  erratic operation, 525-526
  intermittent operation, 524-525
  overheated, 525

  squeaky bearings, 526
outside suspected circuit, 248-249, **249**
PC boards
  cracked and broken, 312-313, **313**
  intermittent problems, 309-310, 312, 662, 664, 666, **311**, **663**
  wiring problems, 309, **310**
phonographs, 428-431
  dead/no turntable movement, 428-429
  howling noise, 434
  miscellaneous problems, 431
  motor does not rotate, 432
  motor rotates/no turntable motion, 429
  motor runs fast, 432
  motor runs hot, 432
  no muting, 434
  pick-up arm won't move, 433
  record arm keeper loose, 433-434
  rumble noise, 434
  slow rotation, 432
  speed erratic, 433
  speed problems, 429-430, **429**
  speed won't change, 433
  tone arm problems, 433
  tone arm skips on record, 434
  turntable runs slow, 432
  turntable won't rotate, 432-433
  two records drop together, 434
  won't play last record, 431, 433
  won't reject records, 430-431, **430**, **432**
pinpointing defective circuit, 244-248, **245**, **247**
power supplies, television, 677-678

radio circuits using oscilloscope, 377-378
radios, 493-519
  AM reception weak, 513
  AM/FM reception intermittent, 514
  dial cord problems, 512, **512**
  erratic FM noise, 514
  FM reception poor, 515
  FM reception weak, 514
  intermittent problems with, 672, 674-675, **673**
  loud tuning hum, 515
  no AM/FM normal, 513
  no FM/AM normal, 513
  no tuning in digital, 515
solid-state circuits, 455-467
speakers, 303-304, **304**
stereo decoder circuits, 952
symptoms,
  determining, 237-238
  different types of, 239-240
televisions
  AGC problems, 582
  audio, 676-677, **676**
  black hum bars, 878-879, **879**
  black raster/no picture/no sound, 574, **574**
  bow in picture, 582
  bunching lines, 583, **583**
  chassis shutdown, 583-584, 881
  dead, 573
  diagonal horizontal dark bars, 581, **581**
  excessively bright raster, 576, **576**
  firing lines in the raster, 584, **584**
  fuses keep blowing, 588, 874-876, 922-923, **589**, **875**, **923**
  high-voltage arcover, 585-586, **585**
  high-voltage problems, 908, **909**
  horizontal foldover, 586

horizontal white line, 465-467, 576-577, 594, 884, **466**, **577**, **594**, **884**
hot chassis, 187-188, **187**, **188**, **189**
hum bars in picture, 586, **587**
HV shutdown, 586-588, 919-920, **587**, **920**
insufficient height, 577, **578**
intermittent horizontal problems, 927
intermittent picture/good sound, 575-576
intermittent problems, 658-661, 676-681, 744-746, **676**, **745**
multiple problems, 743
narrow picture, 578
no color, 579-580, **579**
no red in picture, 589
no sound, 580, **580**
no sound/no picture/no raster, 240, 242, **241**
picture rolls up/down, 890-891, **891**
poor focus, 578, **579**
poor width in chassis, 249, **249**
power supplies, 677-678
raster, 680-681, **680**
red color missing, 464-465, **464**
rolling pictures, 577
sides pulled in on each side of picture, 916-917
snowy picture/fair sound, 575, **575**
tuners, 677, **678**
vertical foldover, 590, **590**
vertical raster, 678-680, **679**
weak color, 589
weak sound, 590
white lines in raster, 927-929, **928**, **929**
white raster/rush in sound, 575

tough repair problems, 741-761
tuners, 677, **678**
VCRs, 437-443
 cannot remove cassette, 850, **851**
 cassette will not load properly, 851, **851**
 dead, 243, 854-855, **243**, **855**
 no tape motion, 852, **853**
 poor recording, 855, 857, **857**, **858**
 pulling of tape, 850, **850**
 shuts down after loading, 852
 no audio in playback, 858-859
 poor picture in playback, 859
tube tester, 336, **336**
tubes
 gas-filled regulator, 272
 magnetron, 764-765, 769-770, **764**, **765**, **766**
 picture, 613, 615-637, **613**
tuner-subber, 340, **340**
tuners
 computer, 946-948, **946**, **947**
 testing, 611-612, **611**, **612**
 troubleshooting intermittent problems, 677, **678**
 varactor, 713, **714**
tuning
 digital, 507, 509-510, **508**, **509**
 permeability, 505-506, **505**
 varactor, 506-507, **506**
 varicap, 506-507, **506**
turntable (*see* phonograph)
turntable motor, 770, 805-807, **771**, **806**
TZL International, 1009

U

UCM Corp., 1009
Ultimate Sound, 1009
unijunction transistor, 16
 symbol, **29**

Union Electronics Distributor, 740
unit of capacitance, 3
unit of inductance, 5
unit of resistance, 10
universal carbon resistor, 11
universal video analyzer, 349
universal video generator, 349

V

vacuum-tube voltmeter (VTVM), 84-85, 319, **85**
Vance Baldwin Inc., 740
varactor diode, 18, **20**
 symbol, **29**, **32**
varactor tuner, 713, **714**
varactor tuning, 506-507, **506**
variable capacitors, 504, **504**
 symbol, **30**
variable dc power supply, 365-367, **365**, **366**
variable inductor, symbol, **30**
variable power supply, 337-338, **338**, **339**
variable resistors, 10, **11**, **12**
 replacing, 478, 480
 symbol, **30**
variable transformer, symbol, **30**
varicap tuning, 506-507, **506**
VCRs, 437-443, 825-840
 Beta format, 826, **830**
 Beta tape-loading path, 835, **836**
 block diagram, 826, **827**
 built into televisions, 725, **726**
 capstan motor, 440, 847, **440**
 cleaning, 437, **437**
 cylinder or drum motor, 440, **441**
 drive belts, 437, 840, **438**, **841**
 loading motor, 438, **439**
 locating components in RCA/GE guide, 728, **729**

VCRs, *continued*
  motor, 521
  PC board, 306, **306**
  preventive maintenance and lubrication, 836-838
  problems with, 849-860
  reel table drive mechanism, 838-839, **840**
  servo and control systems, 847
  supply and take-up reels, 442-443, **443**
  tape heads, 830, **831**
    cleaning, 832-833, **832**, **833**, **834**, **839**
  troubleshooting,
    cannot remove cassette, 850, **851**
    cassette will not load properly, 851, **851**
    dead, 854-855, **855**
    no audio in playback, 858-859
    no tape motion, 852, **853**
    poor recording, 855, 857, **857**, **858**
    pulling of tape, 850, **850**
    shuts down after loading, 852
    poor picture in playback, 859
  VHS format, 826, **830**
  VHS tape-loading paths, 833-835, **835**
  worm gear, 438, **439**
vertical circuits, 883-900
  checking by the numbers, 890, **890**
  imported, 736-737
  insufficient vertical raster, 899-900, **899**
  servicing directly coupled, 696-697, **697**
  servicing without schematics, 886-887
  voltage injection, 893-894, **895**
  voltage supply sources, 885-886, **885**

vertical countdown ICs, 888-889, **889**
vertical foldover, 895-898, **896**
vertical input sync signal, 890-891
vertical output transistors
  replacing, 606-607, **607**
  testing, 237, 605-606, **237**, **605**, **606**
vertical sweep, intermittent, 891-892, **892**
vertical sweep waveforms, 382-385, **384**, **385**
vertical waveforms, 887-888, **888**
video amplifier, 219-220, **220**
video circuits
  camcorder, 844-845, **845**, **846**
  servicing directly coupled, 699, **700**
video head tester, 347
video tracker, 349
VIPUR power supply, 938-940
voice coil, 14
volt-ohm-millimeter (VOM), 83-84
volt-ohmmeter, 318-319, 775, **318**
  repairing, 967-968, **968**, **969**
voltage
  ac, 2
  ac power line problems, 185-186, **186**, **187**
  bias, 452, 454, **452**, **453**
  dc, 6
  derived, 230
  different sources of, 869-870
  hot and cold ac resistance tests, 202, **203**
  measuring, 251, 253, **252**, **253**
  measuring transistor, 79, **80**

  measuring with DMM, 62-63, 75-78, **62**, **63**, **75**, **76**, **77**, **78**
  power cords, 186
  random measurements for signaltracing, 367-368, **367**
  scan-derived secondary, 913
  source markings on diagrams, 79, **79**
  supply tests, 78-79, **78**
  tests and testing, 446-447, 692, 715, **446**, **447**, **716**
    diodes, 100-101, **100**, **101**
    ICs, 454-455, **454**
    motors, 523
    semiconductors, 97-99, **97**, **98**, **99**
voltage controlled oscillator (VCO), 389-390, **390**
voltage doubler, 267-268, **268**
voltage injection, vertical, 893-894, **895**
voltage regulators, 271-272, **271**, **272**
  IC, 272-273, **273**
  line, 870, 872, **871**
  symbol, **32**
  testing, 599, **599**, **600**
  transistor, 872-874
  zener diode, 872-874, **873**
voltage-controlled oscillator (VCO), 813
voltage-doubler power circuits, 862-863, **863**
voltmeter
  FET, 85, 319
  vacuum tube, 319
  vacuum-tube, 84-85, **85**
volume controls, 672

W

Wald Sound, 1009
wall adapter, 280, **280**
wattage, 25-26
watts, 25-26

waveform and circuit analyzer, 350
waveforms, 256, 373-393, 913-916, **256**, **257**, **914**, **915**, **916**
 bias oscillator, **546**
 CD, 385
 EFM signal, 386, **386**
 focus coil, 389, **389**
 focus error, 388, **388**
 HF, 386, **386**
 horizontal sweep, 379, 381-382, **380**, **381**, **382**
 measuring on television, 657, **657**
 oscilloscope, 364, **365**
 PLL VCO, 389-390, **390**
 RF signal, 386, **386**
 sine, 362
 SLED motor, 390-392, **391**, **392**
 spindle motor, 392, **393**
 square, 361, 374, **361**, **374**
 television sweep, 378-379, **379**
 tests in audio circuits, 374-377, **374**, **375**, **376**
 tracking coil, 387, **388**
 tracking error, 387, **387**
 vertical, 605, 887-888, **888**
 vertical sweep, 382-385, **384**, **385**
WEKA Publishing Inc., 1010
Whirlpool Corp., 789
White Westinghouse, 789
Winegard Company, 1010
wow and flutter meter, 345-346, **346**

X

X-ray radiation, 199-200

Y

Yamaha Electronic Corp., 1010
yoke, 718
York Electronics, 1010

Z

Zeff Advanced Products, 1010
zener diodes, 18, 128, 271-272, 277, **20**, **271**, **272**, **277**
 resistance tests, 255
 symbol, **29**, **32**, **476**
 testing voltage, 101, **101**
 testing with DMM, 68-69
 voltage regulating, 872-874, **873**
Zenith Corp., 1010
zoom, erratic operation, 853, **853**, **854**

1071

Index

About the author

Homer L. Davidson is an electronics technician who owned and operated a successful small appliance repair business for 38 years before retiring to write full-time. He is a regular contributor to *Electronic Servicing and Technology* magazine and the author of 18 TAB/McGraw-Hill books.

Other Bestsellers of Related Interest

Maintaining and Repairing VCRs, 3rd Edition
Robert L. Goodman
This is a 50,000-copy bestseller—the #1 guide for technicians and advanced do-it-yourselfers—covering the latest VCR technology from General Electric, RCA, JVC, Zenith, Panasonic, Sony, and more. Goodman offers professional tips and techniques for keeping VCRs operating trouble-free—and for fixing them when they don't—using pinout diagrams and exploded-view photographs to illustrate procedures.
$19.95 Paper, ISBN 0-8306-4080-0, #023970-3
$39.95 Hard, ISBN 0-8306-4079-7, #023969-X

The Complete Book of Oscilloscopes, 2nd Edition
Stan Prentiss
This revised edition of the bestselling Complete Book of Oscilloscopes provides an up-to-date look at all of the latest oscilloscope equipment and advanced testing procedures developed during the last five years. Professional technicians and electronics hobbyists will find detailed information on all types of oscilloscopes and their applications.
$19.95 Paper, ISBN 0-8306-3908-X, #157781-5

Troubleshooting and Repairing Compact Disc Players, 2nd Edition
Homer L. Davidson
Updated to cover all the latest CD player makes and models, this bestselling guide provides all the information professionals need to diagnose and fix the most common problems in today's players.
$24.95 Paper, ISBN 0-07-015670-0

Troubleshooting and Repairing Computer Monitors
Stephen J. Bigelow
The technician's guide to maintaining, aligning, troubleshooting, and repairing both CRT and flat-panel computer monitors.
$42.95 Hard, ISBN 0-07-005408-8

Troubleshooting and Repairing Consumer Electronics Without a Schematic
Homer L. Davidson
Indispensable for electronics technicians, students, and advanced hobbyists, this hands-on guide comes to the rescue for all those times when no schematic diagram is available.
$22.95 Paper, ISBN 0-07-015650-6
$34.95 Hard, ISBN 0-07-015649-2

Troubleshooting and Repairing Solid-State TVs, 2nd Edition
Homer L. Davidson
A complete workbench reference for electronics technicians and students. Packed with case study examples, troubleshooting photos, and diagrams for every kind of TV circuit, this popular guide is just what technicians need to diagnose and repair virtually any TV malfunction.
$26.95 Paper, ISBN 0-8306-3893-8, #157677-0
$36.95 Hard, ISBN 0-8306-3894-6, #157678-9

How to Order

Call 1-800-822-8158
24 hours a day,
7 days a week
in U.S. and Canada

Mail this coupon to:
McGraw-Hill, Inc.
P.O. Box 182067,
Columbus, OH 43218-2607

Fax your order to:
614-759-3644

EMAIL
70007.1531@COMPUSERVE.COM
COMPUSERVE: GO MH

Shipping and Handling Charges

Order Amount	Within U.S.	Outside U.S.
Less than $15	$3.50	$5.50
$15.00 - $24.99	$4.00	$6.00
$25.00 - $49.99	$5.00	$7.00
$50.00 - $74.49	$6.00	$8.00
$75.00 - and up	$7.00	$9.00

EASY ORDER FORM— SATISFACTION GUARANTEED

Ship to:
Name _____
Address _____
City/State/Zip _____
Daytime Telephone No. _____

Thank you for your order!

ITEM NO.	QUANTITY	AMT.

Method of Payment:
☐ Check or money order enclosed (payable to McGraw-Hill)
☐ VISA   ☐ DISCOVER
☐ AMERICAN EXPRESS Cards   ☐ MasterCard

Shipping & Handling charge from chart below	
Subtotal	
Please add applicable state & local sales tax	
TOTAL	

Account No. ☐☐☐☐☐☐☐☐☐☐☐☐☐

Signature _____ Exp. Date _____
Order invalid without signature

In a hurry? Call 1-800-822-8158 anytime, day or night, or visit your local bookstore.

Code = BC15ZZA